STP 1091

Plants for Toxicity Assessment

Wuncheng Wang, Joseph W. Gorsuch, and William R. Lower,
editors

ASTM
1916 Race Street
Philadelphia, PA 19103

Library of Congress Cataloging-in-Publication Data

Plants for toxicity assessment / W. Wang, J. W. Gorsuch, and W. R. Lower, editors
    p. cm.—(STP; 1091)
    Papers from the First Symposium on Use of Plants for Toxicity Assessment, held in Atlanta, Ga., Apr. 19–20, 1989, and sponsored by ASTM Committee E-47 on Biological Effects and Environmental Fate and its Subcommittee E47.11 on Plant Toxicity.
    "ASTM publication code number (PCN) 04-010910-16"—T.p. verso.
    Includes bibliographical references and index.
    ISBN 0-8031-1397-8
    1. Plants, Effect of pollution on—Congresses. 2. Plant indicators—Congresses. 3. Toxicity testing—Congresses. 4. Biological monitoring—Congresses. I. Wang, Wuncheng. II. Gorsuch, J. W. (Joseph W.), 1947–   . III. Lower, W. R. IV. Symposium on Use of Plants for Toxicity Assessment (1st: 1989: Atlanta, Ga.)   V. ASTM Committee E-47 on Biological Effects and Environmental Fate. Subcommittee E47.11 on Plant Toxicity. VI. Series: ASTM special technical publication; 1091.
QK750.P56   1990
615.9′07—dc20
                                                  90-40925
                                                     CIP

Copyright © 1990 by the AMERICAN SOCIETY FOR TESTING AND MATERIALS. All rights reserved. No part of this publication may be reproduced, stored in a retrieval system, or transmitted, in any form or by any means, electronic, mechanical, photocopying, recording, or otherwise, without the prior written permission of the publisher.

NOTE

The Society is not responsible, as a body,
for the statements and opinions
advanced in this publication.

Peer Review Policy

Each paper published in this volume was evaluated by three peer reviewers. The authors addressed all of the reviewers' comments to the satisfaction of both the technical editor(s) and the ASTM Committee on Publications.

The quality of the papers in this publication reflects not only the obvious efforts of the authors and the technical editor(s), but also the work of these peer reviewers. The ASTM Committee on Publications acknowledges with appreciation their dedication and contribution of time and effort on behalf of ASTM.

Printed in Baltimore, Md.
September 1990

Foreword

The First Symposium on Use of Plants for Toxicity Assessment was held in Atlanta, Georgia, on 19–20 April 1989. ASTM Committee E-47 on Biological Effects and Environmental Fate and its Subcommittee E47.11 on Plant Toxicity sponsored the event.

The symposium chairmen were: Wuncheng Wang, Illinois State Water Survey; Joseph W. Gorsuch, Eastman Kodak Company; and William R. Lower, University of Missouri. They have also served as editors of this volume.

Contents

Overview   1

REGULATORY PERSPECTIVES

Keynote Address: Plants—Keystone to Risk Assessment—F. BENENATI   5

Development of Guidelines for Testing Pesticide Toxicity to Nontarget Plants for Canada—K. FREEMARK, P. MAC QUARRIE, S. SWANSON, AND H. PETERSON   14

COMPARATIVE TOXICOLOGY

Use of Algae versus Vascular Plants to Test for Chemical Toxicity—J. S. FLETCHER   33

Comparison of Short- and Long-Term Toxicity Test Results for the Green Alga, *Selenastrum capricornutum*—D. J. VERSTEEG   40

Chemical Effects on the Germination and Early Growth of Terrestrial Plants—J. W. GORSUCH, R. O. KRINGLE, AND K. A. ROBILLARD   49

Comparison of the Seagrass *Thalassia testudinum* and Its Epiphytes in the Field and in Laboratory Test Systems—J. R. CLARK AND J. M. MACAULEY   59

Multispecies Methods of Testing for Toxicity: Use of the *Rhizobium*-Legume Symbiosis in Nitrogen Fixation and Correlations Between Responses by Algae and Terrestrial Plants—C. T. GARTEN, JR.   69

PLANTS AND XENOBIOTIC UPTAKE

Characterization of Xenobiotic Uptake Utilizing an Isolated Root Uptake Test (IRUT) and a Whole Plant Uptake Test (WPUT)—M. A. KRSTICH AND O. J. SCHWARZ   87

Bioaccumulation of Mercury Compounds in Two Aquatic Plants (*Elodea densa* and *Ludwigia natans*): Actions and Interactions of Four Abiotic Factors—F. RIBEYRE AND A. BOUDOU   97

Plants and Air Pollution

Use of Native and Cultivated Plants as Bioindicators and Biomonitors of Pollution Damage—L. H. WEINSTEIN, J. A. LAURENCE, R. H. MANDL, AND K. WÄLTI ... 117

Ecological Effects Evaluation of Two Phosphorus Smoke-Producing Compounds Using Terrestrial Microcosms—D. A. TOLLE, M. F. ARTHUR, K. M. DUKE, AND J. CHESSON ... 127

Use of Physiological and Biochemical Markers for Assessing Air Pollution Stress in Trees—C. J. RICHARDSON, T. W. SASEK, AND R. T. DI GIULIO ... 143

A Bioindicator System Assessing Air Quality Within Minnesota— K. W. KROMROY, M. F. OLSON, D. F. GRIGAL, P. S. TENG, D. R. FRENCH, AND G. H. AMUNDSON ... 156

The *Vicia* Leaf Tip Cell-Micronucleus Bioassay and Air Pollution Monitoring— CHEN GUANGRONG, JIN BO, LI MING, AND WENG XINGGUO ... 170

General Phytotoxicology

Seed Germination and Root Elongation Toxicity Tests in Hazardous Waste Site Evaluation: Methods Development and Applications—G. LINDER, J. C. GREENE, H. RATSCH, J, NWOSU, S. SMITH, AND D. WILBORN ... 177

Survey of Rocky Mountain Arsenal for Phytotoxic Substances—D. L. SIROIS ... 188

Protocol for Evaluating Soil Contaminated with Fuel or Herbicide—T. H. LILLIE AND R. W. BARTINE ... 198

Effects of Chromium in Freshwater Algae and Macrophytes—M. BASSI, M. G. CORRADI, AND M. A. FAVALI ... 204

Evaluation of Protocols for the Assessment of Phytotoxicity—D. L. SIROIS ... 225

Time-Dependent Toxicity Assessment of Herbicide Contaminated Soil Using the Green Alga *Selenastrum capricornutum*—M. W. THOMAS, B. M. JUDY, W. R. LOWER, G. F. KRAUSE, AND W. W. SUTTON ... 235

Use of *Thalassia* and Its Epiphytes for Toxicity Assessment: Effects of a Drilling Fluid and Tributyltin—J. M. MACAULEY, J. R. CLARK, AND A. R. PITTS ... 255

A Plant Toxicity Test with the Moss *Physcomitrella patens* (Hedw.) B.S.G.— E. L. MORGAN, YUEH-CHIN A. WU, AND R. C. YOUNG ... 267

Millet Root Elongation in Toxicological Studies of Heavy Metals: A Mathematical Model—WUNCHENG WANG AND G. ELSETH ... 280

New Approaches

The Role of Chlorophyll Fluorescence as a Bioassay for Assessment of Toxicity in Plants—D. MILES 297

Chlorophyll Fluorescence of a Higher Plant as an Assay for Toxicity Assessment of Soil and Water—B. M. JUDY, W. R. LOWER, C. D. MILES, M. W. THOMAS, AND G. F. KRAUSE 308

***Tradescantia* Sister-Chromatid-Exchange (SCE) Bioassay for Environmental Mutagens**—YAN PENG AND TE-HSIU MA 319

Activation of Promutagens by a Unicellular Green Alga—K. R. SAUSER AND S. J. KLAINE 324

Some Approaches to Rapid and Pre-Symptom Diagnosis of Chemical Stress in Plants—R. R. VELAGALETI, D. KRAMER, S. S. MARSH, N. G. REICHENBACH, AND D. E. FLEISCHMAN 333

Image Analysis System for Rapid Data Processing in *Tradescantia*-Micronucleus Bioassay—JIANHUA XU, WENJIE XIA, XUDONG JONG, WEICHI SUN, GUANGHENG LIN, AND TE-HSIU MA 346

Author Index 357

Subject Index 359

Overview

Both the higher and lower plants are essential parts of a healthy, balanced ecosystem. Both are primary producers, the oxygen-generating organisms of the biosphere, and are considered the "lungs of the earth." Nevertheless, plants have received relatively little attention in ecotoxicological studies during the past four decades in contrast to the overwhelming emphasis on faunal studies. It is not uncommon for the public to become agitated about fish kills, while plant (forest) die-off in mountains has aroused little concern until the 1980s. It is perhaps fitting to coin the expression "not seeing is not disturbing."

Only recently has the general public been made aware through the mass media of the grim reality that plant life has been suffering from adverse impacts caused by cultural (large expanses of the rain forest being cleared in the Amazon region of South America) and natural activities (the destruction of forests by volcanic activity) and that the decline of plant life affects all other life forms. Forest declines in western Europe and North America due to air pollution, especially the northeastern part of North America, are well-documented. In turn, the effects of contaminants on plants can directly, as well as indirectly, affect man. For example, in parts of the world soil pollution is so severe that farmers harvesting grain sell it on the open market rather than consume it themselves. Grain purchased by the same farmer at the market, however, might have been rejected by farmers of other regions, creating a trade-off of contaminated food. The conviction of the symposium committee that plants are a vital part of a healthy environment, and that there has been little attention paid to plant life by many scientific sectors, convinced them of the need for a symposium that would direct the attention of scientists and non-scientists to these lapses. This symposium provided a forum to promote (1) the gathering and dissemination of information and (2) the development of standard practices to assess the impact of chemicals and other xenobiotics upon the plant communities and to use plants to biologically assess toxicity.

The First Symposium on Use of Plants for Toxicity Assessment was sponsored by ASTM Committee E-47 on Biological Effects and Environmental Fate and its Subcommittee E47.11 on Plant Toxicology, and held in Atlanta, Georgia, on 19–20 April 1989. The symposium attracted some 112 researchers from Canada, China, France, Italy, the Netherlands, the United Kingdom, and the United States. Scientists from academia, industry, consulting laboratories, and governments were represented. The attendees' disciplines spanned ecotoxicology, biochemistry, ecology, plant physiology, and genetics, to name only a few. The platform and poster sessions included topics ranging from freshwater to seawater environments, from higher to lower plants, and from air pollution to water pollution to soil pollution. Themes covered many aspects of ecotoxicology as well as genotoxicology.

The papers indicated that research needs to focus on standardizing bioassays to evaluate the impact of new and existing chemicals, to evaluate the impact of fugitive emissions on forests and other plant ecosystems, and to evaluate the introduction of biological control agents. The papers further indicated that plant toxicologists and regulatory scientists need to reach an agreement on precise test methods, test endpoints, and data interpretation.

This ASTM Special Technical Publication is an outgrowth of the symposium. It contains 29 refereed papers divided into six groups: Regulatory Perspectives, Comparative Toxicology, Plants and Xenobiotic Uptake, Plants and Air Pollution, General Phytotoxicology, and New Approaches. Frank Benenati of the U.S. EPA presented the keynote speech, describing the role of plant toxicology in risk assessment under the guidance of the Toxic Substances Control Act. He noted that relatively few Premanufacturing Notifications (PMNs) have any plant data submitted with them and that the data are primarily algal inhibition studies. There is concern that the plant environment is not being fully investigated or protected.

The use of plants to assess the toxicity of air, water, and soil has a promising future. To be successful the momentum that began at this symposium must be maintained through the participation of scientists around the world. We must encourage ourselves to seek answers to the many problems that face us. The papers presented herein can serve as the catalyst that will provide for more opportunities to advance the use of plants to biologically assess the impact of environmental contamination on all life.

The symposium committee thanks Rick Cardwell, former chairman of Committee E-47, for his support of this project; the authors, who contributed and shared their findings with the attendees; and the reviewers, whose time-consuming efforts and constructive comments resulted in much-improved papers. ASTM staff members are acknowledged for their assistance in organizing this ground-breaking symposium as well as their efforts in producing this timely publication. They include Dorothy Savini, Kathy Greene, Monica Armata, and Rita Harhut.

Wuncheng (Woodrow) Wang
Symposium Chairman and Editor
Illinois State Water Survey
Peoria, IL 61652

Joseph W. Gorsuch
Symposium Co-Chairman and Editor
Eastman Kodak Company
Rochester, NY 14652-3617

William R. Lower
Symposium Co-Chairman and Editor
University of Missouri
Columbia, MO 65203

Regulatory Perspectives

Frank Benenati[1]

Keynote Address: Plants—Keystone to Risk Assessment

REFERENCE: Benenati, F., "**Keynote Address: Plants—Keystone to Risk Assessment,**" *Plants for Toxicity Assessment, ASTM STP 1091,* W. Wang, J. W. Gorsuch, and W. R. Lower, Eds., American Society for Testing and Materials, Philadelphia, 1990, pp. 5-13.

ABSTRACT: This keynote speech provides an overview of the plant test data submitted in the past ten years to the U.S. Environmental Protection Agency under the Toxic Substances Control Act (TSCA). Only 155 of 12 403 Premanufacture Notices required for new chemicals contained plant data. Most of these notices (149) contained only algal data. An additional 23 algal assays were required for existing chemicals subject to test rules and testing consent orders. Plant test data from over 400 studies (again mostly algal) were also submitted in response to information gathering rules. The plant data, while less frequently submitted than fish or aquatic invertebrate data, demonstrated that the algal assay is a good tool for screening chemicals for toxicity. However, whether or not this test adequately screens chemicals for phytotoxicity is questioned. The use and limitations of plant data in risk assessment are also discussed along with the need for standardization and validation of plant test methods.

KEY WORDS: Toxic Substances Control Act (TSCA), plant testing, phytotoxicity, risk assessment, algal assay, screening

I assume that most of us attending this symposium agree that plants are important and accept as a truth that without plants life on this planet would not exist as we know it. So, if plants are so important, why are we so hard pressed to convince others of their importance in environmental risk assessment?

The one exception to this dilemma is man's continued interest in food chain contamination. There is a shared concern for the presence of radionuclides, pesticide residues, and other contaminants in crops and plant products. In support of this, many of us have conducted laboratory and field studies to evaluate contamination of crops and forage and the resultant effects on livestock. These studies may also have included a visual assessment of damage to vegetation as a diagnostic aid in identifying the causal agent(s).

A lesser concern, which does not seem to fare as well, is testing for direct effects on plants. I believe the reported nonsensitivity of plants to chemical stress by a number of researchers [e.g., *1-3*] has decreased the perceived importance of plant toxicity testing. Our presence here today suggests that we believe the opposite. I anticipate that the papers presented at this symposium will reinforce our belief.

My purpose today is twofold: (1) to provide a summary and characterization of the plant test data submitted to the U.S. Environmental Protection Agency (EPA) during the past ten years in response to requirements of Sections 5, 4, and 8 of the Toxic Substances Control Act (TSCA); and (2) to discuss what plant scientists need to do to enhance the role of plants in risk assessment.

---

[1] Section Chief, Chemical Testing Branch, Office of Toxic Substances, U.S. Environmental Protection Agency, Washington, DC 20460.

TSCA Section 5

Premanufacture Notices (PMNs) are required for new chemicals (those not on the TSCA Chemical Inventory) and are submitted under TSCA Section 5. Since the identity of many PMN chemicals is confidential business information, my analysis is limited to an overview of the frequency of testing, type of tests conducted, endpoints reported, and relative test sensitivity.

The number of PMNs submitted to EPA has gone from 8 in 1979 to over 2600 in 1988 (Table 1). The percentage of PMN submissions received in any given year that included environmental effects data of any kind ranged from 4 to 25%. Only 1 to 3% of the PMNs included plant data. While the frequency of plant testing is somewhat disappointing to plant scientists, some interesting observations can be drawn from these sparse data.

Of the 155 PMNs submitted with plant data, the traditional algal assay was the test of choice for 149. All but 18 of these algal tests used *Selenastrum capricornutum* as the test species. *Lemna minor* acute toxicity (1), early seedling growth (4), and plant uptake (1) were the tests of choice for the 6 PMNs submitted with other plant data. A variety of crop and weed species were tested. Apparently, the terrestrial plant test data were not developed specifically for the PMN. Rather, the data appear to have been developed for registration of a pesticide or for a patent and submitted as available data.

For all but 22 of these 149 algal tests, a 96-h EC_{50} was reported. For 9 of these 22, a 96-h NOEL was reported; while for the remaining 13, a variety of endpoints were reported. In recent years, almost all studies have reported a 96-h EC_{50}. In addition, for almost all algal studies, the effect of the test substance on biomass (cells/mL) and/or population growth (change in cell number over time) were the evaluated criteria. Chlorophyll *a* fluorescence was also routinely measured, but as an indirect determinant of biomass. For only a few tests was a direct measure of cell function (assimilation of 14-C) reported. Studies to determine if a chemical was algicidal or algistatic were seldom conducted.

Another interesting observation is that 36 of the algal tests (29%) reported a 96-h EC_{50} equal to or less than 1 mg/L. More than half (20) of these were between 0.01 and 0.1 mg/L (Table 2). As such, the test system appears responsive and able to detect effects at concentrations that may occur in the environment.

I next compared sensitivity of the endpoints from the algae tests to the traditional endpoints from the *Daphnia* acute toxicity (48-h EC_{50}) and fish acute toxicity (96-h LC_{50}) tests

TABLE 1—*PMN environmental effects testing.*

Year	No. of PMNs Submitted	No. with Environmental Data, %	No. with Plant Data, %
1979	8	2 (25)	0
1980	300	21 (7)	2 (<1)
1981	600	70 (12)	3 (<1)
1982	850	32 (4)	3 (<1)
1983	1300	116 (9)	6 (<1)
1984	1200	125 (10)	24 (2)
1985	1450	114 (8)	38 (3)
1986	1700	90 (5)	21 (1)
1987	1800	120 (7)	38 (2)
1988	2645	168 (6)	18 (<1)
1989[a]	550	34 (6)	2 (<1)
Total	12403	892 (7)	155 (1)

[a] As of March 1989.

TABLE 2—*Algal toxicity (96-h EC_{50}) values.*

Toxicity, mg/L	No. of PMNs
0.01–0.1	20
0.11–1.0	16
1.1–10.0	35
10.1–100	38
>100	16

reported for those chemicals tested for both plant and animal toxicity. Based upon these data, algae were more sensitive than *Daphnia* or fish to 78 of the 155 PMN chemicals tested (50%). Another 15 of the chemicals (10%) were equally toxic; while only 47 (30%) were more toxic to *Daphnia* or fish. Data were inadequate for the remaining 15 chemicals (10%).

These data, the seeming proof that the algal assay is the keystone test in a chemical screening base set, might not be all that they appear. The types of new chemicals we usually receive environmental data on are either charged (e.g., cationic or anionic surfactants) or are anticipated to be toxic at a low dose based upon their chemical structure (e.g., structural analogue to a pesticide or known toxic chemical). The analysis of sensitivity might be slightly biased in favor of plants, since about 33% of the PMNs submitted for a class of chemicals to which plants are consistently more sensitive to than fish or *Daphnia* contained algal or plant data. On average, though, only 17% of the PMNs submitted with environmental data also contained plant data. Although some of these chemicals were tested because the manufacturer or EPA thought the chemical might be phytotoxic or have herbicidal properties, it does not appear that plant testing was only conducted when there was cause. Rather, I believe that many of the plant data submissions resulted from voluntary compliance with the "base set" testing philosophy supported by many environmental scientists.

The apparently greater sensitivity of algae might also be explained by comparing the type of test the algal assay is with the type of test the fish and *Daphnia* acute toxicity tests are. The 72 to 96-h algal assay is often considered an acute test, but is it? It measures several cell divisions of an algal population. Time and dose needed to cause a lethal effect alone do not make a test "acute". For comparison, the fish and *Daphnia* acute tests measure response to a lethal dose over only a brief (48 to 96 h) portion of the young organism's life. This difference alone might suggest why the algal test appears to be more sensitive than the other two tests.

Over the last two years, the EPA has received about 20 PMNs for genetically engineered organisms that have included plant testing data. These PMNs are not accounted for in Tables 1 and 2. The plant testing submitted with these PMNs has been conducted primarily to evaluate if nitrogen fixation could be enhanced by genetically altered *Rhizobium*. The testing has included laboratory growth chamber and limited (small-scale) field trials with a variety of crop species.

TSCA Section 4

Manufacturers have also conducted plant testing under Section 4 of TSCA. Section 4 covers existing chemicals, those on the TSCA Chemical Inventory. Plant testing, as with section 5, has focussed on algal toxicity. To date, 23 algal assays have been completed while 3 are underway. Of the completed tests, 96-h EC_{50}s could be determined for only 9 (Table 3). Volatility or solubility of the remaining 14 chemicals prevented determination of EC_{50}s (Table 4). As with PMNs, *Selenastrum* was the species of choice, and cell density and chlo-

TABLE 3—*Successful algal testing under Section 4.*

96-h EC_{50}, mg/L	Chemical	Ref.
0.1–1.0	ortho-Phenylenediamine	4
	para-Phenylenediamine	4
	Di-n-butyl phthalate	5
	Butyl benzyl phthalate	5
1.1–10.0	meta-Phenylenediamine	4
	Bisphenyl A	6 & 7[a]
	Octylphenol[b]	8
10.1–100	Diethyl Phthalate	5
>100	Dimethyl Phthalate	5

[a] *Skeletonema nostratum.*
[b] Tetramethylbutyl phenol.

rophyll *a* fluorescence were the most frequently reported evaluation criteria. The chemicals with algal testing under way are: 2,6-dichloro-4-nitroanaline, octamethylcyclotetrasiloxane, and di-*tert*-butyl phenol.

During the past ten years, EPA has proposed other plant testing under Section 4 but has not required such testing in either a final rule or consent order (Tables 5 and 6). Both routine (e.g., seed germination/root elongation, early seedling growth) and less routine (e.g., plant life cycle, aquatic macrophyte) testing have been proposed but then withdrawn. The reasons for this have been either that after re-evaluation, the EPA was not convinced that either exposure or release was sufficient to warrant testing under Section 4(a)(1)(B) of TSCA (the chemical "may present a risk due to significant exposure or substantial release"), or that adequate test standards were not available to require manufacturers to do this testing. A Section 4 test rule or consent order must specify the test standards and reporting requirements for required tests.

Note that the terrestrial plant test guidelines published in Chapter 40 of the Code of Federal Regulations, Section 797, have not been converted to test standards. Most of the other

TABLE 4—*Section 4 chemicals with 96-h $EC_{50}s$ not determined.*

Problem	Chemical	Ref.
Volatility	3,4-Dichlorobenzotrifluoride	9
	Dichloropropane	10 & 11
Solubility	Tetrabromobisphenyl A	12
	Chlorowax 500 C	13
	Phthalates:	7
	Butyl-2-ethylhexyl	
	Di (heptyl, nonyl, undecyl)	
	Di (*n*-hexyl,*n*-octyl,*n*-decyl)	
	Di-2-ethylhexyl	
	Dihexyl	
	Diisodecyl	
	Diisononyl	
	Diisooctyl	
	Ditridecyl	
	Diundecyl	

TABLE 5—*Terrestrial plant testing proposed but not finalized.*

Chemicals	Reasons for Drop
Seed Germination & Root Elongation	
• Dichloromethane	• Insufficient exposure to support testing
• Nitrobenzene	• No substantial release
Early Seeding Growth	
• Monochlorobenzene • 1,2-Dichlorobenzene • Dichlorobenzene • 1,2,4-Trichlorobenzene • Dichloromethane	• Insufficient exposure to support testing
• Nitrobenzene	• No substantial release
Plant Life Cycle Test	
• Dichloromethane • Nitrobenzene	• Standard methods not available
Plant Uptake/Translocation	
• Dichloromethane • Nitrobenzene • 1,1,1-Trichloroethane	• Insufficient exposure to support testing

environmental effects guidelines were converted in 1985. As a consequence, the incorporation of plant testing other than with algae in test rules continues to be difficult.

TSCA Section 8

Industry has also submitted about 400 plant studies under Section 8(d) and 6 under Section 8(e) of TSCA. Section 8(d) provides EPA with the authority to require the submission of known health and safety studies. Section 8(e) requires any person who manufactures, processes, or distributes in commerce a chemical that may cause substantial risk to notify the Administrator of EPA.

TABLE 6—*Aquatic plant testing proposed but not finalized.*

Chemicals	Reasons for Drop
Fresh and Saltwater Algae	
• Hydroquinone • Quinone	• Insufficient exposure to support testing
Lemna Minor	
• Biphenyl • Tetrachlorobenzene • 1,2,4-Trichlorobenzene	• Insufficient hazard data to support *Lemna* as more sensitive
Fresh and Saltwater Macrophyte	
• Dichloromethane • Nitrobenzene • Hydroquinone • Quinone	• Standard methods not available

EPA has used most 8(d) rules to support rule-making under Section 4 of TSCA. The submitted studies and any published studies are evaluated by EPA to determine if additional testing is necessary. EPA has issued 8(d) rules for over 300 Section 4 chemicals. These rules cover 15 chemical categories, 1 mixture, and 114 individual chemicals. EPA has also used Section 8(d) of TSCA to collect information on over 150 additional chemicals in support of other regulatory programs within EPA (e.g., Office of Solid Waste, Office of Pesticides).

The quality of the 8(d) data is extremely mixed; some studies are quite old and of limited value for assessing risk. For example, in one study conducted in the 1950s, 70 different oil emulsions were each sprayed on cucumber. The plants were only visually scored for phytotoxicity 14 days later. Many other 8(d) plant studies were conducted on test substances with physical or chemical properties that raise questions on study design and data adequacy. In a few instances, 8(d) studies have been considered adequate, and additional testing has not been required under Section 4. The termination of rule-making for 3,4-dichloroaniline is a recent example [14].

About 75% of the 8(d) studies were conducted with algae. The remaining laboratory studies were either early seedling growth (20%) or seed germination tests (5%). The study with oil emulsions, though, accounted for 70 records, suggesting that the seedling growth test was not a more popular choice than the seed germination test. Of special note, a few nonlaboratory plant studies were also submitted.

To date, industry has submitted two 8(e) notices that include plant test data. The notices were for 4-chloro-3-nitroaniline and 4-nitrobenzamide. In both instances, the initial submissions were made due to concern for possible health effects. The results of environmental effects testing, including phytotoxicity screening (seed germination, root growth, and early seedling growth testing), were reported in subsequent notices. The submitter considered the potential for adverse environmental effects to be low, given the physical chemical properties, relative toxicity, and the use and disposal practices for these chemicals.

Plants and Risk Assessment

The algal assay has been and will continue to be the test of choice for assessing phytotoxicity. The data submitted under TSCA support its usefulness and importance in environment risk assessment. We do not know, however, whether another plant test method or even evaluation of other assessment criteria in the algal assay might be better for assessing phytotoxicity. I have never been convinced that the measurement of cell numbers or population growth in the algal assay adequately screens chemicals for phytotoxic effects, since a wide variety of herbicides test negative by this method [15,16]. Of greater concern, a variety of other biological processes unique to plants are not evaluated in this test; thus we can not conclude that an effect on algal biomass is predictive of effects unique to other algae or higher plants. For example, *Champia parvula* (a marine red alga) and *Arabidopsis thaliana* (a herbaceous member of the Cruciferae family) have both shown effects on sexual reproduction from chemical concentrations much lower than reported for criteria evaluated in traditional assays [17,18].

As mentioned earlier, several researchers have concluded that plants are nonsensitive to chemical stress and that assessment of effects on animal life (i.e., fish and aquatic invertebrates) will protect plants since animals are more sensitive. The basis for such claims, though, have been studies that in many instances tested nonsensitive species, such as *Chlorella vulgaris,* and selected nonsensitive or severe endpoints for quantification of effects, such as 100% death. Undoing the impact of such publications continues to be a slow process. The screening level data upon which these conclusions are based were likely developed for the purpose of identifying if existing chemicals could eradicate hardy weed species and be potentially marketed as herbicides. If the purpose had been to determine at what concentration a chemical might significantly affect more sensitive plant species or cultivars or might cause subtle effects, the results and conclusions reached might have been markedly different. Subtle effects such as decreased seed production, altered fruit size and shape, or a changed ratio of male and female plants may be as important as plant necrosis. For example, 6.4 $\mu g/m^3$ of atmospheric fluoride has been reported to decreae the weight of tomatoes by almost 50% and the number of fruits per plant by about 25% [19]. The tomato, though, is considered a fluoride tolerant species. The farmer would sustain economic loss even though his crop otherwise might appear normal and healthy. We can each provide examples of plant response to chemical stress that demonstrate plant sensitivity and real-world adverse effects.

Since several papers on using plants to assess health effects are being presented at this symposium, there are several comments I would like to make on the applicability of plant testing in health risk assessment. I believe the acceptance of plant and nonmammalian testing as surrogates for man will gain in importance in chemical risk assessment. Concern for animal rights provides a strong incentive for developing alternative test methods. However, the ability of these alternative tests to correctly identify known human (or mammalian) mutagens and nonmutagens needs to be demonstrated before such test data will gain acceptance for risk assessment. My understanding is that, to date, these tests have not been validated as to their predictability of response in mammalian systems. That is: Do chemicals that test negative in the plant micronucleus test also test negative in the mouse micronucleus test? Similarly: Does a positive result in such a test provide an adequate basis to suggest that more traditional mammalian testing is warranted? Most importantly: Do such data support the "may present an unreasonable risk" finding that EPA must make before it can require additional testing under Sections 4 and 5 of TSCA? If it is any consolation to plant scientists, the use of fish for screening chemicals for carcinogenicity is on an "equal footing" with using plants for screening for mutagenicity. However, the application of such tests (both plant and animal) to monitoring emissions near points of discharge (as sentinel species) appears promising and worth pursuing. The current EPA regulatory testing philosophy for mutagenicity and carcinogenicity is limited primarily to *Sulmonella, Drosophila,* and rodent testing. No plant test guidelines for health evaluation have been published by EPA for use as test standards under Section 4 or as guidance under Section 5. To date, no plant test data for evaluating health effects have been submitted with a PMN or required under a Section 4 test rule. If such data were to be received, it would be considered along with any animal data to assess risk; however, I doubt that it alone would be considered sufficient for risk assessment.

Issues

Plant scientists must think about the steps needed to advance the application of our science in risk assessment. We need to determine what additional testing is appropriate for those chemicals that test highly toxic in the algal assay or in other plant toxicity screening tests. Many environmental scientists believe that the algal assay as conducted by most laboratories simply screens for biological activity. Just as many believe it screens for phytotoxicity. Some believe that if positive results are shown in this test, then chronic animal (fish or aquatic invertebrate testing) is appropriate. Do you agree? What additional plant testing is needed to adequately define a NOEL to plants and more importantly to plant communities?

Another issue that has been inadequately addressed is whether the algal assay is a good enough screen for phytotoxicity. Many believe that the early seedling growth test might be a more effective screening test. Probably each of us has a "pet" test that we believe to be the best.

How do we advance our science? ASTM provides us with an opportunity not only to exchange ideas and present results of our research at symposia but to discuss and develop consensus test methods. Our science, as most, suffers from a literature that reflects past practice. It is hard to find two studies that follow the same methodology that were not published by the same author. If we are to see plant testing advance to the same level as animal testing, we need to work towards developing consensus testing and risk assessment methodology. For those of us working to better define the boundary between plant and animal sciences, our task is not just demonstrating the reproducibility or sensitivity of our science but also its applicability. Your participation on ASTM Subcommittee E47.11 on Plant Toxicity, which works towards these ends, is encouraged.

Acknowledgment

I would like to thank Vince Nabholz for his making the PMN data summaries available for my analysis, Jim Bradshaw for searching the TSCATS database for Section 8 submissions, Harriet Colbert for searching the SMART and TSCATS databases for Section 4 actions with plant testing, and Barbara Ostrow for compiling information on the number of Section 8(d) actions.

References

[1] Kenaga, E. E. and Moolenaar, R. J., "Fish and *Daphnia* Toxicity as Surrogates for Aquatic Vascular Plants and Algae," *Environmental Science and Technology,* Vol. 13, 1979, pp. 1479–1480.
[2] Jensen, A., "Marine Ecotoxicological Tests with Seaweeds," in *Ecotoxicological Testing for the Marine Environment,* Vol. 1, G. Persoone, E. Jaspers, and C. Claus, Eds., State University of Ghent and the Institute for Marine Scientific Research, Bredene, Belgium, 1984, pp. 181–183.
[3] Kenaga, E. E., "The Use of Environmental Toxicology and Chemistry Data in Hazard Assessment: Program Needs, Challenges," *Environmental Toxicology and Chemistry,* Vol. 1, 1982, pp. 69–79.
[4] Erco, Inc., Toxicity data for *para-, meta-,* and *ortho-*phenylenediamine to *Selenastrum capricornutum* in a letter dated 27 Oct. 1985 submitted to EPA by E. I. Du Pont de Nemours & Co., Wilmington, Del., 1985.
[5] Springborn Bionomics, "Toxicity of Fourteen Phthalate Esters to the Freshwater Alga *(Selenastrum capricornutum),*" Report BP-84-1-4, Chemical Manufacturers Association, Washington, D.C., 1984.
[6] Alexander, H. C., Dill, D. C., Milazzo, D. P., and Boggs, G. U., *Bisphenol A: Algal Toxicity Test,* Dow Chemical USA, Midland, Mich., 1985.
[7] Springborn Bionomics, "Toxicity of Bisphenol A to the Marine Alga *Skeletonema costatum,*" Report BW-85-8-1832, Society of Plastics Industry, New York, N.Y., 1985.
[8] Analytical Biochemistry Laboratories, "Acute Toxicity of Octylphenol to *Selenastrum capricornutum,*" Report 31913 for Chemical Manufacturers Association, Washington, D.C., 1984.
[9] Springborn Life Sciences, "Toxicity of 3,4-Dichlorobenzotrifluoride to the Freshwater Alga *Selenastrum capricornutum,*" Report 88-04-2686 for Occidental Chemical Corp., Niagara Falls, N.Y., 1988.
[10] Hughes, J. S., "1,2-Dichloropropane: The Toxicity to *Skeletonema costatum,*" Report 0460-03-1100-2, Malcolm Pirnie, Elmsford, N.Y., 1988.
[11] Hughes, J. S., "1,2-Dichloropropane: The Toxicity to *Selenastrum capricornutum,*" Report 0460-03-1100-1, Malcolm Pirnie, Elmsford, N.Y., 1988.
[12] Springborn Life Sciences, "Toxicity of Tetrabromobisphenol A to the Freshwater Green Alga *Selenastrum capricornutum,*" Report 88-10-2828, Brominated Flame Retardant Industry Panel, West Lafayette, Ind., 1988.
[13] Thompson, R. S. and Madley, J. R., *"Toxicity of a Chlorinated Paraffin to the Green Alga Selenastrum capricornutum,"* Imperial Chemicals Industries PLC Brixham Laboratory, Devon, U.K., 1983.
[14] U.S. Environmental Protection Agency, "Termination of Rulemaking for Certain Members of the Aniline Category," *Federal Register,* Vol. 53, No. 161, 1988, pp. 31814–31817.

[15] Decleire, M., DeCat, W., and Bastin, R., "Possibilité de détection d'herbicides dans l'eau par l'inhibition du verdissement des cotylédons de concombres étiolés et excisés," (Application of Six Bioassays for Detection of Herbicides in Water), *Parasitica,* Vol. 29, 1973, pp. 185-193.
[16] Garten, C. T. and Frank, M. L., *"Comparison of Toxicity to Terrestrial Plants with Algal Growth Inhibition by Herbicides,"* Publication No. 2301, Environmental Sciences Division, Oak Ridge National Laboratory, Oak Ridge, Tenn., 1984.
[17] Thursby, G. B. and Steele, R. L., "Comparison of Short- and Long-Term Sexual Reproduction Tests with the Marine Red Alga *Champia parvula,*" *Environmental Toxicology and Chemistry,* Vol. 5, 1986, pp. 1013-1018.
[18] Ratsch, H. C., "Growth Inhibition and Morphological Effects of Several Chemicals in *Arabidopsis thaliana* (L.)," *Environmental Toxicology and Chemistry,* Vol. 5, 1986, pp. 55-60.
[19] Pack, M. R., Hill, A. C., Thomas, M. D., and Transtrum, L. G., "Response of Tomato Fruiting by Hydrogen Fluoride as Influenced by Calcium Nutrition," *Journal of the Air Pollution Control Association,* Vol. 16, 1966, pp. 541-54.

Kathryn Freemark,[1] Patricia MacQuarrie,[2] Stella Swanson,[3] and Hans Peterson[3]

Development of Guidelines for Testing Pesticide Toxicity to Nontarget Plants for Canada

REFERENCE: Freemark, K., MacQuarrie, P., Swanson, S., and Peterson, H., "Development of Guidelines for Testing Pesticide Toxicity to Nontarget Plants for Canada," *Plants for Toxicity Assessment, ASTM STP 1091,* W. Wang, J. W. Gorsuch, and W. R. Lower, Eds., American Society for Testing and Materials, Philadelphia, 1990, pp. 14–29.

ABSTRACT: Guidelines are currently being developed to outline requirements for nontarget plant testing for pesticide registration in Canada. The Saskatchewan Research Council (SRC), under contract to Environment Canada, evaluated international regulatory requirements and test protocols, identified research needs, and recommended guidelines for Canada, by reviewing the scientific literature and databases, and consulting with regulators and scientists involved in phytotoxicity testing. Highlights of the SRC reports and a review by a federal working committee are presented to stimulate additional research, and to solicit critical review from scientists and regulators prior to formal drafting of Canadian regulatory guidelines.

KEY WORDS: plants, algae, pesticides, regulations, phytotoxicity, review, Canada

Use of herbicides is extensive in Canada. In 1986, 22.9 million hectares of farmland were treated with herbicides [*1*], representing almost a threefold increase since the early 1970s. Correspondingly, concerns about the potential for adverse impact of phytotoxic chemicals on nontarget organisms directly, or indirectly through critical alterations in habitat, have increased [*2*].

In Canada, the federal government regulates the registration, classification and labelling of pesticide products, while provincial governments regulate their actual use through licenses, permits, and related regulatory techniques [*3*]. At the federal level, the use of pesticides is regulated primarily by the Pest Control Products (PCP) Act, R.S.C. 1970, C.P-10, as amended, which is administered by Agriculture Canada. Environment Canada and the Department of Fisheries and Oceans are responsible for advising Agriculture Canada on aspects of environmental chemistry and fate and environmental toxicology (including ecological effects) of pesticides being considered for new or continuing registration under the PCP Act (Fig. 1). Data requirements for nontarget phytotoxicity are not formally addressed in the guidelines for pesticide registration and are currently handled on a case-by-case basis only. Accordingly, Environment Canada initiated work towards the development of regulatory guidelines. The present paper provides highlights of those efforts in order to stimulate additional research and solicit comment from scientists and regulators prior to formal drafting of regulatory guidelines.

---

[1] Canadian Wildlife Service, Environment Canada, Ottawa, Ontario, Canada.
[2] Environmental Protection, Environment Canada, Ottawa, Ontario, Canada.
[3] Saskatchewan Research Council, Saskatoon, Saskatchewan, Canada.

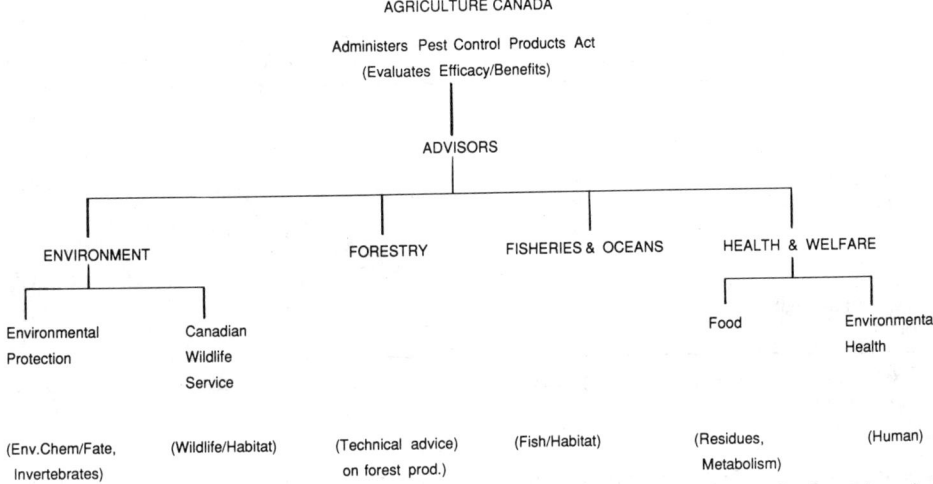

FIG. 1—*Federal departmental structure and responsibilities in the review of pesticides for registration in Canada.*

Methods

Staff (S. Swanson, H. Peterson) of the Saskatchewan Research Council (SRC) were contracted by Environment Canada to evaluate regulatory requirements and test protocols, identify research needs, and generate recommendations for Canadian guidelines. The scientific literature on nontarget plant toxicity testing was reviewed. Various databases of the Chemical Information System (CIS) were searched to obtain additional information to support recommendations for testing pesticide toxicity to aquatic plants. Current regulatory requirements for nontarget plant testing were discussed with representatives of the Organization for Economic Cooperation and Development (OECD), the European Economic Community (EEC), the Danish National Agency for Environmental Protection, and the U.S. Environmental Protection Agency (EPA). A questionnaire was designed and mailed to 64 scientists involved in aspects of phytotoxicity testing in agriculture, environment, forestry, and the pesticide industry.

A federal working committee involving Environment Canada, the Department of Fisheries and Oceans, Forestry Canada, and Agriculture Canada was established to provide a regulatory and scientific forum for exchange of information and review of the project.

Results

Highlights of the SRC contracts are presented below. Full details are contained in Environment Canada reports [4,5].

International Requirements and Test Protocols

The requirements for nontarget plant testing of pesticides vary among international agencies and their member counries from which information could be obtained (see also Table 1). Standard test protocols also vary, particularly for algal growth inhibition tests (see below).

European Economic Community (EEC)—The EEC has published a Directive [6] stipulating that member countries are expected to have pesticide registration requirements that

TABLE 1—*Summary of nontarget plant test requirements for pesticides by international agency and country (OECD = Organization for Economic Cooperation & Development, EEC = European Economic Community, USA = United States of America, UK = United Kingdom, AUS = Australia).*[a]

	OECD	EEC	USA	UK	AUS
Aquatic					
Algal Growth	R	R	R	NR	RI
Vascular Growth	RI	RI	CR	NR	RI
Field	RI	NR	CR	NR	RI
Terrestrial					
Germination/Emergence	CR	R	R	NR	RI
Vegetative Vigor	RI	RI	CR	NR	RI
Field	RI	NR	CR	NR	RI

[a] R = required; CR = conditionally required; NR = not required; RI = requirement implied but no specifics given.

include at least an algal growth inhibition test and a test on a higher plant. EEC test methods [7] are identical with those of the OECD (see below). An improved version of the higher plant test is currently being evaluated.

Food and Agriculture Organization (FAO)—In a report on international harmonization of pesticide registration requirements [8], the FAO supported the need for toxicity data on a variety of nontarget organisms. However, nontarget plants were not mentioned specifically. The recommendation on test species only stated that careful selection was needed in order to justify broad environmental predictions being made from the results of the test program.

International Standards Organization (ISO)—The ISO provides standard test protocols for general chemical testing. The only nontarget plant protocol is for algal growth inhibition [9]. The test involves one algal species (*Scenedesmus subspicatus* or *Selenastrum capricornutum*), at least five chemical concentrations, and a 72-h test duration.

Organization for Economic Cooperation and Development (OECD)—The OECD recommends a three-tiered system of basic, confirmatory, and definitive testing for pesticides and other chemicals [10,11]. Criteria for progressing beyond the basic tier include: persistence, significant bioaccumulation, high acute toxicity, lack of threshold responses, mutagenicity, changes in production volume, use or disposal pattern, increases in toxicity due to environmental modification, and the ratio of the no-observed-effect concentrations (NOEC) to the expected environmental concentration [12].

The only phytotoxicity test required at the basic tier is an algal growth inhibition test. The test uses one species from a recommended list of three, at least five concentrations of the test substance, and a 72-h test duration. At the confirmatory tier, recommended aquatic plant tests include: growth tests on additional algal species, and acute toxicity, growth, and reproduction tests on *Lemna* and other aquatic vascular species. A standard protocol for the *Lemna* test is currently under development by OECD. Recommended terrestrial plant tests include: germination, growth, and partial life-cycle tests on both monocots and dicots. The standard protocol for the growth test involves testing a minimum of three species, one from each of three categories (cereals, brassicas, and legumes/lettuce/cress), using three concentrations of the test substance mixed with soil, and a test duration of no fewer than 14 days after emergence of control seedlings. Tests at the definitive tier have not been fully developed; general categories include tests with confined natural communities, aquarium tests with artificial communities, and compartment tests with separate trophic levels.

Pesticides are regulated under the authority of separate legislation in OECD member countries [13]. Nontarget plant testing is required by most member countries (e.g., Denmark,

Federal Republic of Germany, Finland, Netherlands, Sweden, Switzerland, Turkey, United States, but not Canada or the United Kingdom), although the extent of test requirements beyond the basic tier varies among countries and is often developed on a case-by-case basis.

United States of America—In the United States, the EPA requires a three-tier system of pesticide testing based on general use pattern or hazard [14]. The EPA is the only agency that formally includes field testing as a requirement.

Tier 1 testing is required for terrestrial nonfood (e.g., grasslands), aquatic nonfood, and forestry uses. Other uses (e.g., terrestrial food crop) are exempt except when (*i*) phytotoxicity problems are observed during use and existing data are inadequate to address the problem, or (*ii*) the product poses hazards to endangered or threatened species, or (*iii*) a special review is initiated on the product. In practice, potential hazard to endangered or threatened plant species is a common justification for requiring nontarget plant testing for exempt use patterns. Progression to Tier 2 is based on the level of detrimental effects observed in Tier 1 tests. Progression to Tier 3 field testing is based on the potential hazard posed by the use pattern to terrestrial or aquatic plants tested in Tier 2. Field tests may also be required on a case-by-case basis for terrestrial plants, or for aquatic plants, when the pesticide is to be used within a forest system or is expected to enter water by direct application or direct discharge of treated water.

Test protocols needed to produce the required data for nontarget plants are fully described elsewhere [15]. For terrestrial plants, the Tier 1 test includes protocols for seed germination, seedling emergence, and vegetative vigor tests using 10 species (6 dicots, 4 monocots) and a single test concentration applied to germination/filter paper or in soil or acid-washed quartz sand (germination, emergence) or sprayed (vigor). Technical grade of the active ingredient (as defined in [14]) is used with a test duration of five days for the germination test and at least two weeks for the emergence and vigor tests. The Tier 2 protocol is a further seed germination, seedling emergence, and vigor test using at least five test concentrations and at least the species showing phytotoxic effects in Tier 1. The Tier 3 terrestrial test is a field test using 10 species representing monocots, dicots, vascular cryptogams, bryophytes or hepatophytes, and gymnosperms and the same dosages as those used in Tier 2 but with the end-use product (as defined in [14]). The duration of field tests has to be sufficient to assess multiple applications (if specified on the product label) with observations continuing at least two weeks after the last application (up to a maximum of four weeks) to note recovery or death.

For aquatic plants, Tier 1 includes algal growth inhibition protocols for four species (a green, a freshwater diatom, a marine diatom, a blue-green), and a protocol for the *Lemna gibba* (duckweed) growth inhibition test. A single concentration of the technical grade of the active ingredient is applied in the medium with a test duration of five days for the algae and 14 days for the *Lemna*. Tier 2 tests are conducted using at least five test concentrations on at least those species showing toxic effects in Tier 1. The Tier 3 field tests use the end-use product at the same test concentrations as those in the previous tier. Representatives of monocots, dicots, vascular cryptogams, each algal division and bryophytes or hepatophytes must be tested. The duration of the test must be sufficient to assess multiple applications (if specified on the product label), with observations continuing for at least two weeks after the last application (up to a maximum of four weeks) to note recovery or death.

United Kingdom—The Ministry of Agriculture, Fisheries and Food provides guidelines for pesticide registration that include consideration of flora, but no specific tests are identified [16]. A scheme is used for evaluation of chemical hazard to nontarget aquatic organisms including plants [17]. The scheme involves five steps, each generating more complex data. Progression from the initial step to further testing depends upon the use pattern (e.g., aquatic herbicides) and "margins of safety" for the most sensitive species of fish, invertebrates, and algae.

TABLE 2—*Summary of questionnaire responses from scientists.*

	Agriculture:Government		Environment:Government		Environment:Academic		Forestry:Government		Forestry:Academic		Industry	
No. of Questionnaires Sent[a]	5		21		19		6		7		6	
No. of Respondents[a]	2		10		6		3		3		2	
1. Currently Used Tests (U.S. EPA, OECD, etc.):	adequate	inadequate	adequate	inadequate	adequate	inadequate	adequate	inadequate	adequate	inadequate	adequate	inadequate
- Scope	1	1	5	3	1	4	3	0	1	2	2[b]	0
- Reproducibility	2	0	5	Equal 1	4[b] Equal 1	1	1 Equal 1	0	2 Equal 1	0	0	0
- Advantages versus Disadvantages[c]	1	0	7	0	2	1	0	0	0	2	1	0
2. Endpoints	Germination/Growth		Growth, primary production, mortality, germination		Growth, Physiological parameters, Morphology		Go beyond germination		Death or Necrosis Survival, Growth		Germination, Growth Development	
3. Indicator Species[d]	Sensitive crop species Regional species		Algae (>1 species)* Macrophyte (Lemna, plus others)* Softwoods, Hardwoods Commercial Crops Monocots, Dicots*		Algae (>1 species)* Macrophytes (>1 species)* Crop species Conifers Hardwoods		Major crop trees* Major weed trees/shrubs* Species specific to the pesticide		Conifers (>1 species) Dicots (>1 species) Key browse species Key crop species Adjacent agronomic crops		No need for "Canadian" Species site-specific in special cases	
4. Tier Approach	Agree 1	Disagree 0	Agree 6	Disagree 1	Agree 3	Disagree 0	Agree 1	Disagree 1	Agree 1	Disagree 1	no opinion given	
5. Multi-Species Tests as Part of Guidelines:	Agree	Disagree	Agree	Disagree	Agree	Disagree	Agree	Disagree	Agree	Disagree	Agree	Disagree
- Field Tests	2	0	6[b]	0	3	3	2	0	2[b]	0	2 but need more research	
- Mesocosms	0	0	3[b]	2	5[b]	1	1[b]	1	0	1[b]	1[b]	0
- Species Battery	0	0	7	0	6[b]	0	3[b]	0	0	0	1[b]	0

[a] not all respondents answered all questions
[b] several provisos attached
[c] adequate = advantages outweigh disadvantages
  equal = advantages and disadvantages in balance
  inadequate = disadvantages outweigh advantages
[d] list of all suggestions (some made more than once marked by *)

Australia—The Department of Primary Industry provides general requirements for protection of treated crops and for assessment of hazard to nontarget crops and plants [*18*]. More extensive testing is required for pesticides that are persistent, volatile, or bioaccumulate; have repeated treatments; are applied in or near waterways; or are applied from aircraft or by other methods that might be conducive to drift.

Japan—The Minister of Agriculture, Forestry and Fisheries grants pesticides registrations. Application for registration must be accompanied by a document stating experimental results including efficacy and phytotoxicity [*19*]. Details of phytotoxicity tests were not available at the time this review was conducted.

Evaluation of Test Requirements

Twenty-six scientists responded to the questionnaire on existing test requirements and related test protocols (Table 2).

Species selection is a major problem area for all tests currently required (Table 3) and was the issue most frequently cited by representatives of international agencies. Recommended species are not always available (e.g., *Lemna gibba*), are not necessarily representative of the environments of concern, and are not always the most sensitive species to the pesticide in

TABLE 3—*Summary of nontarget plant toxicity tests.*

EXISTING TEST	ADVANTAGES	DISADVANTAGES	ALTERNATIVE METHODS	RESEARCH REQUIRED
AQUATIC PLANTS				
Algal Growth Inhibition	Inexpensive Simple Reproducible Quick Sensitive	Inadequate scope: too few species, difficult to extrapolate to field Nutrient-sufficient	Microplate Technique	More validation and more sensitivity checks Expand to multi-species
			Species Battery Approach	Species availability and sensitivity Inter-laboratory validation
Lemna Growth	Inexpensive Simple Quick Sensitive Reproducible	Inadequate scope: submergent and emergent species needed	Submergent or Emergent Species Tests	Species selection Methods development re: culture conditions, endpoints Inter-laboratory validation
Field Test (EPA only)	Indicates effects under field conditions.	Very little experience with test No uniform approach	Microcosm	Standardize design
			Mesocosm	Utility as screening tool prior to field testing More standardization
			Field Tests	Develop design based on rigorous sampling & unambiguous endpoints by reference to other areas of research
TERRESTRIAL VASCULAR PLANTS				
Germination, Growth & Vigour	Inexpensive Simple	Little experience with tests; questionable reproducibility and sensitivity Endpoints variable: need to use several	Improved Growth & Vigour Tests	Investigate better endpoints, measurement techn. Develop new species (from wild) to use in tests
			Pollen Germination Tests	Develop standard protocol Inter-laboratory validation with species used in germination and growth tests
Field Tests (EPA only)	Indicates effects under field conditions	Very little experience with test No uniform approach	Microcosm	Standardize design
			Mesocosm	Develop design
			Field Tests	Develop design based on rigorous sampling & unambiguous endpoints by reference to other areas of research

question. In addition, the selection of strains or cultivars within species can be a problem. The EPA emphasized that species should have an existing toxicity database, be readily available, not endangered, and be indicative of the environment. The OECD also looks for species that have already been used extensively in toxicity tests and that are indicative of the environment. The difficulty inherent in species selection was reflected in the variety in questionnaire responses on suitable indicator species for Canada (Table 2). Some respondents disliked limiting tests to a few indicator species, preferring instead to choose species appropriate for the pesticide and use pattern.

In evaluating the adequacy of currently used tests, scientists responding to the questionnaire were equally divided on the issue of scope (Table 2). Some respondents who thought that the scope of tests was generally adequate emphasized the importance of careful species selection. Those who thought that the scope was inadequate were often concerned about the limited number and representativeness of species being tested (particularly for the forest setting) and the need to test for effects at different stages in the life cycle or during different metabolic or mitotic stages. In addition, there was general doubt about the relevance of these tests to the field, particularly given confounding effects of interspecific variability in toxic response. Respondents strongly supported the species battery approach in testing despite existing difficulties in species selection. Some emphasized that the species battery should supplement, not replace, the requirement for field testing. The OECD is promoting the development of multi-species tests as alternatives to current tests of one or two species. In the absence of well-developed multi-species tests, the OECD recommends a series of tests with different species rather than more precise testing with only one species [10]. A dissenting view on the utility of multi-species tests, at least for aquatic systems, was expressed by Gisey [20] on the basis that the tests are not amenable to the type of protocol development and precision required in formal testing schemes.

Most of the questionnaire respondents who felt qualified to comment on the tiered system of testing thought it was a good approach. Provisos stressed the importance of well-defined triggers or progression criteria. Progression criteria generated much comment and criticism from representatives of international agencies. Criteria which were quite specific (e.g., EPA), as well as criteria which were quite general (e.g., OECD), were considered to have drawbacks. The EPA and OECD both emphasized the importance of proper judgment when assessing the need for additional testing.

All but one questionnaire respondent were satisfied with the reproducibility of tests currently recommended. Provisos again emphasized the need for experienced personnel, as well as the importance of rigorously controlled conditions and careful monitoring, the importance of axenic cultures, and the need for a standard source collection that is fully identified taxonomically. The most commonly cited advantages of the current tests were that they were (in order of frequency of response): inexpensive, simple, reproducible, quick, sensitive, and required minimal space and equipment. The most commonly cited disadvantage was the difficulty in extrapolating to the field.

Despite the paucity of standard protocols, scientists and regulators generally agree that microcosm (i.e., multi-species laboratory tests), mesocosm tests (i.e., large-scale indoor/outdoor multi-species tests) or field tests are needed for a full evaluation of the environmental hazard posed by pesticides and to establish the relationship between single-species laboratory tests and effects in the environment [2,21,22]. Questionnaire respondents felt that field tests could be made reproducible enough to be used in the registration process. Emphasis was placed on proper study design. Most of the opposition to requiring field testing came from academics who questioned the reproducibility and practicality of including field tests. Respondent opinions were more divided on the use of mesocosms within the registration framework. Mesocosms were generally regarded as good screening tools to be followed up by field tests. Because the present state of knowledge does not allow for a confident evaluation of the merits and drawbacks of particular protocols, there is considerable variation among agencies with respect to the actual amount of multi-species or field testing being required. While the EPA formally includes field testing in its regulation, requests for field testing were relatively rare at the time of review.

Both the scientific literature review and questionnaire responses revealed several major gaps in our knowledge of pesticide effects and how to measure those effects: effects of mixtures of pesticides [21], degradation products, and, to a lesser extent, surfactants and adjuvants [23,24] and methods for extrapolating standardized test results to assessment of effects in the field [25–27].

Algal Tests—*Selenastrum capricornutum* was the most common algal species used to meet test requirements. Advantages of using *Selenastrum* included fast growth rates, availability, ease of culture, and the extensive database already developed. Disadvantages included inappropriate morphology for accurate volume measurements, considerable mucilage production, and clonal variability. Sensitivity of algal species to toxicants can vary considerably and unpredictably within and among taxonomic groups [28–31]. An alternative to current tests is to test a battery of species comprised of one species from each algal division, and species that are also environmentally important (Table 3). Although the species battery approach was strongly supported by questionnaire respondents, the database for such testing is very small, and the rationale for species selection within the battery is not well developed. In a test of 19 miscellaneous chemicals on 13 freshwater algal species, Blanck et al. [28] found that no generally sensitive or generally insensitive species could be identified. When this same dataset (with three additional species) was examined using multivariate analysis, two algal groups with different sensitivity patterns were recognized, the Cyanobacteria and the Chlorococcales [32]. The authors suggested that one species from within each of these groups should form part of a battery. However, they emphasized that the composition of other "sensitivity" groups is not yet known.

The selection of appropriate endpoints in algal toxicity tests was also a matter of considerable discussion in the scientific literature [33]. Questionnaire respondents most commonly recommended growth as an algal endpoint. The measurement of algal growth, expressed as number of cells per mL, was the most common endpoint in current tests. Indirect measurements of growth such as chlorophyll-a, ATP, protein, RNA, and DNA have also been used. These indirect measurements are comparable only if the growth is balanced (i.e., a doubling of the biomass results in a doubling of any chemical constituent). Therefore, when such indicators are used, calibrating them against direct growth determination is desirable. The advantage in using biomass indicators lies in the potential for automation of testing. For example, the microplate testing method measures ATP formation/degradation in very small volumes on plates with a large number of replicates that can be read simultaneously [34]. Alternatives to using growth or growth indicators are to use less biochemically complex processes such as fluorescence changes [35], nutrient uptake [36,37], oxygen evolution [38] and pH change of the medium [39]. Several of these alternatives are excellent for studying specific problems and some, such as the microplate technique [34], have considerable potential for automation but require additional research (Table 3).

Physical and chemical constraints of test systems must be taken into account if algal toxicity tests are to be properly interpreted. Lack of knowledge about test parameters such as pH, nutrient conditions, and complexing capacity of organics has invalidated a large percentage of published toxicity work [40]. For example, algal cells grown under nutrient-limiting conditions respond differently than those grown under nutrient-sufficient conditions [37,41]. Physical and chemical parameters of the test system that can affect the behavior and toxicity of the test chemical must be specified and controlled [36,42].

Aquatic Vascular Plant Tests—Aquatic vascular plant tests are best developed for duckweed *(Lemna gibba, L. minor)*. *Lemna* was the species most commonly recommended as an indicator species by questionnaire respondents. Commercial availability of *Lemna gibba* has been a problem in the past. Endpoints suggested for *Lemna* included: wet and dry weights, chlorophyll content, frond counts, and mortality. Other endpoints that have been used include: frond area, color [*43-45*], and root length [*46,47*]. Davis [*47*] reported that the best endpoints were frond death or frond count; root length was poor because of the difficulty in obtaining accurate measurements. As with other tests, it is important to consider factors such as pH and organic constituents in the growth medium as these affect the chemical behavior and effects of pesticides [*44*]. Static replacement tests were acceptable when replication was adequate [*47*].

Few tests with rooted, aquatic vascular plants have been reported [*48-50*]. Standard protocols have not been developed. Issues to be resolved include selection of suitable indicator species, method of application, test conditions, and duration of the tests (Table 3).

Aquatic Plant Microcosm, Mesocosm, and Field Tests—Methods for microcosm, mesocosm, and field tests have been developed for aquatic systems but require additional research (Table 3). Microcosm protocols are being standardized [*51-56*]; the results to date on reproducibility, sensitivity, and comparability to field results are encouraging [*53,57-59*]. Mesocosm methods often involve the use of various enclosures (e.g., limnocorrals) within natural or artificial water bodies [*60-62*], or the use of artificial ponds or model streams [*63-66*]. These studies help to bridge the gap between laboratory and field [*57*] but are more complex, expensive, and long-term than microcosms. Field studies are uncommon [*52,57-59,67,68*], particularly of pesticide effects on nontarget plants [*69*].

Terrestrial Plant Germination/Emergence/Vigor Tests—Existing terrestrial plant test protocols use economic species almost exclusively. The relationship between economic species and species that are important in the wild is not well understood. Questionnaire respondents suggested a wide variety of indicator species for Canada (Table 2). The forestry sector provided the most detailed lists which included common "crop" species such as Eastern white pine *(Pinus strobus)*, jackpine *(P. banksiana)*, sitka spruce *(Picea sitchensis)*, and common "weed" or browse species such as trembling aspen *(Populus tremuloides)*, alder (*Alnus* spp.), and raspberry (*Rubus* spp.). Recommendations from other respondents were more general (e.g., representatives of dicots, monocots, crop species). Selection of more functionally important species is difficult because of a lack of data comparing sensitivities to phytotoxicants among species. Variability of sensitivity within ecotypes or cultivars also complicates the choice of test species. Differences in tolerance could result from differential rates of metabolism, plant age, or localization of the detoxification mechanisms in a particular plant organ, such as the root, shoot, or leaf [*70*]. In terrestrial plant tests, the source of the seed must be known, well documented, and reliable, and the relative sensitivity of the different strains known prior to testing. Efficacy data may be useful for determining the relative sensitivity and recovery rates of terrestrial vascular species [*71*].

Existing terrestrial vascular plant tests have not been widely used and evaluated. Results of toxicity tests using vascular plants are more accurate when more than one endpoint is used and the results averaged to determine the toxicant effects [*43*]. Recommended endpoints for vascular species tests varied substantially among questionnaire respondents. The only relatively common recommendation was that "growth" should be measured. Other suggestions included percent germination and time of emergence, height, fresh or dry weight, death, survival and physiological endpoints such as CO_2 fixation, carbon assimilation, O_2 evolution, chlorophyll, and key enzymes. Other measurements include leaf area, leaf color, and CO_2 evolution [*72*].

The route of exposure in terrestrial plant tests can be problematical. The main methods of application in standard protocols are foliar [15] or soil incorporation [11, 15]. Foliar application ignores effects via root uptake almost entirely. The support medium can affect toxicity via pesticide-medium interactions, particularly if soil is used [43].

Alternatives to the terrestrial vascular plant tests currently used include petri dish germination and growth assays [43], improved root elongation tests [73,74], a photosynthesis inhibition test [75], pollen toxicity tests *in vitro* [76,77] and *in vivo* [76], and cell culture techniques [78-81]. Some of these methods may be suitable as test requirements with additional research (Table 3).

Terrestrial Microcosm, Mesocosm and Field Tests—Microcosm methods are not common and need standardization [82,83]. Mesocosm and field studies are extremely rare. All of these tests require additional research (Table 3).

Recommendations for Canadian Guidelines

The SRC recommended that all nontarget plant tests be conducted with the end-use product to improve comparability between laboratory and field tests, to incorporate toxic effects of product adjuvants, and to improve the solubility of the test substance. All tests should require a minimum of five test concentrations as specified by OECD and ISO protocols. A minimum of three replicates at each test concentration was recommended. Dose-response analyses using clearly identified endpoints are required for all tests to determine detrimental effect levels.

Minimum test requirements for pesticides included a battery of algal growth inhibition tests (for all uses), terrestrial plant seed germination, seedling emergence and vegetative vigor tests (for all uses except for control of aquatic weeds only), and an aquatic vascular plant test using *Lemna* (for aquatic uses and for uses that are expected to pose a hazard to aquatic plants).

Additional test requirements included tests with rooted, submerged vascular plants (for aquatic uses of persistent products or for uses expected to pose a hazard to aquatic plants), pollen germination tests (for uses where flowering is important or in progress), and terrestrial or aquatic field testing (for uses expected to pose a hazard to nontarget plants based on previous toxicity testing, environmental chemistry, and fate of the product, and the ecology of the receiving environment). All of these tests require additional research to develop standard protocols.

Algal Tests—The species battery should include a minimum of one species from each algal division present in the receiving environment. Selection of algal species was based on criteria outlined by Cracker [72] including taxonomic diversity, availability of cultures and good taxonomic characterization of strains, ease of culture in media suitable for toxicity testing, sufficiently rapid growth to allow estimation of density after 72 h, suitable morphology for counting, and susceptibility to phytotoxicants. Information was obtained on test results for 97 pesticides and 68 algal species [5]. These data were insufficient for quantitative comparisons because variability in species selection among studies resulted in limited databases for individual species. Ten algal species were recommended for testing including *Ankistrodesmus falcatus, Anabaena flos-aquae, Chlorella pyrenoidosa, C. vulgaris, Kirchneriella subsolitaria, Microcystis aeruginosa, Oocystis minuta, Scenedesmus bijuga, S. obliquus, S. quadricauda*, as well as *Selenastrum capricornutum*. No diatoms were recommended because more research is needed into their culture. *Selenastrum* was included because of the extensive database already developed, although it should not be the only species tested.

Currently used standard protocols were considered satisfactory, provided the test duration is no more than 72 h and initial cell numbers are as low as possible (10^4 cells/mL or less). Short duration and small population size should minimize alteration of the test conditions caused by the production of algal metabolites and changes in the medium.

Aquatic Vascular Plant Tests—*Lemna minor* was recommended to meet minimum test requirements because of the lack of availability of *L. gibba* at the time of the review. The current EPA protocol was satisfactory. A range of test concentrations should be tested using frond counts, frond death or dry weight as endpoints.

Based on the limited information available, the rooted, vascular plant *Potamogeton pectinatus* was recommended as a test species. A standard test protocol for this species needs to be developed. The following ecologically important genera have also been used in toxicity testing but their culturing and testing techniques require additional research: *Elodea, Ceratophyllum, Myriophyllum, Vallisneria, Ruppia, Zannichellia,* and *Typha.*

Terrestrial Vascular Plant Tests—Terrestrial plant species used in the relevant OECD and EPA test protocols were recommended with the proviso that their utility as representatives of wild species needed research.

The EPA and OECD terrestrial plant test protocols were adequate. Several endpoints should be used because endpoints for these tests are not well-researched. The seed germination test should be extended to include growth. The test substance should be applied via foliar spray and via incorporation in the support medium to represent both routes of exposure in the field. An inert support medium, such as acid-washed sand or glass beads, should be used to avoid adsorption problems related to the use of soil.

Pollen Germination Test—Test species need to be selected based on existing research expertise [84–88]. Standard protocols for *in vitro* and *in vivo* tests need to be developed.

Microcosm, Mesocosm, and Field Tests—Several available aquatic microcosm designs are suitable for consideration as part of test guidelines [54]. Microcosm tests generally agree well with field results and provide a first approximation of how an ecosystem will respond to a toxicant. Design should not be so standardized that the ecology of the receiving environment cannot be accommodated in the test conditions.

Mesocosms are very useful research tools because they help to bridge the gap between laboratory single-species and microcosm studies, and the real world. However, their complexity, and the lack of standardization, limits their usefulness as part of formal testing requirements.

Field studies should become a part of test guidelines after they have been developed to include rigorous sampling design and clear, unambiguous end points. Field tests have to be of sufficient duration to show ecosystem resilience and pesticide effects at critical times (e.g., fruit set). Endpoints will be less ambiguous if deleterious effects are defined prior to testing. Definition of deleterious effects will depend on an understanding of the ecosystem being studied, in part from results of laboratory single-species tests, microcosm and mesocosm studies, and previous fieldwork. More experience with nontarget plant testing should make defining deleterious effects easier in the future.

Federal Working Committee

The interdepartmental committee reviewed the SRC report [4] and recommended that the report be distributed with the following technical comments for review by the pesticide industry and the regulatory and scientific communities concerned with the study of environmental effects of pesticides:

> Data requirements should be tailored according to the use pattern (e.g. field vs greenhouse), and type of action (e.g. herbicidal, insecticidal, or fungicidal) for major groups of pesticides. It may be

feasible to limit data requirements to herbicides. Other pesticides (e.g. insecticides, fungicides) are not expected to exhibit phytotoxicity at the concentrations likely to reach nontarget areas. Data to show a lack of phytotoxicity during the screening process for pesticidal activity conducted by industry may be sufficient for regulatory purposes. Data normally generated during the industry screening process may also be adequate for evaluating herbicide toxicity to nontarget plants. The feasibility of this approach needs to be determined. Vegetative vigour, seedling emergence, and seed germination studies may be more appropriately requested as advanced studies. Data requirements might also be tailored according to the results of laboratory environmental chemistry and fate studies.

Basic (minimal) data requirements for herbicides should include testing with *Lemna* because inherent differences in the physiology of algae and aquatic macrophytes suggest that one group cannot realistically be used an an indicator for the other. Aquatic macrophytes are significant producers in a number of aquatic systems. Monitoring studies have shown that a portion of many pesticides applied terrestrially enters aquatic environments.

All laboratory studies should be conducted with technical grade of the active ingredient. A representative number of these studies should be repeated with the product intended for use. Field tests and ecosystem level fate and impact experiments, if required, should be conducted with the product intended for use. Test concentrations should be confirmed (using analytical techniques) at appropriate times during the course of a test, or justification for not doing so should be provided. For aquatic plant tests, rather than testing five concentrations, it may be sufficient to show that no effect occurs at the water solubility limit of the product, or at an environmental concentration of the pesticide under a worst-case scenario for the proposed use pattern. The response of appropriate controls should also be measured. Control treatments may include treatment with a toxicant of known phytotoxicity to the test plant; a treatment which duplicates the exposure treatment, but contains no technical grade of the active ingredient; or a treatment which duplicates the exposure treatment, but contains no technical grade of the active ingredient or carrier (e.g. solvent).

Discussion

Nontarget plant testing guidelines are being developed for Canada for the purpose of outlining data requirements for pesticide registration under the Pest Control Products Act. The current project has been useful for generating recommendations on test requirements for review by the scientific and regulatory communities prior to drafting of guidelines for Canada.

The project has also stimulated research in both the public and private sectors. Specific problems with existing test requirements and protocols, such as the development of the microplate technique for pesticides, and multi-species tests, are currently being addressed by Environment Canada and the Saskatchewan Research Council. Hazard and risk assessment methods need to be improved, particularly with respect to generating ecologically-relevant toxicity data and estimating expected environmental concentrations (cf. [89]). Phytotoxic impacts in the field need to be identified and quantified in order to improve the design of laboratory, mesocosm, and field tests. Different approaches, such as biomonitoring and ecological modelling, should be examined for their utility in predicting and assessing phytotoxicant impacts on plants and their associated wildlife.

Acknowledgments

This paper benefitted immeasurably from reviews by staff of Environmental Protection and the National Wildlife Research Centre of the Canadian Wildlife Service, Conservation and Protection, Environment Canada, Ottawa. Useful comments were also provided by two anonymous reviewers.

References

[1] "Census of Canada: Agriculture 1986," Statistics Canada, Ottawa, Canada, 1987.
[2] Sheehan, P. J., Baril, A., Mineau, P., Smith, D. K., Harfenist, A., and Marshall, W. K., "The Impact of Pesticides on the Ecology of Prairie Nesting Ducks," Technical Report Series No. 19, Canadian Wildlife Service, Conservation and Protection, Environment Canada, Ottawa, Canada, 1987.
[3] Standing Senate Committee on Agriculture and Forestry, "The Canadian Regulatory Process for Pesticides," Twelfth Report, Proceedings Issue No. 33, Canadian Government Publishing Centre, Ottawa, Canada, Sept. 1988, pp. 6–30.
[4] Swanson, S. and Peterson, H., "Development of Guidelines for Testing Pesticide Toxicity to Non-target Plants," Environmental Protection Series, Environment Canada, Ottawa, Canada, in press.
[5] Swanson, S., "Aquatic Plant Toxicity Testing: Recommendations for Test Species," unpublished manuscript, Environment Canada, Ottawa, Canada, 1989.
[6] European Economic Community (EEC), "Council Directive Amending for the Sixth Time, Directive 67/548/EEC on the Approximation of the Laws, Regulations and Administrative Provisions Relating to the Classification, Packaging and Labelling of Dangerous Substances," *Official Journal of the European Communities,* L 259/10, 1979.
[7] European Economic Community (EEC), "Methods for the Determination of Ecotoxicity: Algal Inhibition Test," EEC Directive, 79/831, Annex V, Part C, 1987.
[8] Food and Agriculture Organization of the United Nations (FAO), "Report of the Second Government Consultation on International Harmonization of Pesticide Registration Requirements," Rome, Oct. 1982.
[9] International Standards Organization (ISO), "Draft Method—Water Quality: Algal Growth Inhibition Test," ISO/TC 147, Nederlands Normalisatie-Instituut, Delft, The Netherlands, 1987.
[10] Organization for Economic Cooperation and Development (OECD), "OECD Guidelines for Testing of Chemicals," Paris, 1981.
[11] Organization for Economic Cooperation and Development (OECD), "Second Addendum to OECD Guidelines for Testing of Chemicals," Paris, 1984.
[12] Organization for Economic Cooperation and Development (OECD), "Report on the Assessment of Potential Environmental Effects of Chemicals; the Effects on Organisms Other than Man and on Ecosystems," Volume I, Division of Technology for Society TNO Department of Biology, Study and Information Centre TNO for Environmental Research, Delft, The Netherlands, 1980.
[13] Organization for Economic Cooperation and Development (OECD), "Report of the OECD Workshop on Ecological Effects Assessment," OECD Environment Monographs No. 26, Paris, May 1989.
[14] Federal Insecticide, Fungicide and Rodenticide Act, "Protection of Environment—Data Requirements for Registration (revised as of July 1, 1986)," Code of Federal Regulations 40, Part 158, United States Environmental Protection Agency, Washington, D.C., *Federal Register,* Special Edition, 1986, pp. 44–92.
[15] Holst, R. W. and Ellwanger, T. C., "Pesticide Assessment Guidelines, Subdivision J, Hazard Evaluation: Non-Target Plants," Office of Pesticide and Toxic Substances, United States Environmental Protection Agency, Washington, D.C., 1982.
[16] United Kingdom Ministry of Agriculture, Fisheries and Food, "Pesticides Registration in the United Kingdom," the work of the Pesticides Registration Department at Harpenden, 1984.
[17] Tooby, T. E. in *Proceedings,* 5th Symposium on Aquatic Weeds, European Weed Research Society, 1978, pp. 287–294.
[18] Australian Department of Primary Industry, "Requirements for Clearance of Agricultural Chemicals," Agricultural and Veterinary Chemicals Section, Department of Primary Industry, Canberra, Australia, 1985.
[19] Fukuda, H., "Registration and Inspection of Pesticides in Agriculture," ZEN-NOH Agricultural Technical Center, Division of Pesticides, National Institute of Agro-Environmental Sciences, Kannondai3-1-1, Yatabe-Machi, Tsukuba-Gun, Ibaraki-Ken 305, Japan, 1987, pp. 1–6.
[20] Gisey, J. P. in *Aquatic Toxicology and Hazard Assessments: Eighth Symposium, ASTM STP 891,* R. C. Bahner and D. J. Hansen, Eds., American Society for Testing and Materials, Philadelphia, 1985, pp. 67–77.
[21] *Methods for Assessing the Effects of Mixtures of Chemicals,* Vouk, V. B., Butler, G. C., Upton, A. C., Parke, D. V., and Asher, S. C., Eds., SCOPE 30, SGOMSEC 3, United Nations Environment Programme, International Labour Organization and World Health Organization, International Programme on Chemical Safety, IPSC Joint Symposia 6, Wiley, Toronto, 1987.

[22] Wong, P. T. S. and Couture, P. in *Toxicity Testing Using Microorganisms,* Vol. II, B. J. Dutka and G. Bitton, Eds., CRC Press, Boca Raton, Fla., 1986, pp. 79–100.
[23] Parr, J. F. in *Adjuvants for Herbicides,* Weed Science Society of America, Champaign, Ill., 1978, pp. 93–113.
[24] Sortkjaer, O., *Ecological Bulletin* (Stockholm), Vol. 36, 1984, pp. 75–80.
[25] Cairns, J., Jr., in *Aquatic Toxicology and Hazard Assessment: Sixth Symposium, ASTM STP 802,* W. E. Bishop, R. D. Cardwell, and B. B. Heidolph, Eds., American Society for Testing and Materials, Philadelphia, 1983, pp. 111–127.
[26] Stephan, C. E. in *Aquatic Toxicology and Environmental Fate: Ninth Volume, ASTM STP 921,* T. M. Poston and R. Purdy, Eds., American Society for Testing and Materials, Philadelphia, 1986, pp. 3–10.
[27] Suter II, G. W., Barnthouse, L. W., Breck, J. E., Gardner, R. H. and O'Neil, R. V. in *Aquatic Toxicology and Hazard Assessment: Seventh Symposium, ASTM STP 854,* R. D. Cardwell, R. Purdy, and R. C. Bahner, Eds., American Society for Testing and Materials, Philadelphia, 1985, pp. 400–413.
[28] Blanck, H., Wallin, G., and Wangberg, S., *Ecotoxicology and Environmental Safety,* Vol. 8, 1984, pp. 339–351.
[29] Hollister, T. A. and Walsh, G. E., *Bulletin of Environmental Contamination and Toxicology,* Vol. 9, No. 5, pp. 291–295.
[30] Ibrahim, A., *Aquatic Toxicology,* Vol. 3, 1984, pp. 1–14.
[31] Stratton, G. W., *Archives of Environmental Contamination and Toxicology,* Vol. 13, 1984, pp. 35–42.
[32] Wangberg, S. and Blanck, H., *Ecotoxicology and Environmental Safety,* Vol. 16, 1988, pp. 72–82.
[33] Nyholm, N., *Water Research,* Vol. 19, 1985, pp. 273–279.
[34] Blaise, C. R., Legault, R., Bermingham, N., Van Coillie, R., and Vasseur, P., *Toxicity Assessment: An International Quarterly,* Vol. 1, 1986, pp. 261–281.
[35] Christoffers, D. and Ernst, D. E. W., *Toxicology and Environmental Chemistry,* Vol. 7, 1983, pp. 61–71.
[36] Peterson, H. G., Healey, F. P., and Wagemann, R., *Canadian Journal of Fisheries and Aquatic Science,* Vol. 41, No. 6, 1984, pp. 974–979.
[37] Peterson, H. G. and Healey, F. P., *Journal of Phycology,* Vol. 21, 1985, pp. 217–222.
[38] Turbak, S. C., Olson, S. B., and McFeters, G. A., *Water Research,* Vol. 20, No. 1, 1986, pp. 91–96.
[39] Lusse, B., Schroder, D., and Soeder, C. J., *Archives of Hydrobiology* (Supplement), Vol. 73, No. 1, 1986, pp. 147–152.
[40] Campbell, P. G. C. and Stokes, P. M., *Canadian Journal of Fisheries and Aquatic Sciences,* Vol. 42, 1985, pp. 2034–2049.
[41] Hall, J., "The Interaction of Chronic Copper Toxicity with Nutrient Limitation in Two Chlorophytes," Ph.D. thesis, University of Manitoba, Winnipeg, Manitoba, Canada, 1986.
[42] Soeder, C. J. and Stengel, E., in *Algal Physiology and Biochemistry,* W. D. P. Stewart, Ed., Blackwell, Oxford, 1974, pp. 714–740.
[43] Santelmann, P. W. in *Research Methods in Weed Science,* 2nd ed., B. Truelove, Ed., Southern Weed Science Society, Auburn, Ala., 1977, pp. 79–87.
[44] Lockhart, W. L. and Blouw, A. P., *Canadian Special Publication of Fisheries and Aquatic Sciences,* Vol. 44, 1979, pp. 112–118.
[45] Lockhart, W. L., Billeck, B. N., de March, L. G. E. and Muir, D. C. G. in *Aquatic Toxicology and Hazard Assessment: Sixth Symposium, ASTM STP 802,* W. E. Bishop, R. D. Cardwell, and B. B. Heidolph, Eds., American Society for Testing and Materials, Philadelphia, 1983, pp. 460–468.
[46] Bishop, W. E. and Perry, R. L. in *Aquatic Toxicology and Hazard Assessment: Fourth Symposium, ASTM STP 737,* D. R. Branson and K. L. Dickson, Eds., American Society for Testing and Materials, Philadelphia, 1981, pp. 421–435.
[47] Davis, J. A., "Comparison of Static-Replacement and Flow-Through Bioassays Using Duckweed, *Lemna gibba* G-3," EPA 560/6-81-003, Office of Pesticides and Toxic Substances, United States Environmental Protection Agency, Washington, D.C., 1981.
[48] Correll, D. L. and Wu, T. L., *Aquatic Biology,* Vol. 14, 1982, pp. 151–158.
[49] Davis, E., "Effect of Herbicides on Submerged Seed Plants," W800685 OWRTA-067-ALA(2), Office of Water Research and Technology, Washington, D.C., 1980.
[50] Jones, T. W. and Winchell, L., *Journal of Environmental Quality,* Vol. 13, No. 2, 1984, pp. 243–247.

[51] Brockway, D. L., Smith, P. D., and Stancil, F. E., *Bulletin of Environmental Contamination and Toxicology*, Vol. 32, 1984, pp. 345-353.
[52] Francis, B. M. and Metcalf, R. L., "Screening of Pesticides for Potential Adverse Environmental Effects in Illinois," Institute of Environmental Studies and Department of Entomology, University of Illinois at Urbana-Champaign, 1981.
[53] Hamala, J. A. and Kollig, H. P., *Chemosphere*, Vol. 14, No. 9, 1985, pp. 1391-1408.
[54] Taub, F. B. and Kindig, A. C., "Interlaboratory Testing of an Aquatic Microcosm Protocol," Final Report, Section II of V, Standardized Aquatic Protocol, Department of Health and Human Services, Food and Drug Administration, Rockville, Md., 1986.
[55] Toxic Substances Control Act, "Test Guidelines; Proposed Rule," Code of Federal Regulations 40, Parts 796-797, United States Environmental Protection Agency, Washington, D.C., *Federal Register*, Vol. 52, 1987, pp. 36334-36371.
[56] Toxic Substances Control Act (TSCA), "Revision of TSCA Test Guidelines," Code of Federal Regulations 40, Parts 796-798, United States Environmental Protection Agency, Washington, D.C., *Federal Register*, Vol. 52, 1987, pp. 19056-19082.
[57] Giddings, J. M. and Franco, P. J. in *Validation and Predictability of Laboratory Methods for Assessing the Fate and Effects of Contaminants in Aquatic Ecosystems, ASTM STP 865*, T. P. Boyle, Ed., American Society for Testing and Materials, Philadelphia, 1985, pp. 104-119.
[58] Harrass, M. C. and Taub, F. B. in *Validation and Predictability of Laboratory Methods for Assessing the Fate and Effects of Contaminants in Aquatic Ecosystems, ASTM STP 865*, T. P. Boyle, Ed., American Society for Testing and Materials, Philadelphia, 1985, pp. 57-74.
[59] Neuhold, J. M. in *Aquatic Toxicology and Environmental Fate, ASTM STP 921*, T. M. Poston and R. Purdy, Eds., American Society for Testing and Materials, Philadelphia, 1986, pp. 11-21.
[60] Hamilton, P. B., Jackson, G. S., Kaushik, N. K., Soloman, K. R., and Stephenson, D. L., *Aquatic Toxicology*, Vol. 13, 1988, pp. 123-140.
[61] Solomon, K. R., Smith, K., Guest, G., Yoo, J. Y., and Kaushik, N. K., *Canadian Technical Report of Fisheries and Aquatic Sciences*, Vol. 975, 1980, pp. 1-9.
[62] Van der Werf, B., Schrotenboer, J., Richterm, A. F., Moed, J. R., Hoogveld, H. L. and De Haan, H., *Canadian Journal of Fisheries and Aquatic Sciences*, Vol. 44, 1987, pp. 1649-1652.
[63] deNoyelles, F., Kettle, W. D., and Sinn, D. E., *Ecology*, Vol. 63, No. 5, 1982, pp. 1285-1293.
[64] Kosinski, R. J., *Environmental Pollution*, Vol. 36, 1984, pp. 165-189.
[65] Kosinski, R. J. and Merkle, M. G., *Journal of Environmental Quality*, Vol. 13, No. 1, 1984, pp. 75-82.
[66] Strange, R. J., *Journal of Applied Ecology*, Vol. 13, No. 3, 1976, pp. 889-897.
[67] Klaassen, H. E. and Kadoum, A. M., *Archives of Environmental Contamination and Toxicology*, Vol. 8, 1979, pp. 345-353.
[68] Walsh, G. E., Miller, C. W., and Heitmuller, P. T., *Bulletin of Environmental Contamination and Toxicology*, Vol. 6, No. 3, 1971, pp. 279-288.
[69] Osborn, D., in *Toxic Hazard Assessment of Chemicals*, M. Richardson, Ed., Royal Society of Chemistry, London, 1986, pp. 247-258.
[70] Hatzios, K. K. and Penner, D., *Metabolism of Herbicides in Higher Plants*, Burgess, Minneapolis, Minn., 1982.
[71] Crawford, S. A., *Hydrobiologia*, Vol. 77, 1981, pp. 217-223.
[72] Craker, L. E., "The Determination of Phytotoxicity," United States Army Medical Engineering Research and Development Laboratory, Fort Detrick, Frederick, Md., 1981.
[73] Edwards, N. T. and Ross-Todd, B. M., *Environmental and Experimental Botany*, Vol. 20, 1980, pp. 31-38.
[74] Ratsch, H. C. and Johndro, D., *Environmental Monitoring and Assessment*, Vol. 6, No. 3, 1986, pp. 267-276.
[75] Truelove, B., Davis, D. E., and Jones, L. R., *Weed Science*, Vol. 22, 1974, pp. 15-17.
[76] Church, R. M. and Williams, R. R., *Journal of Horticultural Science*, Vol. 52, 1977, pp. 429-436.
[77] Bristow, P. R. and Windom, G. E., *Plant Disease*, Vol. 71, No. 4, 1987, pp. 326-328.
[78] Davis, F. S., Villasrreal, A., Baur, J. R., and Goldstein, I. S., *Weed Science*, Vol. 20, No. 2, 1972, pp. 185-188.
[79] Zilkah, S. and Gressel, J., *Plant Cell Physiology*, Vol. 18, 1977, pp. 815-820.
[80] Zilkah, S. and Gressel, J., *Plant Cell Physiology*, Vol. 18, 1977, pp. 641-655.
[81] Zilkah, S., Bocion, P. F., and Gressel, J., *Plant Cell Physiology*, Vol. 18, 1977, pp. 657-670.
[82] Gile, J. D., Collins, J. C., and Gillett, J. W., *Environmental Science and Technology*, Vol. 14, No. 9, 1981, pp. 1124-1128.
[83] Gillett, J. W., and Witt, J. M., Eds., "Terrestrial Microcosms," in *Proceedings*, Workshop on Terrestrial Microcosms, Symposium on Terrestrial Microcosms and Environmental Chemistry,

Applied Science and Research Applications, U.S. EPA Project No. 1EA714, National Science Foundation, Washington, D.C., 1979.
[84] Gentile, A. G. and Gallagher, K. J., *Journal of Economic Entomolology,* Vol. 65, No. 2, 1972, pp. 488–491.
[85] Gentile, A. G., Vaughan, A. W., Richman, S. M. and Eaton, A. T., *Environmental Entomology,* Vol. 2, No. 3, 1973, pp. 473–476.
[86] Gentile, A. G., Vaughan, A. W. and Pfeiffer, D. G., *Environmental Entomology,* Vol. 7, 1978, pp. 689–691.
[87] Regupathy, A. and Subramaniam, T. R., *Phytoparasitica,* Vol. 1, No. 2, 1973, pp. 115–116.
[88] Regupathy, A. and Subramaniam, T. R., *Journal of Horticultural Science,* Vol. 49, 1974, pp. 197–198.
[89] J. A. Fava, W. J. Adams, R. J. Larson, G. W. Dickson, K. L. Dickson, and W. E. Bishop, Eds., "Research Priorities in Environmental Risk Assessment," Workshop Report, Society of Environmental Toxicology and Chemistry, Washington, D.C., Aug. 1987.

Comparative Toxicology

John S. Fletcher[1]

Use of Algae versus Vascular Plants to Test for Chemical Toxicity

REFERENCE: Fletcher, J. S., "**Use of Algae versus Vascular Plants to Test for Chemical Toxicity**," *Plants for Toxicity Assessment, ASTM STP 1091*, W. Wang, J. W. Gorsuch, and W. R. Lower, Eds., American Society for Testing and Materials, Philadelphia, 1990, pp. 33–39.

ABSTRACT: The credibility of using an algal plant as a surrogate for vascular plants in chemical toxicity testing was evaluated. Data taken from the PHYTOTOX database were used to estimate what proportion of the chemicals reported to influence vascular plants would also influence algae. The relative sensitivity of algae versus vascular plants to common herbicides was also examined. Analysis of the data suggested that approximately 20% of the time an algal screening test for chemical toxicity may not detect chemicals that elicit a response unique to the growth and development of vascular plants. Another conclusion was that no single plant type (algal, monocotyledon, or dicotyledon plant) can always be viewed as the least or most sensitive test system.

KEY WORDS: PHYTOTOX database, vascular plants, algae, phytotoxicity, screening tests, chemical influence

In several tier testing schemes that have been proposed or are in use, a species of algae has been used in the first tier as a surrogate for the entire plant kingdom [1,2]. For example, "The Guidelines for Testing of Chemicals" formulated by the Organization for Economic Cooperation and Development (OECD) includes an alga growth inhibition test as a part of the basic level (1st tier) of testing. The results from the alga test determines whether or not subsequent testing will be done with vascular plants in a second tier of testing (confirmatory level). Thus, in the OECD test scheme the alga test in the first tier of testing serves as a surrogate for the plant kingdom. This practice is certainly convenient and inexpensive, but its accuracy is questionable in view of the limited research that has been conducted to study the comparative effect of chemicals applied to algae versus vascular plant species.

There are many metabolic processes, such as DNA synthesis, photosynthesis, and respiration, that have been demonstrated for the most part to be enzymatically and functionally identical processes throughout the plant kingdom. Therefore, chemicals that influence any one of these universal processes should have an equivalent influence on either vascular plants or algae. Recognizing how essential these fundamental processes are to the survival and growth of organisms, what follows is that whenever a chemical is shown to inhibit growth as measured by reduced dry weight, fresh weight, etc., one can assume that the inhibition is probably due to the chemicals direct blockage of an underlying metabolic process. This logic is the primary justification for using simple inexpensive algal toxicity tests as a surrogate test for the entire plant kingdom. The flaw with this line of reasoning is that because of the more complex structure and life cycle of vascular plants, they possess numerous physiological fea-

---

[1] Professor, Department of Botany and Microbiology, University of Oklahoma, Norman, OK 70319.

tures, such as leaf abscission and flowering that are not characteristics of algae. Therefore a chemical may inhibit one of these processes that are unique to vascular plants but have no influence on other processes that are shared by both algae and vascular plants. If selective inhibition of processes unique to vascular plants is a common feature of organic chemicals, then algae are not reliable surrogates for phytotoxicity testing.

The credibility of using an algal plant as a surrogate for vascular plants in chemical toxicity testing was evaluated in two ways. First, the published literature was analyzed through the use of PHYTOTOX to estimate what proportion of the chemicals reported to influence vascular plants would also be presumed to influence algae. Secondly, the sensitivities of vascular plant responses to selected chemicals were compared with the sensitivity of *Chlorella* as reported by Decleire et al. [3] or Garten and Frank [4].

Procedure

The analyses reported in this paper were made through the use of PHYTOTOX, a database dealing with the response of vascular plants to organic chemicals. This database is being compiled at the University of Oklahoma from data appearing in the published literature between 1928 and 1988. At the time when this study was conducted the database contained information on approximately 5000 different chemicals tested on at least one of 1500 different plant species. The database includes data pertaining to 153 different plant responses. A detailed description of the database and the procedures used for its development are available in previous publications [5–7].

To determine the extent of common response of vascular plants and algae to individual chemicals, responses were grouped in three categories: unique features to vascular plants, common metabolic processes for both vascular plants and algae, and growth parameters. Records pertaining to each of these categories were recovered from PHYTOTOX and comparisons were made.

A comparison of the sensitivity of algae versus vascular plants was made by comparing the dose-response data reported for *Chlorella* by Decleire et al. [3] or Garten and Frank [4] with data present in PHYTOTOX. The EC_{50} values (effective concentration required to inhibit growth by 50%) reported in the algal studies [3,4] were compared with the minimum concentrations required to cause a 35 to 70% reduction in one of nine important parameters of vascular plant growth and development. Chemical concentrations eliciting a 35 to 70% reduction were used as an approximation of EC_{50} values, since they are almost never determined in phytotoxicity studies conducted with vascular plants, an aspect of the literature that has been discussed in more detail in an earlier paper [6]. The parameters considered were a reduction in: number of living plants, dry mass, fresh mass, harvest yield, seed mass, seed number, root size, shoot size, or stem size.

Results

Analysis of Chemical Response Data in PHYTOTOX

The total number of records present in each of the three examined response categories is shown in Table 1. This compilation of information showed that growth parameters were reported most often (28 757), with unique features being second (11 879), and common metabolic features a distant third (1839).

Table 2 shows the number of different kinds of chemicals which had been used in studies associated with each of the categories listed in Table 1. In growth response studies (Category III), 1770 different chemicals were found to have been tested; approximately two thirds of

TABLE 1—*Three categories of plant response and a listing of PHYTOTOX symptoms and records associated with each.*

(I) Unique Features[a]		(II) Common Features[b]		(III) Growth Parameters	
Symptom	Records	Symptom	Records	Symptom	Records
Abscission	869[c]	DNA Synthesis	105	Cover Range	267
Cold hardiness	38	Ion uptake	86	Dry weight[d]	2 965
Stem curvature	530	Photosynthesis	448	Fresh weight[d]	3 526
Deformed organs	704	Protein synthesis	340	Harvest yield	2 055
Dessication	78	Respiration	721	Kill	5 292
Dormancy	266	RNA synthesis	139	Plant number	5 590
Floral development	188		1839	Plant or organ size[d]	9 062
Seed germination	4 316				28 757
Maturation	535				
Nastic movements	160				
Parthenocarpy	212				
Secondary metabolism	200				
Seed development	115				
Seed number change	223				
Senescence	67				
Flower sex change	130				
Sterility	81				
Stomatal closure	155				
Transpiration	508				
Vascular disruption	31				
Wilt	154				
	11 879				

[a] Physiological features which are characteristic of vascular plants but not algae.
[b] Metabolic processes which are common to both vascular plants and algae.
[c] Each record in PHYTOTOX pertains to the effect of a single dose of one chemical applied to one plant species.
[d] The numbers of records shown for changes in dry weight, fresh weight, and plant size include data pertaining to plant organs as well as for whole plants. For example, root elongation records are included among the 9062 records dealing with changes in plant size.

that number (1097) were tested in unique feature studies (Category I) and only 245 in common feature studies (Category II).

A comparison of lists of chemicals associated with each category was conducted to determine how much overlap existed between chemicals that elicited the responses of the three categories (Table 3). For example, only 2% of the 1097 chemicals that caused response symptoms unique to vascular plants (Category I) were also used in studies showing that they influ-

TABLE 2—*Number of different chemicals in PHYTOTOX associated with each of the categories shown in Table 1.*

Category of Symptom	No. of Chemicals
(I) Physiological feature unique to vascular plants	1097
(II) Metabolic process common to both vascular plants and algae	245
(III) Growth Parameters	1770

TABLE 3—*Frequency (%) of chemicals having an effect on more than one category of response.*

Common with Category	(I) Unique to Vascular Plants	(II) Common to Vascular Plants and Algae	(III) Growth Parameters
	(1097)[a]	(245)	(1770)
I	...	8	25
II	2[b]	...	2
III	41	13	...
I, II	...	...	7
I, III	...	47	...
II, III	11	...	...
None	47	32	66

[a] Number of different kinds of chemicals in the category.
[b] Each of the values indicates what % of the chemicals in the category at the top of the column also influences the category(s) listed in the first column headed with the title "Common with Category".

enced a common metabolic process (Category II). A surprisingly large number of the chemicals in each category were not reported in the other two categories. Sixty-six percent of the chemicals reported to elicit growth responses were not included in either of the other categories. In like fashion, 47 and 32% of the chemicals in the unique and common categories elicited no responses in the other two categories. The greatest amount of overlap between lists occurred for Category II, where 47% of the chemicals reported to influence a common metabolic process also influenced some of the responses in Categories I and III.

Comparison of Vascular Plant and Algae Sensitivities to Selected Chemicals

In the sensitivity comparison (Table 4), attention was given to 33 different chemicals representing 16 of the 20 general classes of herbicides identified by Fletcher et al. [8]. The relative sensitivities reported in Table 4 are summarized in Table 5 by determining the frequencies of each relative response for each of the three types of plants. The summary values appearing in Table 5 show that there is no clear pattern to the sensitivities of the three types of plants. For example, the algal was most sensitive to 10 individual compounds and 6 herbicide classes, but it was also least sensitive to 16 compounds and 10 herbicide classes. The monocotyledon and dicotyledon plants showed equivalent sensitivities to most of the compounds examined; and their relationships to the algal sensitivities were mixed, sometimes more sensitive, sometimes less.

TABLE 4—*Relative sensitivity of algal, monocotyledon, and dicotyledon plants to specific chemicals, from different classes of herbicides.*

Herbicide Class		Relative Sensitivity[a]		
No.	Chemical Name	Algae	Monocotyledon	Dicotyledon
I	Haloalkanoic acids			
	Dalapon	d	d	d
II	Phenoxyalkanic acids			
	2,4-D	c	a	b
III	Aromatic Acids			
	Chloramben	c	a	...
	Dicamba	c	b	a

TABLE 4—(continued)

	Herbicide Class	Relative Sensivitity[a]		
No.	Chemical Name	Algae	Monocotyledon	Dicotyledon
IV	Amides			
	Naptalam	c	a	b
V	Nitrites			
	Bromoxynil	c	a	b
VI	Amilides			
	Propanil	a	...	c
	Benefin	d	d	d
	Propachlor	c	...	a
	Alachlor	c	a	b
VII	Nitrophenols			
	DNOC	c	a	b
VIII	Nitrophenylethers	...	...	...
IX	Nitroanilines			
	Trifluralin	d	d	d
	Nitralin	c	d	d
	Benfluralin	c	d	d
	Isopropalin	c	d	d
X	Carbamates			
	Propham	d	d	d
	Chlorpropham	a	d	d
XI	Triocarbamates			
	EPTC	c	d	d
	Triallate	c	b	a
	Moninate	c	a	...
XII	Ureas			
	Linuron	d	...	d
	Fluometuron	d	d	d
	Monuron	a	d	d
	Fenuron	a	c	...
XIII	Triazines			
	Terbutryn	a	c	b
	Atrazine	a	c	b
	Simazine	a	b	c
XIV	Pyridines			
	Picloram	c	d	d
	Diquat	a	d	d
	Paraquat	a	d	d
XV	Pyridazines	...	...	...
XVI	Pyrimidines			
	Bromacil	a	d	d
XVII	Unclassified heterocyclic nitrogen compounds			
	Amitrole	d	...	d
XVIII	Heterocyclic compounds with heteroatoms other than nitrogen	...	...	...
XIX	Organoarsenic compounds			
	Cacodylic acid	c	d	d
XX	Organophosphorous compounds	...	...	...

[a] For categories of Relative Sensitivity:
a = Most sensitive—lowest EC_{50}.
b = Intermediate sensitivity—intermediate EC_{50}.
c = Least sensitive—highest EC_{50}.
d = Equivalent sensitivity—equal EC_{50}.

TABLE 5—*Frequency of relative sensitivities shown in Table 4, summarizing the responses of different types of plants to specific herbicides and different classes of herbicides.*

Type of Plant	Relative Sensitivity				Total
	Most	Intermediate	Least	Equivalent	
Algal	$10^a(6)^b$	0	16(10)	7	33
Monocotyledon	7(7)	2	3(2)	16	28
Dicotyledon	3(3)	7	2(2)	18	30

[a] Number of specific herbicides to which the plant group (algae) had the most sensitive response.
[b] Number of herbicide classes to which the plant group (algae) had the most sensitive response.

Discussion

For algal phytotoxicity data to be a valid indication of how vascular plants react to a chemical two things must hold:
1. The chemical must invoke the same mode of action on the same process in both plant types, or it must influence a multitude of processes whereby at least one of these processes is impacted upon in both test systems.
2. The sensitivity of algae and vascular plants should be similar or correction factors must be known to permit accurate use of algal data in predicting vascular plant dose-response curves. The analyses conducted in this study address both of these issues.

A comparison of different plant responses and their associated chemicals showed that approximately 50% of the chemicals listed in PHYTOTOX which elicited a response unique to vascular plants (Table 1) were not reported to influence processes recognized as algal functions. This lack of evidence suggests that these chemicals (≈ 500) do not influence algae. If this is true, one can argue that these chemicals and any newly manufactured chemicals with similar chemical structures would not show toxicity in an algal screening test. If such false testing occurred in an actual screening program at the same frequency as nonoverlapping chemicals appeared in our analysis of data from PHYTOTOX (Table 3), it would occur approximately 500 times out of 2,500 or 20% of the time. This reasoning is highly speculative, since the frequency is based on the lack of data for 500 chemicals, but it serves to illustrate the possible inaccuracy of algal screening tests and emphasizes the need for additional research to resolve this issue.

The sensitivity comparisons conducted in this study between algal and vascular plants clearly shows that no single plant type can always be viewed as the least or most sensitive test system.

Acknowledgments

Although the research described in this article has been funded wholly or in part by the U.S. Environmental Protection Agency agreement CR 813094010, it has not been subjected to the Agency's review and therefore does not necessarily reflect the views of the Agency, and no official endorsement should be inferred.

References

[1] Maki, A. W. in *Analyzing the Hazard Evaluation Process*, K. L. Dickson, A. W. Maki, and J. Carines, Eds., American Fisheries Society, Washington, D.C., 1979, pp. 83–100.
[2] "OECD Guideline For Testing of Chemicals," Organization for Economic Cooperation and Development, Paris, 1981.

[3] Decleire, M. S., de Cat, W., and Bastin, R., *Reveu de l'Agriculture,* Vol. 28, No. 1, 1975, pp. 27–38.
[4] Garten, C. T. and Frank, M. L., Environmental Sciences Division Publication 2361, Oak Ridge National Laboratory, Oak Ridge, Tenn., 1984.
[5] Royce, C. L., Fletcher, J. S., Risser, P. R., McFarlane, J. C., and Benenati, F. E., *Journal of Chemical Information and Computer Sciences,* Vol. 24, 1984, pp. 7–10.
[6] Fletcher, J. S., Muhitch, M. J., Vann, D. R., McFarlane, J. C., and Benenati, F. E., *Environmental Toxicology and Chemistry,* Vol. 4, 1985, pp. 523–532.
[7] Fletcher, J. S., Johnson, F. L., and McFarlane, J. C., *Environmental Toxicology and Chemistry,* Vol. 7, 1988, pp. 615–622.
[8] Fletcher, W. W. and Kirkwood, R. C., *Herbicides and Plant Growth Regulators,* Granada, London, 1982, p. 15

Donald J. Versteeg[1]

Comparison of Short- and Long-Term Toxicity Test Results for the Green Alga, *Selenastrum capricornutum*

REFERENCE: Versteeg, D. J., "**Comparison of Short- and Long-Term Toxicity Test Results for the Green Alga, *Selenastrum capricornutum*,**" *Plants for Toxicity Assessment, ASTM STP 1091,* W. Wang, J. W. Gorsuch, and W. R. Lower, Eds., American Society for Testing and Materials, Philadelphia, 1990, pp. 40–48.

ABSTRACT: Algae are important components of many aquatic communities that can be sensitive to effects of aquatic pollutants. Standard algal toxicity tests quantify effects on population growth over 3 to 14 days. These population growth tests are useful for environmental hazard assessment but have limitations. These limitations include test duration, pH, nutrient and metabolite changes during toxicity testing, inability to test at the community level, potential interaction between algal media and test materials, and poor control or understanding of the effects of sample volatilization, sorption, biodegradation, photolysis, or hydrolysis on test material toxicity. These test limitations can be addressed in short-term (<4 h duration) toxicity tests.

Short-term algal toxicity tests were conducted with seven pure compounds and three municipal wastewater treatment plant effluents, and the results were compared to those obtained in a standard four-day population growth test (PGT). The short-term tests involved exposure of the green alga, *Selenastrum capricornutum,* to test materials and quantification of effects on oxygen generation (O_2) in a 30-min exposure and $^{14}CO_2$ fixation in a 35 min exposure. For test materials used in this research, the short-term tests were consistently less sensitive than the PGT, with ratios of the EC_{50} (effective concentrations) for the CO_2:PGT and the O_2:PGT ranging from 1.4 to >161. However, short-term to long-term test ratios were consistently less than 20, indicating that these tests may be appropriate systems to study effects of test material on algae without the limitations of the long-term test. Although the short-term algal test methods have their own set of limitations for environmental hazard testing, the information gained in the short-term approaches can be used to complement information gained in long-term toxicity tests. This research has demonstrated the relative sensitivities of two short-term tests.

KEY WORDS: algae, *Selenastrum capricornutum,* toxicity, short-term tests, cadmium, copper, pentachlorophenol, diethyl phthalate, carbaryl, atrazine, simazine

The traditional algal toxicity test [1,2] is a static, 3 to 14 day *chronic*[2] toxicity test. However, limitations of the experimental protocol can complicate interpretation by the toxicologist and reduce extrapolation of results to environmental situations. These limitations include changes in media pH due to rapid algal growth, algal nutrient and metabolite concentrations, the test material to algal cell ratio, and algal cell densities during exposure [3].

---

[1] Environmental Safety Department, Procter and Gamble Company, Cincinnati, OH 45217.
[2] Algal toxicity tests lasting 3 to 14 days are chronic tests, since the test duration is longer than the reproductive cycle of the organism. To avoid confusion over acute and chronic, the *acute* algal methods presented here will be referred to as *short-term*. Chronic tests, those lasting three days or more, will be referred to as *long-term* population growth tests.

In addition, the apparent toxicity of sample components can be affected through interactions with components of the growth media or through hydrolysis, photolysis, volatilization, sorption, and/or biodegradation of sample components. The effect of these factors on the bioavailability and toxicity of environmental contaminants confounds algal toxicity data interpretation. Finally, long-term population growth test procedures have not been developed for *in situ* testing or simultaneous testing of multiple species [4].

Many of these limitations of long-term population growth tests can be overcome in short-term toxicity tests [3-5]. Due to the short duration of these tests, changes in pH, nutrient and metabolite concentrations, test material to algal cell ratio, and algal cell densities have minimal effects on interpretation of results. Further, short-term test procedures minimize changes in test material concentration and can be conducted with few media components reducing potential interactions with test materials. Finally, short-term methods have been used successfully with natural phytoplankton assemblages [3,4]. Increased utilization of short-term test methods can improve laboratory to field extrapolations especially when used to supplement long-term toxicity test results. These advantages suggest short-term algal test procedures can be an important tool to understand the impact of environmental contaminants on aquatic plants [3-5].

The two short-term test procedures used most commonly are the O_2 generation and CO_2 fixation tests. The duration of these short-term tests typically ranges from a few [3] to 24h [6]; however, toxicity tests of shorter duration are possible. Short-term test procedures need to be fully developed and validated and the relative sensitivity of the short-term test understood. The objectives of this research were to develop short-term algal toxicity test methods (approximately 30 min exposure) and to compare the results of these short-term tests with each other and with the population growth test (four days).

Procedure

Test Materials

Test materials used in these experiments were selected to represent important environmental pollutants with a diversity of chemical properties and modes of toxic action. All test compounds were reagent grade or better and included cadmium sulfate ($CdSO_4$; Fisher Scientific), copper oxide (CuO; Fisher Scientific), diethyl phthalate (Supelco), pentachlorophenol (Fluka AB), carbaryl (Union Carbide), atrazine (2-chloro-4-ethylamino-6-isopropylamino-*s*-triazine; Chem Service), and simazine (2-chloro-4,6-*bis*-ethylamino-*s*-triazine; Chem Service). When possible, test compounds were introduced to toxicity tests in a water carrier. For test compounds with low water solubility, a carrier solvent of methanol (pentachlorophenol) or dimethyl sulfoxide (carbaryl, atrazine, simazine) was used. Carrier solvent concentrations were kept to a minimum and solvent controls were included in the experimental design. Effluents from three municipal wastewater treatment plants (WWTP) without filter sterilization were used as test materials to determine the applicability of these methods to effluents. WWTP A is an 18 million gallon per day (MGD), conventional activated sludge plant receiving wastes from domestic, commercial, and industrial sources. WWTP B is a one MGD extended aeration activated sludge plant receiving waste from domestic and commercial sources. WWTP C is a six MGD trickling filter plant receiving wastes from domestic, commercial, and industrial sources. Effluent samples were grab samples collected prior to chlorination and were utilized in toxicity tests within 24 h of collection. Dilution water for all toxicity tests consisted of charcoal-filtered well water with the following chemical characteristics (mean value reported): hardness, 171 mg/L as $CaCO_3$; alkalinity, 113 mg/L as $CaCO_3$; pH, 8.2; conductivity, 380 μmhos.

Test Procedures

All tests were conducted with the green alga *Selenastrum capricornutum* Printz. Algae were cultured according to the methods of Horning and Weber [2]. In this research, the results of three algal toxicity tests were compared: a four-day population growth test, a CO_2 fixation test, and an oxygen generation test. The three algal tests were each performed with the seven pure test compounds. WWTP effluents were assessed for toxicity to algae in the population growth and the CO_2 fixation test only, since background biological and chemical oxidation could not be entirely corrected for in the oxygen generation test. For each test, definitive test exposure concentrations consisted of a geometric progression of concentrations (concentration ratio 1.5 to 2.0) established based on the results of rangefinding exposures to a logarithmic concentration series. Algal cultures were maintained axenically; however, toxicity tests were conducted without precautions insuring axenic test conditions. Exposure media and test material solutions were checked prior to, and during, each test to insure that competing algae and algal grazer did not influence test results.

Population Growth Test (PGT)

The population growth test followed the procedures of Horning and Weber [2]. Cultures of algae were exposed to a geometric progression of test material concentrations at 24 ± 1°C and an illumination of 86 $\mu E\ s^{-1}\ m^{-2}$ (cool white fluorescent light bulbs; equivalent to 400 ft-c) on a shaker table oscillating at 100 rpm. Light levels in each test was measured with a Li-Cor quantum photometer. After four days of exposure to test materials, cell numbers were quantified by counting on a hemocytometer.

CO_2 Fixation Test

Algae from a culture in logarithmic growth phase were concentrated by centrifugation and resuspended in cold (4°C) 20 mM HEPES-KOH (Sigma Chemical Company) buffer, pH 7.5. Chlorophyll-a was quantified following dimethyl sulfoxide extraction by the method of Shoaf and Lium [7] and adjusted so that the final assay concentration was approximately 15 μg/mL. The test was initiated by adding an aliquot of the algal suspension to a test solution buffered with 20 mM HEPES-KOH to pH 7.5. Light was provided by a Sylvania Spot Grow high-intensity light bulb at 350 $\mu E\ m^{-2}\ s^{-1}$. This light level was used to insure saturation of algal photosynthetic pigments [3]. Algae were exposed to the test material for 30 min at 24°C. At that time, an aliquot of $NaH^{14}CO_3$ (final assay concentration four mM, 0.5 μCi/μmol) was added, and the fixation of $^{14}CO_2$ into acid stable counts was determined at specific time intervals, typically 5 min. Radioactivity was determined on a Beckman LS 7800 liquid scintillation counter. Multiple replicates of a concentration series were conducted simultaneously.

One potential disadvantage of the CO_2 fixation test is the requirement for a buffer. Buffering is necessary to avoid pH shifts during the test which could alter CO_2 concentration and availability to algae. To address this concern, we examined the effect of buffer on the toxicity of cadmium and copper by conducting the population growth test and CO_2 fixation test in the presence of HEPES-KOH buffer. Cadmium and copper were selected as the most probable candidates for HEPES, an organic acid, to affect bioavailability.

Oxygen Generation Test

A solution of algae from a culture in logarithmic growth phase was centrifuged and resuspended in AAP media at approximately 10^6 cells mL^{-1}. The algal suspension was gas purged with a mixture of 500 ppm CO_2 in N_2 to remove oxygen. Test material was added to the algal culture and the culture placed into an airtight Plexiglas exposure chamber. Algae were exposed to approximately 250 μE m^{-2} s^{-1} at 24°C, and oxygen generation was continuously monitored with an oxygen probe (Yellow Springs Instrument Company) for 30 min. Continuous monitoring of oxygen limited the number of tests which could be conducted simultaneously; thus multiple test concentrations were used but individual test concentrations were not replicated.

Statistics

Toxicity was assessed as a decrease, relative to appropriate controls, in population numbers in the population growth test, CO_2 fixation into acid stable counts in the CO_2 fixation test, and O_2 produced in the oxygen generation test. In the four-day test and CO_2 fixation test, effective concentrations (EC values) reducing the biotic parameter by 50% and the associated 95% confidence intervals were calculated using the nonlinear multiple regression analysis model of SAS Institute. Due to the lack of replication, the oxygen generation test could not be assessed by nonlinear multiple regression. The EC_{50} concentrations for the oxygen generation test were estimated by graphical interpolation [8]. The EC_{50} statistic was selected as an indicator of the toxic concentration to compare data from different treatments and does not necessarily indicate a toxicity threshold.

Results

Results of the three toxicity tests on the seven pure materials and three municipal wastewater treatment plant effluent samples appear in Table 1. Pure test compounds produced dose-related effects in the three tests. Dose-response curves for cadmium and simazine are shown and are representative of the data obtained (Fig. 1). Cadmium dose-response curves were similar in the CO_2 fixation test and the population growth test generating EC_{50} concentrations of 0.13 mg/L and 0.18 mg/L cadmium, respectively. The slope of the dose-response curve was less steep in the O_2 generation test and the EC_{50} concentration was 4.7 mg/L. Likewise, for simazine, the CO_2 fixation test and the population growth test had similar slopes and the slope in the O_2 generation test was less. The EC_{50} concentrations were 0.10 mg/L, 0.22 mg/L, and 0.49 mg/L in the population growth test, CO_2 fixation test, and O_2 generation test, respectively.

Based on the EC_{50} values, the order of sensitivities of the three toxicity tests to pure test materials were: population growth > CO_2 fixation > oxygen generation (Table 1). This relative sensitivity of the three tests was consistent among all pure test materials except for carbaryl, where the O_2 generation test was similar in sensitivity to the CO_2 fixation test.

For effluents, the CO_2 fixation test was less sensitive than the four-day test. The population growth test detected two of the effluents as toxic and the CO_2 test detected only one effluent as toxic.

For pure compounds, the ratio of the EC_{50} concentrations of the CO_2 fixation test to the population growth test ranged from 1.4 to 68, with six of the seven ratios less than 20 (Table 2). Copper had the greatest CO_2:PGT ratio. The ratio of effect concentrations in the O_2 generation test and the population growth test ranged from 4.0 to >161, with four of the six ratios less than 20; again, copper had the greatest short- to long-term ratio (Table 2). For effluents, only one ratio could be calculated due to the low toxicity of these effluents to the alga. The ratio obtained for WWTP B effluent agreed well with ratios on pure materials.

The effect of the buffer, HEPES-KOH, on the toxicity of cadmium and copper were determined in the CO_2 fixation and the population growth tests. HEPES had little effect on the

TABLE 1—*Comparison of the toxicity (EC_{50} concentrations) of pure test compounds and effluents to the alga,* Selenastrum capricornutum, *as assessed by the four-day population growth test, the CO_2 fixation test, and the O_2 generation test.*

Test Material	Population Growth Test, mg/L	CO_2 Fixation Test, mg/L	O_2 Generation Test, mg/L
Pure Compounds:			
Cadmium	0.13[a]	0.18	4.70[b]
	(0.111–0.151)	(0.102–0.317)	
Copper	0.03	1.90	>4.50
	(0.021–0.037)	(1.66–2.17)	
Diethyl Phthalate	30.1	>100	ND
	(21.1–42.9)		
Pentachlorophenol	0.07	0.85	1.21
	(0.044–0.122)	(0.751–0.952)	
Carbaryl[c]	1.04	18.1	17.2
	(0.686–1.57)	(7.59–43.0)	
Atrazine	0.05	0.10	0.38
	(0.042–0.067)	(0.059–0.160)	
Simazine	0.10	0.22	0.45
	(0.085–0.107)	(0.167–0.281)	
WWTP Effluents:[c]			
WWTP A	>100	>100	ND[d]
WWTP B	7.05	86.8	ND
	(3.32–14.98)	(52.4–144.59)	
WWTP C	20.6	>100	ND
	(10.59–39.99)		

[a] EC_{50} with 95% confidence intervals in parentheses.
[b] EC_{50}; 95% confidence intervals could not be estimated.
[c] EC_{20} concentrations reported. For carbaryl, EC_{50} values exceeded solubility limit. For effluents, EC_{50} values frequently exceeded 100% effluent.
[d] Effluents were not assessed with the O_2 generation test due to difficulties in controlling O_2 uptake by bacteria and oxidizable compounds in the effluent. These effects occurred at different levels at each effluent concentration eliminating the specificity of algae O_2 generation.

toxicity of cadmium to algae in either the population growth test or the CO_2 fixation test (Table 3). The EC_{50} concentration increased from 0.12 mg/L to 0.18 mg/L cadmium with 0 and 20 mM HEPES, respectively, in the population growth test. In the CO_2 fixation test, the cadmium EC_{50} levels ranged from 0.04 to 0.18 mg/L but did not show a relationship with HEPES concentration. With copper, the EC_{50} level in the population growth test increased in a dose-dependent manner from 0.03 to 0.23 mg/L with an increase in buffer concentration from 0 to 20 mM HEPES (Table 3). Copper EC_{50} concentrations in the CO_2 fixation test were unaffected by the level of HEPES used. The EC_{50} concentrations ranged from 1.3 to 1.9 mg/L at the three buffer concentrations.

Discussion

In this research, short-term tests were less sensitive than the standard four-day population growth test. This was expected as the CO_2 fixation and oxygen generation tests assess short-term effects on photosynthesis. Due to the timing of the test, compounds indirectly affecting photosynthesis and/or requiring more than 35 min to elicit an effect will have high short- to long-term test ratios. In long-term tests, test materials have additional time and sites to exert a toxic action. In support of this rationale, the compounds known specifically to inhibit elec-

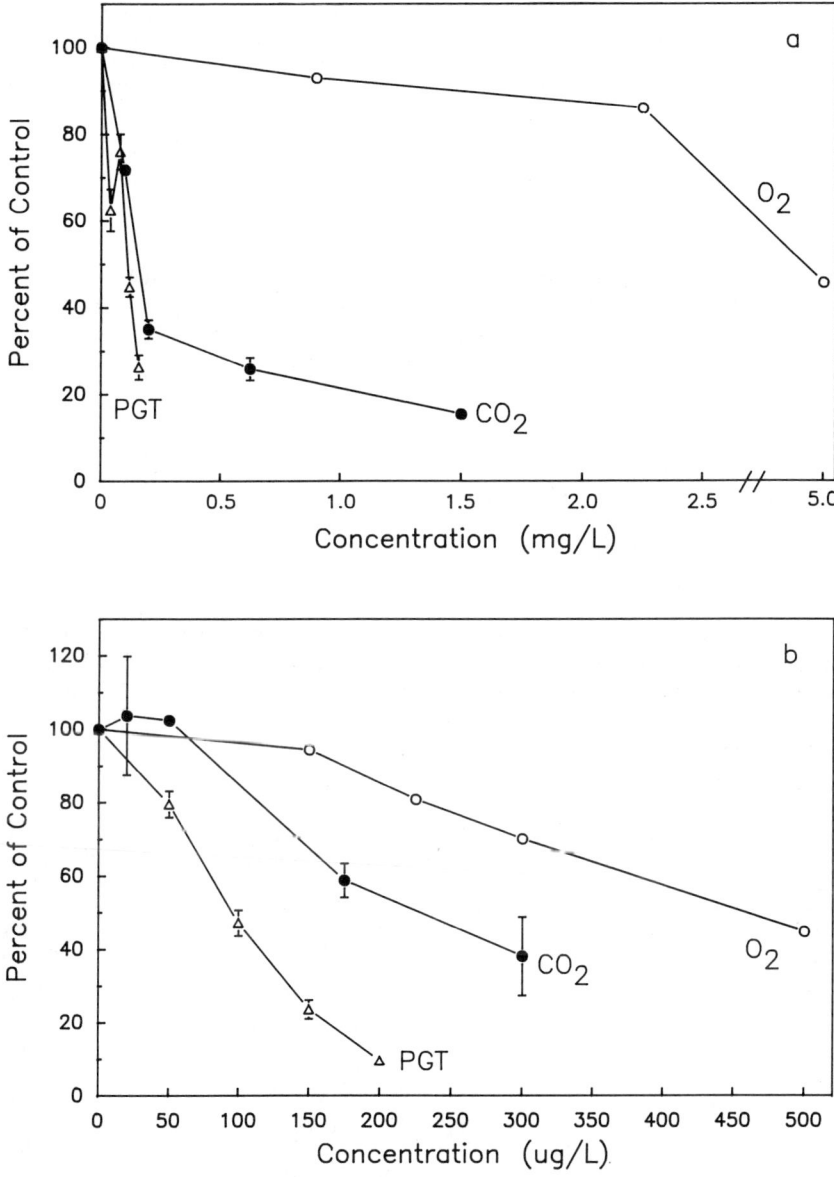

FIG. 1—*Effect of* (a) *cadmium and* (b) *simazine on the alga,* Selenastrum capricornutum, *in the population growth test, CO_2 fixation test, and O_2 generation test.*

tron transport during photosynthesis, atrazine and simazine [9], had short- to long-term ratios at the low end of the range. Reduced sensitivity, however, does not eliminate the utility of a short-term test method. Acute toxicity tests have been used successfully to address a wide variety of environmental questions. As with acute to chronic ratios for other classes of aquatic organisms [10], ratios for algae were compound specific and generally ranged from 2 to 20.

TABLE 2—*Comparison of the sensitivity to pure test compounds and effluents of the alga, Selenastrum capricornutum, in three toxicity tests. Ratios of the EC_{50} concentrations of the CO_2 fixation test to population growth test (PGT) and the O_2 generation test to population growth test.*

Test Material	Ratios	
	CO_2:PGT	O_2:PGT
Pure Compounds:		
Cadmium	1.4	36
Copper	68	>161
Diethyl phthalate	>3.3	ND[a]
Pentachlorophenol	12	17
Carbaryl[b]	17	17
Atrazine	1.8	7.3
Simazine	2.3	4.0
WWTP Effluents:[b]		
WWTP A	no ratio	ND
WWTP B	12	ND
WWTP C	>4.9	ND

[a] Not determined.
[b] Ratios based on EC_{20} concentrations.

The CO_2 fixation test was generally more sensitive than the oxygen generation test to test materials. This sensitivity is due at least in part to the use of a 30 min incubation period prior to quantification of effects. With more sensitive and rapid oxygen sensing probes, a procedure similar to the CO_2 fixation test could be employed. This improved procedure would be expected to generate short- to long-term test ratios more comparable to the CO_2

TABLE 3—*Comparison of the toxicity (EC_{50} concentration) of cadmium and copper to the alga, Selenastrum capricornutum, at three levels of HEPES-KOH buffer as assessed by the four-day population growth test and the CO_2 fixation test.*

Test Material	HEPES, mM	Population Growth Test, mg/L	CO_2 Fixation Test, mg/L
Cadmium	0.0	0.13[a] (0.111–0.151)	ND[b]
	0.2	0.12 (0.106–0.144)	0.13 (0.096–0.178)
	2.0	0.13 (0.119–0.135)	0.04 (0.032–0.062)
	20.0	0.18 (0.154–0.221)	0.18 (0.102–0.317)
Copper	0.0	0.03 (0.021–0.037)	ND
	0.2	0.06 (0.055–0.076)	1.6 (1.49–1.82)
	2.0	0.09 (0.080–0.100)	1.3 (0.938–1.72)
	20.0	0.23 (0.139–0.383)	1.9 (1.66–2.17)

[a] EC_{50} with 95% confidence limit in parentheses.
[b] No data; CO_2 fixation test required buffer to control pH.

fixation test. However, the inability to use the oxygen generation approach in samples containing appreciable chemical or biological oxygen demand is a drawback.

The use of buffer did not affect the toxicity of copper or cadmium in the CO_2 fixation test. HEPES did reduce copper toxicity in the long-term population growth test, which suggests that copper-HEPES binding is slow or that HEPES has a protective effect on a cellular function other than photosynthesis. The high copper CO_2:PGT copper ratio appears to be due to a relatively slow (i.e., >35 min) expression of toxicity. This is further supported by the copper O_2:PGT ratio of >161. The use of HEPES as a buffer in short-term toxicity tests is supported by the observation that most short- to long-term test ratios were less than 20. This indicates that HEPES had minimal, if any, effect on the toxicity of these test materials to algae in the short-term test. Use of a buffer to stabilize pH in algal toxicity tests allows assessment of toxicity at several pHs, reduces the effect rising pH (due to CO_2 depletion) may have on test compound bioavailability, and can facilitate the interpretation of data on environmental samples (e.g., effluents and leachates) where pH may not be stable.

The use of short-term algal photosynthetic tests is not new. These tests have been used to study the biochemistry of photosynthesis [11], primary production in aquatic and marine habitats [12], and in toxicity testing [3–6,13,14]. Although some studies report comparable sensitivities between short- and long-term methods [6,13], the key advantage of short-term algal tests is not necessarily sensitivity but their speed and flexibility for assessing the interaction of environmental factors on toxicity. The effects of media components, cell to toxicant ratio, changes in sample composition due to sorption, volatilization, etc., and a variety of environmental factors on the toxicity of pure test compounds or environmental samples to an alga or a natural phytoplankton community can be assessed quickly [3,4,6]. Long-term test approaches reduce the level of control and understanding of the influence these factors have on test results.

Summary

This and other studies have demonstrated the utility of short-term algal tests to address environmental questions in a timely manner on a variety of environmental samples. In general, short-term (approximately 30 min) toxicity tests are less sensitive than the long-term population growth test. Long-term tests are still needed for environmental hazard assessment. However, the advantages of short-term toxicity tests can greatly improve the toxicologist's ability to interpret and apply results of long-term algal toxicity tests. Short-term algal tests should be used to provide information complementary to long-term test data. Like other "acute" toxicity tests, short-term algal tests have specific advantages over long-term toxicity tests and should be utilized for these advantages.

Acknowledgments

The assistance of Michael Lewis and Daniel Woltering in reviewing and commenting on this manuscript is gratefully acknowledged.

References

[1] Miller, W. E., Greene, J. C., and Shiroyama, T., "The *Selenastrum capricornutum* Printz Algal Assay Bottle Test: Experimental Design, Application, and Data Interpretation Protocol," EPA-600/9-78-018, U.S. Environmental Protection Agency, Corvallis, Ore., July 1978.

[2] "Methods for Estimating the Chronic Toxicity of Effluents and Receiving Waters to Freshwater Organisms—Draft," W. B. Horning and C. I. Weber, Eds., EPA-600/4-85-014, U.S. Environmen-

tal Protection Agency, Environmental Monitoring and Support Laboratory-Cincinnati, Ohio, May 1985.
[3] Giddings, J. M., Stewart, A. J., O'Neill, R. V., and Gardner, R. H., "An Efficient Algal Bioassay Based on Short-Term Photosynthetic Response," in *Aquatic Toxicology and Hazard Assessment: Sixth Symposium, ASTM STP 802*, W. E. Bishop, R. D. Cardwell, and B. B. Heidolph, Eds., American Society for Testing and Materials, Philadelphia, 1983, pp. 445-459.
[4] Lewis, M. A. and Hamm, B. G., "Environmental Modification of the Photosynthetic Response of Lake Phytoplankton to Surfactants and Significance to a Laboratory-Field Comparison," *Water Research*, Vol. 20, No. 12, 1986, pp. 1575-1582.
[5] Blaise, C., Van Collie, R., Bermingham, N., and Coulombe, G., "Comparison of the Toxic Responses of Three Biological Indicators (Bacteria, Algae, and Fish) Exposed to Effluents from Pulp and Paper Mills," *Revue Internationale des Sciences de l'Eau*, Vol. 3, No. 1, 1987, pp. 9-17.
[6] Turbak, S. C., Olsen, S. B., and McFeters, G. A., "Comparison of Algal Assay Systems for Detecting Waterborne Herbicides and Metals," *Water Research*, Vol. 20, 1986, pp. 91-96.
[7] Shoaf, W. T. and Lium, B. W., "Improved Extraction of Chlorophyll *a* and *b* from Algae using Dimethyl Sulfoxide," *Limnology and Oceanography*, Vol. 21, 1976, pp. 926-928.
[8] Walsh, G. E., Deans, C. H., and McLaughlin, L. L., "Comparison of the EC_{50}'s of Algal Toxicity Tests Calculated by Four Methods," *Environmental Toxicology and Chemistry*, Vol. 6, No. 10, 1987, pp. 767-770.
[9] Moreland, D. E., "Mechanism of Action of Herbicides," *Annual Review of Plant Physiology*, Vol. 31, 1980, pp. 597-638.
[10] Kenaga, E. E., "Predictability of Chronic Toxicity from Acute Toxicity of Chemicals in Fish and Aquatic Invertebrates," *Environmental Toxicology and Chemistry*, Vol. 1, No. 4, 1982, pp. 347-358.
[11] Tolbert, N. E., Harrison, M., and Selph, N., "Aminooxyacetate Stimulation of Glycolate Formation and Excretion by *Chlamydomonas*," *Plant Physiology*, Vol. 72, 1983, pp. 1075-1083.
[12] Lind, O. T., *Handbook of Common Methods in Limnology*, The C.V. Mosby Company, St. Louis, 1974.
[13] Walsh, G. E., "Effects of Herbicides on Photosynthesis and Growth of Marine Unicellular Algae," *Hyacinth Control Journal*, Vol. 10, 1972, pp. 45-48.
[14] Ross, P. E. and Henebry, M. S., "Use of Four Microbial Tests to Assess the Ecotoxicological Hazard of Contaminated Sediments," *Toxicity Assessment: An International Journal*, Vol. 4, No. 1, 1989, pp. 1-21.

Joseph W. Gorsuch,[1] *Robert O. Kringle,*[2] *and Kenneth A. Robillard*[3]

Chemical Effects on the Germination and Early Growth of Terrestrial Plants

REFERENCE: Gorsuch, J. W., Kringle, R. O., and Robillard, K. A., "**Chemical Effects on the Germination and Early Growth of Terrestrial Plants**," *Plants for Toxicity Assessment, ASTM STP 1091,* W. Wang, J. W. Gorsuch, and W. R. Lower, Eds., American Society for Testing and Materials, Philadelphia, 1990, pp. 49–58.

ABSTRACT: A rapid, simple, and cost-effective testing procedure is described for determining the potential of a chemical to affect germination and early growth of terrestrial plants. Radish (*Raphanus sativus* L. var. Champion 708), lettuce (*Lactuca sativa* L. var. 525 Ithaca M.T.O.), and perennial ryegrass (*Lolium perenne* L. var. Manhattan) are routinely used in this test because of their economic and ecological importance, quick and high rate of germination, and relatively uniform root and hypocotyl/coleoptile growth. Plants are exposed to the chemical for seven days, after which germination, root length, and plant height values are determined. Average values for treated plants are compared with the corresponding values for the control (untreated) plants. An observed ratio (treated versus control means) of <90% is considered to be a significant indication of adverse chemical effect.

Test results for 26 commercial chemicals are compared with information in the literature. For these commercial chemicals, reduction in root length was the most sensitive plant response, followed by reduction in the plant height. By comparison, germination was not noticeably affected. Ryegrass and radish were found to be equally tolerant species, and lettuce was the most sensitive plant species. Where data for other plant-effect methods are available, comparisons show that this method is generally of equal or greater sensitivity in identifying the potential for chemical effects.

KEY WORDS: lettuce, radish, ryegrass, phytotoxicity (bioassay), germination, early growth effects

The potential for a chemical, or mixture of chemicals, to inhibit the growth of terrestrial plants is an important environmental consideration. In recent years various seed germination screening procedures have been useful in determining the phytotoxicity of chemicals, leachates, and effluents. Some of these procedures use petri dishes [1–4], liquid-shaking cultures [5–6], and blotter-sandwich techniques [7–12]. The bioassay described in this paper, which is most similar to the blotter-sandwich techniques, is designed to determine the dose-response characteristics of a chemical, leachate, or effluent on the germination and early growth of several representative terrestrial plants during a 7-day exposure period. Wang [4] and Ratsch and Johndro [6] have described advantages of the plant seed germination and root elongation tests, which require a minimum of time, space, equipment, and cost. In our laboratory we have found all the above to be true, and we found the results to be biologically and statistically sensitive.

The most time-consuming feature of most germination and root elongation studies is measuring the roots and shoots (hypocotyls or coleoptiles). Our method uses a test system that

---

[1] Health and Environment Laboratories, Eastman Kodak Company, Rochester, NY 14652-3617.
[2] Eastman Pharmaceuticals, Great Valley Corporate Center, Malvern, PA 19355.
[3] Chemicals Quality Services Division, Eastman Kodak Company, Rochester, NY 14652-3615.

provides for germination in a vertical plane, allowing the roots and shoots to grow straight and therefore making these measurements easier. Three plant species are routinely used in the test: ryegrass, radish, and lettuce. These species were selected because of their economic and ecological importance (representing monocotyledons and dicotyledons), rapid and high rate of germination, and relatively uniform hypocotyl/coleoptile and root growth.

An evaluation was performed of our Germination and Early Growth Effects Test using 26 commercially important, large-volume chemicals. The seven-day No Observed Effect Concentration (NOEC) values were derived for each species. NOEC is defined for this procedure as the highest test chemical concentration that produces less than 10% difference in any of the 3-endpoints measured and where there are no abnormal physical observations in the treated plants when compared to the control plants. The statistical validity of our 10% measured difference criterion was evaluated by using a method for the ratio of two normally distributed random variables [13]. Our results were compared to literature values for the same species, regardless of the phytotoxicity method used, to assess the relative sensitivity of the procedure described here.

Materials and Methods

Apparatus

Seed-pack growth pouches (manufactured by Northrup, King, and Company) were used in this test (Fig. 1). Separator racks (Fig. 2), containing the prepared growth pouches, kept the growth pouches vertical.

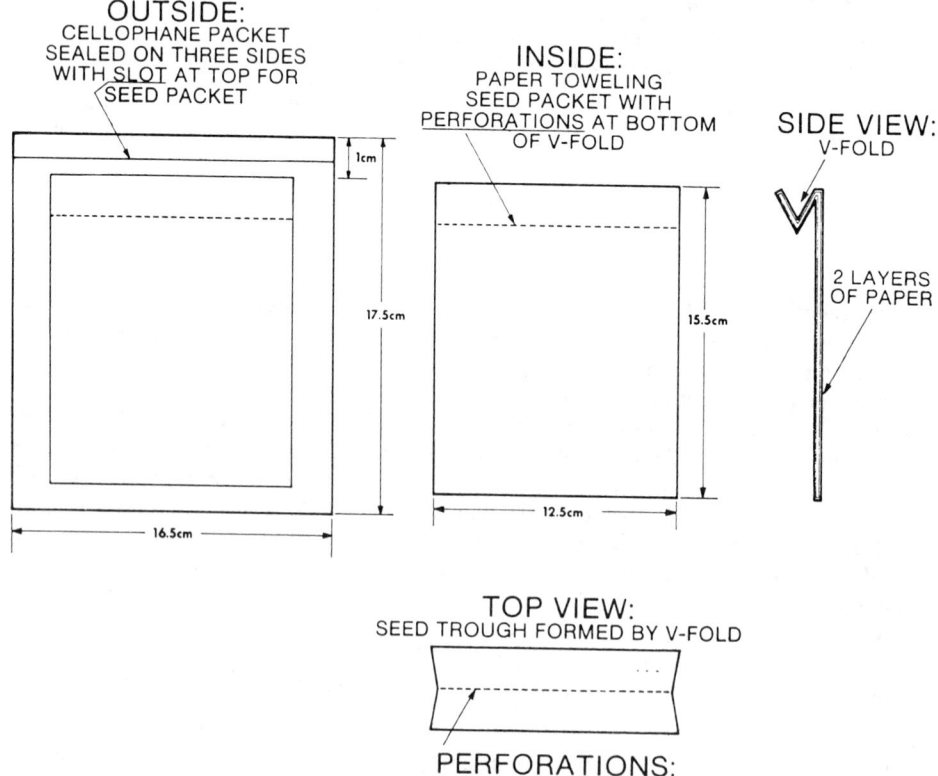

FIG. 1—*Seed-pack growth pouch.*

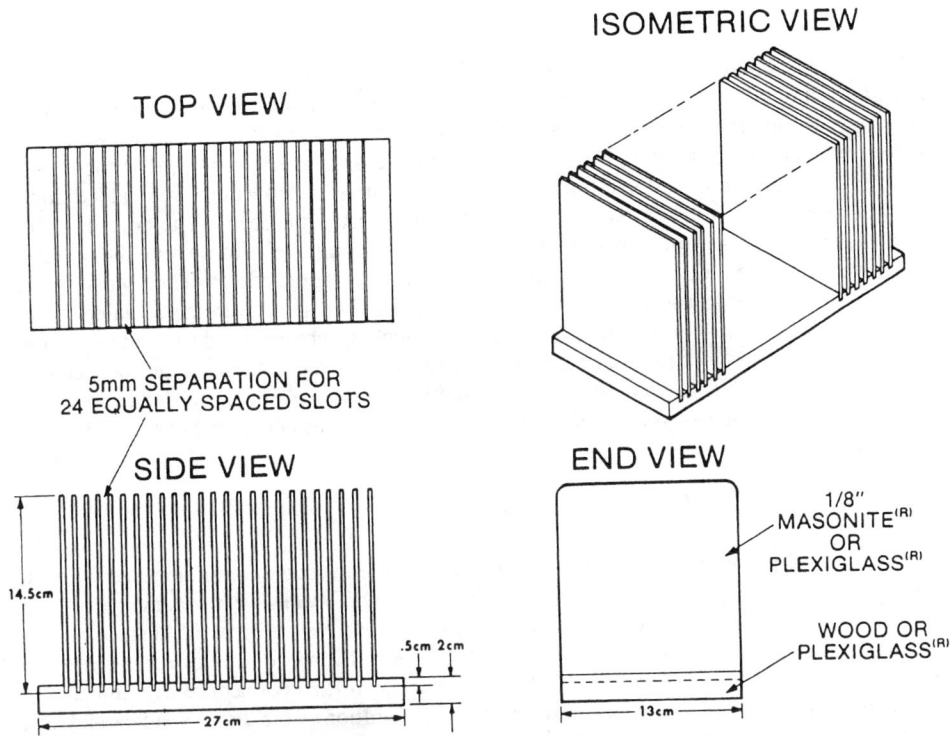

FIG. 2—*Germination growth pouch separator rack.*

Test Species

The seeds used in this study, obtained from Harris Moran Seed Company, Rochester, New York, were taken from the same batch for each plant species. The seed identity was obtained from the container in which they were sold. Radish (*Raphanus sativus* L., cv. Champion 708), lettuce (*Lactuca sativa* L., cv. 525 Ithaca, M.T.O.), and perennial ryegrass (*Lolium perenne* L., cv. Manhattan) seeds were used in this test.

Test Solution Preparation

Chemicals evaluated in this study were known to be readily soluble at 100 mg/L [14], the maximum test chemical concentration routinely used. A 100 ± 1 mg quantity of the test article was added to 1 L of distilled water. This mixture was mechanically stirred at room temperature (22 ± 2°C) until dissolution was complete. The pH of this solution was checked and, if outside the range of 6.5 to 8.0, adjusted to 7.0 ± 0.1 with either aqueous hydrochloric acid (HCl) or sodium hydroxide (NaOH) solutions. Fifty millilitres of the above solution were transferred to another flask and diluted with 450 mL of distilled water. Nominal concentrations of 100 and 10 mg/L were used in describing these two test solutions, except for the metal salts, which were expressed as the nominal concentration of the metal ion. When lower concentrations were needed to achieve the NOEC, they were prepared by further 1 to 9 serial dilutions with distilled water. No testing was done at concentrations greater than 100 mg/L.

Test Procedure

Radish, lettuce, and ryegrass seeds were each dispersed in growth pouches (Fig. 1). Each replicate contained 20 seeds of one plant type. Four replicates (80 seeds for each plant type) were used for each treatment level (e.g., 100 mg/L, 10 mg/L, etc.). Two replicates were used for controls (40 seeds for each plant type). With the seeds uniformly arranged in the trough of each pouch, 20 mL of test solution or control solution (distilled water) were added down the back side of the paper toweling which served to moisten the seeds. The bottom of the pouch was slowly raised until the pouch was horizontal, completely immersing the paper in the solution; this step lessened the likelihood of chromatographic effects. The pouches were placed in a chamber at room temperature (22 ± 2°C). The seeds were germinated in the dark, as occurs in soil. After seven days the growth pouches were removed from the separator racks, and the number of seeds germinated, the number of roots, the average root lengths, the average plant heights, and the general condition of the plants were determined and recorded. No attempt was made to determine the amount of test chemical absorbed by each plant species.

Calculations

Measured values of the replicates for each species were averaged and compared with the corresponding average values for the controls. The NOEC value was obtained from the results of the relative mean germination response, the relative mean plant height, and the relative mean root length. A value of <90% of the control value for any of these properties was considered indicative of a significant adverse effect. Mottled cotyledons or plant shoots, desiccated plants, loss of hair roots, and marked discolorations were also considered adverse effects. The highest concentration causing no significant adverse effect was denoted the NOEC. In tests where there were no adverse effects, or where the treated plants exhibited enhanced growth or germination response (>110% of the control value) at the highest concentration tested and there were no adverse effects, that concentration was designated the NOEC.

Statistical Validation of 10% Measured Difference Criterion

The method used to statistically test the hypothesis that 10% measured differences were indeed statistically significant is based on the ratio of two normally distributed random variables [13]. The percent growth (W) of the treated group values, relative to control values, was estimated by

$$W = \frac{\overline{X}}{\overline{C}} \cdot 100\%$$

where

$\overline{X}$ = mean of treated group replicates, and
$\overline{C}$ = mean of control group replicates.

Estimates of the relative percent growth were obtained for root length and plant height at each chemical concentration. An estimate of the variance, with three degrees of freedom, was calculated for root length and plant height for each species at each chemical concentration. For each control group a single degree of freedom estimate of the variance was calculated for root length and plant height for each species. The control estimates of the variance for each plant type were averaged over 22 test sets to obtain a pooled estimate of the variability of the control.

The 95% confidence interval for the relative percent growth (W) was determined by finding the values that satisfy the following inequality:

$$AW^2 - BW + C \leq 0$$

where

$A = \overline{C}^2 - t_c^2 S_c^2/(n_c - 1)$
$B = (\overline{C})(\overline{X})$
$C = \overline{X}^2 - t_x^2 S_x^2/(n_x - 1)$

where

n_c = number of control replicates (= 2),
n_x = number of treatment replicates (= 4),
$\overline{C}$ = mean of n_c control replicates,
$\overline{X}$ = mean of n_x treatment replicates,
S_c^2 = maximum likelihood estimate of variance of control replicates,
S_x^2 = maximum likelihood estimate of variance of treatment replicates,
t_c^2 = upper 2.5% point of t-distribution with degrees of freedom associated with S_c^2,
t_x^2 = upper 2.5% point of t-distribution with degrees of freedom associated with S_x^2 and
t = upper 2.5% point of t-distribution with degrees of freedom associated with the corresponding variance.

In theory, this interval will contain the true percent effect due to treatment when compared with a control 95% of the time. The estimated 95% confidence intervals were used to test the hypothesis that the treatment level had no adverse effect on plant growth when compared with a control group. If the 95% confidence interval includes the value 100%, then there is not enough statistical evidence to conclude that the growth of the treated plant group was different from that of the control group. If the 95% confidence interval includes only values greater than 100%, then one could conclude that the growth of the treated group was enhanced compared to that of the control group. Similarly, if the 95% confidence interval includes only values below 100%, then one could conclude that the growth of the treated group was inhibited compared to that of the control group. The germination responses were not validated by this method.

Results

Estimation of NOEC Values

The No Observed Effect Concentration (NOEC) values for the 26 commercial chemicals tested using this procedure are given in Table 1. The NOEC values are based on adverse effects noted from any of the three endpoints: germination response, plant height, and root length. The NOEC value is set at the highest concentration tested where enhanced growth is observed and no adverse effects noted. The NOEC values are expressed in milligrams per litre (mg/L) of the nominal chemical concentration and without water of hydration for potassium ferrocyanide. The metal salt NOEC values are expressed as the nominal metal ion concentration.

TABLE 1—*NOECs (mg/L) for the 26 commercial chemicals.*

Name	Formula	Chemical Abstract Service (CAS) Registry Number	Ryegrass	Radish	Lettuce
Acetone	$(CH_3)_2CO$	(67-64-1)	100	100	100
Acetonitrile	CH_3CN	(75-05-8)	100	100	10
Allyl Alcohol	$CH_2=CHCH_2OH$	(107-18-6)	1	$<10^a$	1
Ammonium Sulfate	$(NH_4)_2SO_4$	(7783-20-2)	100	100	100
Aniline	$C_6H_5NH_2$	(62-53-3)	10	10	10
Benzoic Acid, Sodium Salt	C_6H_5COONa	(532-32-1)	10	10	10
Beryllium Chloride	$BeCl_2$	(7787-47-5)	11.3^b	11.3^b	1.1^b
Cobaltous Chloride	$CoCl_2$	(7646-79-9)	4.5^b	4.5^b	4.5^b
Copper Sulfate	$CuSO_4$	(7758-98-7)	0.26^b	0.26^b	$<0.26^{a,b}$
Diethanolamine	$(HOCH_2CH_2)_2NH$	(111-42-2)	100	100	10
Ethyl Alcohol	C_2H_5OH	(64-17-5)	10	100	10
Ethylenediamine Dihydrochloride	$NH_2CH_2CH_2NH_2 \cdot 2HCl$	(333-18-6)	100	100	10
(Ethylenedinitrilo)-tetraacetic Acid Ferric Sodium Salt	$C_{10}H_{12}FeN_2NaO_8$	(15708-41-5)	$<10^a$	100	100
Methyl Alcohol	CH_3OH	(67-56-1)	100	10	100
Nickel Chloride	$NiCl_2$	(7718-54-9)	0.45^b	$<4.5^{a,b}$	0.45^b
Pentachlorophenol	C_6Cl_5OH	(87-86-5)	0.1	0.1	0.1
Phenol	C_6H_5OH	(108-95-2)	0.1	1	10
Potassium Bromide	KBr	(7758-02-3)	100	100	10
Potassium Dichromate	$K_2Cr_2O_7$	(7778-50-9)	3.5^b	3.5^b	3.5^b
Potassium Ferrocyanide	$K_4Fe(CN)_6 \cdot 3H_2O$	(13943-58-3)	10	10	$<10^a$
Resorcinol	$C_6H_4\text{-}1,3\text{-}(OH)_2$	(108-46-3)	10	10	100
Sodium Cyanide	$NaCN$	(143-33-9)	$<10^a$	1	1
Sodium Hypochlorite	$NaClO$	(7681-52-9)	$<10^a$	100	10
Sodium Nitrite	$NaNO_2$	(7632-00-0)	100	100	10
Thymol	$5\text{-}CH_3C_6H_3\text{-}2\text{-}CH(CH_3)_2\text{-}1\text{-}OH$	(89-83-8)	10	10	10
Zinc Chloride	$ZnCl_2$	(7646-85-7)	48^b	4.8^b	48^b

[a] Lowest concentrations tested; 100 mg/L was the highest concentration tested.
[b] Value represents metal ion.

Comparison to Literature Values

Results from the Germination and Early Growth Effects Test were compared with values reported in eight literature citations for some chemicals using the same plant species (Table 2). The literature values were for four metals (chromium, copper, nickel, and zinc) and two organic compounds (phenol and aniline on radish).

TABLE 2—*Comparison with results of other test procedures reported in the literature (values in mg/L).*

Compound	Test Description, (Endpoint)	Plant Type	Literature Value[a]	(NOEC)
Phenol	96-Hour Germination Test (EC_{50}) [4][d]	Radish	230	1.0
Cr	14-Day Root Elongation Test (EC_{50}) [15]	Ryegrass	2.0	3.5
Cu			0.02	0.26
Ni			0.18	0.45
Zn			1.6	48[b]
Cu	5-Day Illuminated Solution Culture (Toxicity Limit) [16]	Lettuce	0.5	0.26[c]
Zn			2.16	48[b]
Cr	4-Week Seedling Test (EC_{50}) [17]	Ryegrass	2.5	3.5
Cu			10.8	0.26
Ni			100	0.45
Zn			1000[b]	48[b]
Cu	4-Week Seedling Test (NOEC) [18]	Radish	4	0.26
Ni	4-Week Seedling Test (NOEC) [19]	Ryegrass	30	0.45
Aniline	16-Hour Imbibed, 5-Day Germination Followed by 14-Day Seedling Test (NOEC) [20]	Radish	330	10
Cu	16-Day Seedling Test Solution Toxicity Level [21]	Lettuce	10-60	0.26[c]
Cu	36-Day Seedling Test (Solution Toxicity Level from Line Extrapolation) [21]	Ryegrass	200	0.26
Ni			1000	0.45
Zn			1000	48[b]

[a] Metal salt values are expressed as the metal ion concentration.
[b] Highest concentration tested.
[c] Lowest concentration tested.
[d] Number in bracket is reference for that data set.

Species Sensitivity

The relative sensitivities of the three plant species to the 26 commercial chemicals are presented in Tables 3 and 4. Two sets of comparisons were made: (1) for the plant species (Table 3) and (2) for the plant growth of root length and plant height (Table 4). The values are expressed as the number of test chemicals and percentage of test compounds that resulted in a particular response. Literature values were not available to make similar comparisons.

Statistical Validity of the 10% Measured Difference Criterion

Consideration of the 95% confidence intervals for relative percent growth on 427 sets of data revealed a 2.8% false negative rate for the 10% criterion. This criterion is considered to be an excellent predictor of plant growth effects.

TABLE 3—*Relative sensitivity: comparison by plant species.*[a]

Concentration, mg/L	No. of Compounds with NOEC Values at a Given Concentration		
	Ryegrass	Radish	Lettuce
100	10	11	6
10	8	9	14
1	6	5	4
0.1	2	1	2

[a] Twenty-six commercial compounds were tested on each plant species.

Discussion and Conclusions

We have demonstrated that the Germination and Early Growth Effects Test procedure is reliable, sensitive, and accurate for establishing the dose response and the NOEC of chemicals on several terrestrial plants. This procedure requires little time or expense for materials, yet provides useful information on the effects of chemicals. The vertical plane of growth and the clear pouch allow for easy measurement of roots. The clear pouch allows for easy electronic or photometric measurements of roots and photography of roots if further documentation is desired. By containing the test solutions, the disposable pouches reduce human exposure to most chemicals.

This method works best with plants with small seeds, as used in this study. Larger seeds (beans and corn) can, however, be used in the growth pouches by enlarging the holes in the trough. Cucumber seeds work well in the growth pouches but, as with the other large seeds, crowding and subsequent overlapping of roots were avoided by using no more than ten seeds per pouch.

The majority of the NOEC values determined using the Germination and Early Growth Effects Test procedure were less than or equal to those values recorded in the literature using other plant effect procedures. Only the values reported by Wong and Bradshaw [15] for a 14-day root elongation test on ryegrass and Berry [16] using lettuce were less than the NOEC values determined using this procedure. None of the seven seedling tests with exposure periods of 14 to 36 days (two to five times the length of the Germination and Early Growth Effects Test procedure) [18-21] reported NOEC or similar endpoint values as low as the Germination and Early Growth Effects Test NOEC values.

Occasionally lower exposure concentrations used in the Germination and Early Growth

TABLE 4—*Relative sensitivity: comparison of root and height growth.*[a]

	Values are Expressed as Percentage of Compounds		
	Ryegrass	Radish	Lettuce
Roots Most Sensitive	50	46	54
Height Most Sensitive	4	4	11
Equally Sensitive	46	50	35
Total	100	100	100

[a] The germination response was not appreciably affected by 25 of the 26 commercial compounds. Only the 100 mg/L allyl alcohol solution inhibited germination response.

Effects Test procedure had greater inhibitory effects on the percent growth of treated plants than higher concentrations. For example, ryegrass exposed to nominal 100 mg/L and 10 mg/L solutions of sodium ion (as nitrite) had percent growth values for plant height, compared to the control values, of 95% and 87.3%, respectively. This difference could be due to either experimental variability or plant architecture [22].

For these commercial chemicals, lettuce was the most sensitive plant species used in this test (Table 3). The relative sensitivities of ryegrass and radish were similar. Reduction in root length was the more sensitive plant response (in 46 to 54% of the studies) when compared with reduction of plant height (Table 4). The greater sensitivity of root growth to chemical exposure has been observed by others [11,23]. Reduction in plant height was a less sensitive response in approximately 50% of the studies, equally sensitive in 35 to 50% of the studies, and more sensitive in 4 to 11% of the studies. Although germination is a critical stage in the development of terrestrial plants, it was not adversely affected by 25 of the 26 commercial chemicals used and therefore appears not to be a sensitive indicator of phytotoxicity of chemicals to plants. This conclusion is also consistent with what other researchers have found [11,12,23].

Acknowledgments

We wish to thank Mr. Barry Zacharias for his technical assistance.

References

[1] Skinner, C. G. and Shive, W., "Stimulation of Lettuce Seed Germination by 6-(Substituted) Purines," *Plant Physiology,* Vol. 34, 1959, pp. 1–3.
[2] Prat, S. and Sladecek, V., "An Inexpensive Bioassay Aimed at the Agricultural Disposal of Waste Waters," *Hydrobiologia,* Vol. 23, 1964, pp. 246–252.
[3] Horowitz, M., "A Rapid Bioassay for PEBC and its Application in Volatilization and Adsorption Studies," *Weed Research,* Vol. 6, 1965, pp. 22–36.
[4] Wang, W., "The Use of Plant Seeds in Toxicity Tests of Phenolic Compounds," *Environment International,* Vol. 11, 1985, pp. 49–55.
[5] Kuboi, T. and Fujii, K., "A New Method for Seedling Assay of Phytotoxic Substances: Liquid Shaking Culture," *Soil Science and Plant Nutrition,* Vol. 30, 1984, pp. 209–218.
[6] Ratsch, H. C. and Johndro, D., "Comparative Toxicity of Several Test Chemicals to Lettuce Using Two Root Elongation Test Methods," *Journal of Environment Monitoring Assessment,* Vol. 6, 1986, pp. 267–276.
[7] Myhill, R. R., and Konzak, C. F., "A New Technique for Culturing and Measuring Barley Seedlings," *Crop Science,* Vol. 7, 1967, pp. 275–276.
[8] Rajanna, B. and De La Cruz, A. A., "Mirex Incorporation in the Environment: Phytotoxicity on Germination Emergence, and Early Growth of Crop Seedlings," *Bulletin of Environmental Contamination and Toxicology,* Vol. 14, 1978, pp. 77–82.
[9] Konzak, C. F., Polle, E., and Kittrick, J. A., "Screening Several Crops for Aluminum Tolerance," in *Proceedings,* Workshop on Plant Adaptation to Mineral Stress in Problem Soils, sponsored by Office of Agriculture, Technical Assistance Bureau, Agency for International Development (Washington, D.C.), Beltsville, Md., 22-23 Nov. 1976, pp. 311–327.
[10] Edwards, N. T. and Ross-Todd, B. M., "An Improved Bioassay Technique Used in Solid Waste Leachate Phytotoxicity Research," *Environmental and Experimental Botany,* Vol. 20, 1980, pp. 31–38.
[11] Ratsch, H. C., "Interlaboratory Root Elongation Testing of Toxic Substances on Selected Plant Species," EPA 600/53-83-051, U.S. Environmental Protection Agency, Environmental Research Laboratory, Corvallis, Ore., 1983.
[12] U.S. Environmental Protection Agency, "Environmental Effects Test Guidelines," EPA 560/6-82-002, Office of Toxic Substances, Washington, D.C., 1982.
[13] Fieller, E. C., "Some Problems in Interval Estimation," *Royal Statistical Society,* Vol. B16, 1954, pp. 175–183.

[14] Ewell, W. S., Gorsuch, J. W., Kringle, R. O., Robillard, K. A., and Spiegel, R. C., "Simultaneous Evaluation of the Acute Effects of Chemicals on Seven Aquatic Species," *Environmental Toxicology and Chemistry,* Vol. 5, 1986, pp. 831–840.

[15] Wong, M. H., and Bradshaw, A. D., "A Comparison of the Toxicity of Heavy Metals, Using Root Elongation of Rye Grass, *Lolium perenne,*" *New Phytologist,* Vol. 91, 1982, pp. 255–261.

[16] Berry, W. L., "Dose-Response Curves for Lettuce Subjected to Acute Toxic Levels of Copper and Zinc," in *Proceedings,* 15th Annual Hanford Life Sciences Symposium, ERDA Symposium Series 42, Biological Implication of Metals in the Environment, Richland, Wash. 29 Sept.–1 Oct. 1975, pp. 365–369.

[17] Dijkshoorn, W., van Broekhoven, L. W., and Lampe, J. E. M., "Phytotoxicity of Zinc, Nickel, Cadmium, Lead, Copper, and Chromium in Three Pasture Plant Species Supplied with Graduated Amounts from the Soil," *Netherlands Journal of Agriculture,* Vol. 27, 1979, pp. 241–253.

[18] Garten, C. T., and Frank, M. L., "Comparison of Toxicity to Terrestrial Plants with Algal Growth Inhibition by Herbicides," ORNL/TM-9177, Oak Ridge National Laboratory, Oak Ridge, Tenn., 1984, 55 pp.

[19] Khalid, B. Y., and Tinsley, J., "Some Effects of Nickel Toxicity on Ryegrass," *Plant and Soil,* Vol. 55, 1980, pp. 139–144.

[20] Nozzolillo, C., "The Effect of Imbition of Solutions of Aniline on Germination, Growth, and Red Pigmentation of Seedlings," *Canadian Journal of Botany,* Vol. 49, 1971, pp. 2113–2117.

[21] Davis, R. D., and Beckett, P. H. T., "Upper Critical Levels of Toxic Elements in Plants: II—Critical Levels of Copper in Young Barley, Wheat, Rape, Lettuce, and Ryegrass, and of Nickel and Zinc in Young Barley and Ryegrass," *New Phytologist,* Vol. 80, 1978, pp. 23–32.

[22] Evans, G. C., *The Quantitative Analysis of Plant Growth, Studies in Ecology,* Vol. 1. University of California Press, Berkeley and Los Angeles, 1972.

[23] Wang, W., "Use of Millet Root Elongation for Toxicity Tests of Phenolic Compounds," *Environment International,* Vol. 11, 1985, pp. 95–98.

James R. Clark[1] and John M. Macauley[1]

Comparison of the Seagrass *Thalassia testudinum* and Its Epiphytes in the Field and in Laboratory Test Systems

REFERENCE: Clark, J. R. and Macauley, J. M., "**Comparison of the Seagrass *Thalassia testudinum* and Its Epiphytes in the Field and in Laboratory Test Systems**," *Plants for Toxicity Assessment, ASTM STP 1091*, W. Wang, J. W. Gorsuch, and W. R. Lower, Eds., American Society for Testing and Materials, Philadelphia, 1990, pp. 59–68.

ABSTRACT: *Thalassia testudinum* and associated epiphytes from field plots were compared with plants from laboratory microcosms to determine if laboratory observations obtained from plants undergoing seasonal growth patterns were characteristic of plants in natural systems. The measurements selected for characterizing *Thalassia* health were chlorophyll and protein content of leaves and carbohydrate content of rhizomes. For epiphytes, we measured standing crop and chlorophyll content. Mean values for field and laboratory data were statistically analyzed for two experiments conducted over six-week intervals and for one experiment extended to twelve weeks. *Thalassia* plants in the laboratory followed similar trends of field plants during the six-week experiments, but the laboratory plants differed significantly from field plants at twelve weeks. Chlorophyll content of epiphyte communities colonizing *Thalassia* leaves was significantly different in the laboratory compared to field communities, but trends from test initiation to the six-week sampling were consistent between field measurements and laboratory test systems.

KEY WORDS: seagrass, *Thalassia testudinum*, epiphytes, lab to field comparisons, field validation, chlorophyll, protein, carbohydrate

Laboratory test systems offer an opportunity to assess responses of marine plants exposed to xenobiotic chemicals under controlled conditions. However, plants tested under laboratory conditions should be compared to plants in a natural setting to ensure that observed responses provide realistic data on the potential for chemical effects using characteristic growth cycles and plant vigor [*1–3*]. This is a special concern when measures of chemical effect on plants also are sensitive to a variety of other stresses [*4,5*].

We were interested in determining the extent to which seasonally prominent growth cycles, observed in the field, were represented during laboratory studies. The degree of correspondence would not necessarily provide a basis to accept or reject exposure-response tests with toxicants; rather, it would provide one measure of the extent to which laboratory data could be applied to field situations. The results of our comparisons, although not comprehensive, provide some insight into important factors for consideration.

We compared changes in *Thalassia* chlorophyll, protein, and carbohydrate content and epiphyte chlorophyll and standing crop in untreated laboratory microcosms over 6- or 12-week test intervals with changes in plants from the field over similar times. Comparisons from three experiments, conducted at different times of the growing season, are presented.

---
[1] U.S. Environmental Protection Agency, Environmental Research Labortory, Gulf Breeze, FL 32561.

Methods

Characterizations of *Thalassia testudinum* and its epiphytes in the field are from samples collected in a seagrass bed that we have studied since 1983 [6]. Laboratory data were taken from tests designed to determine effects of contaminants on *Thalssia*-dominated ecosystems. Only results from untreated (control) microcosms are presented for comparisons with field data. Test I was conducted from 14 April to 8 July 1986, Test II from 14 October to 5 December 1986, and Test III from 14 May to 29 June 1987.

Field Samples

Field samples were taken from a *Thalassia testudinum* bed located in an embayment 200 m west of the Environmental Research Laboratory, Gulf Breeze (ERL/GB), in Santa Rosa Sound, Florida. We studied a 10 by 10 m plot, described by Macauley et al. [6], in the center of the bed where *Thalassia* density was most uniform. Five replicate *Thalassia* and 15 epiphyte samples were collected from the site at the beginning of each laboratory experiment and within three to twelve days of test termination. *Thalassia* leaves within a 0.25 m^2 plastic frame placed randomly at five sites within the 10 m^2 plot were clipped at the sediment surface, placed in a plastic bag, and taken to the laboratory. There, epiphytes were removed from three leaves of each sample and epiphyte and leaf samples processed as described below. In addition, two 14-cm-diameter cores were taken from the plot, intact *Thalassia* plants were removed from the cores, and placed in plastic bags. At the laboratory, five intact plants were selected for leaf protein and rhizome carbohydrate analyses.

Laboratory Samples

Divers collected intact samples of *Thalassia* and substrate for laboratory tests from a *Thalassia* bed in Santa Rosa Sound approximately 1.0 km east of ERL/GB. The site was selected for laboratory test source material to avoid additional disturbances in our long-term monitoring bed [6]. The density and distribution of plants and water quality were similar at both sites. Cores were taken to a substrate depth of 10 cm, placed directly in clear acrylic cylinders (15.9 cm ID, 50 cm height) without disrupting rooted plants, and transported to ERL/GB for testing. There *Thalassia* plants were maintained in flowing seawater that was pumped from Santa Rosa Sound and delivered to each cylinder (microcosm) at the rate of 7 L/h, providing one complete water exchange each hour. Two cores were sacrificed immediately (test initiation) and 5 intact *Thalassia* plants selected for leaf protein and rhizome carbohydrate analyses and 15 leaves selected for epiphyte samples. Protein and carbohydrate samples were accidentally destroyed during processing of Test III samples. Further details on sample collection and laboratory microcosms are provided by Morton et al. [7]. Four 400-W, halogen-halide lamps provided light for the laboratory test system. This light was supplemented by high-intensity fluorescent bulbs (GE Power Groove) rated at 250 W. A cycle of 12 h light to 12 h darkness was used for laboratory tests to conform to standard testing schemes. Lights that were suspended 46 cm above the microcosms provided an average photon flux of 225 $\mu E/m^2/s$ to the water surface of each microcosm, as measured by a LICOR LI 188B photometer. Laboratory light represents 35% of the 650 $\mu E/m^2/s$ measured during a sunny July day at noon in the field plot at a depth of 1.25 m.

Laboratory experiments were 6 weeks' duration with an additional sample taken during Test I at 12 weeks. At termination, one *Thalassia* leaf was collected from each of 15 microcosms for determination of epiphyte parameters and *Thalassia* leaf chlorophyll analyses. An intact *Thalassia* plant also was removed from each of 15 microcosms for quantitative determination of leaf protein and rhizome carbohydrate.

Thalassia Samples

Epiphytic material was removed from *Thalassia* leaves with a razor blade and collected for subsequent analyses. Chlorophylls a and b were quantified using the SCOR/UNESCO equations [8] and chlorophyll concentrations calculated per gram dry weight of leaf material according to the procedures of Macauley et al. [6]. Whole *Thalassia* plants were separated into leaf and rhizome components, lyophilized, and ground in a Wiley tissue mill. Leaf protein and rhizome carbohydrate concentrations were determined using the Lowry procedures and carbohydrate analysis as described by Dawes [4]. Values are reported as mg per gram of dry tissue.

Epiphyte Samples

Epiphytes scraped from *Thalassia* leaves were processed for chlorophyll extraction and determination of epiphyte ash-free dry weight (AFDW) [9]. Epiphyte standing crop was estimated as AFDW per cm^2, calculated from the area of leaf tissue sampled [6]. Phaeophytin-corrected chlorophyll a was calculated according to the method of Weber [10] and expressed as μg per cm^2 of *Thalassia* leaf surface and per mg AFDW of epiphyte material.

Data Analyses

Means and standard deviations were computed for each set of replicate samples. For each test, means were compared using the general linear models procedures and Duncans multiple comparison procedures [11] to demonstrate significant differences between laboratory and field samples at test initiation and completion and to discern significant changes in laboratory or field communities over the duration of the test. Tests for significant differences were conducted at $\alpha = 0.05$.

Results

Thalassia

Chlorophyll Analyses—In Tests I and III, chlorophylls a and b increased for *Thalassia* in the field during the 6-week test period as they did for laboratory plants (Fig. 1). When sampled at 12 weeks in Test I, leaves from laboratory-held plants had significantly greater chlorophyll compared to other field or laboratory samples. For Test II, chlorophyll content of laboratory-held plants was greater than that of plants sampled in the field at test initiation, although this difference was not significant. No significant difference between laboratory and field values or time-0 and 6-week sample values were recorded for Test II.

Protein Content—*Thalassia* from Test I had significantly less leaf protein after 6 weeks in the laboratory relative to field plants at that time (Fig. 2). However, no significant change over the 6-week test period occurred within either field and laboratory plants. By 12 weeks, protein content of leaves from field and laboratory plants were nearly equal; laboratory plants had a significant increase from 6-week protein concentrations. In Test II, conducted during the fall, there were no changes in leaf protein values for field plants over the 6-week test period, whereas laboratory-held plants significantly increased leaf protein content. During Test III, leaf protein values for field plants showed no significant difference from test initiation through the 6-week sampling period, although mean protein values were less at 6 weeks. Laboratory plants had leaf protein values significantly lower than field plants.

Carbohydrate Content—No significant differences were seen between laboratory and field plants at 6-week sampling periods for any of the tests. When held in the laboratory for 12 weeks during Test I, rhizome carbohydrate content significantly decreased from previous

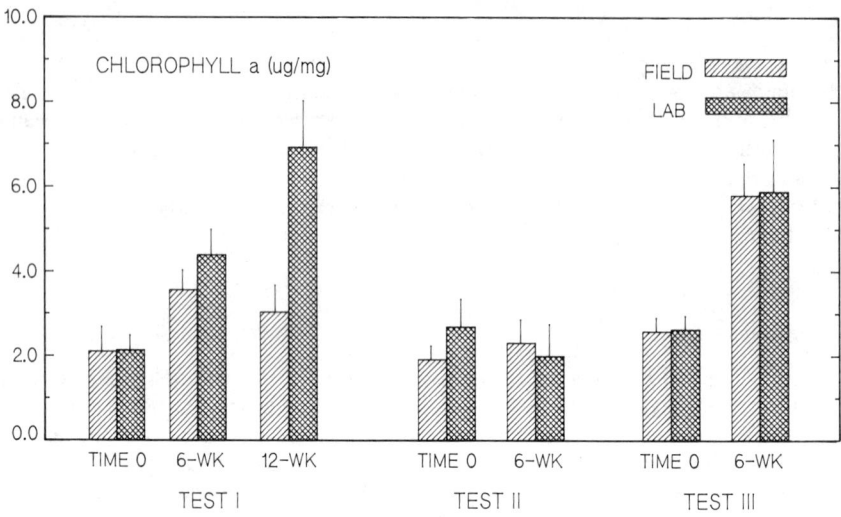
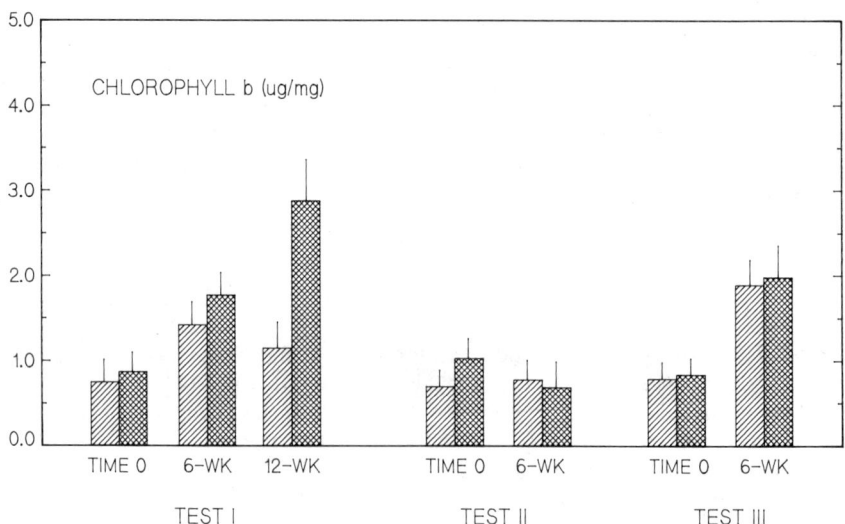

FIG. 1—*Chlorophyll* a *and* b *content of* Thalassia testudinum *leaves sampled from a monitoring plot in Santa Rosa Sound, Florida (field), and from laboratory test systems (lab). Means of 15 replicates are plotted for each bar; vertical line is ± 1 SD. Test I conducted 4/14/86 to 7/8/86; Test II from 10/14/86 to 2/5/86; Test III from 5/14/87 to 6/29/87.*

levels while field values significantly increased. Carbohydrate values for field plants sampled in May 1986 and 1987 were similar (565 and 620 mg/g). However, by the end of June 1986 values increased (although not statistically significant) to 729 mg/g whereas by June 1987 values significantly decreased to 390 mg/kg.

Epiphytes

$AFDW/cm^2$—Epiphyte standing crop, measured as $AFDW/cm^2$, was similar in laboratory and field samples at the beginning of each test and when sampled at 6 weeks for Tests I and

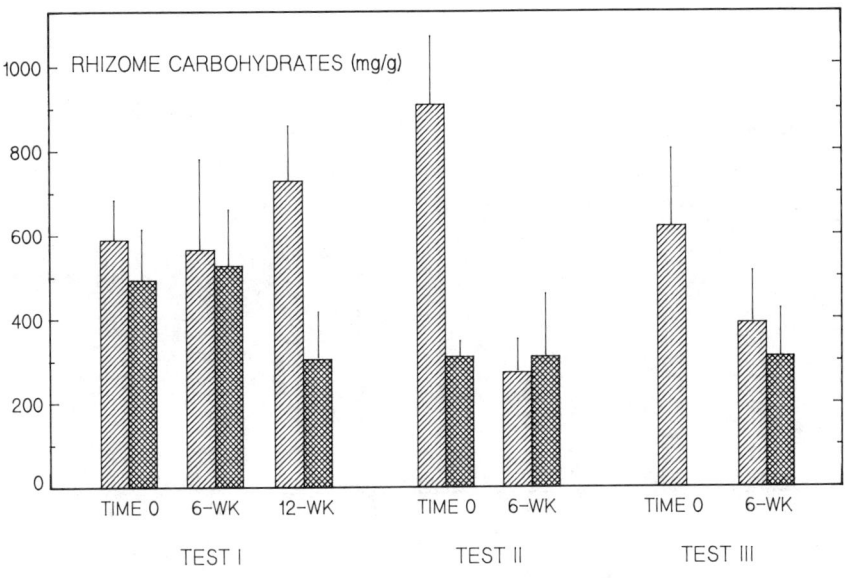
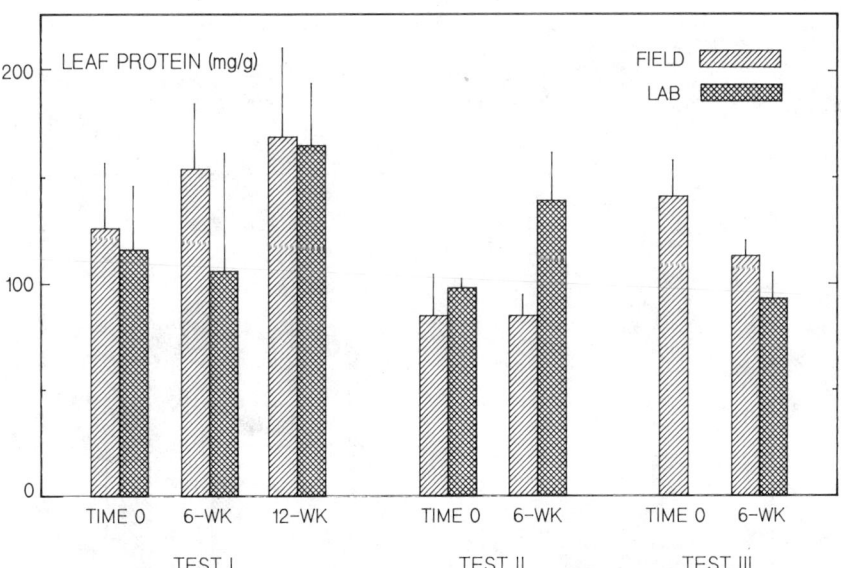

FIG. 2—*Leaf protein and rhizome carbohydrate content of* Thalassia testudinum *sampled from a monitoring plot in Santa Rosa Sound, Florida (field), and from laboratory test systems (lab). Means of 5 replicates (field) or 15 replicates (lab) are plotted for each bar; vertical line is ± 1 SD.*

III (Fig. 3). No significant changes occurred in laboratory or field standing crops sampled at 12 weeks (Test I). A seasonal increase in epiphyte biomass that occurs during fall and winter [6] occurred in field samples taken at 6 weeks during Test II, but not in the laboratory test system.

Chlorophyll a/cm²—Epiphyte chlorophyll *a* per cm² of *Thalassia* leaf was significantly greater in laboratory systems compared to field samples at test initiation and each sample

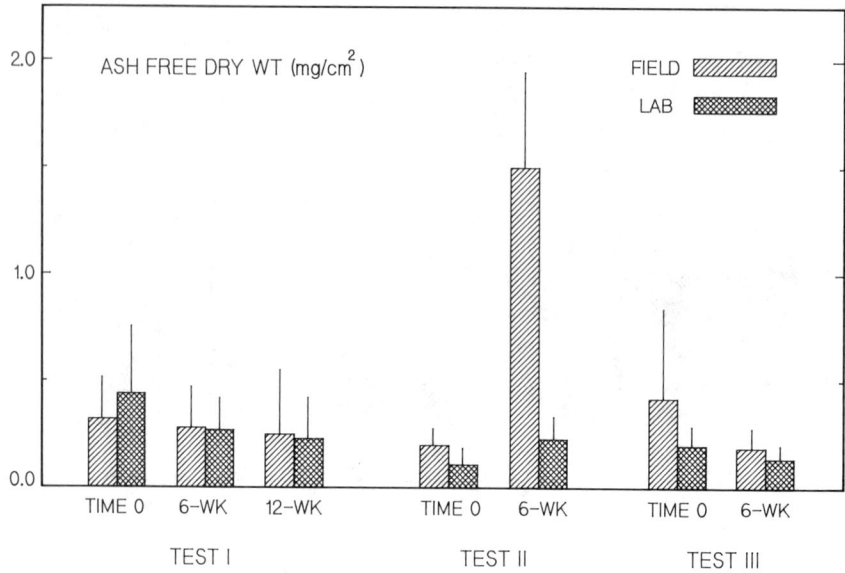
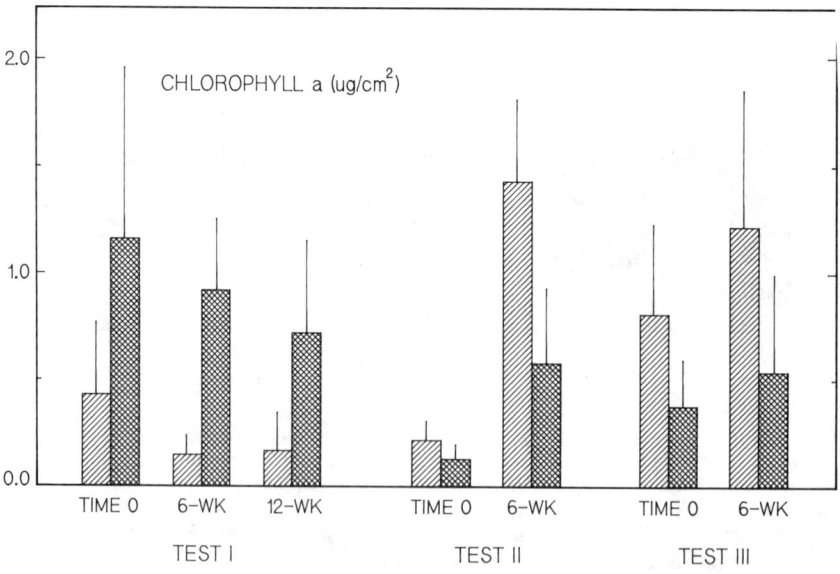

FIG. 3—*Epiphyte biomass (AFDW/cm², algal standing crop (chlorophyll a/cm²), and epiphyte chlorophyll content (chlorophyll a/mg AFDW) for epiphytic communities sampled from* Thalassia testudinum *leaves in a monitoring plot in Santa Rosa Sound, Florida (field), and from laboratory test systems (lab). Means of 15 replicates are plotted for each bar; vertical line is ±1 SD.*

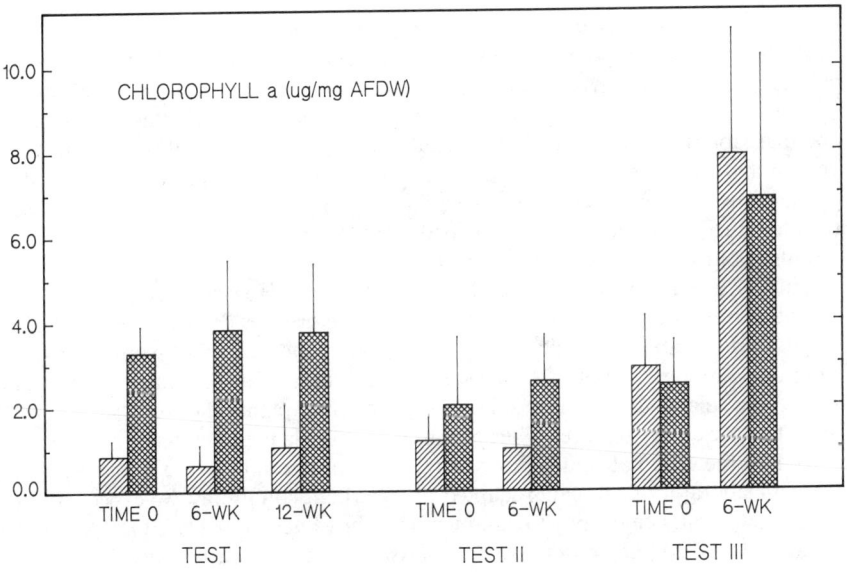

FIG. 3—(continued)

date during Test I. Both field and laboratory values decreased during the first 6 weeks of the experiment. Values of chlorophyll a/cm^2 increased during the course of Tests II and III, and significant differences between field and laboratory samples were recorded at 6 weeks for both of these experiments.

Chlorophyll a/g AFDW—This parameter was calculated from existing data to reflect relative abundance of algae in the epiphytic communities. Chlorophyll a/g AFDW was significantly greater in laboratory communities relative to field communities at test initiation and each subsequent sample date after initiating Tests I and II. No consistent trend at 6 or 12 weeks was seen in either field or laboratory data. No significant differences between laboratory and field samples at test initiation or 6 weeks occurred in Test III, although both 6-week values were significantly greater than those measured at test initiation.

Discussion

Thalassia

The annual cycle of *Thalassia* growth in northwestern Florida has been characterized as having a low, over wintering standing crop, and a spring growth period beginning in April or May, leading to maximum standing crop in June or July [6]. During Tests I and III, conducted during the spring-summer growth phase, we observed an increase in leaf length, indicating that laboratory-held plants followed the same trends during the interval between test initiation and the 6-week sampling period. This similarity gave us confidence that laboratory test results would reflect responses of plants in the spring-summer growth phase. The test conducted during the fall, Test II, provided similar confidence in test results characteristic of plants about to enter the over wintering phase.

Day length (12 h) was constant during the laboratory tests. However, temperatures in the laboratory were similar to field temperatures ($\pm 2°C$). Our greatest concerns centered on laboratory light and the ability of microcosms to provide conditions that would allow us to observe seasonal growth cycles for *Thalassia* in the laboratory. Such conditions would allow us a system to test plants undergoing rapid, spring growth or plants allocating photosynthate to rhizomes to store energy reserves. The photon flux available to plants inside the test system was 35% of that measured at midday on several midsummer days in the *Thalassia* bed we monitored. As laboratory-held plants adapted to these light conditions, we expected chlorophyll content, especially chlorophyll *b*, and leaf protein values to increase [4,12]. However, Wiginton and McMillan [13] reported that *Thalassia* showed the least propensity to increase chlorophyll *b* content compared to several other seagrasses common to the Gulf of Mexico. After 6 weeks in the laboratory, we observed a significant increase beyond concomitant field measurements only for *Thalassia* chlorophylls *a* and *b* in Test I and leaf protein in Test II. Lack of consistent response suggests that the laboratory test system was adequate for tests of 6-week duration. Significant increases in chlorophyll *a* and *b* and a decrease in rhizome carbohydrate for *Thalassia* held in the laboratory demonstrate that results from a 12-week test would not be representative of plants under field conditions.

Epiphytes

None of the epiphyte test parameters showed consistent laboratory-field relationships over all experiments, and no two parameters demonstrated similar relative differences for all tests. During Tests I and II, laboratory communities were dominated by algae (as indicated by the ratio of chlorophyll *a* to AFDW) compared to field communities, although these differences were not significant during Test III. We suspected that the absence of large grazers, omnivores, and physical turbulence in laboratory microcosms might lead to increased epiphyte biomass or algal dominance relative to field samples [14,15]. However, biomass differences between laboratory and field samples were not significant at 6-weeks during Tests I and III. The seasonal increase in epiphyte biomass in *Thalassia* beds in fall and winter [6] did not occur in the laboratory during Test II, although field samples had greater AFDW and chlorophyll *a*. Significantly greater chlorophyll *a* in laboratory samples relative to field for Test I occurred during a time of unexplained seasonal low in epiphyte standing crop at the field site [6]. This difference was reversed during the experiment conducted the following spring (Test III).

It is unreasonable to expect strict correspondence between laboratory and field communities of epiphytes subjected to widely different environmental conditions for 6 weeks. Epiphyte communities are composed of microbial autotrophs and heterotrophs with rapid turnover capabilities [16,17]. Test results can be influenced by environmental factors that occur only days or hours before sampling, and species composition may change once habitat features are altered [17]. Although we did not address species composition in our test procedures (because of the time demands), we recognize that different chlorophyll and biomass relationships among epiphytic species added variability to our results.

Summary

Because plants in our test system were exposed to less light intensity and turbulence than those in the field, we questioned whether laboratory-held plants reflected conditions of plants in the field. Changes in chlorophyll, protein, and carbohydrate content of *Thalassia* plants during 6-week laboratory tests were similar to trends observed in the field. However, after 12 weeks in the laboratory, chlorophyll had increased significantly and rhizome carbohydrate significantly decreased, most likely the result of prolonged exposure to low light intensities. The standing crop and chlorophyll content of laboratory epiphyte communities were considerably different than that of field communities. Nevertheless, relative increases or decreases in these parameters from test initiation to the 6-week sampling period were similar for laboratory and field communities. Laboratory test systems can provide useful information on probable effects of stress (pollutant or physio-chemical) on *Thalassia* and its epiphytes when test conditions relative to field communities are evaluated.

Acknowledgments

This work is contribution No. 629 of the U.S. EPA Environmental Research Laboratory at Gulf Breeze, FL. We thank the ERL/GB Dive Team for collecting samples, A. Pitts and K. Roadman for assistance in sample processing and data management, and V. Coseo for typing the manuscript.

References

[1] Giddings, J. M. and Eddlemon G. K., *International Journal of Environmental Studies*, Vol. 13, 1979, pp. 119–123.
[2] Kemp W. M., Boynton W. R., Cunningham J. J., Stevenson J. C., Jones, T. W., and Means J. C., *Marine Environmental Research*, Vol. 16, 1985, pp. 255–280.
[3] Short, F. T., *Aquatic Botany*, Vol. 27, 1987, pp. 41–57.
[4] Dawes, C. J., *Marine Botany*, John Wiley, New York, 1981.
[5] Phillips, R. C. and Mcroy, C. P., *Handbook of Seagrass Biology: An Ecosystem Perspective*, Garland STPM Press, New York, 1980.
[6] Macauley, J. M., Clark, J. R., and Price, W. A., *Aquatic Botany*, Vol. 31, 1988, pp. 277–287.

[7] Morton, R. D., Duke, T. W., Macauley, J. M., Clark, J. R., Price, W. A., Hendricks, S. J., Owsley-Montgomery, S. L., and Plaia, G. R. in *Community Toxicity Testing, ASTM STP 920*, J. Cairns, Jr., Ed., American Society for Testing and Materials, Philadelphia, 1986, pp. 199–212.
[8] Strickland, J. D. H. and Parsons, T. R., *A Practical Handbook of Seawater Analysis*, 2nd ed., Bulletin 167, Ottawa Fisheries Research Board of Canada, 1972.
[9] American Public Health Association, American Water Works Association, Water Pollution Control Federation, *Standard Methods for the Examination of Water and Wastewater*, 16th ed., American Public Health Association, Washington, D.C., 1985.
[10] Weber, C. I., "Biological Field and Laboratory Methods for Measuring the Quality of Surface Waters and Effluents," U.S. EPA 670/4-73-001, Washington, D.C., 1973.
[11] *SAS User's Guide: Statistics*, SAS Institute, Cary, N.C., 1982.
[12] Ferguson, R. L., Thayer, G. W., and Rice, R. R. in *Functional Adaptations of Marine Organisms*, J. Smith, Ed., Academic Press, Ann Arbor, Mich., 1980, pp. 9–69.
[13] Wiginton, J. R. and McMillan, C., *Aquatic Botany*, Vol. 6, 1979, pp. 171–184.
[14] Kirkman, H., *Aquatic Botany*, Vol. 4, 1978, pp. 367–372.
[15] Sand-Jansen, K., *Aquatic Botany*, Vol. 3, 1977, pp. 55–63.
[16] Humm, H. J., *Bulletin of Marine Science of the Gulf and Caribbean*, Vol. 14, 1964, pp. 306–341.
[17] Weitzel, R. L., *Methods and Measurements of Periphyton Communities: A Review, ASTM STP 690*, American Society for Testing and Materials, Philadelphia, 1979.

C. T. Garten, Jr.[1]

Multispecies Methods of Testing for Toxicity: Use of the *Rhizobium*-Legume Symbiosis in Nitrogen Fixation and Correlations Between Responses by Algae and Terrestrial Plants

REFERENCE: Garten, C. T., Jr., "**Multispecies Methods of Testing for Toxicity: Use of the *Rhizobium*-Legume Symbiosis in Nitrogen Fixation and Correlations Between Responses by Algae and Terrestrial Plants**," *Plants for Toxicity Assessment, ASTM STP 1091*, W. Wang, J. W. Gorsuch, and W. R. Lower, Eds., American Society for Testing and Materials, Philadelphia, 1990, pp. 69–84.

ABSTRACT: Responses of the *Rhizobium*-legume symbiotic relationship to long-term (5- to 7-week) and short-term (2-week) exposures of copper sulfate, 2,4-dichlorophenoxyacetic acid, and streptomycin sulfate were examined in bush beans and clover. Toxic effects were evaluated by comparing plant biomass (yield), nodulation success, nitrogen fixation rate as indicated by acetylene reduction, and plant nitrogen content in controls and in plants exposed to various concentrations of the chemicals. Plants with long-term continuous root exposure were affected more than plants with short-term exposure, as indicated by reductions in nitrogen fixation rates and plant growth. Although rates of acetylene reduction (nitrogen fixation) were depressed, plant biomass and the numbers of root nodules produced were simpler, less expensive indices of exposure.

In a second test, the responses of algae (*Selenastrum capricornutum* and *Chlorella vulgaris*) and terrestrial plants (radishes, barley, bush beans, and soybeans) to 21 different herbicides were compared to evaluate the use of a short-term (96-h) algal growth inhibition test for identifying chemicals potentially toxic to terrestrial plants. Two test end points were evaluated. (1) the highest concentration resulting in no statistically significant reduction in the growth rate of algae or terrestrial plants, and (2) the concentration at which algal growth or terrestrial plant biomass was reduced by 50%. Test results were correlated for the two algae and for all pairs of vascular plants. However, results of the algal tests did not correlate with those of the terrestrial plants. Furthermore, according to tests with *Selenastrum capricornutum*, there was only a 50% chance of successfully identifying herbicide levels that reduced terrestrial plant biomass. For the herbicides and species examined, short-term toxicity tests with algae were not good indicators of toxicity to terrestrial plants.

KEY WORDS: nitrogen fixation, legumes, algae, herbicides, pesticides, toxicity testing, multispecies toxicity testing, symbiosis, terrestrial plants.

Toxicity tests with plants are now commonly used to evaluate the environmental hazards of chemicals; however, a multispecies approach to toxicity testing remains uncommon. Relatively few studies of multispecies toxicity tests involve assessing effects on parasitism, grazing, or symbiotic relationships [1], partly because such tests tend to be time-consuming and expensive. However, these disadvantages may be offset by the sensitivity and environmental relevance of multispecies test methods.

[1] Environmental Sciences Division, Oak Ridge National Laboratory, P.O. Box 2008, Oak Ridge, TN 37831-6038.

In a symbiotic relationship, chemical effects on one symbiont may indirectly affect the other member(s). Because of its importance to agriculture, the *Rhizobium*-legume symbiosis is potentially useful as a multispecies toxicity test system and has already been used to research the effects of various pesticide chemicals on nitrogen fixation (Table 1). Testing to assess effects of pesticides on nitrogen fixation has yielded mixed results. When used at recommended rates for field use, some pesticides have been found to exert an adverse effect on nitrogen fixation by legumes, whereas others have had no apparent adverse effects.

Toxicity tests with terrestrial plants may involve various stages of the life cycle or even an entire life cycle. However, for economic reasons, sometimes what is needed is an inexpensive short-term test (lasting hours or days) to indicate chemical toxicity to a terrestrial plant. Correlations between the outcome of toxicity tests with algae and tests with terrestrial plants may provide such data, if short-term algal tests can be used to predict the phytotoxicity of chemicals in longer-term terrestrial plant tests.

For example, Decleire et al. [14] found that growth inhibition in oat shoots was predictable from algal (*Chlorella*) toxicity tests for approximately 43% of the 61 herbicides they tested. Therefore further study of correlations between the outcome of short-term tests with algae and longer-term toxicity tests with terrestrial plants was warranted. In this work, the toxicity of 21 herbicides to algae (*Selenastrum capricornutum* and *Chlorella vulgaris*) and to terrestrial plants (radishes, barley, and bush beans or soybeans) were compared to evaluate the accuracy with which a 96-h algal growth inhibition test could identify chemicals that would be toxic in a 4-week test using terrestrial plants.

This work was undertaken to evaluate the utility of two different approaches to multispecies methods of toxicity testing with plants. First, sensitivity of the *Rhizobium*-legume symbiosis in nitrogen fixation by beans and clover was examined as a multispecies test involving interdependencies between different species. Second, the use of correlations between the outcomes of toxicity tests in algae and terrestrial plants was examined as a multispecies test method involving the similarity or dissimilarity of responses of independent, unrelated species to herbicides. These methodologies differ in many respects, especially the

TABLE 1—*Effects of various pesticides on N-fixation by* Rhizobium-*legume symbioses.*

Common Name	Primary Action	Host	Observed Effects in One or More Host Species[a]	Type Test[b]	Reference
Aldicarb	insecticide	alfalfa, clover	R, P, N, NF	G	[2,3]
Bentazon	herbicide	bean	NF	G	[4]
Carbaryl	insecticide	alfalfa, clover	N, NF	G	[3]
Carbofuran	insecticide	clover, alfalfa	R, P, N, NF	G	[2,3]
Chloropropham	herbicide	soybean	N, P	F	[5]
DDT	insecticide	peas, clover, soybean	N, P, NF	G	[6,7]
Lindane	insecticide	clover, groundnut	N, P	F, G	[8,9]
Phosmet	insecticide	clover	NF	G	[3]
Simazine	herbicide	groundnut	N, P	G	[9]
Terbufos	insecticide	clover	N, NF	G	[3]
2,4-D	herbicide	alfalfa, peas, clover, beans	R, N, P	G	[10,11]
2,4,5-T	herbicide	clover	N, P	G	[12,13]

[a] A statistically significant response to chemical exposure (relative to experimental controls) is indicated by the following abbreviations: "R" – *Rhizobium* viability, "N" – nodulation success, "P" – plant yield, "NF" – N-fixation.

[b] The type of test is indicated as follows: "G" – greenhouse or laboratory test, "F" – field test.

degree of association between the species used in testing. Nevertheless, both are presented here as examples of a multispecies approach to using plants in toxicity assessments.

Materials and Methods

Effects of Test Chemicals on Nitrogen Fixation by Beans and Clover

Test Species Selection and Test Conditions—Details of the experimental methods are published elsewhere [15] and, for the sake of brevity, are summarized here. Nitrogen fixation by legumes is a property of the *Rhizobium*-host relationship when both the host plant and the bacteria are healthy. Bush beans (*Phaseolus vulgaris*, var. Tennessee Greenpod and Blue Lake) and white clover (*Trifolium repens*) were chosen as host legumes in the present work because of their prior use in chemical toxicity tests involving measures of nitrogen fixation (Table 1). Bush bean seeds were obtained from a commercial source and were inoculated with *Rhizobium* bacteria by mixing them with a moist peat inoculant. Seeds were then germinated between moist paper towels for six days before planting. Clover seeds were inoculated by mixing them in a small vial with a pinch of moist peat inoculant immediately before planting. *Rhizobium* inoculants were obtained from the Nitragen Company (Milwaukee, Wisconsin).

Beans and clover were grown separately in plastic pots (12-cm diameter and 12-cm deep) filled with coarse sand (>0.595 mm). The sand was washed and dried at 70°C before use. Pots were irrigated from below with a N-free nutrient solution containing the following salts (in mg/L): K_2SO_4, 801; KH_2PO_4, 272; $CaCl_2$, 416; $MgSO_4 \cdot 7H_2O$, 493; Fe Sequestrene®, 8.3; H_3BO_3, 2.9; $MnCl_2 \cdot 4H_2O$, 1.8; $ZnSO_4 \cdot 7H_2O$, 0.22; $H_2MoO_4 \cdot H_2O$, 0.02; and $CuSO_4 \cdot 5H_2O$, 0.08. The nutrient solution also contained the test chemical. Beans were irrigated with N-free nutrient solutions for two weeks after planting to encourage establishment of the symbiosis with *Rhizobium*. Thereafter, combined nitrogen [KNO_3 and $Ca(NO_3)_2 \cdot 4H_2O$] was added to the nutrient solution to maintain a NO_3^- concentration of 1.4 mmol (90 mg/L). The nitrogen supplement helped to maintain healthy bush bean plants and was not detrimental to nitrogen fixation. White clover was grown only in N-free nutrient solution. The tests were conducted in a greenhouse, where natural daylight was supplemented by plant lights suspended above each greenhouse bench (16-h photoperiod). Photosynthetically active radiation, air temperature, and the pH of the nutrient medium were routinely measured to ensure optimum conditions for the establishment of the *Rhizobium*-host relationship [15].

Test Chemicals and Exposure Regimes—The chemicals tested were copper [Cu^{2+} (as $CuSO_4 \cdot 5H_2O$)], 2,4-dichlorophenoxyacetic acid (2,4-D), and streptomycin sulfate. On the basis of available data from published studies (Table 1), concentrations that would yield observable effects on plant growth or nitrogen fixation were chosen. Results of seed germination tests were also used for guidance in selecting concentrations to be tested [15]. Root elongation in germination tests with both bush beans and clover was inhibited at exposure concentrations of 2,4-D as low as 1 mg/L. By comparison, streptomycin had no detectable effect on seed germination at concentrations up to 100 mg/L. Both clover and bush bean germination was severely affected by Cu^{2+} at 100 mg/L.

The following concentrations (in mg/L) were finally selected for testing: 2,4-D (0, 0.01, 0.1, and 1.0), Cu^{2+} (0, 0.1, 1, and 10), and streptomycin sulfate (0, 1, 10, and 100). The antibiotic streptomycin sulfate was chosen for testing because it should primarily affect the *Rhizobium* bacteria rather than the host plant. Streptomycin sulfate is an antibiotic that is effective against gram-negative bacteria, including *Rhizobium* [16], because of its ability to inhibit protein synthesis. For *Rhizobium trifolii* strains not previously exposed to streptomycin, growth is severely inhibited by 50 to 100 mg/L [17]. The herbicide 2,4-D was tested as a representative chemical with strong phytotoxic properties [10,11]. The growth of beans in nutrient solutions is decreased by 2,4-D concentrations as low as 0.15 mg/L [*18*]. Fletcher et al. [*12*] found that 2,4-D severely inhibited nodulation and growth of white clover on mineral agar at concentrations as low as 0.05 ppm. By comparison, the growth of *Rhizobium trifolii* and *Rhizobium phaseoli* is inhibited by 2,4-D salts at concentrations of $\approx$ 100 to 200 mg/L [*11,12*].

For long-term (5- to 7-week) exposures, beans and clover were irrigated with nutrient solution containing the test chemical from the time of planting until harvest. The irrigation system is described in detail elsewhere [*15*]. Bean plants were harvested twice between 3 and 5 weeks after planting, and clover was harvested twice between 4 and 7 weeks after planting. For short-term exposures (2 weeks), plants were not exposed to the test chemical until after the first sampling period, which occurred at least 3 weeks after the planting of beans and 4 weeks after the planting of clover. After the first set of measurements, which was to ensure similarity between plants assigned to the various treatment groups, the test chemical was added in the desired concentration to the nutrient solution (except for controls). Plants were then tested again at 1 and 2 weeks after the test chemical had been added.

Response Variables—Plant biomass (yield), nodulation success, nitrogen concentration (milligrams of nitrogen per milligram of dry biomass), and nitrogen fixation rate (estimated by acetylene reduction assay) were measured to determine the effect of the chemicals on the *Rhizobium*-legume association in beans and clover. Total root and shoot dry weight (plant biomass) was measured after drying plants at 70°C for 2 days. Nodulation success was determined by counting the number of nodules and the total fresh weight of nodules per plant. Plant nitrogen concentration was determined by standard methods of Kjeldahl analysis [*19*].

Estimates of nitrogen fixation rate were made by using the acetylene (C_2H_2)-reduction technique [*20,21*]. Acetylene reduction (i.e., ethylene production) in beans was measured in experiments with freshly detached root systems [*15*]. The root system was removed from each plant, washed with distilled water, blotted dry between paper towels, and placed in a 500-mL gas-tight, clear, plastic jar for exposure to acetylene (Fig. 1). Each jar was injected with C_2H_2 to give a partial pressure of 5 kPa. There is $\approx$30% reduction in the C_2H_4 produc-

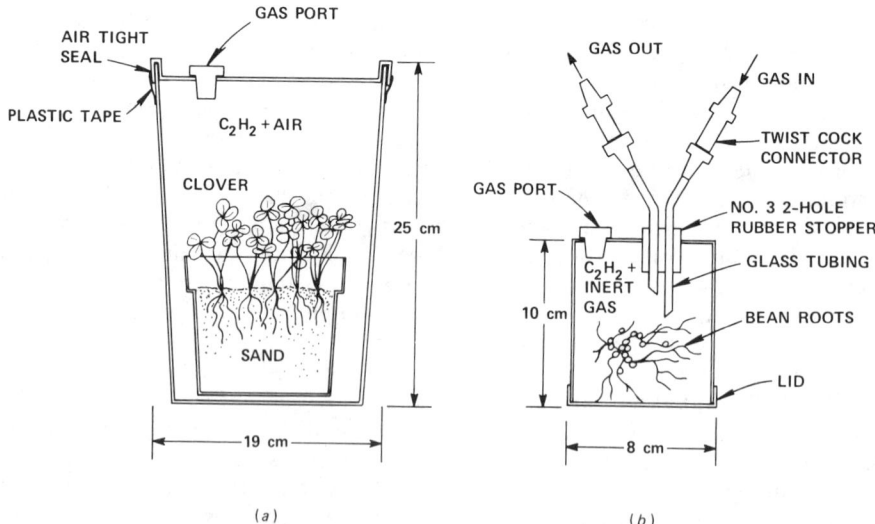

FIG. 1.—*Apparatus used for the acetylene reduction assay to estimate nitrogen fixation rates in potted clover* (a) *and detached nodulated bean roots* (b).

tion rate when the C_2H_2 reduction assay on nodulated bean roots is conducted in air rather than in inert gas (1 part CO_2:500 parts O_2:2000 parts Ar) [15]. The nitrogen in air competes with C_2H_2 for active sites on the nitrogenase enzyme and apparently accounts for this difference in results from the two incubation techniques. Gas samples were withdrawn from the jars after a 1-h incubation period and analyzed for C_2H_4 by gas chromatography. Roots were removed from each jar, nodules were detached, and the fresh weight of nodules from each root system was determined to the nearest milligram. At low rates of C_2H_4 production (1 to 5 nmol/mg fresh weight of nodules), the C_2H_4 production by bean root nodules was approximately linear with time. The C_2H_4 production versus incubation time was nonlinear at higher C_2H_4 production rates; because of this, incubation times with nodulated bush bean roots were limited to 1 h for acetylene reduction assays [15].

Acetylene reduction in clover was measured by enclosing entire pots in gas-tight plastic containers (7-L volume) and injecting C_2H_2 to give a partial pressure of 2.5 kPa (Fig. 1). The nodule nitrogenase system was saturated by 2 to 10 kPa of acetylene [20,21]. Gas samples were withdrawn hourly from each plastic container over a 5-h incubation period. The samples (2 cm^3) were analyzed for C_2H_4 by gas chromatography (Perkin Elmer 3920 gas chromatograph equipped with an H-flame photometric detector and Porapak R column at 50°C; carrier gas was He at 50 mL/min). Potted clover plants were not irrigated for 3 to 4 h before testing, because C_2H_2 reduction by roots is reduced if the growth medium is too wet [22,23]. Ethylene production was linearly related to incubation time in the 7-L test chamber; therefore a 5-h incubation time was chosen for acetylene reduction assays using clover [15].

Statistical Analysis—Results from each test were analyzed by a two-way, completely randomized, factorial analysis of variance (ANOVA). The sources of variation included treatment (concentration of the test chemical), time of testing (weeks after planting), treatment by time interaction, and experimental error. Three or four replicates were tested for each treatment combination. Statistical significance was indicated by $P < 0.05$.

Correlations Between Algal Tests and Terrestrial Plant Tests

Terrestrial Plant Tests

Test Species Selection and Test Conditions—Radishes (*Raphanus sativus*, var. Champion), barley (*Hordeum vulgare*), and soybeans (*Glycine max*, var. Prize) or bush beans (*Phaseolus vulgaris*, var. Blue Lake) were planted in 10 by 10 by 9 cm plastic pots containing a sand-soil mixture. After germination, the plants were thinned to leave the largest six per pot. Three replicate pots at each of the concentrations tested were available for each plant species and for each herbicide tested. Plants were grown in a greenhouse with supplemental lighting. All pots were maintained at the same moisture level by watering to a constant weight. After 4 weeks growth, the aboveground portions of six plants per treatment level were harvested, dried at 70°C for 2 days, and weighed to determine the dry weight of the individual plants.

Test Chemicals and Exposure Regimes—The herbicides that were tested are listed in Table 2. Pure (>98%) compounds or commercial formulations containing a single active ingredient were bought from chemical companies or obtained as gifts courtesy of manufacturers. The chemicals tested represented a wide array of compounds that have single or multiple modes of phytotoxicity, including interference with cell division, inhibition of protein synthesis, disruption of the cell membrane or phosphorus metabolism, inhibition of oxidative phosphorylation or nucleic acid metabolism, blockage of photosynthesis, inhibition of the Hill reaction of photosynthesis, and disruption of enzyme reactions.

TABLE 2—*Common names, chemical names, sources, and percent of active ingredient in the form or herbicide formulation used for terrestrial plant and algal toxicity testing.*

Common Name	Chemical Name of Active Ingredient	Source	Percent Active Ingredient
2,4-D	(2,4-dichlorophenoxy) acetic acid	Eastman-Kodak	99
Monuron	3-(p-chlorophenyl)-1,1-dimethylurea	Hopkins	98[a]
Dalapon	2,2-dichloropropionic acid (Na salt)	Diamond Shamrock	74[a]
Bromacil	5-bromo-3-*sec*-butyl-6-methyluracil	DuPont	21.8[a]
Linuron	3-(3,4-dichlorophenyl)-1-methoxy-1-methylurea	DuPont	41.8[a]
Benefin	N-butyl-N-ethyl-a,a,a-trifluoro-2,6-dinitro-*p*-toluidine	Elanco	16.3[a]
Sodium TCA	Sodium trichloroacetate	Hopkins	41.5[a,b]
Dicamba	2-methoxy-3,6-dichlorobenzoic acid (Na salt)	Velsicol	22.5[a,b]
Alachlor	2-chloro-2',6'-diethyl-N-(methoxymethyl)-acetanilide	Monsanto	45.1[a]
Diquat	6,7-dihydrodipyridol (1,2-a:2',1'-c) pyrazdiium ion	Ortho	19.3[a,c]
Naptalam	Sodium 2-[(1-naphthalenylamino) carbonyl] benzoate	Uniroyal	23.5[a]
DSMA	Disodium methanearsonate	Drexel	21.9[a]
Bromoxynil	3,5-dibromo-4-hydroxybenzonitrile	Aldrich	98
Picloram	4-amino-3,5,6-trichloropicolinic acid	Aldrich	99
Simazine	2-chloro-4,6-*bis*(ethylamino)-*s*-triazine	Ciba-Geigy	41.7[a]
Cacodylic acid	Hydroxydimethylarsine oxide	Aldrich	98
Triallate	S-(2,3,3-trichloroallyl) diisopropyl-thiocarbamate	Monsanto	45[a]
Amitrole	3-amino-1,2,4-triazole	Aldrich	95
DNOC	Dinitro-*o*-cresol	Aldrich	90
Cu^{2+}	Cupric sulfate (pentahydrate)	J. T. Baker	25.5[c]
Propham	Isopropyl carbanilate	PPG Industries	41.4[a]

[a] Herbicide and analysis provided courtesy of the manufacturer.
[b] Based on acid equivalent.
[c] Cation only.

Terrestrial plants were grown in a nonsterile sand-soil medium made from 3.5 kg of clean, dry, coarse sand (99% between 0.2 and 2 mm in diameter); 0.9 kg of air-dried silt-loam soil (sieved to pass a 2-mm screen); and 90 g of sand that contained 0.0005, 0.005, 0.05, or 5 mg of herbicide per gram of sand. Sand containing each test herbicide was prepared by concentrating the chemical onto 100 g of coarse sand at temperatures <40°C through the use of a rotary evaporator. Sand bearing the herbicide was then thoroughly mixed with the sand-soil medium in a rotary mixer. Final properties of the sand-soil mixture were as follows: textural composition, 15% soil fines (<0.02 mm in diameter); pH 5.3; and maximum water-holding capacity, 20 mL/100 g of mixture. The resulting concentrations in the mixed medium were 0 (control), 0.01, 0.1, 1.0, 10, and 100 mg of herbicide per kilogram of sand-soil mixture.

Response Variables and Statistical Analysis—The toxicity of the 21 test chemicals to terrestrial plants was evaluated in terms of (1) the no-observed-effect concentration (NOEC); that is, the highest test concentration at which the mean individual dry weight of the exposed plants, after 4 weeks of growth, was not significantly less than that of the controls; and (2) the concentration (c') at which mean plant dry weight was reduced by 50% or more in relation to controls [24]. Differences between plant dry weights (aerial portions only) for the concentrations tested were analyzed by a one-way ANOVA.

Algal Tests

Test Species Selection and Test Conditions—Procedures used in the algal toxicity tests were based on a test protocol described by the Organization for Economic Cooperation and Development (OECD) [25]. Each herbicide was tested with two species of algae: *Chlorella vulgaris* and *Selenastrum capricornutum* (purchased from Carolina Biological Supply Company). The growth medium for the algae, described elsewhere [24,25], was sterilized by autoclaving for 20 min at 120°C and 100 kPa before use. *Chlorella vulgaris* and *Selenastrum capricornutum* inocula were 3- or 4-day-old cultures grown in the same medium used in the tests. The cultures were in a logarithmic phase of growth at the time of their use as inocula.

The test vessels were 250-mL Erlenmeyer flasks with aluminum foil covers. Three replicates were used for each concentration of each herbicide tested. Initially, each flask contained 100 mL of test solution and $\approx 1 \times 10^6$ algal cells. The quantity of inoculum used was determined from microscopic counts of the number of algal cells in the inoculum at the start of the test.

Test Chemicals and Exposure Regimes—Each herbicide (Table 2) was added to the medium after sterilization to avoid possible thermal decomposition. The nominal concentrations used in the algal tests were 0 (control), 0.01, 0.1, 1.0, 10 and 100 mg of active ingredient per liter of growth medium. The amount of active ingredient in herbicide formulations was determined from data supplied by the manufacturer (Table 2). A stock solution (pH 7.7) containing 100 mg of a given herbicide in 1 L of sterile medium was used for the highest treatment level and for preparing more dilute mediums.

The culture flasks were incubated on a rotary shaker (125 rpm) in an environmental chamber at 20°C under constant illumination (4300 lx [400 fc]). Samples (4.5 mL) were taken at 24 and 96 h in the first seven tests. Samples were also taken at 1 h during subsequent tests because, for some herbicides, algal inhibition was observed at high treatment concentrations in relation to controls within the first 24 h.

Response Variables and Statistical Analysis—The end-point of the algal toxicity test was inhibition of growth rate. Growth in each culture was estimated based on fluorometric measurements of chlorophyll-*a*. Chlorophyll-*a* was measured *in vivo* and after being extracted from the algal cells with basic methanol. The sensitivity of the *in vivo* measurements was normally equal to or greater than determinations based on extraction with methanol; hence *in vivo* measurements were used. When necessary, fluorometer readings were corrected for background fluorescence contributed by the test chemical. The estimated growth rate in each algal culture was calculated from fluorometric readings taken at the beginning and the end of the 96-h test [24]. The percentage inhibition at each treatment concentration was then calculated from the average growth rate of algae in the presence of the test chemical and the average growth rate of controls. The coefficient of variation for average growth rate of algal controls over the 21 chemicals tested was less than 10%. Following analysis of variance, Dunnett's test [26] for comparisons involving a control mean was used to determine whether the growth rates of algae at the treatment concentrations used were significantly different from controls [24].

Results and Discussion

Effects of Test Chemicals on Nitrogen Fixation

Statistically significant sources of variation in measurements of bean plant biomass, nodulation success, or acetylene reduction in long-term and short-term exposure tests with Cu^{2+}, streptomycin sulfate, and 2,4-D are presented in Table 3. Both chemical concentration and date of testing (plant age) affected the various responses. Plant yield, nodule mass per plant,

TABLE 3—*Summary of statistically significant sources of variation (plant age, chemical concentration, and interaction) in response variables of bush beans exposed to three test chemicals in nutrient solution and grown under long-term (5-week) and short-term (2-week) exposure conditions.*[a]

Toxicant	Response Variable	Long-Term Exposure			Short-Term Exposure		
		Age	Concentration	Interaction	Age	Concentration	Interaction
Cu^{2+}	Top biomass	●●●	●●●	●●●	●●●	●	NS
	Root biomass	●●●	●●●	●●	●●●	NS	NS
	Nodule numbers	●●	●●●	NS	●●	●	NS
	Nodule mass per plant	●●●	●●●	●●	●●●	NS	NS
	Nitrogen fixation[b]	●	●	●	●●●	NS	NS
	N concentration	●	●	NS	NS	NS	NS
Streptomycin sulfate	Top biomass	NS	NS	NS	●●	●●●	●
	Root biomass	NS	●●	NS	●●	●●●	NS
	Nodule numbers	NS	●●●	NS	NS	NS	NS
	Nodule mass per plant	NS	●●●	NS	●●	●	NS
	Nitrogen fixation[b]	NS	NS	NS	●●●	NS	NS
	N-concentration	NS	NS	NS	NS	●	NS
2,4-D[c]	Top biomass	●●●	●●●	●●●	●●●	●●	●●●
	Root biomass	NS	●●●	●●	●●●	NS	NS
	Nodule numbers	●●●	●●●	●	●●●	NS	NS
	Nodule mass per plant	●●●	●●●	●●●	●●●	NS	NS
	Nitrogen fixation[b]	NS	●●●	●●●	NS	●●	NS
	N-concentration	NS	NS	NS	●	NS	NS

[a] ● = statistically significant effect at 95% level; ●● = statistically significant effect at 99% level; ●●● = statistically significant effect at 99.9% level; and NS = differences between treatment levels were not statistically significant.
[b] Nitrogen fixation rate was assayed by the acetylene reduction technique [20,21].
[c] There were also varietal differences in bush beans for many response variables [15].

and C_2H_4 production rate increased with plant age in bush beans [15]. Statistically significant interactions between time of testing and concentration level occurred in tests with Cu^{2+} and 2,4-D, indicating that the response to changing concentration levels for some response variables depended on the time of testing. These interactions in ANOVA complicated the straightforward presentation of concentration effects on nitrogen fixation more frequently in long-term exposure tests than in short-term exposure tests (Table 3).

By design, short-term exposures were not useful for detecting effects on the early stages of bean growth and nodulation. In addition, concentration effects in the short-term exposure tests were more often not statistically significant as a source of variation in response variables related to nitrogen fixation (Table 3). As a specific example, in the long-term exposure test, bean plant biomass, nodulation success, and ethylene production were significantly less than those of controls at 10 mg of Cu^{2+} per liter (Fig. 2). By comparison, short-term exposures at the same Cu^{2+} concentration affected fewer response variables. Furthermore, in short-term exposure tests, the amounts of root biomass, nodule mass, acetylene reduction, and plant nitrogen concentration in plants exposed to 10 mg of Cu^{2+} per liter were not significantly different from those in controls (Table 3).

Statistically significant sources of variation in the measurements of clover biomass, nodule numbers, acetylene reduction, and whole-plant nitrogen concentration in long-term and

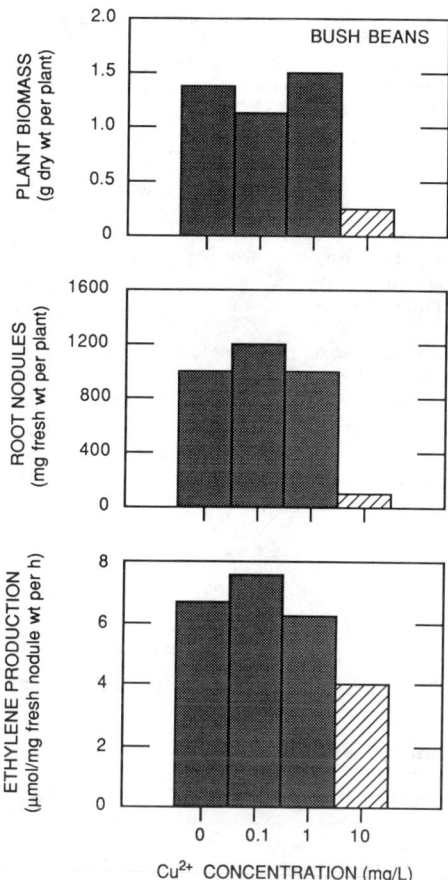

FIG. 2.—*Plant biomass, nodulation success, and rate of symbiotic nitrogen fixation (ethylene production) in bush beans following long-term exposure to concentrations of Cu^{2+} in nutrient solution. Solid bars show Cu^{2+} concentrations at which each response variable was not significantly different ($P < 0.05$) from the control; hatched bars are different from the control.*

short-term exposures with Cu^{2+}, streptomycin sulfate, and 2,4-D are presented in Table 4. White clover was more sensitive to the three test chemicals than bush beans were. In long-term exposures to Cu^{2+} at 10 mg per liter, clover plants died a few days after germination. Bush beans grew and survived under long-term exposure to 10 mg of Cu^{2+} per liter, but were stunted. In comparison with controls, exposed clover had significantly reduced plant biomass and nitrogen fixation rates under long-term exposure to 1 mg of streptomycin per liter, whereas the same response variables in bush beans were unaffected by concentrations as high as 100 mg of streptomycin per liter.

Unlike Cu^{2+} and streptomycin, 2,4-D did not affect all response variables in white clover (Table 4). In long-term exposures, root nodule numbers were not significantly different from those of the controls at any of the 2,4-D concentrations tested, whereas the nitrogen fixation rate (ethylene production) on a pot basis was significantly different from that of controls, but only at the highest test concentration (Fig. 3). Clover biomass per pot appeared to be the most sensitive measure of toxicity in the long-term exposure tests. For example, the NOEC

TABLE 4—*Summary of statistically significant sources of variation (plant age, chemical concentration, and interaction) in response variables of white clover exposed to three test chemicals in nutrient solution and grown under long-term (7-week) and short-term (2-week) exposure conditions.*[a]

Toxicant	Response Variable	Long-Term Exposure			Short-Term Exposure		
		Age	Concentration	Interaction	Age	Concentration	Interaction
Cu^{2+}	Clover biomass	●●●	●●●	●●●	●●●	●●	●
	Nodule numbers	NS	●●●	NS	NS	NS	NS
	C_2H_4 production rate per gram DW	●●●	●●●	NS	●●●	●●	NS
	C_2H_4 production rate per pot	●●●	●●●	●●●	●●●	●●●	●●
	N concentration	NS	●●●	NS	●	NS	NS
Streptomycin sulfate	Clover biomass		●●●		●●	NS	NS
	Nodule numbers		●●●		NS	NS	NS
	C_2H_4 production rate per gram DW		●●●		NS	●	NS
	C_2H_4 production rate per pot		●●●		●	●●	●
	N concentration		●●●		●●	●	NS
2,4-D	Clover biomass	●●●	●●●	●●●	●●●	●●●	●●●
	Nodule numbers	NS	NS	●●●	●●●	●●●	●●●
	C_2H_4 production rate per gram DW	●●	●	NS	●●●	●●	●●●
	C_2H_4 production rate per pot	●●	●●●	●	●●●	●●●	●●●
	N concentration	NS	●●	NS	●●●	●●●	●●●

[a] ● = statistically significant effect at 95% level; ●● = statistically significant effect at 99% level; ●●● = statistically significant effect at 99.9% level; and NS = differences between treatment levels were not statistically significant.

for clover biomass was less than 0.01 mg of 2,4-D per liter, whereas nodule numbers and C_2H_4 production in exposed plants were not different from those of the controls for the same treatment concentration (Fig. 3).

As was the case with bush beans, chemical additions in the short-term exposure tests with white clover apparently were too late to enable early detection of effects on the different response variables. In the short-term tests, both Cu^{2+} and streptomycin sulfate had somewhat less effect on response variables than was observed in the long-term exposure tests. Clover biomass was not totally inhibited at 10 mg of Cu^{2+} per liter, unlike in the long-term exposure test. Streptomycin sulfate also had no significant effect on clover biomass and nodulation in the short-term exposure tests, but it had a significant effect on these same response variables in the long-term exposure test (Table 4).

Sample sizes necessary to detect a statistically significant difference ($P<0.05$) between treatment means of 10, 25, or 50% of their absolute values were calculated from each ANOVA according to methods presented by Clarke [27]. Sample sizes used in the long- and short-term exposures (three to four replicate bean plants or replicate pots of clover per concentration) were large enough to detect differences of ≈50% or more in the treatment means

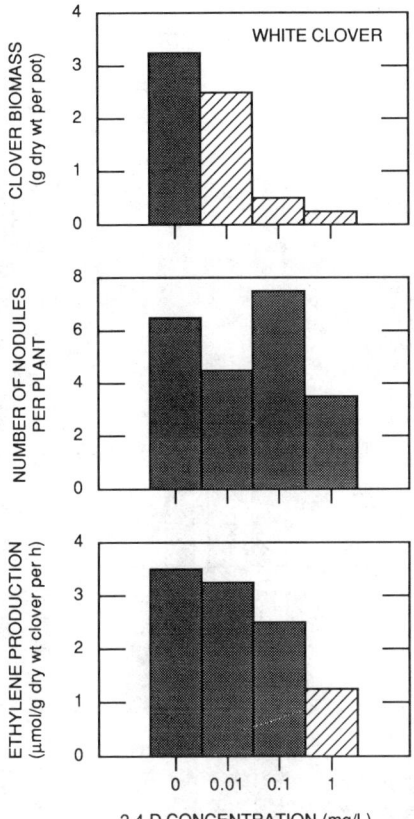

FIG. 3.—*Clover biomass, nodulation success, and rate of symbiotic nitrogen fixation (ethylene production) in potted white clover following long-term exposure to concentrations of 2,4-D in nutrient solution. Solid bars show 2,4-D concentrations at which each response variable was not significantly different ($P < 0.05$) from the control; hatched bars are different from the control.*

of many response variables. In general, the best sensitivity that can be expected with these multispecies toxicity tests is an ≈25 to 50% difference between treatment means. Due to the variability between replicates, prohibitively large sample sizes are needed to detect differences of 10% or less in all response variables [15].

Correlations Between Algal and Terrestrial Plant Toxicity Tests

The no-observed-effect concentration (NOEC; the highest test concentration at which the growth rate in exposed plants was not significantly different from that of the control) and the concentration at which the algal growth rate was reduced by 50% or more in comparison with that of the controls (c') was determined in *Selenastrum* and *Chlorella* for each of the 21 herbicides tested [24]. As an example of the type of data used in determining the NOEC and c', Fig. 4 shows the growth rate of *Selenastrum* as a percentage of the control algal culture for each of the 21 herbicides tested. A comparison of the NOECs showed that *Selenastrum* was more sensitive than *Chlorella* to 11 test chemicals, less sensitive than *Chlorella* to

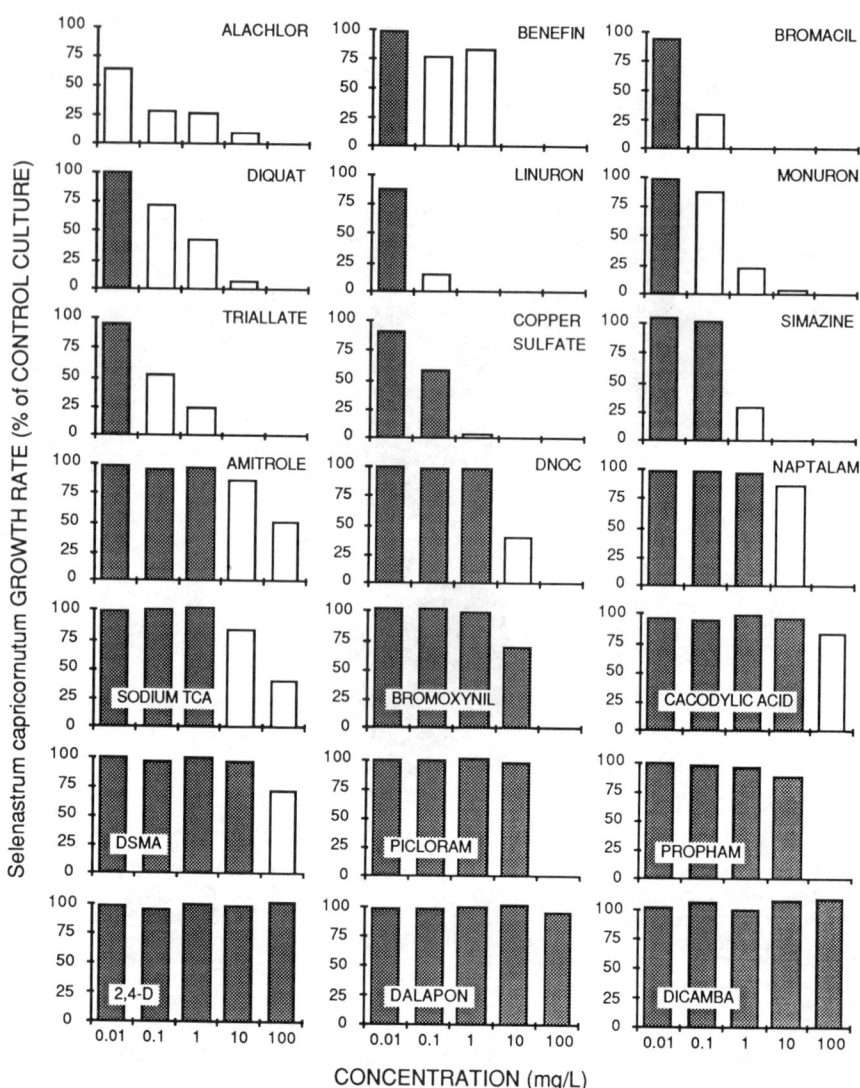

FIG. 4.—*Relative growth rates of* Selenastrum capricornutum *exposed to various concentrations of 21 herbicides (see Table 2) in a short term (96-h) toxicity test. Growth rates are in relation to control cultures that did not contain herbicides. Shaded columns show those concentrations at which growth rates were not different from the control; open columns are significantly different ($P < 0.05$) from the control.*

2 chemicals, and equally sensitive to 8 chemicals (Fig. 5). Therefore the probability of agreement between NOECs from the two algae for the same chemical was 0.38. The NOEC was usually a more sensitive measure of algal growth inhibition than c'. In tests with *Chlorella vulgaris*, the concentration at which algal growth was inhibited by 50% or more (c') could be specified only at >100 mg/L for more than half of the herbicides used. Three of the test chemicals (2,4-D, dalapon, and dicamba) did not significantly reduce growth rates of either species of algae at any concentration tested.

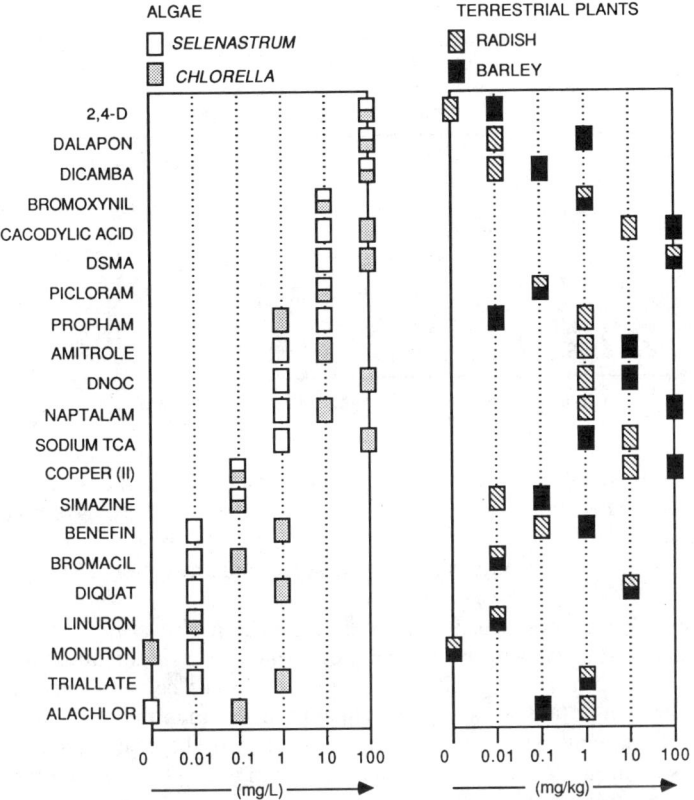

FIG. 5.—*No observed-effect-concentrations (see text) in two species of algae (*Selenastrum capricornutum *and* Chlorella vulgaris*) and two species of terrestrial plants (radish and barley) for 21 different herbicides.*

The concentration at which the growth of radishes, barley, and beans was inhibited by 50% or more was generally about ten times greater than the NOEC for each herbicide. Hence, for these species, the NOEC was more sensitive than c' as a measure of phytotoxicity [*24*]. Bush beans were substituted for the original test species (soybeans) in most of the tests because of poor germination in soybean controls during the first month of testing. Although phytotoxicity varied depending upon the plant species and the test chemical used, the growth of radishes, barley, and dry beans was inhibited when concentrations of 2,4-D, bromacil, monuron, dicamba, simazine, and picloram were ≤ 1 mg/kg [*24*].

Radishes and barley had the same NOEC for 8 of the 21 chemicals tested (38% agreement between test outcomes) (Fig. 5). Radishes were less sensitive than barley to only 2 of the 21 chemicals tested. NOECs in radishes generally could not be used to make accurate predictions about NOECs in barley. For example, chemicals with a 1-mg/kg NOEC in radishes had NOECs ranging from 0.01 to 100 mg/kg in barley (Fig. 5). When the agreement criterion was relaxed to include treatment concentrations within a factor of ten of the predicted NOEC, the probability of agreement was 0.86.

Overall, the degree of association between the algal and terrestrial tests was such that there

TABLE 5—*Statistically significant (P < 0.05) Pearson product moment correlation coefficients (upper right triangle of matrix) and Spearman rank correlation coefficients (lower left triangle of matrix) between no-observed-effect concentrations for 21 herbicides (N = 21 for each correlation) in a 4-week test with terrestrial plants (radishes, barley, and beans) and a 96 h test with algae* (Selenastrum capricornutum *and* Chlorella vulgaris).[a]

Test Species	Correlation Coefficients Between No-Observed-Effect Concentrations				
	Radish	Barley	Beans	*Selenastrum*	*Chlorella*
Radish	1.00	0.54	NS	NS	NS
Barley	0.78	1.00	0.49	NS	NS
Beans	0.65	0.58	1.00	NS	NS
Selenastrum	NS	NS	NS	1.00	0.60
Chlorella	NS	NS	NS	0.80	1.00

[a] NS denotes that the correlation is not significantly different from zero.

was less than 50% probability of the NOEC from one test matching the NOEC from another. For example, *Selenastrum capricornutum* had the same relative NOEC as radishes and barley for less than 30% of the herbicides tested [24]. Agreement between the observed NOECs in *Chlorella vulgaris* and those in radishes and barley was not any better (24 to 43% agreement) [24].

The usefulness of the most sensitive algal species tested *(Selenastrum capricornutum)* as a "screen" to identify the concentration at or above which herbicides reduced terrestrial plant growth was evaluated in terms of either the NOEC or c'. To be an effective screen, *Selenastrum* must have an NOEC or a c' value equal to or less than that observed for the most sensitive terrestrial plant tested. This was true for only 11 of the 21 herbicides tested. Therefore there was about a 50% chance that the short-term algal phytotoxicity test with *Selenastrum* would identify the level of response to the herbicide by one or more of the terrestrial plants tested.

Correlation coefficients for the NOECs for the terrestrial plant and algal toxicity tests were calculated on the basis of the results of the tests with the 21 herbicides (Table 5). Significant correlations between NOECs from different species were found within a particular type of test (e.g., between NOECs for radishes and barley or between NOECs for *Chlorella* and *Selenastrum*). However, the NOECs from tests of herbicides on algae were not significantly correlated with results from the terrestrial plant tests.

Conclusion

Toxicity tests using the *Rhizobium*-legume symbiosis involve interspecies dependencies (i.e., the interactive associations between plant host and bacterium) that can change in response to toxic chemicals. In the *Rhizobium*-legume tests, longer exposure times (4- to 5-week tests) resulted in more detectable changes. The *Rhizobium*-legume toxicity test, however, required considerable amounts of maintenance, time, and greenhouse space. This test also did not appear to be particularly sensitive to differences in the concentrations of the test chemicals. Response variables related to nitrogen fixation for treatments with 4 to 12 replicates were statistically distinguishable only when differences between means were 25 to 50% [15]. Plant biomass and measures of nodulation success were the simplest and least expensive indices of response to the test chemicals.

Independent correlative testing with algae and terrestrial plants is a predictive approach to help identify chemical toxicity in one species on the basis of the response by a second, unrelated species. However, the test systems used were physically different in several respects including the application of test chemicals and composition of the growth media. These differences undoubtedly affect the bioavailability of toxicants and partly contribute to very different test outcomes. There was generally poor agreement between results from the two types of tests; the results of the 96-h algal growth inhibition test did not correlate well with the results from the longer-term terrestrial plant test. Overall, there was only about a 50% chance that an algal test would successfully detect herbicide levels that reduced terrestrial plant yield. For the herbicides and species examined, short-term toxicity tests with algae were not good indicators of toxicity to terrestrial plants [24].

Acknowledgments

I wish to thank M. L. Frank and B. G. Blaylock for their contributions to this work. Appreciation is also extended to A. J. Stewart and R. J. Norby, ORNL, for their helpful reviews of the manuscript. This research was sponsored by the U.S. Environmental Protection Agency under Interagency Agreements EPA 78-D-X0387 and 40-1067-80 with the U.S. Department of Energy, under Contract DE-AC05-84OR21400 with Martin Marietta Energy Systems, Inc. This paper is Publication No. 3349, Environmental Sciences Division, Oak Ridge National Laboratory.

References

[1] Hammons, A. S., Giddings, J. M., Suter, G. W., II, and Barnthouse, L. W., *Methods for Ecological Toxicology: A Critical Review of Laboratory Multispecies Tests*, Ann Arbor Science, Ann Arbor, Mich., 1981.
[2] Lin, S., Funke, B. R., and Schulz, J. T., *Plant and Soil*, Vol. 37, 1972, pp. 489–496.
[3] Smith, C. R., Funke, B. R., and Schulz, J. T., *Soil Biology and Biochemistry*, Vol. 10, 1978, pp. 463–466.
[4] Bethlenfalvay, G. J., Norris, R. F., and Phillips, D. A., *Plant Physiology*, Vol. 63, 1979, pp. 213–215.
[5] Kapusta, G. and Rouwenhorst, D. L., *Agronomy Journal*, Vol. 65, 1973, pp. 112–115.
[6] Appleman, M. D. and Sears, O. H., *Journal of the American Society of Agronomy*, Vol. 38, 1946, pp. 545–550.
[7] Pareek, R. P. and Gaur, A. C., *Plant and Soil*, Vol. 33, 1969, pp. 297–304.
[8] Braithwaite, B. M., Jane, A., and Swain, F. G., *Journal of the Australian Institute of Agricultural Sciences*, Vol. 24, 1958, pp. 155–157.
[9] Misra, K. C. and Gaur, A. C., *Indian Journal of Agricultural Science*, Vol. 44, 1974, pp. 838–840.
[10] Payne, M. G. and Fults, J. L., *Journal of the American Society of Agronomy*, Vol. 39, 1947, pp. 52–55.
[11] Carlyle, R. E. and Thorpe, J. D., *Journal of the American Society of Agronomy*, Vol. 39, 1947, pp. 929–936.
[12] Fletcher, W. W., Dickenson, P. B., and Raymond, J. C., *Phyton* (Argentina), Vol. 7, 1956, pp. 121–130.
[13] Fletcher, W. W., Dickenson, P. B., Forrest, J. D., and Raymond, J. C., *Phyton* (Argentina), Vol. 9, 1957, pp. 41–46.
[14] Decleire, M., de Cat, W., and Bastin, R., *Revue de l'Agriculture*, Vol. 1, 1975, pp. 27–38.
[15] Garten, C. T., Jr., Suter, G. W., II, and Blaylock, B. G., "Development and Evaluation of Multispecies Test Protocols for Assessing Chemical Toxicity," ORNL/TM-9225, Oak Ridge National Laboratory, Oak Ridge, TN, June 1985.
[16] Levin, R. A. and Montgomery, M. P., *Plant and Soil*, Vol. 41, 1971, pp. 669–676.
[17] Zelazna-Kowalska, I., *Plant and Soil*, Special Vol. 1971, 1971, pp. 67–71.
[18] Taylor, D. L., *Botanical Gazette*, Vol. 107, 1946, pp. 597–611.

[19] Bremner, J. M., in *Methods of Soil Analysis*, C. A. Black, Ed., American Society of Agronomy, Madison, Wisc., 1965, Chapter 83, pp. 1149–1178.

[20] Hardy, R. W. F., Holsten, R. D., Jackson, E. K., and Burns, R. C., *Plant Physiology*, Vol. 43, 1968, pp. 1185–1207.

[21] Hardy, R. W. F., Burns, R. C., and Holsten, R. D., *Soil Biology and Biochemistry*, Vol. 5, 1973, pp. 47–81.

[22] Fishbeck, K., Evans, H. J., and Boersma, L. L., *Agronomy Journal*, Vol. 65, 1973, pp. 429–433.

[23] Tu, C. M. and Hietkamp, G., *Communications Soil Science and Plant Analysis*, Vol. 8, 1977, pp. 81–86.

[24] Garten, C. T., Jr., and Frank, M. L., "Comparison of Toxicity to Terrestrial Plants with Algal Growth Inhibition by Herbicides," ORNL/TM-9177, Oak Ridge National Laboratory, Oak Ridge, Tenn., Oct. 1984.

[25] Organization for Economic Cooperation and Development, "OECD Guideline for Testing of Chemicals: Algal Growth Inhibition Test," Number 201, Organization for Economic Cooperation and Development, Paris, May 1981.

[26] Kirk, R. E., *Experimental Design: Procedures for the Behavioral Sciences*, Brooks/Cole, Belmont, Calif., 1968.

[27] Clarke, G. M., *Statistics and Experimental Design*, American Elsevier, New York, 1969.

Plants and Xenobiotic Uptake

Michael A. Krstich[1] *and Otto J. Schwarz*[1]

Characterization of Xenobiotic Uptake Utilizing an Isolated Root Uptake Test (IRUT) and a Whole Plant Uptake Test (WPUT)

REFERENCE: Krstich, M. A. and Schwarz, O. J., "**Characterization of Xenobiotic Uptake Utilizing an Isolated Root Uptake Test (IRUT) and a Whole Plant Uptake Test (WPUT),**" *Plants for Toxicity Assessment, ASTM STP 1091,* W. Wang, J. W. Gorsuch, and W. R. Lower, Eds., American Society for Testing and Materials, Philadelphia, 1990, pp. 87–96.

ABSTRACT: An isolated root uptake test (IRUT) and a whole plant uptake test (WPUT) were used to characterize the uptake and bioaccumulation of naphthol in two varieties of fescue (*Festuca arundinaceae* Schreb. var. KY-31 and var. Auburn) and one variety of clover (*Trifolium pratense* L. var. Redland). Whole plant uptake tests (WPUT) run for 24, 48, and 72 h revealed that the amount of radiolabel translocated to the shoots of both species was directly related to the length of exposure time. ^{14}C-naphthol concentration in the roots of whole plants was directly proportional to naphthol concentration in solution between 0.1 and 0.3 μM. The transpiration stream concentration factor (TSCF) in both varieties of fescue decreased with increasing naphthol concentration in solution. ^{14}C-naphthol elution from the apparent free space (AFS) was very rapid, occurring in less than 1 min. The AFS volume of clover and KY-31 fescue was determined to be approximately 10% of the total root volume. Isolated root tissue, incubated in 10 μM naphthol for 24 h, contained less than 1% of the radiolabel as recoverable parent compound.

KEY WORDS: xenobiotic, naphthol, isolated root, IRUT, TSCF, AFS, clover, fescue, whole plant uptake

The fate and transport of xenobiotic chemicals in a terrestrial environment are of major concern since many are known toxic substances. Polynuclear aromatics (PNA) consitute a group of xenobiotic compounds with a wide range in molecular weights, octanol:water partition coefficients (K_{ow}), and water solubilities (K_s). Introduced into the environment as byproducts of synfuel production [1], xenobiotics such as naphthalene, naphthol, and antracene may enter the foodchain by accumulating in soil, leaching into groundwater, and accumulating in terrestrial plants. This paper reports on a combination of isolated tissue uptake studies and whole plant uptake studies designed to help characterize the accumulation of these chemicals in plants, which indicates their potential for transfer through the foodchain.

The uptake of xenobiotic chemicals by roots and their subsequent translocation to the shoots has been investigated [1–9]. The use of isolated root systems to study the uptake of xenobiotic chemicals has been demonstrated to be a valuable bioassay for understanding the fate of xenobiotics used in agriculture [10–13] and those of industrial origin [14,15].

The relationship between chemical uptake by isolated roots and chemical uptake and translocation by whole plants has received little attention. Martin and Edgington [16] studied the uptake of several xenobiotics in soybean and barley plants and potato tuber tissue

---
[1] Waste Management Institute and Botany Department, University of Tennessee, Knoxville, TN 37996.

and noted similar levels of fenapanil and fenarimol uptake in both whole plants and isolated tissue. McFarlane et al. [15] studied the uptake, distribution, and metabolism of four organic chemicals by soybean plants and excised barley roots and reported similar patterns of uptake rate constants for both plants and isolated roots.

Differences in uptake and elution rates over time of various chemicals in isolated roots have indicated the presence of multiple compartments within root tissue and their relative importance in the uptake and retention of xenobiotics [10,17,18]. The largest compartment may represent the portion of the root that is accessible to chemicals that are actively adsorbed or that passively diffuse into the lipophilic membranes of the symplast [19]. This chemical accumulation has been described by Shone et al. [5] as the root concentration factor (RCF). The smaller compartment is thought to represent the intercellular area in the root including the cell wall, which contains the readily diffusible portion of the chemical in solution and is described as the apparent free space (AFS) [19,20]. Briggs et al. [20] defined the apparent volume of free space as equalling that amount of solute in the free space divided by the amount of solute per unit volume of external solution when free space and external solution are in equilibrium. The percent volume that the AFS constitutes of the total volume of the root has been estimated to range from 10 to 28% [20–22].

Plant uptake of xenobiotics can occur by an active mechanism as in the case of 2,4-D [10], a passive mechanism as illustrated for carbendazim [11] and Monuron (3-p-chlorophenyl-1,1-dimethylurea) [10], or a combination of the two mechanisms as shown for naphthol [23]. Certain xenobiotics such as naphthol have been found to accumulate in plant roots against a concentration gradient [23]. After they pass through the endodermis, these chemicals may (1) accumulate in the roots as the parent compound with little translocation to the shoots, (2) accumulate in the shoots as parent compound, (3) accumulate in roots or shoots as metabolites of the parent compound, and/or (4) be released through the external surfaces of the plant as volatiles or exudates either in parent compound or metabolite form.

The metabolism of xenobiotic chemicals has been reported to result in predominantly polar conjugates and insoluble residues as stable end-products in a wide variety of plant species [24]. Oxidation of highly lipophilic compounds to phenols and alcohols increases their water solubility and permits their conjugation via glycosidic bond formation [25]. Glutathione conjugation is also an important detoxifying step in plant herbicide metabolism [26]. Hofmann [27] found yeast to oxidize anthracene and naphthalene and noted that the predominant metabolite of naphthalene metabolism was 1-naphthol. Although oxidations appear to be the major first step in plant xenobiotic transformation, other forms of attack (e.g., hydrolysis and photoconversion) may also be important [25].

Because the relationship between the amount of chemical concentrated in the roots and the amount that may subsequently accumulate in the shoots cannot be easily predicted from the xenobiotic's physical and chemical properties, research on xenobiotic fate in plants depends on measurements of compound accumulation. To study aspects of xenobiotic chemical/plant interactions, an isolated root uptake test (IRUT) was coupled with a whole plant uptake test (WPUT) and used to measure and characterize naphthol uptake, its accumulation in shoots, and the extent of its metabolism.

Procedures

Tall fescue (*Festuca arundinaceae* Schreb.) and red clover (*Trifolium pratense* L.) were used to study the application of IRUT and WPUT to naphthol uptake, accumulation, and metabolism. Experiments were conducted to describe the (1) concentration of naphthol in the roots and shoots of whole plants, (2) concentration of naphthol in the apparent free space (AFS) of roots, and (3) metabolism of naphthol within the root. The IRUT protocol described by Krstich and Schwarz [23] was used in the AFS study with ^{14}C-naphthol and in the metabolite study with nonradiolabeled naphthol. ^{14}C-naphthol was used with WPUT to determine the naphthol concentration in shoots and the transpiration stream concentration factor (TSCF).

The ^{14}C-naphthol (Amersham) (MW 146) used in all studies had a specific activity of 58 mCi mM^{-1} and a chemical purity of ca. 97%. A 0.1 mM, carrier-free, ^{14}C-naphthol methanolic stock solution was prepared for use in all incubation studies. Fifty microliters of this stock solution per 50 mL Knops nutrient solution (KNS) [28], adjusted to a pH of 5.5 with KOH, provided a 0.1 μM ^{14}C-naphthol incubation solution with ca. 1290 dpm ^{14}C counts per 100 μL aliquot. A 50% counting efficiency provided adequate counts for uptake studies. All ^{14}C-naphthol analyses were conducted using a Beckman LS 3800 Liquid Scintillation System at ambient temperature. Root and shoot sample preparation consisted of tissue digestion in 1 mL of a quarternary ammonium hydroxide compound (Unisol) for 7 days at 37°C, followed by the addition of 0.5 mL methanol, 0.1 mL H_2O_2 (roots) or 0.3 mL H_2O_2 (shoots), and 5 mL of a toluene-based complement (Unisol). Incubation solutions were analyzed for ^{14}C-naphthol by transferring 100 μL aliquots to 1 mL Unisol and adding 0.5 mL methanol and 5 mL complement.

Where appropriate, standard analysis of variance was determined using a general linear model (GLM) [29] with mean separation by the Duncan Multiple Range test. The confidence interval for all analyses was set at 0.05 level of significance.

Whole Plant Uptake Test (WPUT)

Ninety-day-old plants, two varieties of fescue (KY-31 and Auburn) and one variety of clover (Redland) grown under greenhouse conditions in a soil medium, were transferred into 60 mL amber glass bottles containing 50 mL KNS (pH 5.5). Three plants of each species for each time period, naphthol concentration, and control were used as replicates. The root system of each plant was first rinsed clean of any soil and trimmed to ca. 3 cm average length. A polyurethane foam plug placed in the mouth of the jar supported the plant. Shoots and roots trimmed before transfer provided a compact plant and ensured that most tissue during the test would be young and rapidly growing. Transfers were placed in a fume hood (25 ± 1°C) containing mixed fluorescent and incandescent lighting (25 W/m^2 measure in visible region as photosynthetically active radiation [PAR]) (12 h alternating light and dark periods) and allowed to grow for ca. 14 to 21 days. All plants, both treated and controls, were run simultaneously under the same conditions of relative humidity and air velocity (hood blower running). Transpirational water loss was the same for all treatments. During this time period, clover shoots were trimmed at the crown periodically before treatments so as to maintain 3 to 4 stems per plant, while fescue was trimmed so as to maintain ca. 5 cm top growth. Individual fresh plant weights, before treatments, were ca. 1 g for fescue and 2 g for clover with ca. 1:1 root-to-shoot ratio by weight. KNS was changed twice a week. Plants were transferred after 14 to 21 days to a second set of jars containing 50 mL KNS and various concentrations of ^{14}C-naphthol (50 mL KNS in the 60 mL bottle with plug resulted in ca. 5 cm^3 head space). All tissue was collected from each plant, with each plant representing one replication of a single concentration and time period.

Two studies were conducted. One study consisted of measuring ^{14}C-naphthol uptake from a 0.1 μM ^{14}C-naphthol solution at three time periods: 24, 48, and 72 h. Plants were transferred to new jars containing KNS with ^{14}C-naphthol at 24 and 48 h to provide a more constant dose of the test chemical over the experimental period and to minimize the development of microbial growth. The second study consisted of measuring ^{14}C-naphthol uptake at 24 h from 0.1, 0.2, and 0.3 μM ^{14}C-naphthol solutions.

All plants were cut into roots and shoots of ca. 0.1 g sample sizes, the weight of each sample was recorded, and the total weight of each plant determined by summing the individual sample weights. Total plant weights ranged from 2.5 to 4.0 g, with individual samples ranging in weight from 0.1 to 0.15 g. Samples were placed in 1 mL Unisol and digested for 7 days at 37 ± 1°C. Clover shoots were cut into leaves and stems before analysis. Total ^{14}C concentration (naphthol and metabolites) for each plant part was calculated. Naphthol concentration in the solution was calculated by measuring the naphthol concentration at the beginning and end of the experiment and determining the average concentration. The final amount of KNS remaining in each jar was measured and the amount transpired determined. The transpiration stream concentration factor (TSCF) was calculated according to the formula [3]

$$\text{TSCF} = \frac{\text{nmol radiolabel (shoot)/mL H}_2\text{O transpired}}{\text{nmol radiolabel (solution)/mL solution}}$$

Apparent Free Space Study

Excised roots (0.1 g fresh weight) from KY-31 fescue and Redland clover were incubated in 50 mL KNS containing 0.1 μM ^{14}C-naphthol for ca. 24 h at 25 ± 1°C. Roots were separated from the solution by vacuum filtration using a Buchner funnel fitted with #2 Whatman filter paper, rapidly rinsed twice with ca. 10 mL of 1.0 μM unlabeled naphthol in KNS, and placed in 20 mL scintillation vials. Vials were wrapped in foil to minimize photodegredation during the elution period. Stepwise elution of ^{14}C-naphthol from the roots was measured by adding 5 mL KNS to each vial and removing a 100 μL aliquot at each designated sampling time. Time intervals for sample collection were 1, 5, 10, 15, 20, 25, 30, 45, 60, 90, and 120 min. Following aliquot removal, the solution was decanted, the root tissue lightly blotted while in the vessel with absorbant tissue, and 5 mL KNS added to the vial. All vials were maintained on a shaker at 25 ± 1°C during the entire elution period. Three ^{14}C-naphthol concentrations (0.1, 0.2, and 0.3 μM) were tested. AFS was determined according to the formula [20]

$$\text{AFS} = \frac{\text{nmol readily diffusible chemical/g fresh root weight}}{\text{nmol chemical/mL ambient solution}}$$

Metabolite Study

Root tissue from KY-31 fescue and Redland clover was subjected to the IRUT protocol using nonradiolabeled naphthol. Roots were incubated at 25 ± 1°C in KNS of 10 or 100 μM naphthol for 24 h and rinsed twice with ca. 10 mL KNS. Root tissue was ground in a glass tissue homogenizer using 5 mL methanol in 1 mL increments and decanting the supernatant between each increment. The combined supernatant (ca. 5 mL) was filtered through a 0.2 μm nylon Acrodisc, stored on ice, and analyzed within 6 h on a Waters HPLC. HPLC was performed with Waters Model 501 pumps equipped with a Model 680 gradient controller and a U6K injector with a 2 mL fixed sample loop, Model 481 UV-Visible detector, and Model 740 data module. The analytical column was a reverse-phase Nova Pak C_{18}, 3.9 mm by 15 cm column (Waters Associates). A 50 μL sample was injected for each run. The mobile phase was 40/60 (v/v) water:methanol at a flow rate of 0.6 mL·m^{-1}. Retention time was ca. 6.8 min. Duplicate injections of samples were bracketed by injections of known amounts of standard. Percent parent compound retention in root tissue was calculated by dividing the naphthol concentration in root tissue by the amount of naphthol lost from the incubation solution. To estimate the efficiency of nahpthol extraction in this system, a correction factor was calculated by comparing the amount of ^{14}C-naphthol extracted using the tissue grinder with the amount determined to be in the root tissue using the total tissue digestion method (Unisol).

Results and Discussion

Increasing ^{14}C-naphthol concentration in KNS from 0.1 to 0.3 μM for the WPUT increased uptake of ^{14}C-naphthol by roots of whole plants of clover and both varieties of fescue during the 24-h exposure period (Table 1). This was not a linear response. Increasing ^{14}C-naphthol from 0.1 to 0.3 μM resulted in a sixfold increase of radiolabel in Auburn, a fourfold increase in KY-31, and a fivefold increase in clover. Leaves of both varieties of fescue showed a significant reduction in radiolabel accumulation as the concentration of ^{14}C-naphthol increased from 0.2 to 0.3 μM. The reason for this decrease is not readily apparent as transpirational water loss remained relatively constant at each treatment level. Since the total amount of radiolabel found in the leaves of both varieties of fescue tested decreased at the higher levels of naphthol exposure (i.e., between 0.2 and 0.3 μM [Table 1]), perhaps the chemical reached levels that in some way inhibited or disrupted the transport tissues involved. This would result in a reduced total chemical delivery to the leaves. Another possibility is a rapid increase in the catabolic metabolism of naphthol triggered at the higher concentrations resulting in ^{14}C isotope loss from the leaves as CO_2. Since the determination of a total naphthol carbon mass balance was not possible with the experimental design used, tracking of volatilized chemical losses of naphthol and/or its metabolites from aerial plant organs remains to be accomplished. By contrast, clover leaves showed a nearly linear increase in radiolabel from 0.23 to 0.64 nmol g^{-1} in response to increasing naphthol levels from 0.1 to 0.3 μM. The stems of clover contained low levels of radiolabel similar to concentrations found in fescue leaves. Martin and Edgington [16] also reported differences in the translocation of individual xenobiotics between barley and soybean.

The TSCF as determined under our experimental conditions reflects the concentration of radiolabel in shoots with respect to the amount of water transpired and the average concentration of compound in solution between $t = 0$ and $t = 24$ h. Different levels of naphthol concentration in KNS for a 24-h exposure period resulted in varying effects on the TSCF for both fescue and clover. Figure 1 illustrates the decrease in the TSCF for Auburn fescue, 0.188 to 0.005, and KY-31 fescue, 0.051 to 0.005, as the concentration of naphthol in solution increased from 0.1 to 0.3 μM, indicating saturation of the xenobiotic transport system. Clover showed no significant change in the TSCF as naphthol concentration increased from 0.1 to 0.3 μM, indicating that the transport system was able to readily accommodate the

TABLE 1—*Distribution of ^{14}C-isotope given as naphthol in roots and shoots of clover and two varieties of fescue after 24 h exposure in the whole plant uptake test (WPUT) (radiolabel in nmol g^{-1}).*

^{14}C Naphthol (μM)	Species[a]						
	Auburn		KY-31		Clover		
	Leaf	Root	Leaf	Root	Leaf	Stem	Root
0.1	0.09b	8.9c	0.08a	11.7b	0.23b	0.03b	2.8c
0.2	0.22a	35.4b	0.07a	34.0a	0.33ab	0.07a	6.2b
0.3	0.04b	58.2a	0.04b	44.7a	0.64a	0.07a	13.5a

[a] Values in same column followed by the same letter are not significantly different at $p = 0.05$.

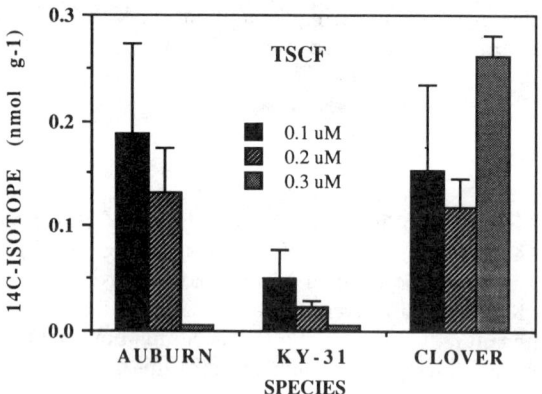

FIG. 1—*Transpiration stream concentration factor (TSCF) versus level of ^{14}C-isotope in the incubation solution.*

increased chemical load. Differences in the TSCF between species was significant at the 0.2 and 0.3 μM concentrations (Table 2). KY-31 showed a significantly lower TSCF than both Auburn fescue and clover at 0.2 μM. However, at 0.3 μM, KY-31, and Auburn had similar TSCFs, while clover had a significantly higher TSCF.

A number of authors have determined TSCF for xenobiotic chemicals of greatly differing lipid/water solubilities [3,6,8,9]. The TSCF values obtained here are in reasonable agreement with those reported by these authors for chemicals possessing similar octanol/water partition coefficients. Clearly the reduction in the TSCF values for both varieties of fescue indicates saturation of the transport system at the 0.3 μM exposure level. Clover, however, showed no such limitation at the test level, suggesting a greater potential for acropetal transport and accumulation in the aerial portion of the plant of naphthol and/or its metabolites.

Figure 2 illustrates the uptake of ^{14}C-naphthol from a 0.1 μM ^{14}C-naphthol/KNS by roots of whole plants of clover and two varieties of fescue at 24, 48, and 72 h. KNS containing 0.1 μM ^{14}C-naphthol was replaced at 48 and 72 h. ^{14}C-naphthol uptake by the roots of all three plants increased with increasing levels of chemical in the bathing solution. Naphthol uptake increased from 7.1 to 29.4 nmol g^{-1} fresh root weight for Auburn, 6.6 to 21.4 for KY-31, and 2.9 to 11.0 for clover, as root exposure to a 0.1 μM ^{14}C-naphthol in KNS increased from 24 to 72 h respectively. Accumulation of carbon-14 in the shoots of the test plants is illustrated in Fig. 3. Shoots of clover are the combined values for both stems and leaves. The

TABLE 2—*TSCF for two varieties of fescue and clover at three concentrations of naphthol in KNS determined after 24 h incubation.*

Species	Concentration (μM)[a]		
	0.1	0.2	0.3
Auburn	0.1875a	0.1308a	0.0050b
KY-31	0.0510a	0.0238b	0.0046b
Clover	0.1519a	0.1179a	0.2606a

[a] Means in same column followed by the same letter are not significantly different at $p = 0.05$.

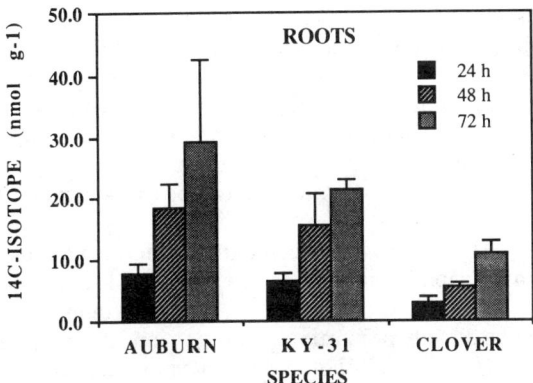

FIG. 2—*Uptake of ^{14}C-isotope as a function of exposure time by roots of whole plants of clover and two varieties of fescue.*

pattern of translocation of radiolabel to the shoots is similar to the pattern of uptake of ^{14}C-naphthol by the roots. The data indicate that the amount of radiolabel in the shoots is cumulative with respect to exposure time. The accumulation profiles for the two varieties of fescue and clover are similar to those described for Asulam (methylsulfanilcarbamate) in bean and maize [8]. Interestingly, Asulam is described as being relatively lipophobic, reducing the likelihood of its accumulation in the lipid fractions of plant membranes, yet it accumulated to the greatest extent in the roots with less than 10% translocated to shoots after 96 h.

The relative distribution of a xenobiotic chemical within a whole plant is the result of a complex set of interactions involving the physical/chemical properties of the compound and the plant species in question. Martin and Edgington [16] found major differences in patterns of translocation between plant genera. Three xenobiotics having octanol/water partition coefficients (log P) ranging from 0.67 to -0.48 exhibited primarily apoplastic transport in barley and an ambimobile transport pattern when applied to soybean. They reported that compounds with larger octanol/water partition coefficients (i.e., compounds only slightly water

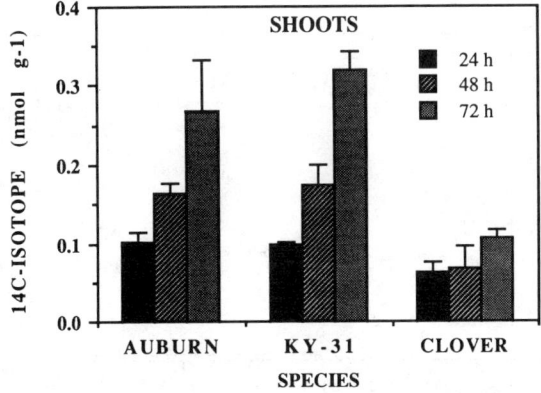

FIG. 3—*Accumulation of ^{14}C-isotope given as naphthol, with increasing incubation time in the shoots of clover and two varieties of fescue.*

soluble) may be transported in the symplast due primarily to their lipid solubility. Because our experimental approach involved only root application of chemical over relatively short time intervals (in this experiment 24 h) the relative importance of apoplastic, symplastic, or ambimobile transport could not be determined. Naphthol is slightly water soluble with a log P of 0.45 [30], putting it in the range of chemicals with relatively high lipid solubility thought to primarily utilize symplastic transport. However, it was apparent that naphthol or its metabolites are not fixed at its point of entry and can be transported to the aerial portion of the test species.

Apparent free space (AFS) values have been estimated to range from 10 to 28% [20-22]. Given that the amount of naphthol in the AFS is a function of AFS volume and the concentration of chemical in the bathing solution when at or near equilibrium with the roots, the amount eluted should be directly proportional to the chemical concentration in the bathing solution. Increasing levels of ^{14}C-naphthol in the incubation solution (0.1, 0.2, and 0.3 μM) resulted in increased amounts of radiolabel eluted during the first minute (11, 21, and 25 pmol g^{-1} fresh root weight respectively). Calculations of AFS for KY-31 based upon the three incubation concentrations were 10, 9, and 11% respectively. Verification that the amount eluted during the first minute respresents the chemical contained within the AFS was calculated using a simple mean square distance formula which states that $x^2 = 2Dt$, where x = distance (ca. 0.01 mm between epidermis and endodermis (cortex)), t = time, and D = diffusion coefficient (ca. 10^{-6} cm s^{-1} for naphthol). It was calculated that naphthol could diffuse from the AFS into the external solution in less than 1 min. These results coincide with the findings of Shone et al. [18], who noted a rapid loss to water (less than 1 min) of triazines from roots of intact barley plants.

Figure 4 illustrates the elution pattern for KY-31 fescue and clover roots that were previously incubated in a 0.1 μM ^{14}C-naphthol solution for 24 h. There is a rapid efflux of radiolabel during the first minute for both species, 8.3 pmol g^{-1} for fescue and 3.8 pmol g^{-1} for clover. The total amount of radiolabel eluted during this initial release accounted for less than 1% of the total amount of radiolabel contained in the root tissue. This initial efflux was followed by a sharp reduction in the rate of radiolabel eluted, which also mirrored the findings of Shone et al. [18] for the triazines. This gradual rate of radiolabel release from the root tissue remained relatively constant over the 2 h duration of the experiment, resulting in total release of ^{14}C-isotope that was still less than 1% of the total amount of radiolabel contained

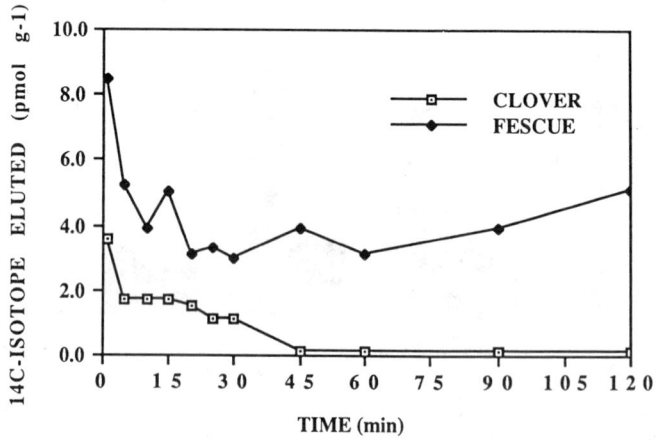

FIG. 4—*Elution of ^{14}C-isotope given as naphthol from isolated roots of clover and fescue.*

in the root tissue. These results are in contrast with the findings for some other xenobiotics [13,31,32] which show rapid elution of most of the chemicals tested. The much reduced rate of ^{14}C-isotope loss observed here may be attributed to several factors, one of which is naphthol's lipophilic nature that should allow it to readily partition into membranes. In addition, naphthol may be rapidly metabolized upon uptake to substances that are tightly bound to cellular constituents.

HPLC analysis of naphthol in the incubation vessel containing KNS without root tissue, at the beginning and end of the 24-h incubation period, indicated no loss or transformation of naphthol. The extraction procedure for the metabolite study, using methanol solvent and a glass tissue grinder, was contrasted with the use of a liquid tissue digester as to the efficiency of naphthol extraction from root tissue. Based on the stability of naphthol in the test system and the ca. 90% naphthol recovery rate from root tissue digested in Unisol, naphthal recovery from root tissue using methanol and grinder was 14% for clover and 17% for fescue. Edwards et al. [7] reported a 8 to 32% recovery rate of ^{14}C-anthracene from soybean using an acetone extraction, while Briggs et al. [6] reported an 85% recovery rate for nonionized chemicals from acetone extracted barley roots and shoots.

Most of the naphthol absorbed by the root was not present in root tissue as extractable parent compound. For KY-31 fescue roots, of the 390 nmol naphthol lost from the 10 μM solution, only 3.2 nmol remained as recoverable parent compound in the root tissue, indicating that over 99% of the chemical taken up into the root tissue was present as something other than extractable naphthol. In the 100 μM naphthol incubation solutions, clover and fescue roots were determined to contain 275 nmol of naphthol, or nearly 15% of the naphthol removed from solution by the roots. Naphthol, reported to be a metabolic conversion product of naphthalene [27,33], apparently does not accumulate to any great extent in KY-31 fescue and clover. Based upon naphthol recovery rates attained after 24 h incubation, it is reasonable to assume that extensive binding precluding extraction under reasonably strong conditions or catabolic loss of the parent compound has occurred in both fescue and clover. Edwards et al. [7] reported that soybeans could catabolize the xenobiotic anthracene to CO_2 and other undetermined metabolites.

Conclusions

We have studied the uptake, bioaccumulation, translocation, and parent compound retention of ^{14}C-naphthol by using isolated roots and whole plants of two varieties of fescue and one variety of clover, both species useful as primary cover on industrial and municipal waste disposal sites. The data clearly indicate that ^{14}C-naphthol is taken up by whole plant roots of these species at levels directly proportional to naphthol concentration in the range of 0.1 to 0.3 μM. It was determined that although less than 1% of the radiolabel accumulated in the roots was translocated to the shoots even after 72 h, there was a direct increase in radiolabel accumulation (i.e., parent compound and metabolites) in the shoots with increasing exposure time from 24 to 72 h. Exposure of whole plants to increasing levels of ^{14}C-naphthol in solution (0.1 to 0.3 μM) for 24 h resulted in a declining TSCF for both varieties of fescue, indicating saturation of the transport system at higher chemical exposure levels.

Our results also clearly indicate that although naphthol is taken up by the roots of these species, it is not retained primarily as parent compound but as tightly bound species or metabolites. The efflux of ^{14}C-isotope from isolated root tissue indicated that the AFS of both fescue and clover roots was approximately 10% of the total root volume. Results of elution studies indicated that efflux of radiolabel chemical (assumed to be naphthol) from the AFS occurred in less than 1 min. Calculations using the mean square distance formula verified this observation.

The implications of these data show that the test species can physically and chemically interact with the xenobiotic naphthol and therefore, as with anthracene and soybeans [7], provide naphthol an entrance into the terrestrial foodchain. These data also suggest that these cover species may prove useful in helping to reduce ambient levels of xenobiotics through catabolic metabolism.

References

[1] Schwarz, O. J. and Eisele, G. R. in *Proceedings, Fifth Life Sciences Symposium*, Butterworth, Boston, 1982, pp. 441-462.
[2] Crafts, A. S. and Yamaguchi, S., *Am. J. Bot.*, Vol. 47, 1960, pp. 248-255.
[3] Shone, M. G. T. and Wood, A. V., *J. Exp. Bot.*, Vol. 23, 1972, pp. 141-151.
[4] Shone, M. G. T., Bartlett, B. O., and Wood, A. V., *J. Exp. Bot.*, Vol. 25, 1974, pp. 390-400.
[5] Shone, M. G. T., Bartlett, B. O., and Wood, A. V., *J. Exp. Bot.*, Vol. 25, 1974, pp. 401-409.
[6] Briggs, G. E., Bromilow, R. H., and Evans, A. A., *Pest. Sci.*, Vol. 13, 1982, pp. 495-504.
[7] Edwards, N. T., Ross-Todd, B. M., and Garver, E. G., *Environ. Exp. Bot.*, Vol. 22, 1982, pp. 349-357.
[8] Gortz, J. H. and Van Oorschot, J. L. P., *Pest. Biochem. and Physiol.*, Vol. 21, 1984, pp. 45-52.
[9] Rigitano, R. L. O., Briggs, G. G., Bromilow, R. H., and Chamberlain, K., *Movement of Pesticides in Plants, Physicochemical and Biophysical Panel Symposium*, 1986, pp. 62-63.
[10] Donaldson, T. W., Bayer, D.E., and Leonard, O. A., *Plant. Physiol.*, Vol. 52, 1973, pp. 638-645.
[11] Leroux, P. and Gredt, M., *Pest. Biochem. and Physiol.*, Vol. 5, 1975, pp. 507-514.
[12] Lichtner, F. T., *Plant Physiol.*, Vol. 71, 1983, pp. 307-312.
[13] Devine, M. D., Bestman, H. D., and Vanden Born, W.H., *Plant Physiol.*, Vol. 85, 1987, pp. 82-86.
[14] McFarlane, J. C. and Wickliff, C., *Envir. Monit. and Assess.*, Vol. 5, 1985, pp. 385-391.
[15] McFarlane, J. C., Nolt, C., Wickliff, C., Pfleeger, T., Shimabuku, R., and McDowel, M., *Environ. Toxic. and Chem.*, Vol. 6, 1987, pp. 847-856.
[16] Martin, R. A. and Edgington, L. V., *Pest. Biochem. and Physiol.*, Vol. 16, 1981, pp. 87-96.
[17] Barrett, M. and Ashton, F. M., *Weed Sci.*, Vol. 31, 1983, pp. 43-48.
[18] Shone, M.G.T., Clarkson, D.T., Sanderson, J., and Wood, A. V., *Ion Transport in Plants*, Academic Press, London, 1973, pp. 571-582.
[19] Bidwell, R. G. S., *Plant Physiology*, Macmillan, New York, 1974.
[20] Briggs, G. E., Hope, A. B., and Robertson, R. N., *Electrolytes and Plant Cells*, Wiley, New York, 1961.
[21] Butler, G., *Physiol. Plantarum*, Vol. 6, 1953, pp. 617-635.
[22] Epstich, E., *Plant Physiol.*, Vol. 30, 1955, pp. 529-535.
[23] Krstich, M. A. and Schwarz, O.J., *Environ. Monit. and Assess.* (in press)
[24] Lamoureux, G. L. and Rusness, D. G., *Sulfur in Pesticide Action and Metabolism*, American Chemical Society, 1980, pp. 133-164.
[25] Cole, D., *Progress in Pesticide Biochemistry and Toxicology*, Vol. 3, 1983, pp. 199-243.
[26] Shimabukuro, R. H. and Walsh, W. C., *American Chemical Society Symposium Series*, Vol. 97, 1979, pp. 3-34.
[27] Hofmann, K. H., *J. Basic Microbiol.*, Vol. 26, 1986, pp. 109-111.
[28] Nitsch, J. P., *Amer. J. Bot.*, Vol. 38, 1951, pp. 566-577.
[29] SAS Institute Inc., *SAS User's Guide: Statistics*, Cary, N.C., 1985.
[30] Eadsforth, C. V., *Pest. Sci.*, Vol. 17, 1986, pp. 311-325.
[31] Moody, K., Kust, C. A., and Buchholtz, K. P., *Weed Sci.*, Vol. 18, 1970, pp. 214-218.
[32] Orwick, P. L. and Schreiber, M. M., *Weed Res.*, Vol. 16, 1976, pp. 139-144.
[33] Durmishidze, S. V., Chrikishvili, D. I., Beriashvili, T. V., Maisuradze, Ts. M., and Gugunishvili, B. Sh., *Prikladnaya Biokhimiya i Mikrobiologiya.*, Vol. 21, 1985, pp. 395-400.

F. Ribeyre[1] *and A. Boudou*[1]

Bioaccumulation of Mercury Compounds in Two Aquatic Plants (*Elodea densa* and *Ludwigia natans*): Actions and Interactions of Four Abiotic Factors

REFERENCE: Ribeyre, F. and Boudou, A., "**Bioaccumulation of Mercury Compounds in Two Aquatic Plants (*Elodea densa* and *Ludwigia natans*): Actions and Interactions of Four Abiotic Factors,**" *Plants for Toxicity Assessment, ASTM STP 1091*, W. Wang, J. W. Gorsuch, and W. R. Lower, Eds., American Society for Testing and Materials, Philadelphia, 1990, pp. 97–113.

ABSTRACT: The research presented is based on an experimental study of bioaccumulation of two mercury compounds ($HgCl_2$ and CH_3HgCl) by two species of rooted macrophytes (*Elodea densa* and *Ludwigia natans*). Contamination of the experimental model (water, natural sediment, and macrophytes) is carried out by introducing contaminants into the water column. The actions and interactions of four abiotic factors (temperature, pH, photoperiod, and light intensity) are studied by setting up a complete experimental design corresponding to 72 different experimental conditions. Results relate to the evolution of mercury concentrations in the water over 28 days' exposure, plant growth (weight and stem length), and quantification of total mercury bioaccumulation (concentration and content) in the whole plants.

KEY WORDS: aquatic ecotoxicology, bioaccumulation, mercury compounds, rooted macrophytes, *Elodea densa, Ludwigia natans*

Research carried out in our laboratory studies bioaccumulation and transfers of mercury compounds within continental aquatic systems. Two chemical forms of the metal are compared: an inorganic form (mercuric chloride [$HgCl_2$]) and an organic form (methylmercury chloride [CH_3HgCl]).

From appropriate methodologies, we selected ecotoxicological models as multicompartment systems. They come between monospecific approaches, carried out in the laboratory and requiring a considerable degree of reductionism, and *"in situ"* studies, on the ecosystem or ecocomplex scale. Ecotoxicological models present a level of complexity which is closely dependent on their biological component—for instance, the diversity or the number of trophic links between the different species [1,2].

Our research program has been based on two main stages. The first yielded a linear model of trophic transfers (i.e., an experimental trophic chain) [3,4]. Since 1982, the second stage has been based on the development of an interactive model (experimental ecosystem), taking into account several factors relating to the three fundamentals of ecotoxicology: abiotic, biotic, and contamination factors [5]. In establishing our methodology, the study of interactions was the most important consideration and largely determined the structure of the model. In the first step, priority was given to analyzing actions and interactions of abiotic and contamination factors rather than to the biological component of the model. Interac-

[1] Research Scientist and Professor in Ecotoxicology, respectively, Fundamental Ecology and Ecotoxicology Laboratory, U.R.A. CNRS 1356, Bordeaux I University, 33405 Talence Cedex, France.

tions can be defined as the nonadditivity of factor effects; they express synergy or antagonism between factors. The model consists of a three-compartment system: water, natural sediment, and rooted macrophytes. We analyzed the processes of bioaccumulation and transfer of mercury compounds from two contamination sources (water and sediment). Three abiotic factors were initially selected: temperature, daily period of light or photoperiod, and pH of the medium. A complementary factor, light intensity, was also considered in relation to the structure of the equipment.

In this paper we present results relating to the study of mercury accumulation by two species of macrophytes: *Elodea densa* and *Ludwigia natans;* the initial contamination derived from the water source.

Material and Methods

Structure of Ecotoxicological Model

Many preliminary experiments were carried out to determine precisely the characteristics of each of the three compartments of the system. Various trials were carried out using different sediments with contrasting structural and geochemical properties: sands, pure clays, natural sediments, and so on. We selected a sediment from a research station on the banks of the Garonne River, upstream from Bordeaux. Geochemical and granulometric features were determined by the Institut de Géologie du Bassin d'Aquitaine (Bordeaux I University). The sediment was a very homogeneous silt, rich in clays (70 to 80%) and with a low level of organic carbon particles (2%). Water content was about 50% of fresh weight. The natural level of total mercury was 0.124 ± 0.012 mgHg kg^{-1} (fresh weight). Similar comparative studies were carried out for the water compartment, as a result of which we selected tap water from ground water supplies, which was first dechlorinated by aeration.

The choice of the biological component had to be made in relation to the overall aims of our research and had to fulfill two major requirements: (1) it had to establish links between the two compartments of the biotope (water and sediment); and (2) it had to belong to the same trophic level as producers so that at a later stage it would be possible to set up alimentary transfers between this level and a primary consumer. We turned to rooted macrophytes, which are in direct contact with the sediment via their root system and with the aquatic phase through their upper part (stems, leaves, buds and, in some cases, adventive roots) [6]. In a natural environment these plants would be a food source for many herbivorous species such as molluscs or fishes. Several trials were carried out to determine which plant species best matched our requirements. The main features that we considered were morphometric characteristics, ability to adapt to abiotic conditions, method of implantation (propagation by cutting preferred), growth rate, homogeneity of batches, and the ease with which they can be obtained (quantity, quality, availability, etc.) [7]. Two species were selected: *Elodea densa* and *Ludwigia natans*. The former had a rapid growth rate, was particularly sensitive to the physicochemical conditions of the environment, and had leaves suitable as a trophic source for herbivores, especially fish. The latter had a slow growth rate and was very robust and easy to dissect.

The basic structure of the model is the experimental unit (EU) consisting of a glass tank (24 by 24 by 30 cm) containing 3.5 kg of fresh sediment (5 cm in depth) and 11.5 L of water (18.5 cm in height). In each unit ten *Elodea densa* and four *Ludwigia natans* cuttings were planted (apical parts of main stems); each was 12 cm long, on average.

Experimental Design and Specific Equipment

A complete experimental design was drawn up, including all combinations of parameter levels. Three levels were decided on for the temperature and photoperiod: 16, 20, and 24°C and 8, 12, and 16 h of light per day. The upper and lower values were selected in relation to the ecophysiological tolerance of the two species studied, *Elodea densa* in particular. By using three levels, any nonlinearity in responses could be revealed. In the case of the pH, only two levels were used, one corresponding to the pH of the nonregulated experimental medium (8.0 ± 0.2), the second to an acidic condition (6.0 ± 0.2).

To control the abiotic factors, specific equipment has to be set up. Experimental units are placed in larger tanks (140 by 65 by 30 cm), which are themselves in enclosed containers. Each tank has thermoregulation equipment, which is very efficient due to the large volume of water that is constantly being stirred by submerged pumps (±0.2°C). In this way contact between the regulation system and the medium in the EUs is avoided. Light is produced by two neon tubes (Sylvania F36W/GRO) positioned at 45 cm from the surface of the experimental units and operated by a timer switch. Light intensity at the surface was between 1700 lx (edge of larger tank—low intensity level) and 2400 lx (center of larger tank—high intensity level) (Quantum Sensor LI190SB). The pH regulation of the water column is a complex matter, mainly because of the presence of natural sediment and organisms. We decided to use the technique of controlled additions of acid solutions (H_2SO_4) using an automatized regulation system [8].

The resulting experiment produced 72 different experimental conditions. Three exposure periods were selected: 15 and 28 days for *Elodea* and 20 days for *Ludwigia*.

Contamination of Experimental System

In any experimental procedure in ecotoxicology it is of paramount importance that the contamination methods should be properly controlled, because they determine the results obtained and hence affect the validity of conclusions. In our approach, where a large number of experimental units must be carried out simultaneously, the method best suited to our requirements was cyclical additions of constant amounts of mercury. Mercury compounds were introduced into the aquatic phase by means of twice-daily additions (morning and evening) of 5 mL mercury solution at 0.27 mgHg L^{-1} ($HgCl_2$ or CH_3HgCl). Mercury concentrations in the water tend to diminish rapidly after each addition due to metal losses through adsorption to the tank walls, water-sediment and water-organism transfers, and volatilization. The amount of this decrease is influenced by the experimental conditions pertaining in each unit—for example, temperature, pH, plant biomass, and growth rate of buds. Therefore mercury concentration in the water column was measured frequently in order to monitor its evolution throughout the experiment.

It is fairly straightforward to show up relationships between different abiotic conditions and accumulation in the organisms via water source if the metal concentration in the water is constant throughout the exposure period; when this concentration fluctuates, it is much more difficult. From the various approaches to considering these phenomena in order to correct differences in bioaccumulation due to these differences in water source contamination levels for the different experimental conditions, we selected an index representing contamination pressure in each EU and its evolution with time. To calculate this index, mercury concentrations in the water are measured at intervals. For each interval of time between two measurements, we calculate the product of "average mercury concentration during the period considered" × "corresponding time interval". Hence, for a given experimental period (e.g., 28 days), the index is the sum of previous products and is expressed as "concentration × days" equivalent (μgHg $L^{-1}\cdot$d or ppb·d). In order to compare bioaccumulation in the 72 experimental conditions, we first consider mercury concentrations or contents measured in the plant; secondly, in order to have as a basis for calculation the same contamination pressure in each EU, measurements are derived from the corresponding values from the index. Although this estimate is based strictly on hypotheses of proportionality and linearity in relation to mercury accumulated by the plants, we can proceed to analyses that complement those carried out directly on the measurements taken.

Analysis of Experimental System

The morphometric criteria selected for this experiment to characterize each plant and its growth are fresh weight (whole plant) and overall length of stems. The ratio between fresh weight and dry weight is 10% in *Elodea densa* and 13% in *Ludwigia natans*.

Quantification of mercury bioaccumulation in the whole plant is based on two complementary criteria: concentration (ngHg g^{-1}/fresh weight) and content or burden (ng Hg). From these two criteria it is possible to express in particular the effects of growth dilution on the metal accumulated in the plants.

Total mercury determination was carried out by atomic absorption spectrometry without flame (spectrophotometer Varian AA475). Sediment or macrophyte samples were first mineralized by a nitric acid treatment (pure HNO_3) in a pressurized medium (Pyrex glass tube) at 95°C for 3 h. A treatment with bromine salts was applied before addition of stannous chloride [9]. The detection limit is 5 ng Hg. The validity of the analytical method was checked periodically using determination standards (National Bureau of Standards, Washington; International Atomic Energy Agency, Monaco; Kernforschunganlage, Jülich) and intercalibration exercises on the same matrix (i.e., sediment and macrophytes) with other laboratories.

Data treatment was carried out following progressive steps from exhaustive information to synthetic conclusions. In order to quantify the actions of the different controlled factors and their interactions we based our calculations on multiple regression methods [10-12]. The use of orthogonal polynomials simplifies interpretation of the effects of each regressor because of the independence of the regression coefficients [13].[2] As a general rule we adopt an alpha risk equal to 0.01 in order to discuss the statistical significance of the effects observed; F values are calculated with reference to the residual terms that appear after ordination of the corresponding contributions. For the explained variables, no transformation was carried out on the morphometric criteria (weight and length); for the accumulation criteria (mercury concentration and content in the plants), however, logarithmic transformation was used.

Results and Discussion

Characteristics of Contamination Source, Water Source

Figure 1 shows the evolution of mercury concentration in the aquatic phase over the 28-day period for the 36 experimental conditions corresponding to the three main abiotic factors and to the two chemical forms of mercury. The controlled factors have a considerable influence on mercury concentration in the water. Thus when we compare the two extreme results (i.e., from the experimental conditions "16°C, 16 h light per day, pH 6.0, $HgCl_2$" and "24°C, 8 h light per day, pH 8.0, CH_3HgCl"), we find a ratio of 7.3. Several observations can be made from more detailed analysis of these graphs:

1. Mercury concentrations measured in the water were higher in the units contaminated by the inorganic compound, with differences increasing as the temperature rose. For exam-

---

[2] Regressor coding is as follows: Temperature: 16°C (-1), 20°C (0), 24°C ($+1$); Photoperiod: 8 h (-1), 12 h (0), 16 h ($+1$); Light Intensity: low level (-1), high level ($+1$); pH: 6.0 (-1), 8.0 ($+1$); Hg Chemical Form: $HgCl_2$ (-1), CH_3HgCl ($+1$).

ple, the ratio of average concentrations measured in EUs contaminated by $HgCl_2$ and by CH_3HgCl is 1.26 at 16°C and 2.14 at 24°C for the experimental conditions of 8 h light per day, pH 6.0, and 28 days' exposure.

2. The evolution trends of mercury concentrations in the water are fairly close to linearity, although this flattens into a plateau after about 20 days. This stabilization is more marked when the inorganic form of mercury is used and when the temperature is high.

3. An increase in temperature generally gives rise to a decrease in mercury concentration (C) in the water, which is more apparent with the organic form: $C_{24°C}/C_{16°C}$ = 1.5, approximately, for $HgCl_2$ and 2.6 for CH_3HgCl.

4. Mercury concentrations in the water column are generally higher when pH is acid. The differences are more obvious with the organic compound, and much more so when the temperature is high.

5. The photoperiod affects results much less than the other factors considered; its influence is seen most clearly at 16°C, with the highest concentrations occurring with a photoperiod of 16 h per day.

In conjunction with this analysis of the evolution of mercury concentration in the water column, which was measured periodically in each experimental unit, we defined the contamination source using the index "concentration × days" equivalents, calculated for the three observation periods of 15, 20, and 28 days. As an example, we show the influence of the three main abiotic parameters (temperature, photoperiod, and pH) on the contamination level of the water source, expressed as $\mu gHg\ L^{-1}$ × days, after 28 days' exposure (Fig. 2). Like the graph showing the evolution of measured concentrations with time, Fig. 2 shows the wide differences between experimental conditions. We see again that the chemical form of mercury has considerable influence. Differences in the values of "concentration × days" equivalents for the two pH conditions increased with temperature, particularly so in the case of $HgCl_2$. Photoperiod has little effect on the contamination pressure expressed by this index. When considering all readings taken throughout the experiment, the values of this index present a more representative view of the contamination source than do occasional readings of mercury concentration in the water, at the end of the experiment, for example.

The major differences observed between experimental conditions are related to various processes that combine to decrease mercury concentration in the water column of each experimental unit: metal adsorption on the tank walls; water-sediment transfers; mercury volatilization, increased because of abiotic or biotic chemical transformations of the mercury compounds (e.g., production of metal form or dimethylmercury); and bioaccumulation by the plants. The degree of influence of these different processes depends largely on physicochemical conditions in the medium, but also on metal chemical forms introduced into the water column (solubility, binding capacity with inorganic and organic ligands, etc.) and the anatomical and physiological characteristics of the macrophytes (surface contact with the surrounding environment, metabolic exchanges, etc.). All these elements are interactive and move on with time, creating specificities that are significant in varying degrees in relation to the contamination source for each of the 72 experimental conditions.

Mercury accumulation by the plants is very much dependent on the level of contamination in the water source and its evolution. Thus the differences between metal concentrations observed in the water may reduce, or increase, depending on the particular case of mercury transfer between the water column and the macrophytes. This can be most clearly seen in the case of the two extreme conditions (16°C, 16h/24h, pH 6.0, $HgCl_2$ and 24°C, 8h/24h, pH 8.0, CH_3HgCl) where the values of the "concentration × days" equivalents are 52.5 and 6.8 $\mu gHg\ L^{-1}$ × days, respectively. Thus, as we saw earlier, the interpretation of results relating to mercury bioaccumulation by the two species of macrophytes will be based on measured

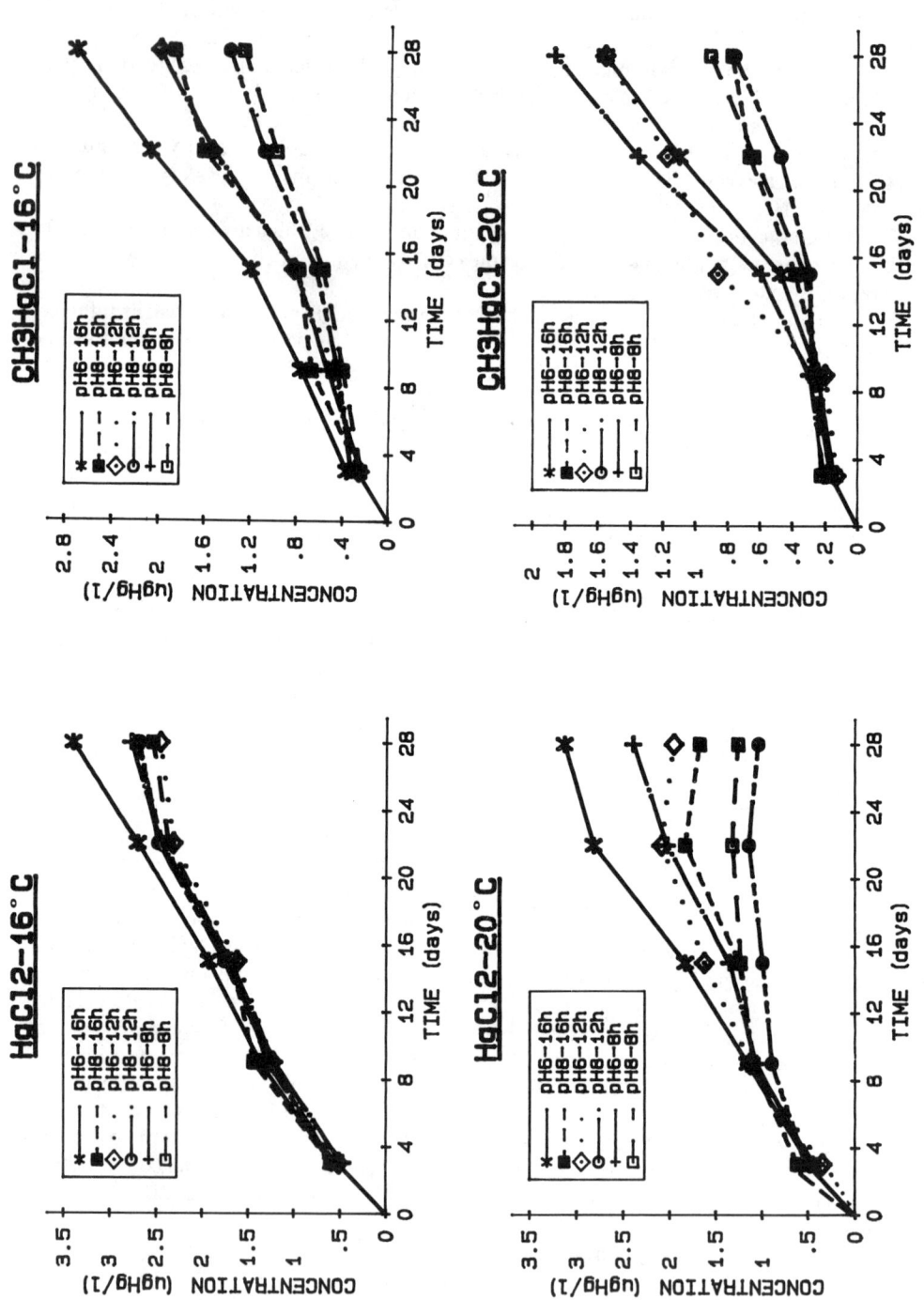

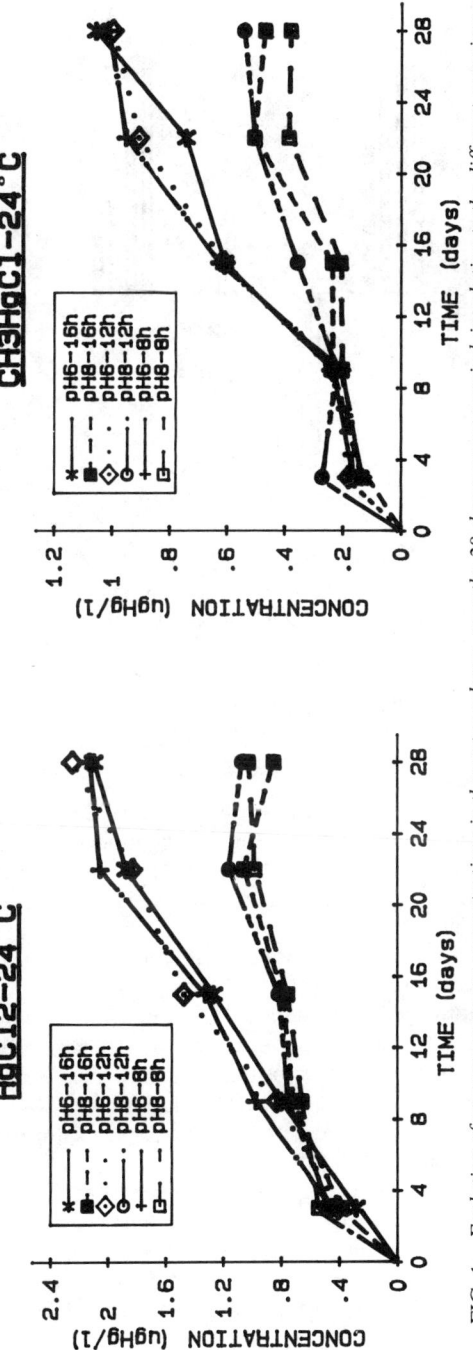

FIG. 1—*Evolution of average mercury concentrations in the water column, over the 28-day exposure period, in relation to the different experimental conditions studied.*

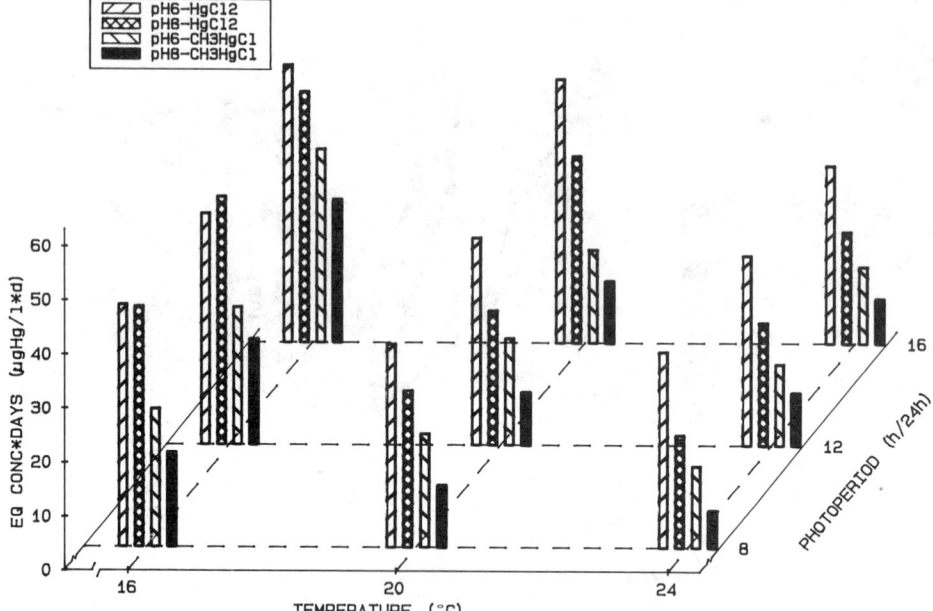

FIG. 2—*Influence of three abiotic factors (temperature, photoperiod, and pH) on the "concentration × days" equivalents calculated for the water source after 28 days' exposure.*

concentrations and contents, and also on corrected values for these two criteria using the index "concentration × days" equivalents.

Plant Growth Analysis, Length and Weight Criteria

The total length values for *Elodea densa*, measured after 28 days' exposure, and the corresponding regressing model, are shown in Fig. 3. The model includes seven terms; despite a very great reduction in the number of regressors (71 terms initially), it represents 87% of all dispersion measured.

Increases in temperature, photoperiod, and light intensity all give rise to an increase in stem length. The pH, on the other hand, has a negative action on the growth of *Elodea densa*, with stem length being significantly greater with pH 6.0. This negative action increases with an increase in temperature (term of interaction in the regression model: "temperature × pH"). It is important to note that the chemical form of mercury has no significant effect on the length of the plants after 28 days' exposure. We should mention that studies carried out in similar conditions, with different contamination levels in the water source, did not show any growth inhibition when compared with the control plants, for similar mercury concentrations in the water to those used in this experiment [7].

Analysis of the total fresh weight criterion led to different conclusions regarding the effect of the five ecotoxicological parameters on plant growth (Fig. 4). Thus, after 28 days, temperature has no significant effect on this criterion. When photoperiod and light intensity are increased, this leads to a weight increase in the plants and this action predominates in relation to that of the pH, which remains negative. No significant interaction between factors was observed.

For the second species, *Ludwigia natans*, there were no major differences between the two growth criteria studied: length and weight (data not shown). This species is less sensitive than *Elodea densa* to the effects of the abiotic factors studied. Indeed, growth was not significantly

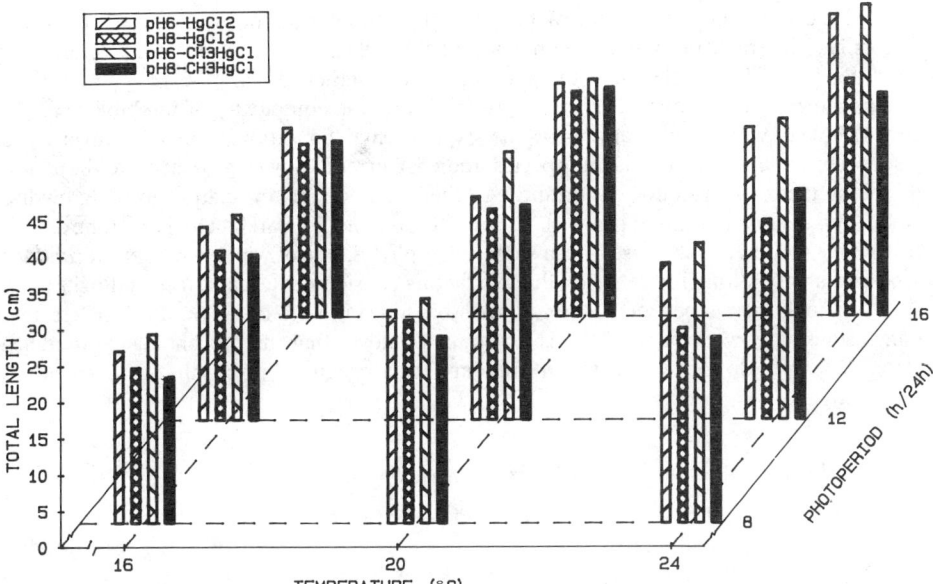

FIG. 3—*Average total length of* Elodea densa *cuttings after 28 days' exposure to the two mercury compounds ($HgCl_2$ and CH_3HgCl–water source) in relation to the 36 abiotic conditions considered. Regression model:* Length (cm) = $29.875 + 5.165$ temp. $- 2.865$ pH $- 1.907$ temp. $\times$ pH $+ 1.745$ photo. $+ 1.235$ light int. $+ 1.214$ temp. $\times$ light int. $- 0.689$ temp.$^2 \times$ pH.

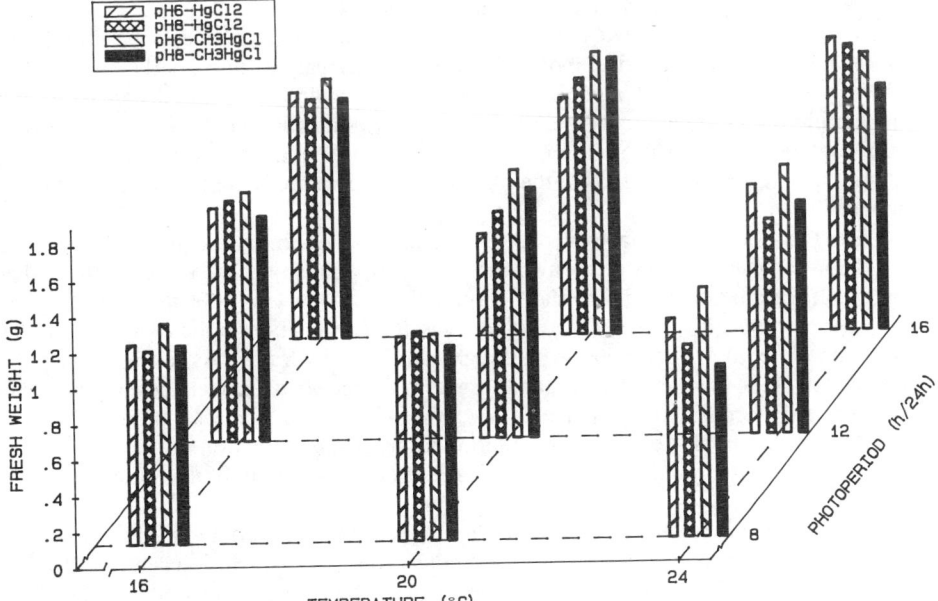

FIG. 4—*Average total fresh weight of* Elodea densa *after 28 days' exposure to the two mercury compounds ($HgCl_2$ and CH_3HgCl–water source) in relation to the 36 abiotic conditions considered. Regression model:* Fresh weight (g) = $1.311 + 0.158$ photo. $+ 0.081$ light int. $- 0.043$ pH.

affected except by the pH (negative action) and the photoperiod (positive action). These two factors also interact, producing an increase in the differences in weight readings between conditions pH 6.0 and pH 8.0 as the photoperiod is lengthened.

This morphometric analysis of plant growth shows the complexity of the biological processes, linked with modifications in the experimental conditions, and the diversity of responses in relation to criteria and species studied. Current knowledge relating to the actions of abiotic factors and, more importantly, to their interactions, in relation to the growth of aquatic plants is very limited [14–17]. The results obtained by Barko et al. [18], for example, from three species of submersed freshwater macrophytes, one of which was *Elodea canadensis,* show a very strong influence by the two factors considered: temperature and light intensity. Although their experimental conditions were not similar to our own, nevertheless, their conclusions agree with ours. There is no doubt whatever that the morphological modifications observed do have effects on mercury bioaccumulation by the two species of macrophytes.

Mercury Bioaccumulation by the Two Macrophyte Species

Analysis of total mercury concentrations measured in *Elodea densa* after 28 days' exposure and the corresponding regression model (Fig. 5) lead us to the following conclusions:

1. Mercury bioaccumulation by the plants is closely dependent on the chemical form of mercury initially introduced into the water column: concentrations in the whole plant are about twice as high when methylmercury is added. These differences, however, are much lower than those observed after contamination via the sediment source. For similar experimental conditions, readings can be almost 50, again in favor of the organic compound [19,20].

2. In most cases, mercury concentration in the plants is higher when pH is 8.0; the ratio between the extremes of measured concentrations ($C_{pH\ 8.0}/C_{pH\ 6.0}$) is almost 2.

3. An increase in temperature gives rise to a decrease in mercury concentrations in *Elodea densa* when metal was added in inorganic form; this decrease is more marked when pH is 6.0 ("temperature $\times$ pH interaction"). In the case of methylmercury, this decrease linked to temperature is less pronounced, and even reversed with alkaline pH ("pH $\times$ Hg chemical form" and "temperature $\times$ Hg chemical form" interactions). In every case, the relationship between mercury concentrations in the plants and temperature is nonlinear (square term in the models).

4. The two other abiotic factors (photoperiod and light intensity) have a negative action in relation to accumulation of the two mercury compounds by *Elodea densa:* the highest concentrations are associated with short photoperiods and low levels of light intensity. However, these actions are relatively weak when considered in relation to the effect of other factors (ratio of about 1.2 between low and high light intensity levels). As can be seen from a comparison of the measured results and those from the regression model, these actions could be completed by considering supplementary terms within the model that are not significant to the alpha threshold = 0.01 (squared term and terms of interaction of second and third orders). In this way it would be possible to make a better assessment of the influence of these two factors, especially photoperiod (e.g., Fig. 5*b*).

The influence of the abiotic factors on total mercury concentrations is less significant in *Ludwigia natans* than in *Elodea densa* (data not shown). This conclusion agrees with observations made when comparing growth in these two species. The photoperiod and light intensity have no significant action. In the regression model selected, only two factors intervene as principal actions: mercury chemical form and pH. The temperature effect can only be seen in association with the two other factors mentioned above (first and second order interactions).

Mercury concentrations in the whole plant, calculated from the "concentration × days" equivalents, enable us to complement the observations above on *Ludwigia natans* and *Elodea densa* after 20 and 28 days' exposure (Fig. 6):

1. The action of the chemical form of mercury is very marked when contamination pressure is considered to be similar for all experimental conditions; hence the ratios C_{CH_3HgCl}/C_{HgCl_2} are between 5 and 8, according to temperature.

2. The effect of temperature is very much modified, however, and even inversed. For both plant species, an increase in temperature gives rise to an increase in mercury concentration in the plants; this increase is greater when the water column is contaminated with the organic form of mercury ("Hg chemical form × temperature" interaction). The relationship between this factor and bioaccumulation rates is very nearly linear.

3. pH always has a strong influence, giving rise to higher contamination levels in basic environment; it interacts in a positive way with temperature and chemical form of mercury, and this is confirmed and amplified with *Elodea densa*.

4. Photoperiod has little effect on bioaccumulation and, as before, has only a negative action on *Elodea densa*. The light intensity effect is not significant, in contrast with conclusions drawn from measured concentrations.

Mercury content, the second criterion selected to study processes of mercury bioaccumulation, represents from an ecotoxicological standpoint the metal quantities that are potentially transferable into other compartments of aquatic systems, especially the herbivore species. We should remember that concentrations can be considered as an indicator of the toxicological risk to the organism or organ considered [1]. An analysis that combines concentrations and contents is fully justified, because organism growth is important during the experiment and, in our case, is influenced by the different experimental factors.

To illustrate this we have selected three examples (data not shown):

1. An increase in pH gives rise to an increase in mercury content in *Elodea densa* and *Ludwigia natans*. This action, however, is much less than that observed in relation to concentrations. This is directly linked with the lowest level of weight increase in the plants during the experiment, at pH 8.0.

2. Temperature does not significantly influence plant weight, so mercury contents and concentrations are similarly influenced by this factor. The increase in bioaccumulation in relation to an increase in temperature (Fig. 6) can thus not be explained directly by the weight criterion. This increase can, however, be linked with an increase in stem length (Fig. 3).

3. The effects of the third abiotic factor, photoperiod, show the oppositions that can be seen between the two accumulation criteria. An increase in photoperiod leads to an increase in mercury content in the plants, and because of the positive action of this factor on weight, there is a decrease in corresponding concentrations.

In our experimental conditions, the bioavailability of mercury compounds in the water column and, secondly, in the sediments, governs the metal uptake by the aquatic plants. Bioavailability is the result of a combination of many phenomena: characteristics of the biotopes (physicochemical factors, importance of suspended matter, geochemical properties of sediments, etc.) and specific features of contamination conditions (inorganic and organic forms of mercury, chemical species, biotransformation processes, etc.) [21–24].

The role of the interfaces between plants and water column, principally leaf surfaces, is also of great importance in relation to mercury fixation on the biological barriers (cell wall, plasmalemma) and to metal absorption. Several studies, essentially based on membrane models, have shown the importance of the physicochemical characteristics of the environment and the importance also of the chemical forms and species of mercury on the binding of the metal onto these cell structures and trans-membrane exchanges [25–27].

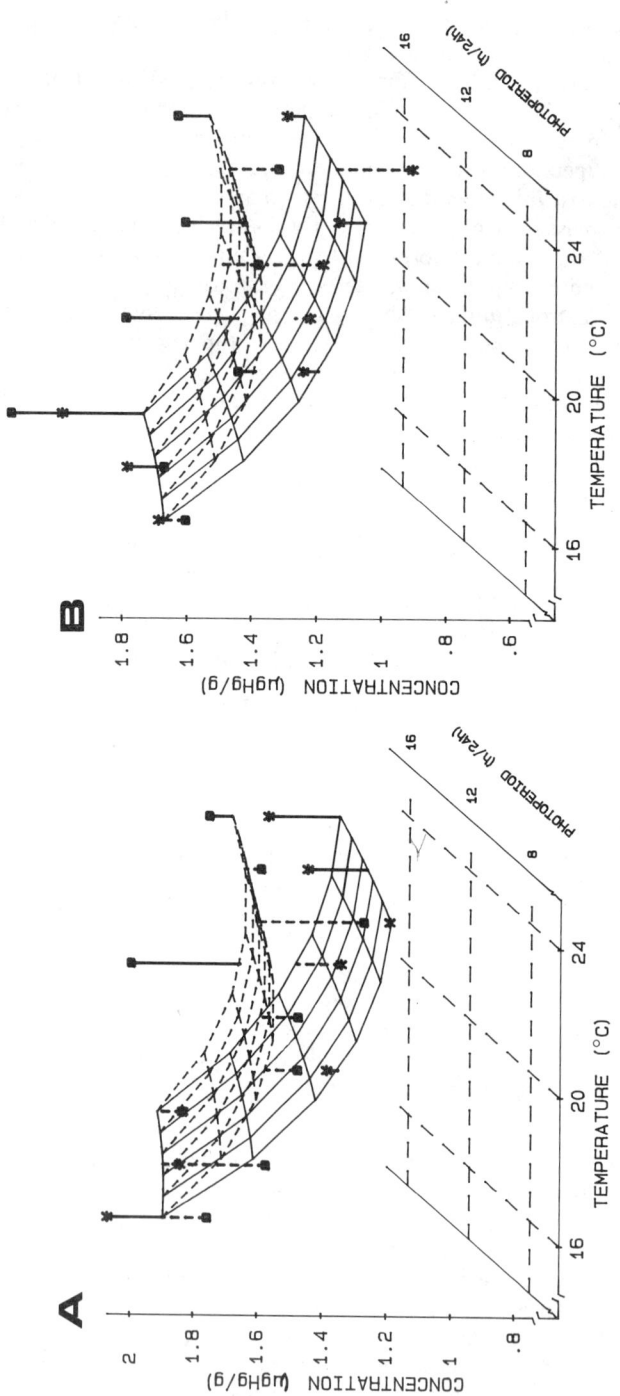

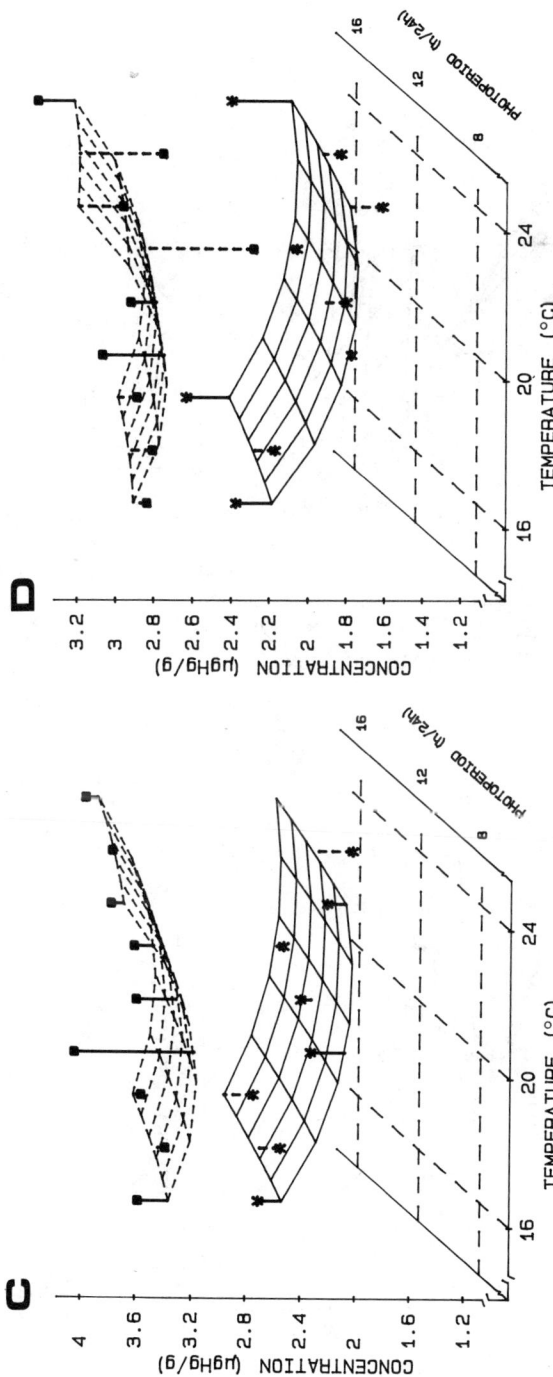

FIG. 5—Average total mercury concentrations in Elodea densa after 28 days' exposure and corresponding theoretical values taken from multiple regression model. (a) $HgCl_2$, low level of light intensity. (b) $HgCl_2$, high level of light intensity. (c) CH_3HgCl, low level of light intensity. (d) CH_3HgCl, high level of light intensity (pH 6.0 *—*; pH 8.0 ■--■). Regression model: log conc. (µg Hg g^{-1}) = 0.207 + 0.118 Hg chem. form + 0.068 pH − 0.049 photo. − 0.043 temp. + 0.033 Hg chem. form × pH − 0.029 light int. + 0.035 temp. × pH + 0.030 Hg chem. form × temp. + 0.016 temp.^{2}.

110 PLANTS FOR TOXICITY ASSESSMENT

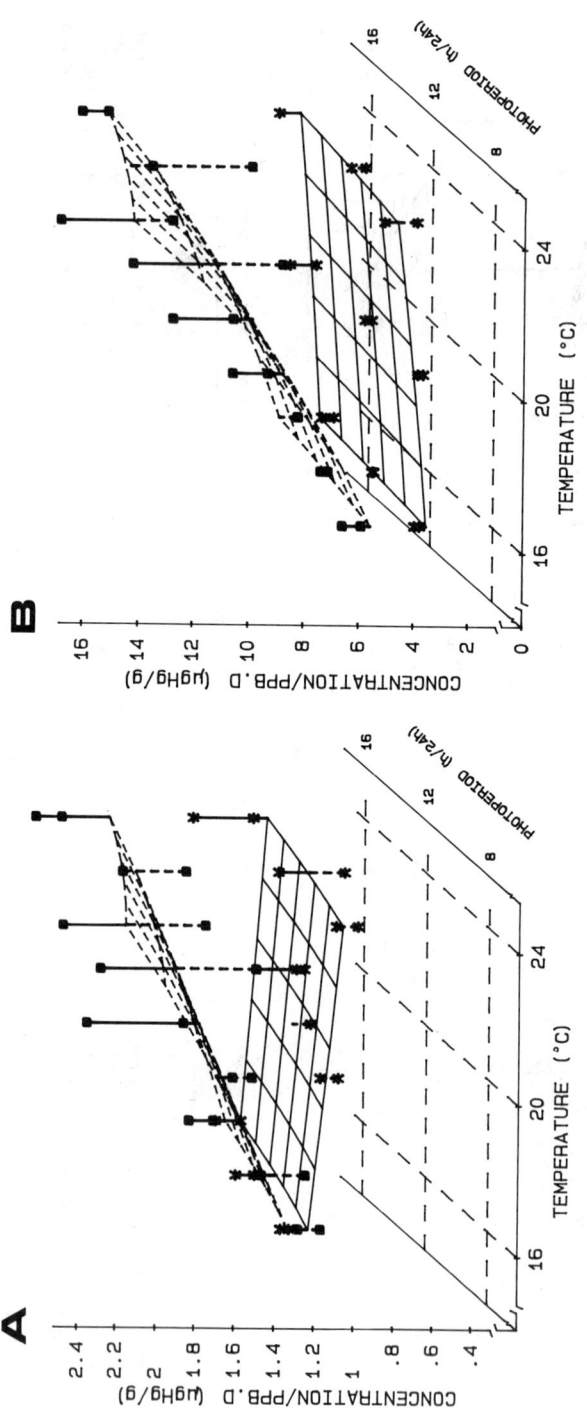

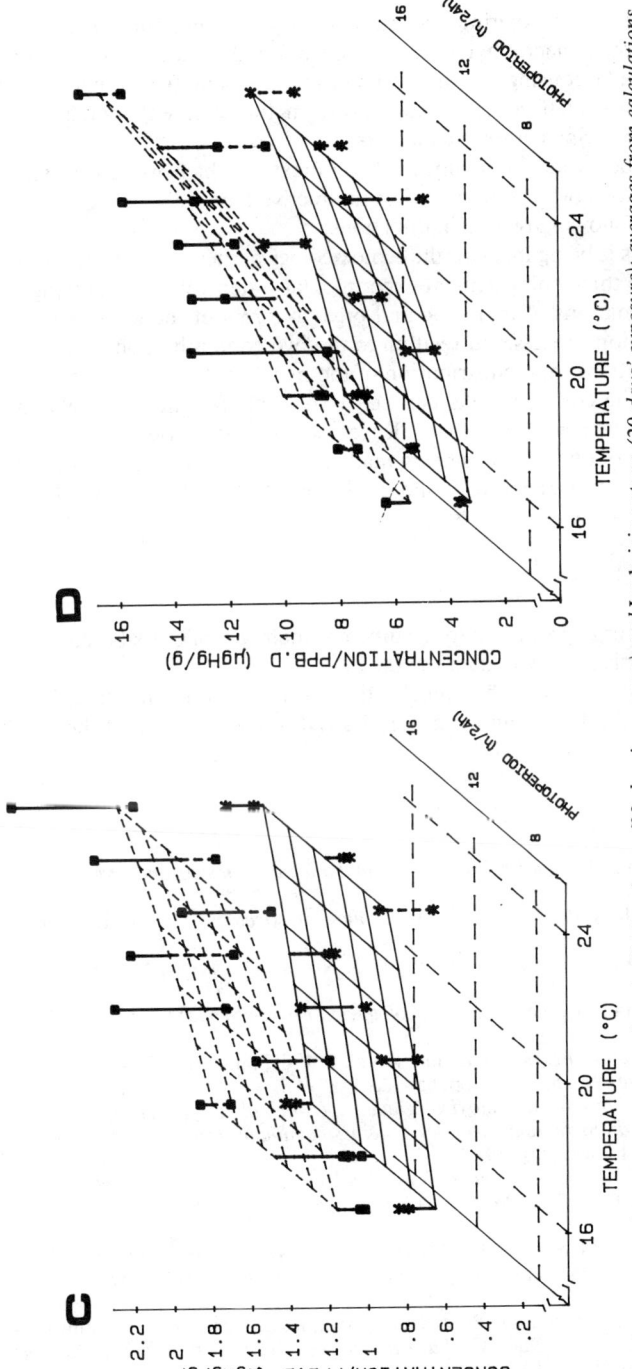

FIG. 6—*Total mercury concentrations in Elodea densa (28 days' exposure) and Ludwigia natans (20 days' exposure): averages from calculations based on "concentration × days" equivalents and values from the corresponding regression models.* (a) Elodea densa, $HgCl_2$. (b) Elodea densa, CH_3HgCl. (c) Ludwigia natans, $HgCl_2$. (d) Ludwigia natans CH_3HgCl (pH 6.0 *——*; pH 8.0 ■--■). *Regression models:* Elodea densa: *log conc.* (μg Hg g^{-1}) = $0.314 + 0.293$ Hg chem.form + 0.152 pH + 0.098 temp. + 0.076 temp. × pH − 0.070 photo. + 0.056 Hg chem.form × pH + 0.059 Hg chem.form × temp. Ludwigia natans: *log conc.* (μg Hg g^{-1}) = $0.312 + 0.377$ Hg chem.form + 0.153 pH + 0.143 temp. + 0.058 Hg chem.form × temp.

Conclusions

The results presented relate to an experimental study of mercury compound bioaccumulation by two species of rooted macrophytes contaminated via the water column. Other experiments have been carried out using this three compartment system (water, natural sediment, rooted macrophytes) to analyze metal transfers from the sediment source and to quantify the actions and interactions of selected factors.

This work is part of a much wider program, in the context of which we aim to set up gradually a multispecies ecotoxicological model of an interactive type.

Our research is currently following two main directions:

1. A more detailed analysis is being made of the ecotoxicological mechanisms that control mercury transfers within the three compartment experimental system. We are studying, for example, chemical transformations of inorganic and organic forms of mercury, especially methylation and demethylation processes in relation to the two contamination sources and to the different abiotic conditions; microdistribution of the metal in the sediment compartment, in order to analyze the complex links between bioavailability and biouptake; and localization of mercury in the macrophyte organs (leaves, stems, buds, roots).

2. The biological structure of our system is being made more complex, with the introduction of herbivorous species (molluscs and fish) and detritivores (intrasedimentary ephemeroptera nymphs).

Acknowledgments

We would like to express our appreciation to the different members of our laboratory, and also to M. Vaillant for his assistance with data treatment.

Our research program is supported by the French Ministry of Environment, the C.N.R.S. (Piren), the Commission of the European Communities (DG12, Brussels), and the E.P.R. (Aquitaine).

References

[1] Boudou, A. and Ribeyre, F., *Aquatic Ecotoxicology: Fundamental Concepts and Methodologies*, CRC Press, Boca Raton, Fla., 1989.
[2] Rand, G. M. and Petrocelli, S. R., *Fundamentals of Aquatic Toxicology: Methods and Applications*, McGraw-Hill, New York, 1985.
[3] Boudou, A. and Ribeyre, F. in *Aquatic Toxicology*, J. O. Nriagu, Ed., Wiley, New York, 1983, Chapter 3, pp. 73–116.
[4] Ribeyre, F., Boudou, A., and Delarche, A., *Ecotoxicology and Environmental Safety*, Vol. 3, 1979, pp. 411–427.
[5] Ribeyre, F., *Ecotoxicology and Environmental Safety*, Vol. 9, 1985, pp. 346–363.
[6] Denny, P., *Biological Revue*, Vol. 55, 1980, pp. 65–92.
[7] Ribeyre, F., "Conception d'un modèle écotoxicologique expérimental et application à l'étude de l'accumulation des dérivés du mercure au sein des systèmes aquatiques continentaux," Doctorate Thesis, No. 957, Bordeaux I University, 1988.
[8] Ribeyre, F. and Boudou, A. in *Aquatic Ecotoxicology: Fundamental Concepts and Methodologies*, A. Boudou and F. Ribeyre, Eds., CRC Press, Boca Raton, Fla., Vol. 2, 1989, Chapter 1-1, pp. 3–46.
[9] Farey, B. J., Nelson, L. A., and Rolph, M. G., *The Analyst*, Vol. 103, 1978, pp. 656–659.
[10] Box, G.E.P., Hunter, W. G., and Hunter, J. S., *Statistics for Experimenters*, Wiley, New York, 1978.
[11] Dagnelie, P., *Principes d'Expérimentation*, Presses Agronomiques de Gembloux, Belgium, 1984.
[12] Tomassone, R., Lesquoy, E., and Millier, C., *La régression-nouveaux regards sur une ancienne méthode statistique*, Masson, Paris, 1983.
[13] Snedecor, G. W. and Cochran, W. G., *Méthodes statistiques*, Association de Coordination Technique Agricole, Paris, 1971.

[14] Baker, D. A., *Annals of Botany,* Vol. 58, 1979, pp. 32–36.
[15] Goodwin, P. B., *Planta,* Vol. 157, 1983, pp. 124–130.
[16] Kendrick, R. E. and Kronenberg, G.H.M., *Photomorphogenesis in Plants,* M. Nijhoff, Dordrecht, The Netherlands, 1986.
[17] Raven, J. A., *Energetics and Transport in Aquatic Plants.* Alan R. Liss, New York, 1985.
[18] Barko, J. W. and Smart, R. M., *Aquatic Botany,* Vol. 10, 1981, pp. 339–352.
[19] Maury, R. and Engrand, P., "Recherches en Ecotoxicologie expérimentale sur la bioaccumulation et les transferts du mercure dans un système "sédiment-eau-macrophytes", " Doctorate Thesis, Bordeaux I University, No. 66 and No. 67, 1986.
[20] Bernhard, M., Brinckman, F. E., and Sadler, P. J., *The Importance of Chemical "Speciation" in Environmental Processes,* Springer-Verlag, Berlin, 1986.
[21] Maury, R., Boudou, A., Ribeyre, F., and Engrand, P., *Aquatic Toxicology,* Vol. 12, 1988, pp. 213–228.
[22] Campbell, P.G.C., Lewis, A.G., Chapman, P.M., Crowder, A.A., Fletsher, W. K., Imber, B., Luoma, S. N., Stokes, P. M., and Winfrey, M., "Biologically Available Metals in Sediments," Report No. 27694, National Research Council Canada, Ottawa, 1988.
[23] Salomons, W. and Forstner, U., *Metals in the Hydrocycle,* Springer-Verlag, Berlin, 1984.
[24] Czuba, M. and Mortimer, D. C., *Canadian Journal of Botany,* Vol. 60, 1982, pp. 657–666.
[25] Bienvenue, E., Boudou, A., Desmazes, J. P., Gavach, C., Georgescauld D., Sandeaux, J., Sandeaux, R., and Seta, P., *Chemico-Biological Interactions,* Vol. 48, 1984, pp. 91–101.
[26] Boudou, A., Desmazes, J. P., and Georgescauld, D., *Ecotoxicology and Environmental Safety,* Vol. 6, 1982, pp. 379–387.
[27] Delnomdedieu, M., Boudou, A., Desmazes, J. P., and Georgescauld, D., *Biochimica et Biophysica Acta,* Vol. 986, 1989, pp. 191–199.

Plants and Air Pollution

L. H. Weinstein,[1,2] *J. A. Laurence,*[1] *R. H. Mandl,*[1] *and Kurt Wälti*[3]

Use of Native and Cultivated Plants as Bioindicators and Biomonitors of Pollution Damage

REFERENCE: Weinstein, L. H., Laurence, J. A., Mandl, R. H., and Wälti, K,, **"Use of Native and Cultivated Plants as Bioindicators and Biomonitors of Pollution Damage,"** *Plants for Toxicity Assessment, ASTM STP 1091,* W. Wang, J. W. Gorsuch, and W. R. Lower, Eds., American Society for Testing and Materials, Philadelphia, 1990, pp. 117–126.

ABSTRACT: Plants are among the most sensitive organisms to pollutants, often responding with distinct, easy-to-recognize symptoms from exposure to specific environmental contaminants. This characteristic makes plants useful as bioindicators and biomonitors of pollutants. Using plants as bioindicators offers several advantages over physical/chemical monitoring systems: plants are easy to grow and maintain and inexpensive to deploy at a great number of sites; plants integrate pollutant exposure with other environmental factors to provide a biological assessment of exposure; and plant samples may be archived for retrospective analysis. Systems in use include indicator gardens, lichen transplants, plant growth and exposure benches, standard grass cultures, field survey of indigenous or cultivated species, and chemical analysis of plant tissue. A case study illustrating the use of bioindicators to assess the level of fluoride pollution in the Rhône valley in Switzerland is presented.

KEY WORDS: air pollution, bioindicator, biomonitor, fluoride, lichen, plant, vegetation

Plants have been used since the early part of the twentieth century for establishing the presence or degree of physical or chemical factors in the environment [1]. Over the years, methods have been developed that employ sensitive plant species as *bioindicators* (i.e., that detect visible, chemical, or physical changes). These include not only the occurrence of foliar lesions, changes in pigmentation, or alteration in growth habits, but also a number of more subtle effects such as physiological, biochemical, or genetic changes. Thus a bioindicator may be defined as a species of plant that responds characteristically to the conditions that occur in a particular region or habitat [2]. In air pollution studies, it is used to define a plant that exhibits a specific symptomology when exposed to a phytotoxicant [3,4]. Bioindicators can be compared to the use of litmus paper to detect the acidity of a liquid. Little quantitative information is provided other than the identification of the toxicant through induction from specific symptoms. Plant *biomonitors* may provide a semi-quantitative or quantitative measure of the amount of the toxicant accumulated by the species, as a pH electrode quantifies acidity. Despite the fact that biomonitors are passive, they might be substituted advantageously for analytical instruments for certain pollutants and under certain conditions. Monitoring of fluoride, chloride, cadmium, lead, etc., can be accomplished at low cost and where

---

[1] Boyce Thompson Institute for Plant Research, Ithaca, NY 14853.
[2] Ecosystems Research Center, Cornell University, Ithaca, NY 14853.
[3] Swiss Aluminum Ltd., Neuhausen am Rheinfall, Switzerland.

there is a lack of available power. Perhaps the most important advantage is the capacity of plants to absorb and integrate doses of pollutant over different environmental and climatological conditions [5–9]. Thus the effect on a living organism can be measured directly under existing conditions in the field. Air monitoring instruments have limited predictive value, because the air sampled is not representative of the whole area, instruments are expensive, and the concentration of a pollutant in the atmosphere provides no information on how a plant will respond under different conditions.

Many of the methods described herein have also been reported earlier [10].

Field Monitoring Systems

Indicator Gardens—Perhaps the simplest system is the indicator garden, consisting of species of plants that respond differentially to pollutants [3,11–15]. Such bioindicator gardens are presently in use in Minnesota to monitor sulfur dioxide and ozone concentrations [12,13]. Indicator gardens are grown in indigenous soil, standard soil, or soil mix, and they utilize species of known and defined sensitivity. For instance, isolines of tobacco, which differ in their sensitivity to ozone, have been used. By comparing the slopes of injury (size of foliar lesions) versus time curves for Bel W-3 (sensitive) to Bel-B (tolerant), air quality can be categorized as poor, moderate, fair, or good [3].

Portable Exposure Benches—Often locations where bioindicators are used are isolated and maintenance (watering, weeding, fertilizing, etc.) becomes a concern. This problem is partly solved by the use of special growth benches [11,16]. These benches not only support potted indicator plants above the ground for protection from animals but provide an automatic watering system (Fig. 1). Exposure benches are used extensively in Europe for supporting indicator trees or other sensitive species.

Lichen Transplants—Disks of corticolous lichen thalli with their bark substrate have been transplanted from areas of clean air to trees growing in polluted air [17,18]. A more successful system, where lichen disks are placed in holes in boards and mounted on posts, is in use in Germany. Lichens may also be transplanted to small blocks of wood that are attached to the vanes of an anemometer, thus ensuring continuous exposure of the lichen transplant to the prevailing winds [11,19,20] (Fig. 2). One disadvantage of using lichens is that they respond slowly to changes in air quality and seldom recover once damaged severely.

Grass Cultures—Standard methods have been devised for determining the accumulation of several air pollutants over short- and long-term periods using ryegrass cultures [11,21] (Fig. 3). The methods are widely used in Germany for monitoring fluoride, sulfur, chloride, lead, cadmium, zinc, copper, nickel, vanadium, and other elements. In general, the cultures consist of ryegrass seedlings growing in a defined medium, provided with a water supply, and mounted above the ground. By removing samples at specific intervals for elemental analysis, the rate of uptake, geographical distribution, and total accumulation of toxicant can be determined.

A valuable aspect of the use of accumulator plants is that they may be harvested, dried, ground, and stored for future analyses, should they be needed. For instance, if concern were to arise over the presence of a specific metal in the environment, archived samples may be analyzed retrospectively to provide information on levels of contamination in the past.

Indigenous or Cultivated Plants In Situ: *A Case Study*

Three aluminum reduction smelters and one chemical plant are more or less evenly distributed in the central Rhône Valley of Switzerland in the canton of Valais (Wallis) (Figs. 4 to 6). The smelters in Martigny and Chippis began operations in 1908; the third, and largest,

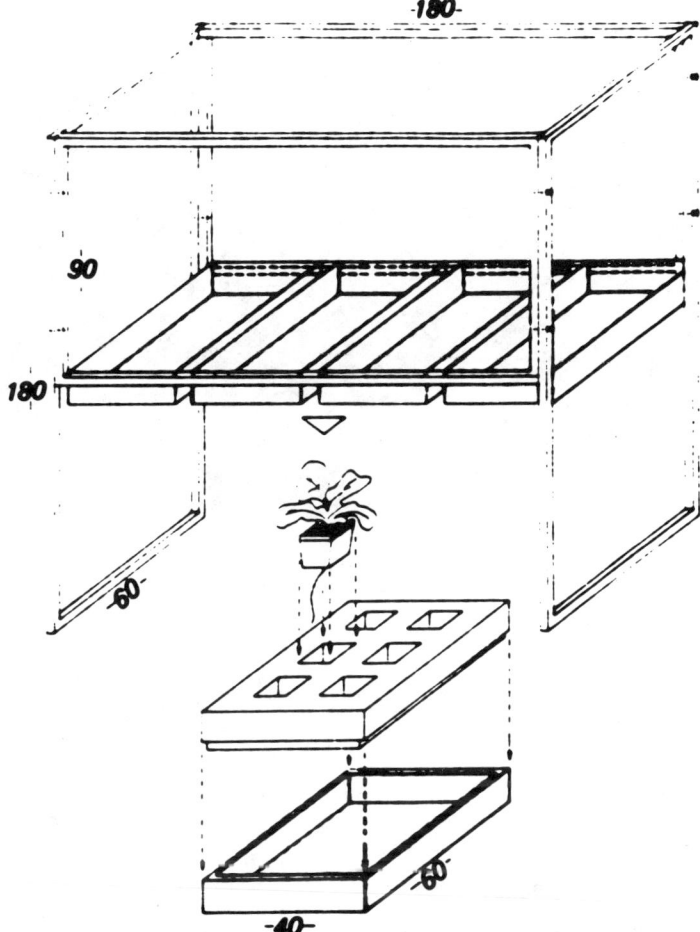

FIG. 1—*Exposure bench for culture of bioindicator plants. Culture bed is supported by a galvanized steel frame and plants are automatically watered. The water reservoir, styrofoam support, and a potted plant are shown* [16].

smelter is located in Steg and began operations in 1962. A fertilizer and chemical factory is located at Visp, east of Steg. North of Martigny to the Lake of Geneva are several industrial areas that include power plants, cement factories, and pharmaceutical plants, and the urbanized areas of Geneva, Lausanne, Montreux, and Monthey.

Aluminum is produced by the reduction of alumina at a temperature above 900°C and in the presence of an electrolyte, usually cryolite, a sodium-aluminum-fluoride mineral. At this high temperature, some of the cryolite is volatilized and hydrogen fluoride (HF) and particulate fluorides are released. Hydrogen fluoride is extremely phytotoxic, being perhaps one hundred to one thousand times more toxic than sulfur dioxide or ozone.

The Rhône Valley, between Monthey and Brig (Figs. 4, 5, and 7), is about 100-km long, with the Alps rising on both sides to 2000 to 4000 m above the narrow valley floor. The prevailing daytime winds are from the north toward Martigny, beyond which the valley turns to the east. The valley is one of the most important and productive agricultural regions of

FIG. 2—*Device for exposing lichen disks to airborne contaminants.*

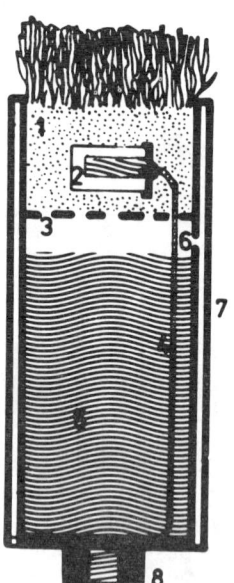
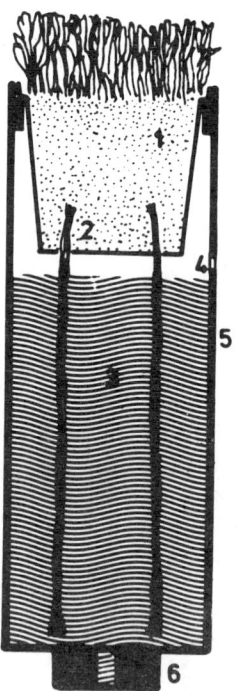

FIG. 3—*Self-watering grass culture containers for biomonitoring of heavy metals, fluoride, etc.* (left) *(1) Standard soil mix. (2) Ceramic cylinder. (3) Filter plate. (4) Wick. (5) Water reservoir. (6) Overflow hole. (7) Double-walled container. (8) Connector flange.* (right) *(1) Standard soil mix. (2) Glass fiber wick. (3) Water reservoir. (4) Overflow hole. (5) Outer container. (6) Connector flange.*

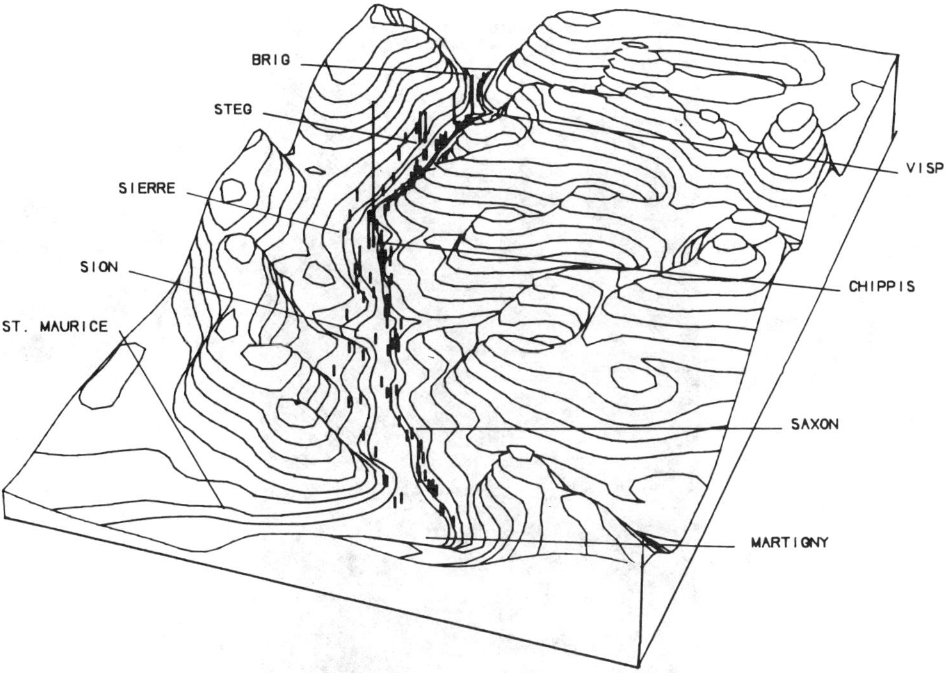

FIG. 4—*Oblique view of the Rhône valley looking to the east and showing the fluoride contents of current-year's needles of Scots pine (1 mm = 5 ppm fluoride on a dry weight basis).*

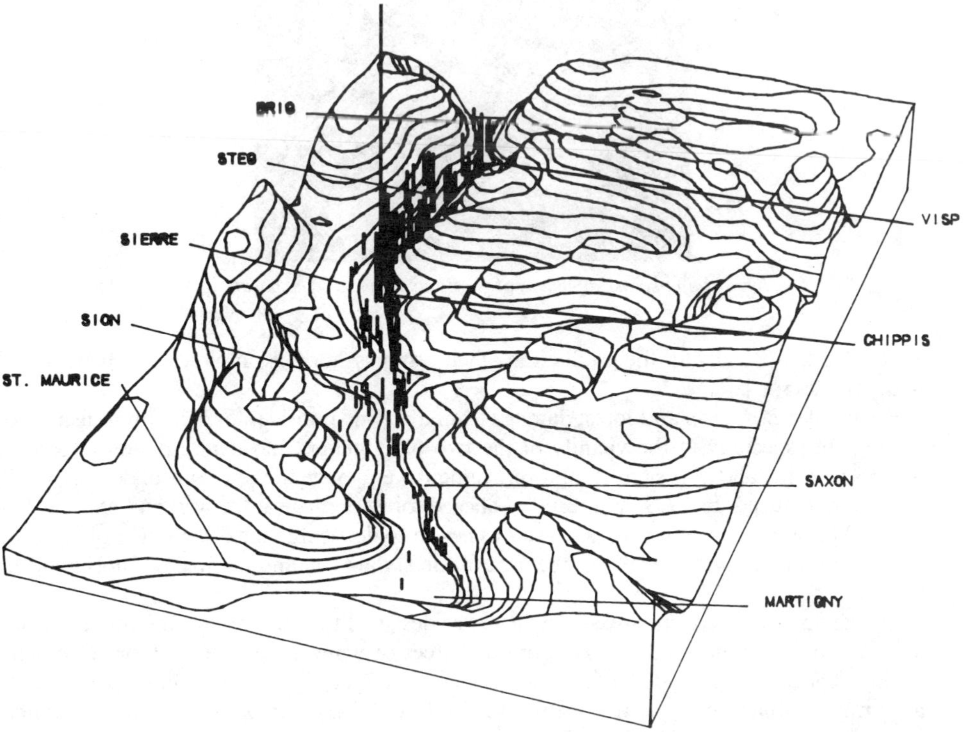

FIG. 5—*Same as Fig. 4, but for 1-year-old needles.*

FIG. 6—*Typical tip necrosis of leaves of gladiolus induced by exposure to airborne fluoride.*

Switzerland, especially for the production of wine grapes, apricots, apples, pears, berries, and many truck crops.

For many years, crop and forest damage from airborne fluoride was a common feature of the region, especially in the vicinity of the two old smelters and before pollution control devices were installed. As significant improvement were made in these control devices, emissions were reduced. In 1979, as a result of alleged forest damage at the west end of the valley east of Martigny (above Saxon), a study was begun to determine the distribution of airborne fluoride in the valley, especially in the area of forest damage, and the role of fluoride in the problem.

Two techniques were employed. The first was designed to measure the amount of fluoride accumulation in foliage of selected crops and trees over several growing seasons *(biomonitoring)*. The second provided a visual assessment of the degree of foliar injury produced on a common ornamental plant sensitive to fluoride *(bioindication)*. Under circumstances

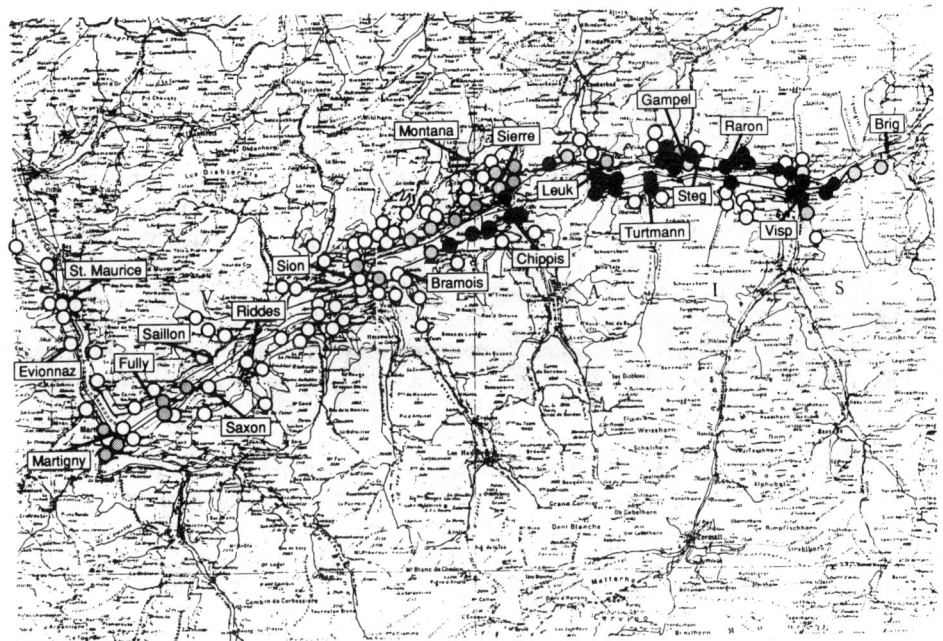

FIG. 7—*Map of the Rhône valley showing locations and degree of injury of gladiolus plantings. Injury is proportional to increasing density of the symbol used.*

where either of these techniques may be used, the choice will probably depend upon the degree of certainty in identification of the pollutant involved, a more quantitative relationship of dose to response, and cost. There have been a number of excellent reviews of these methods [e.g., *11*], and descriptions of earlier research in the Valais utilizing *biomonitoring* and *bioindication* methods [e.g., *23–25*]. Each of the procedures is summarized below.

Biomonitoring

Three species of plants were studied because of their relative importance and distribution throughout the valley: grape, apricot, and Scots pine. With pine, needles from the current and previous years were used. Sampling sites were established at 2-km intervals from Martigny to Brig, a reach of about 75 km. At most sites, samples were collected at five positions: about 1000 m and 500 m elevation on the north slope, on the valley floor, and at about 1000 m and 500 m elevation on the south slope. Foliage of each of the three species of plants was collected where available. Samples were placed in paper bags, labeled, and returned to the laboratory for analysis. Sampling was repeated up to four times during each growing season over a four-year period (1979–1983), representing several thousand analyses. Oven-dried, pulverized samples were analyzed for fluoride. Three-dimensional plots of the summary data for 1980 for the current year's and 1-year-old needles of pipe are shown in Figs. 4 and 5. From these data, several conclusions were drawn:

1. Older needles of pine accumulated at least twice as much fluoride as did new needles, often five times greater. In general, the order of fluoride accumulation was found to be: old pine needles > grape = apricot > new pine needles.

2. Accumulation was greater on the north slope of the valley than the south slope (although this is not obvious from the figures).

3. Fluoride accumulation was greatest near the smelter at Chippis, with concentrations of fluoride as high as 450 ppm on a dry weight basis in 1980 for old needles. As little as about 5 ppm, a value considered to be background, was found in other areas.

4. A second, but much lower, area of fluoride accumulation was evident near the smelter at Steg. A third area of low accumulation occurred near the chemical factory in Visp.

5. A minor peak of accumulation (about two times background values) occurred near Sion where there is a municipal incinerator.

6. There was no evidence of significant fluoride accumulation near the smelter at Martigny.

7. There was no evidence of fluoride accumulation in the regions of pine/spruce forests that lie near Saxon, even though there was severe tree mortality, an observation that caused the study to be initiated.

In addition to the conclusions reached above, it was clear that damage to the forests was caused by factors that did not at that time include airborne fluoride. The syndrome appeared to be closely associated with the more recently recognized complex stress response referred to as forest decline, and probably included high winds, drying conditions, a mixture of airborne contaminants, and the presence of high ambient ozone concentrations.

Bioindication

As we travelled throughout the study area examining vegatation between the sites from which pine, grape, and apricot foliage were collected, it became apparent that the gardeners of the Valais favored the cultivation of gladiolus plants, a species in the lily family that is extremely sensitive to airborne fluoride, and is a favorite indicator species for fluoride monitoring. We therefore extended our study to include the occurrence and degree of foliar injury of dozens of gladiolus plantings distributed throughout the Valais. The symptoms of fluoride injury on gladiolus appear as tip necrosis that usually can be distinguished from injury induced by other pollutants or by other abiotic and biotic agents [22] (Fig. 6). The amount of tip necrosis was estimated visually and recorded as an approximate median value for that site. Sensitive indigenous species, such as common barberry, St. Johnswort, and developing needles of Scots pine, were also inspected and served to confirm our diagnosis. These studies were conducted from 1979 through 1987 during the period from the middle to the end of September. Injury at each site, recorded as characteristic foliar chlorosis and/or necrosis, was then placed into one of four categories: 0 to 1 cm, 2 to 5 cm, 6 to 9 cm, and >10 cm. Representative data for one year are plotted in Fig. 7.

Although the methods of expressing the presence and degree of fluoride pollution are different, essentially the same kinds of information were obtained from each. In fact, because of its much lower cost, the gladiolus inspections can provide nearly as much information as fluoride analyses, assuming that there are a sufficient number of gladiolus plantings available. The following conclusions were drawn from the gladiolus survey for the period shown:

1. Severe necrosis of gladiolus occurred near the smelter at Chippis and, to a lesser degree, near the smelter at Steg.

2. A moderate amount of injury occurred near the chemical factory at Visp.

3. There was little or no injury west of Sion (about 12 km west of the smelter at Chippis). Injury was essentially nonexistent near Saxon.

4. At the highest altitudes, the degree of injury decreased, perhaps due to fact that plants were growing above the inversion layer.

5. Injury was greater on the north slope than on the south slope of the valley.
6. Little or no fluoride injury occurred near the chemical plant and oil refinery in Monthey or near the aluminum smelter at Martigny.

Although *biomonitoring* is by far the more costly method, it can provide a rough estimate of the levels of emissions and the time-course of pollutant uptake and accumulation in plant foliage. Of the two methods, *biomonitoring* is the more useful method for establishing and enforcing air quality standards. *Bioindication,* on the other hand, has the advantages of being rapid and inexpensive, but is less adapted to time-course studies over a growing season, and therefore is less useful for establishing and enforcing air quality standards.

In contrast to air monitoring, both methods provide retrospective information; they are responsive to conditions that have already occurred at the pollution source. On the other hand, air monitoring, especially for airborne fluoride, has disadvantages. For example, continuous fluoride monitoring by physical/chemical means is, at best, expensive and unreliable. The present method of "choice" consists of collecting fluoride in a scrubbing solution followed by measurement with a fluoride specific ion electrode. The equipment is unwieldly and not easily portable. Another problem is that the atmospheric concentration of fluoride is only one of many abiotic factors that determine the incidence and severity of plant injury. While air monitoring only provides a clue to the possibility of an adverse effect on vegetation, *biomonitoring* and *bioindication* methods integrate the exposure with other environmental and biological factors to produce a response.

Because the regulation of air pollution and the assessment of its effects is a complex political and scientific problem, the use of chemical/physical methods of air monitoring is necessary. However, the use of *bioindicators* and *biomonitors* allow the collection of a large quantity of additional and supporting information at relatively low cost to supplement air monitoring and to provide evidence of pollution impact.

Acknowledgments

The authors thank Swiss Aluminum Ltd. for the use of data presented in this paper.

References

[1] Ruston, A. G., *Annals of Applied Biology,* Vol. 7, 1921, pp. 390–403.
[2] Melanby, K. in *Environment and Man, Vol. 7, Measuring and Monitoring the Environment,* J. Lenihan and W. W. Fletcher, Eds., Academic Press, New York, 1978, pp. 1–13.
[3] Feder, W. A. and Manning, W. J. in *Handbook of Methodology for the Assessment of Air Pollution Effects on Vegetation,* W. W. Heck, S. V. Krupa, and S. N. Linzon, Eds., TE-2 Agricultural Committee, Air Pollution Control Association, Pittsburgh, Pa., 1979, Chapter 9, pp. 9-1 to 9-14.
[4] Feder, W. A., *Environmental Health Perspectives,* Vol. 27, 1978, pp. 139–147.
[5] Gilbert, O. L., *New Phytologist,* Vol. 67, 1968, pp. 15–30.
[6] Heck, W. W., Dunning, J. A., and Hindawi, I. J., *Science,* Vol. 151, 1966, pp. 577–578.
[7] LeBlanc, F. and DeSloover, J., *Canadian Journal of Botany,* Vol. 48, 1970, pp. 1485–1496.
[8] Oshima, R. J., *Journal of the Air Pollution Control Association,* Vol. 24, 1974, pp. 576–578.
[9] Treshow, M., *Journal of the Air Pollution Control Association,* Vol. 15, 1965, pp. 266–269.
[10] Weinstein, L. H. and Laurence, J. A. in *Biologic Markers of Air-Pollution Stress and Damage in Forests,* National Academy Press, Washington, D.C., 1989, pp. 195–204.
[11] Arndt, U., Nobel, W., and Schweizer, B., *Bioindikatoren. Möglichkeiten, Grenzen und neue Erkenntnisse,* Eugen Ulmer GmbH & Co., Stuttgart, FRG, 1987.
[12] Kromroy, K. W., Teng, P. S., Olson, M. F., French, D. R., and Lang, D. S., *Environmental Pollution,* Vol. 53, 1988, pp. 439–441.
[13] Kromroy, W. W., Olson, M. F., Grigal, D. F., Teng, P. S., French, D. R., and Amundson, G. A., *Use of Plants for Toxicity Assessment, ASTM STP 1091,* American Society for Testing and Materials, Philadelphia, 1990.

[14] Posthumus, A. C. in *Proceedings, Kuopio Meeting on Plant Damages Caused by Air Pollution*, Sci. Paper Symp., Kuopio, Finland, L. Karenlampi, Ed., 1976, pp. 115-120.
[15] van Raay, A. in *Air Pollution, Proceedings of the First European Congress on Influence of Air Pollution on Plants and Animals*, Centre for Agricultural Publishing and Documentation, Wageningen, Holland, 1968, pp. 319-335.
[16] Arndt, U., Ehrhardt, W., Keitel, A., Michenfelder, K., Nobel, W., and Schlüter, C., *Staub, Reinhaltung Der Luft*, Vol. 45, 1985, pp. 481-483.
[17] Brodo, I. M., *Ecology*, Vol. 42, 1961, pp. 838-841.
[18] Brodo, I. M., *Bryologist*, Vol. 69, 1966, pp. 427-449.
[19] Puckett, K. J. in *Assessing Air Quality with Lichens and Bryophytes*, T. H. Nash, III, and V. Wirth, Eds., Bibliographica Lichenographica, Stuttgart, FRG, 1988.
[20] Schönbeck, H., *Staub, Reinhaltung Der Luft*, Vol. 29, 1969, pp. 17-21.
[21] Scholl, G., *VDI Berichte*, No. 164, 1971, pp. 39-45.
[22] Bolay, A. and Bovay, E., *Agr. Romande IV, Serie A*, Vol. 4, 1965, pp. 43-46.
[23] Zuber, R., Tschannen, W., and Bovay, E., *Rev. Suisse Vitic. Hortic.*, Vol. 13, 1981, pp. 133-138.
[24] Flühler, H., Keller, T., and Schwager, H., *Eidg. Anst. forstl. Versuchswes.*, Vol. 57, 1981, pp. 399-432.
[25] Weinstein, L. H., "Fluoride and Plant Life," *Journal of Occupational Medicine*, 1977, pp. 49-78.

D. A. Tolle,[1] *M. F. Arthur,*[1] *K. M. Duke,*[1] *and J. Chesson*[2]

Ecological Effects Evaluation of Two Phosphorus Smoke-Producing Compounds Using Terrestrial Microcosms

REFERENCE: Tolle, D. A., Arthur, M. F., Duke, K. M., and Chesson, J., "**Ecological Effects Evaluation of Two Phosphorus Smoke-Producing Compounds Using Terrestrial Microcosms,**" *Plants for Toxicity Assessment, ASTM STP 1091,* W. Wang, J. W. Gorsuch, and W. R. Lower, Eds., American Society for Testing and Materials, Philadelphia, 1990, pp. 127–142.

ABSTRACT: An intact soil-core microcosm and static exposure system were used to evaluate the potential ecological effects of two obscurant smokes, red phosphorus/butyl rubber (RP/BR) and white phosphorus/felt (WP/F), used in the field during U.S. Army training exercises. The terrestrial microcosm technique differed from the test protocol prepared for the U.S. Environmental Protection Agency by using a preliminary plant stress-ethylene test to help select the dose range for microcosm testing, and by combining aspects of traditional range-finding and definitive tests into a single microcosm test.

Three plant species (white sweetclover, perennial ryegrass, and wheat) were used for both preliminary and microcosm tests. The stress-ethylene tests indicated that extremely high doses of either smoke would be required to elicit a response in the microcosm test. This information guided selection of microcosm test concentrations. Microcosms were exposed to either RP/BR or WP smoke at target concentrations of 0, 100, 300, 600, and 1500 mg/m^3. These concentrations bracket typical field concentrations.

Minor ecosystem-level effects were detected in microcosms only at the highest smoke concentration and included increased nutrient (Ca) loss in leachate, increased (wheat) or decreased (sweetclover) biomass yield, and increased element (Al, As, Pb, and P) uptake in plant tissue. No negative ecological effects of either smoke were detected at smoke concentrations equal to or below 600 mg/m^3, even for 16 semiweekly exposures over an 8-week period.

It was concluded that: (1) deployment of these two smokes at typical field concentrations is unlikely to cause significant problems to most terrestrial systems; (2) the terrestrial microcosm system appears to be appropriate for evaluating the static, long-term, particle-deposition effects of other aerosols; and (3) the test design produces input for hazard assessment at considerable savings in cost and time compared to the more traditional range-finding plus definitive test strategy.

KEY WORDS: aerosol deposition, ecosystem-level effects, terrestrial microcosm, biomass, element uptake, nutrient loss, red phosphorus/butyl rubber, white phosphorus/felt, stress ethylene, smoke exposure chamber

Introduction

Research on Obscurant Smokes

The U.S. Army Medical Research and Development Command (USAMRDC) is sponsoring research on the environmental effects of the production, use, and disposal of obscurant

---

[1] Environmental Toxicology and Assessment Section, Battelle Columbus Division, Columbus, OH 43201.
[2] Chesson Consulting, Washington, DC 20036.

smokes in order to develop the data base required to establish exposure criteria and standards. Evaluation of ecosystem effects due to the use of red phosphorus/butyl rubber (RP/BR) or white phosphorus/felt (WP/F) smokes during training activities is one component of this research. Previous research has dealt mainly with the chemistry and/or toxicology of these smokes [1-4]. This study is the first attempt to evaluate the effects of phosphorus smokes on a confined portion of a terrestrial ecosystem (herbaceous field) as a single interactive unit and thus fill a major gap in the environmental evaluation of these smokes [5]. A concurrent study investigated effects of phosphorus smokes on single ecosystem components (e.g., separate exposure of plants, sieved soil, or earthworms in artificial soil) [6].

The two smokes (RP/BR and WP/F) are designed to provide a protective screen while moving troops or equipment. Both produce a dense white smoke containing a series of polyphosphoric acids [2] when ignited (RP/BR) [1] or exposed to air (WP/F) [3]. Both smokes also include trace impurities containing elements (e.g., B, As, Cu, Pb, Al, and Ni) [1-3] that are potentially toxic to plants or animals [7].

Background on Microcosm Technique

Microcosms are confined portions of an ecosystem (aquatic or terrestrial) under laboratory control that attempt to mimic the processes and interactions of a larger ecosystem [8]. Terrestrial microcosms with intact soil cores, such as the type used in this study, are thought to be an improvement over artificially constructed terrestrial microcosms, because they contain a *natural* assemblage of soil biota and an undisturbed soil profile [9,10]. Thus, ecosystem-level processes, such as cycling of plant nutrients, element transport and fate, plant-microbial interactions, and soil microsite chemistry occur at rates shown by validation testing to be representative of the field [10].

Battelle has conducted considerable developmental research with terrestrial microcosms since 1979 [5,11-17]. The microcosm resulting from this developmental work is a medium-sized, intact soil core containing producer, decomposer, mycorrhizal, and root components in the soil matrix. Terrestrial microcosm test protocols resulting from this research have been published by the U.S. Environmental Protection Agency (U.S. EPA) as a test guideline for Section 4 of the Toxic Substances Act [18] and by ASTM as Standard E 1197 [19].

Objectives of Study

The objectives of this study were two-fold: (1) to evaluate the applicability of the intact soil-core microcosm as a hazard assessment tool for evaluating ecosystem-level effects of chemicals (including aerosols) used by the Army, and (2) to evaluate the ecological effects (i.e., nutrient loss, element uptake, and biomass yield) of RP/BR and WP smoke using the terrestrial microcosm.

Materials and Methods

Materials and methods are summarized below. A detailed description is provided by Tolle et al. [5]. The white phosphorus (WP) and felt were separated in this study in order to achieve more precise exposure levels. Based on seed germination tests of six different species, three plant species, perennial ryegrass *(Lolium perenne)*, wheat *(Triticum aestivum)*, and white sweetclover *(Melilotus alba)*, were chosen for investigation because they have wide geographic distributions, can be used to reclaim disturbed ground (including Army training areas), are different physiologically and morphologically, and were shown to be capable of

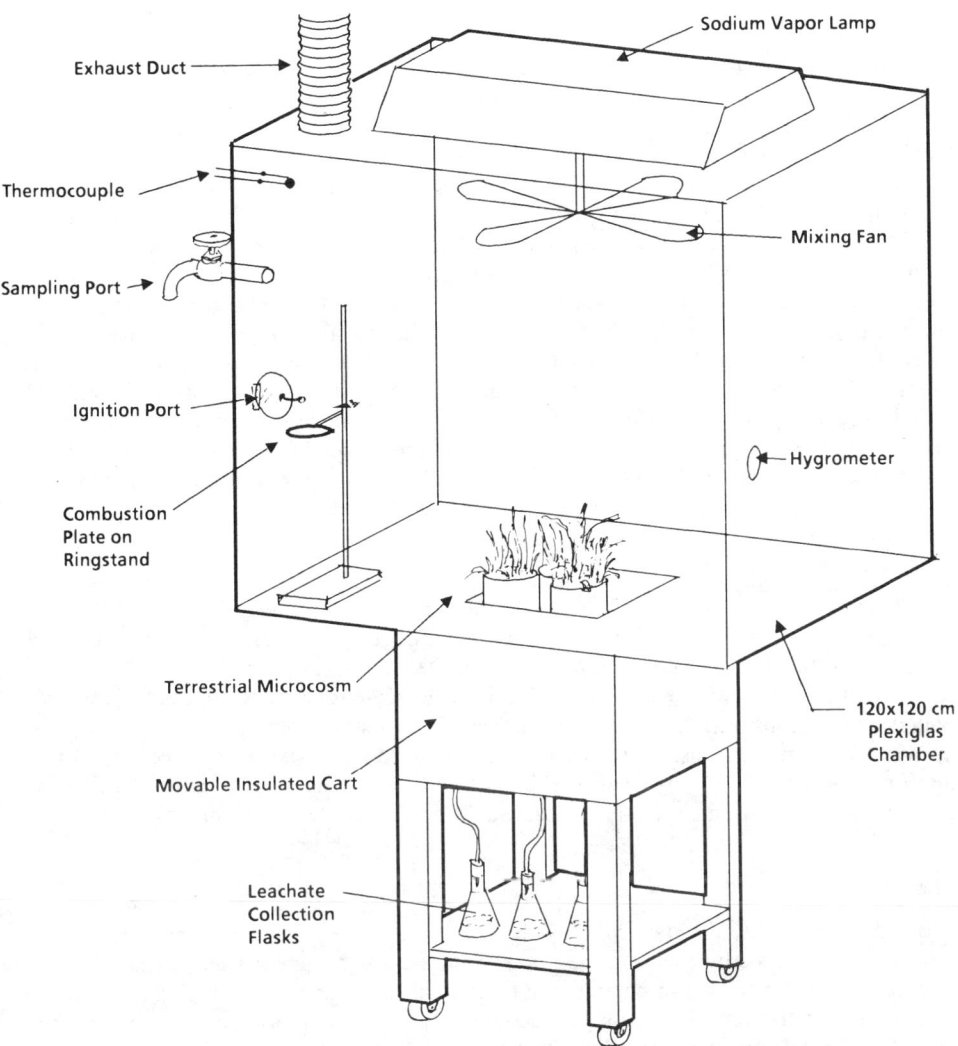

FIG. 1—*Exposure chamber system for dosing microcosms in movable carts.*

field-representative growth in the small surface area (240.5 cm²) of the microcosms. Intact soil-core microcosms were extracted on 21 December 1983 from an undisturbed field site, the existing vegetation was removed, and each microcosm was planted with a mixture of the three plant species. The common soil type for all cores was Crosby silt-loam, which was taken from a fallow agricultural field that had not been farmed or chemically treated for many years and that contained natural herbaceous vegetation (grasses and broadleaves). Microcosms consisted of 60-cm intact soil cores encased in 17-cm-diameter tubes of high-density polyethylene. The microcosms were placed on carts, three per cart (Fig. 1), and kept in a greenhouse under controlled temperature, humidity, and lighting. Additional details on microcosm construction and maintenance are described in the subsection titled Microcosm Test.

Experimental Protocol

Standard toxicological testing procedures usually call for a range-finding test to roughly determine the concentration of the test compound necessary to elicit the response, followed by a definitive test which more accurately determines that concentration. The two test protocols are very similar, the test compound concentrations being the major difference between the two. The cost and time required for the soil core microcosm test to be implemented in this fashion is prohibitive for most applications. For the purpose of these experiments, a modified testing procedure was implemented. The range-finding and definitive tests have been combined into a single expanded range-finding test, hereafter called the *microcosm test*. Here the number of test concentrations (usually five in the standard protocol) is increased and the range of concentrations is determined by a simple preliminary plant toxicity test. Precedent for this approach has been established in the U.S. EPA's Level 1 approach to environmental assessment, where the expanded range-finding testing concept has been validated [20]. In this study, the plant stress-ethylene test was used to determine the lowest concentrations of phosphorus smoke required to elicit plant stress under static acute test conditions. The results were then used to guide the selection of test concentrations for the microcosm test.

Preliminary Stress-Ethylene Tests

The stress-ethylene test is based on the increase in ethylene production by plant tissue as a function of increasing physical or chemical stress. The static, acute, stress-ethylene protocol described by Brusick and Young [20] was used. Details are given by Tolle et al. [5]. The results of the stress-ethylene test indicated little or no acute toxicity due to exposure to very high concentrations of RP/BR or WP/F smokes. The lowest concentration required to elicit an acute ethylene response from the most sensitive test species (white sweetclover) was approximately 20 000 mg/m^3. This preliminary test indicated that the highest dose in the more sensitive subchronic microcosm test should be well under this 20 000 mg/m^3 level.

Microcosm Test

Based on the relatively low acute toxicity to plants of the phosphorus smokes as determined by the stress-ethylene tests, the microcosm test was designed as an expanded range-finding test, with elements in between the traditional short-term range-finding test and the long-term definitive test. For example, the exposure period, monitoring period, and number of leachate collections is intermediate between typical range-finding and definitive microcosm tests, but the number of harvests and ability to limit the suite of elements and nutrients analyzed during later portions of the test are similar to the typical definitive microcosm test [10,14,15,18,19]. The design involved 2-h-long, semiweekly exposures of microcosms to RP/BR or WP smoke over an 8-week period, beginning 94 days after planting. Monitoring of microcosm parameters was conducted during and after the exposure period, for a total of 12 weeks.

The target smoke concentrations were chosen to be between the minimal field concentration (7 mg/m^3) determined by the U.S. Army Dugway Proving Ground in a pilot study for testing smoke munitions and the extremely high levels (18 000 to 20 000 mg/m^3) required to obtain a plant stress-ethylene response in the preliminary tests. Typical field concentrations of both phosphorus smokes are estimated by U.S. Army staff to be in the neighborhood of 50 to 400 mg/m^3. The target dose concentrations chosen for both smokes were 0, 100, 300, 600, and 1500 mg/m^3, which bracketed and exceeded typical field concentrations, yet were far enough apart that deviations of measured aerosol mass concentrations (as much as $\pm 10\%$ from the target dose) from two adjacent target doses were not likely to overlap.

Exposure System—A bottomless, Plexiglas cube (120 cm on a side) was lowered over the microcosm cart to make smoke exposures (Fig. 1). Each of the five exposure chambers was equipped with a low rpm mixing fan, a thermocouple for gas temperature measurement, a hygrometer to determine relative humidity, an exhaust which led to an outside stack, an injection/entry port, and an aerosol sampling port. Ventilation fans provided external cooling to prevent heat build up from the plant growth lamps suspended above each chamber.

Details on the aerosol injection method and measurement of the smoke exposure parameters are described by Tolle et al. [*5*]. Characterization of the smoke exposures involved measurement of the environmental variables, the initial mass concentration, and the deposited phosphorus mass (indicative of actual plant dose).

Microcosm Construction and Maintenance—The design, construction and maintenance of the terrestrial microcosms (Fig. 1) were similar to those used in previous studies [*12,16*]. Intact cores were extracted from an undisturbed herbaceous field using a specially-designed, steel-driving tube [*16*]. Sixty of the 70 cores extracted were chosen for additional work, based on their similar leachability during a 24-h period and absence of surface cracks.

The existing vegetation was removed from each soil-core microcosm; it was then planted with a mixture of the three reclamation plant species. The seeding rate was two times greater than recommended field rates, and the microcosms were later thinned back or replanted as necessary to achieve a uniform number of viable plants per microcosm for each of the three species: perennial ryegrass (9 plants), white sweetclover (5 plants), and wheat (3 plants).

Microcosms were selected randomly and 30 for each phosphorus smoke were evenly distributed between two greenhouse bays. They were watered daily with 150 mL of half-strength Hoagland's nutrient solution until a nutrient stress determined from soil analysis was eliminated. A photoperiod of 16 h light/8 h dark was achieved by the use of metal halide lights, as necessary.

During and after the exposure of microcosms to RP/BR and WP smokes, plants were monitored daily for signs of stress, such as discoloration, wilting, and leaching. After nutrient deficiencies had been corrected with Hoagland's solution, microcosms were watered with 150 mL of reverse osmosis (RO) water on Monday and Wednesday of each week. In addition, on Friday of each week of exposure, the plant leaves were sprayed with 300 mL of RO water to mimic the washing of leaves by rain. The amount of water was based on precipitation during a normal growing season. All cores were monitored daily for insect infestation. During July 1984 (shortly after the exposures to RP/BR and WP were completed), all microcosms were sprayed with malathion to alleviate an aphid infestation.

Soil leachates were monitored for nutrient loss on May 15, June 18, and July 23. RO water was added to the top of each soil-core microcosm and allowed to percolate through the core and Buchner funnel, into an acid-washed collection flask under each funnel. The leachate volume was recorded, leachates from the three microcosms in a cart were pooled, and the pooled leachate was filtered through a 0.45-μm filter and stored at 4°C in the dark.

Monitoring and Analysis of Ecosystem-Level Parameters—Effects of the obscurant smokes on the soil-plant system were monitored through analysis of plant nutrients in soil-core leachate, measurement of total above-ground plant biomass, and analysis of element uptake in above-ground plant biomass.

The first set of soil-core leachates were analyzed for five nutrients. Within 24 h after collection, they were analyzed for ammonium-nitrogen (NH_4-N) via a specific ion electrode and an Orion Research Model 601A digital ion analyzer (Standard Method 417E) [*21*]. These leachates also were analyzed for total organic carbon (TOC) by wet oxidation (consistent with Method 9060) [*22*] using an Oceanography Institute Model 524 carbon analyzer. It was assumed that the TOC values were equal to dissolved organic carbon (DOC), since samples were filtered through a 0.45-μm filter. Calcium (Ca) and potassium (K) were determined using a Model 303 Perkin-Elmer AAS (Standard Method 301.AII) [*23*]. Nitrate-nitrogen

(NO_3-N) was determined using the automated Cd reduction method (Standard Method 605) [23].

Leachates from the second and third leachings were treated similarly, except the number of nutrients analyzed was reduced from five to two. Based on the results (no dose effects on NH_4-N, DOC, or K) from the first leaching, only NO_3-N and Ca were measured in the remaining two leachings.

Plants were harvested at 11 and 81 days after first exposure to the phosphorus smokes, which was, respectively, 105 and 175 days after planting. Crops were cut about 5 cm above the soil surface, and the biomass was oven-dried at 100°C for 72 h. In the first harvest, all plant species were harvested, dried, and weighed together. In the second harvest, each species was dried and weighed separately.

For each harvest, plant biomass was ground in a Wiley mill, mixed, subsampled, and digested with perchloric and nitric acids. Biomass from all species in each of the three microcosms in a given cart (single treatment level) was combined prior to grinding. Biomass from the first harvest was analyzed for 23 elements (Al, B, Ba, Ca, Cd, Co, Cr, Cu, Fe, Mg, Mn, Mo, Na, Ni, P, Pb, Se, Sr, Ti, Tl, V, Y, and Zr) by inductively-coupled-argon-plasma spectroscopy (ICAP), and for As and Se by hybrid generation atomic absorption spectrophotometry (AAS) (Standard Method 301.A VII) [23]. A simultaneous multi-element standard and calibration check was performed at routine intervals during the day to monitor the precision of the method. The instrument was recalibrated if the values detected for the check standard concentrations were more than ±15% in error from the known value. Based on statistical analysis of element uptake data from the first harvest, the biomass from the second harvest was analyzed for five elements (Al, Cr, Mo, P, and Pb) using ICAP and for As by hydride generation AAS.

Statistical Methods—The microcosm exposures to each smoke were designed and performed as two separate studies and were analyzed as such. In all cases involving exposure chambers the microcosm group (or cart) was the unit of replication giving ten independent observations on which to base each dose response curve. A single group measurement from each of the two carts per treatment was obtained from a pooled sample (e.g., leachates from three microcosms per cart were pooled for each one of the three collection dates) or by taking an average of the three individual measurements (e.g., biomass). The pooling reduces the overall error variance and avoids the problem of incorrectly using individual microcosms as independent observations, (i.e., the "psuedo replication" described by Hurlbert [24]).

Quadratic regression curves of the form $y = b_0 + b_1x + b_2x^2$, where y is the response and x is a measure of exposure, were used to describe the dose-response relationship. When the quadratic term was not significant at the 5% level, a linear curve, $y = b_0 + b_1x$, was used to summarize the relationship.

Results and Discussion

Exposure Characterization

Only minor changes in environmental variables were noted during exposures. Ancillary measurements indicated that combustion products (CO ≈ 6 ppm, NO ≈ 20 ppb, and NO_2 < 10 ppb) were all well below values which are physiologically significant to plants, especially considering the short exposure duration (<10 min) of peak concentrations. Also, the aerosol size distribution of the smokes was consistent with measurements reported in other studies of phosphorus smokes. Aerosol samples were analyzed gravimetrically and found to have an aerodynamic mass median diameter (AMMD = 1.0 to 1.2 μm) and a geometric standard deviation σg = 1.5 to 1.7 [5] similar to other published data on phosphorus smokes [1,4].

The aerosol mass concentration in each exposure chamber was measured just after the end of the aerosol generation period to provide a check on the consistency of the smoke generation and produce a measure which can be related to field experience. The measured aerosol mass concentrations were generally within 10% of the target smoke concentration. Analysis of the aerosol mass concentration data showed that, with the exception of the two lowest WP exposures, all four of the exposure levels were significantly ($p < 0.0001$) different from one another.

The mass of aerosol deposited on horizontal surfaces in the exposure chambers was quantified by analyzing the phosphorus mass collected on deposition coupons, using base titration of an aqueous extract. The aerosol mass deposited on the coupons was expected to be similar to the dose deposited on horizontal plant leaves and soil in the microcosms. The deposited phosphorus mass measurements for a given target dose had relative standard deviations on the order of $\pm 30\%$ and $\pm 15\%$, respectively, at the low and high dose levels.

Nutrient Loss

In the first leachate, concentrations of NH_4-N were similar to those in distilled water irrespective of smoke type or exposure dose, and there was no statistically significant effect of smoke exposure on TOC or K (Tables 1 and 2). Thus these three nutrients were not analyzed in subsequent leachates.

TABLE 1—*Loss of Ca and NO_3-N on three separate leaching dates from microcosms exposed to RP/BR smoke.*

		Weight (μg) of Nutrient Loss in Leachate			
		Ca^a		NO_3-N^b	
Leachate	Target Dose (mg/m^3)	Mean[c]	$\pm$ Std. Dev.	Mean[c]	$\pm$ Std. Dev.
First	0	28272.50	6325.07	156.65	202.30
	100	14321.50	6672.97	11.65	2.76
	300	11763.75	4506.04	9.40	2.97
	600	12896.25	6464.72	10.55	4.45
	1500	28112.50	3676.96	17.30	2.26
Second	0	20771.25	5467.70	55.80	47.94
	100	12248.75	7755.19	68.53	62.12
	300	15255.00	8407.50	25.10	12.87
	600	11658.75	1702.36	33.30	32.24
	1500	24207.50	14311.84	68.55	15.63
Third	0	13822.50	1276.33	41.67	39.70
	100	11882.50	2549.12	8.35	5.02
	300	9297.50	1325.83	8.85	0.07
	600	12900.00	3867.87	11.25	0.78
	1500	34552.50	10977.83	18.70	0.14
Cumulative Sum	0	62866.25	13069.10	254.12	194.07
	100	38452.75	16977.28	88.53	69.90
	300	36316.25	2575.64	43.35	15.91
	600	37455.00	8630.24	55.10	37.48
	1500	86872.50	7010.96	104.55	18.03

[a] Ca = Calcium.
[b] NO_3-N = Nitrate-Nitrogen.
[c] N = Two carts per target dose; leachate from three microcosms per cart was pooled for each of the two carts before chemical analysis.
[d] Statistically significant quadratic dose-response, $p < 0.01$.

TABLE 2—*Loss of Ca and NO_3-N on three separate leaching dates from microcosms exposed to WP smoke.*

		Weight (μg) of Nutrient Loss in Leachate			
		Ca^a		NO_3-N^b	
Leachate	Target Dose (mg/m³)	Mean^c	±Std. Dev.	Mean^c	±Std. Dev.
First	0	10810.00	975.81	97.00	123.04
	100	13872.50	130.81	45.00	19.52
	300	15853.75	5891.97	13.35	6.86
	600	10810.00	1774.84	60.25	67.25
	1500	15521.75	6847.27	11.40	3.54
Second	0	24786.25	17912.78	72.28	2.23
	100	17475.00	5536.65	29.50	22.20
	300	30943.75	836.15	22.55	1.77
	600	14355.00	2877.92	89.48	31.50
	1500	18653.75	1833.17	27.55	20.44
Third	0	21175.00	8414.57	15.00	0.71
	100	19845.00	1732.41	15.25	2.76
	300	33208.75	17166.78	17.95	1.06
	600	12171.25	8600.19	11.65	4.60
	1500	29157.00	10441.14	22.60	4.95
Cumulative Sum	0	56771.25	25351.55	184.28	125.97
	100	51192.50	7138.24	89.75	38.96
	300	80006.25	10438.66	53.85	7.57
	600	37336.25	9703.27	161.38	103.34
	1500	63332.50	5427.04	61.55	21.85

^{a} Ca = Calcium.
^{b} NO_3-N = Nitrate-Nitrogen.
^{c} N = Two carts per target dose; leachate from three microcosms per cart was pooled for each of the two carts before chemical analysis.

A quadratic regression with deposition of phosphorus as the dependent variable detected no significant effect of smoke exposure on cumulative (total of three leachates) NO_3-N loss. However, there was a statistically significant effect ($p < 0.01$) of RP/BR smoke on cumulative Ca loss (Fig. 2). These data are presented in Tables 1 and 2.

It appears that exposure of the soil-plant system to relatively high concentrations of RP/BR or WP smokes over an 8-week period had little, if any, detrimental impact on nutrient loss, despite the fact that other studies have demonstrated significant nutrient loss from terrestrial ecosystems due to physical or chemical stresses [25–28]. In an earlier study using soil-core microcosms similar to those reported here, cumulative nutrient losses in leachates correlated well with decreases in plant productivity when soil was amended with toxic levels of fly ash [15].

With the exception of Ca, the nutrient loss data suggest that the obscurant smokes had little short-term (3 months) impact on the soil-plant system under the conditions used in this study (soil type, plant species, microcosm design, etc.). This conclusion is substantiated by biomass yield measurements (see below). The only significant depression in biomass associated with RP/BR was observed with sweetclover exposed to the highest concentration. In general, legumes require considerably more Ca than nonlegumes [29], so a reduction in sweetclover growth could be manifested as increased Ca loss. Conversely, increased yield of sweetclover biomass at the middle dose group may have been associated with Ca uptake and reduced the cumulative Ca loss shown in Fig. 2.

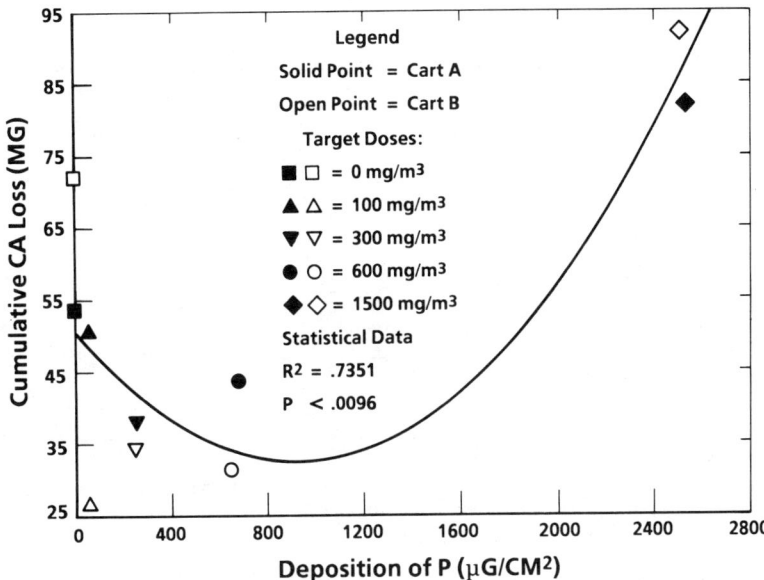

FIG. 2—*Effects of exposure to RP/BR smoke on cumulative (total from three leachates) loss of calcium: Observed values and fitted quadratic regression curve.*

Biomass Yield

For the first harvest, only total biomass was measured. The data showed no statistically significant ($p < 0.05$) effect of exposure to either RP/BR or WP smoke.

Biomass data from the second harvest were obtained for each species (wheat, ryegrass, and sweetclover) individually and for the combined biomass for both harvests (Tables 3 and 4).

TABLE 3—*Biomass[a] yield for individual species from the second harvest of microcosms exposed to RP/BR or WP smoke.*

		Biomass (g)					
		Wheat		Ryegrass		Sweetclover	
Smoke	Target Dose (mg/m^3)	Mean[b]	±Std. Dev.	Mean[b]	±Std. Dev.	Mean[b]	±Std. Dev.
RP/BR	0	6.44	0.63	0.48	0.14	3.71 [c]	1.48
	100	7.03	0.72	0.39	0.20	2.92	0.45
	300	6.37	0.38	0.49	0.14	4.50	0.49
	600	6.27	0.98	0.58	0.18	5.03	1.83
	1500	7.48	1.66	0.66	0.06	1.90	0.33
WP	0	6.72 [d]	1.08	0.59	0.48	4.55 [c]	0.05
	100	6.56	0.45	0.64	0.10	4.81	0.43
	300	6.31	1.71	1.04	0.28	6.13	0.26
	600	5.59	0.57	0.44	0.05	5.75	0.52
	1500	8.70	0.63	0.58	0.08	1.5	0.54

[a] Above-ground dry weight.
[b] N = two carts per target dose; biomass for individual plant species from each of the three microcosms per cart was averaged to give a single value for each cart.
[c] Statistically significant quadratic dose-response, $p < 0.05$.
[d] Statistically significant linear dose-response, $p < 0.05$.

TABLE 4—*Grand total biomass[a] yield for all species from both harvests of microcosms exposed to RP/BR or WP smoke.*

Smoke	Target Dose (mg/m³)	Biomass (g)	
		Mean[b]	±Std. Dev.
RP/BR	0	29.20	0.97
	100	28.58	0.15
	300	30.48	1.98
	600	31.88	3.18
	1500	28.63	1.33
WP	0	31.52	3.67
	100	30.73	1.28
	300	33.00	3.57
	600	31.50	0.89
	1500	28.94	1.14

[a] Above-ground dry weight.
[b] N = two carts per target dose; biomass for all species from both harvests was averaged over each of the three microcosms per cart to give a single value for each cart.

Neither smoke produced a significant effect on total biomass (both harvests) or on ryegrass biomass (second harvest).

A significant effect of both smokes on sweetclover biomass was detected (Table 3). Figure 3 shows the regression analysis for WP. For both smokes, the middle dose group showed improved sweetclover yield relative to controls, while the high dose group showed a reduced

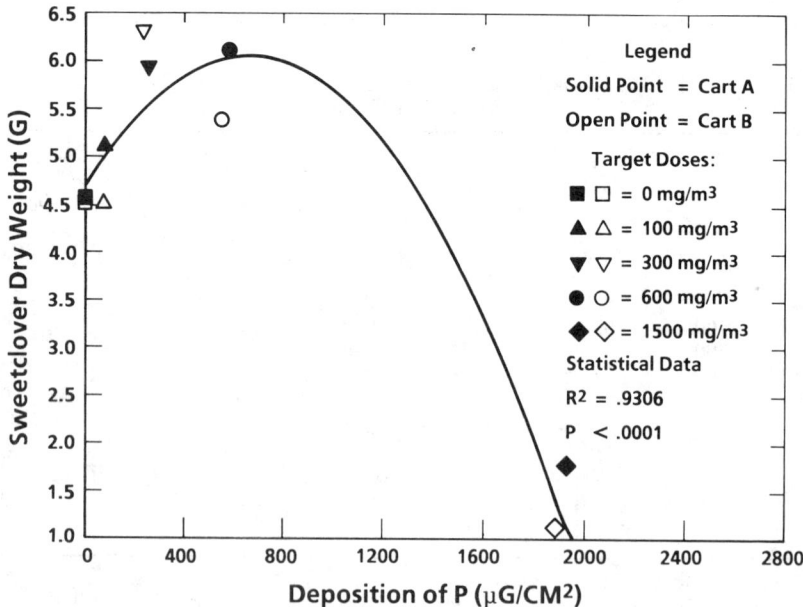

FIG. 3.—*Effect of exposure to WP smoke on sweetclover biomass: Observed values and fitted quadratic regression curve.*

yield relative to controls (e.g., Fig. 3). The sweetclover may have been affected more than the other plant species for two reasons. Firstly, since the sweetclover appeared to have the greatest *horizontal* leaf surface area, it would be expected to receive more smoke aerosol deposition than the other two plant species. This possibility is supported by the preliminary stress-ethylene test, where sweetclover had the steepest dose-response curve of the three species. Data in Tolle et al. [5] indicate that wheat had almost the same total leaf surface area as sweetclover, but that the wheat leaves were oriented *vertically*. Secondly, legumes like sweetclover are extremely sensitive to As [30], which has been measured in unburned WP at a concentration of 84 ppm [3].

Exposure to WP smoke increased wheat biomass significantly relative to the controls (Table 3). The effect may be due to increased As supplied as a contaminant in the WP. Calcium arsenate has been shown to increase wheat biomass [30]. A similar trend was not detected for RP/BR in spite of the fact that more total P was deposited [5] and more As was taken up by the plants (see following section on Element Uptake) at the highest RP/BR dose than at the highest WP dose. It is possible that the As levels in plant tissue (2.55 ppm) at the highest RP/BR smoke treatment level were beyond the stimulatory stage for wheat and beginning to reach toxic levels, while As concentrations in plant tissue (0.9 ppm) at the highest WP smoke treatment level were in the stimulatory range for wheat. Arsenic tissue levels of 0.15 to 0.30 ppm in wheat grain are considered intermediate between deficiency and toxicity [30].

Element Uptake

Statistical analysis of 24 elements in biomass from the first harvest [5], as well as information in the literature on the relative toxicity of the elements to terrestrial biota [7,31,32], resulted in the selection of six elements (P, As, Al, Pb, Cr, and Mo) for analysis in the second harvest. For the first harvest, uptake was regressed on target dose. For the second harvest, uptake was regressed on total P deposition. Chromium and Mo were chosen because of their relatively high toxicity to plants and animals. Zirconium was not analyzed in the second harvest even though there was significant uptake in plants from the first harvest, because of the element's relatively low toxicity.

Statistical analysis of plant concentration data from the second harvest showed significant effects of one or both smokes on uptake of As, P, Al, and Pb (Table 5). Plant uptake of Al and Pb was significantly affected by WP smoke. There was no significant effect ($p > 0.05$) of exposure to either smoke on Cr or Mo uptake or of exposure to RP/BR smoke on Al or Pb uptake. The relationship between element uptake and exposure was adequately described by a straight line for As uptake with RP/BR smoke, P uptake with RP/BR and WP smokes, Al uptake with WP smoke, and Pb uptake with WP smoke (Table 5). However, the relationship between As uptake and exposure to WP smoke followed a quadratic curve (Fig. 4).

Although the levels of elemental impurities were not determined for the RP/BR and WP smoke batches used in this study, uptake of Al, As, and Pb by plants exposed to these smokes is probably a result of their presence as impurities in the unburned phosphorus material in the samples burned. Katz et al. [3] reported the following elemental impurities in WP: Al-20 ppm, As-84 ppm, and Pb-1.27 ppm. Ballou [1] reported that RP also contains trace impurities.

The As levels in plant tissue exposed to the smokes (up to 2.55 ppm) were above the normal (1ppm) level reported by Allaway [32] and may have affected biomass of sweetclover and wheat. Significant declines in sweetclover biomass at high treatment levels for both smokes (Table 3) may be due to As, since Liebig [30] reports that legumes have little or no tolerance to this element. On the other hand, increased wheat biomass at the high treatment

TABLE 5—Element concentrations (μg/g) in plant tissue from the second harvest of microcosms exposed to RP/BR or WP smoke.

Smoke	Element	\multicolumn{2}{c}{0}		\multicolumn{2}{c}{100}		\multicolumn{2}{c}{300}		\multicolumn{2}{c}{600}		\multicolumn{2}{c}{1500}	
		Mean[a]	±Std. Dev.	Mean[a]	±Std. Dev.	Mean[a]	±Std. Dev.	Mean[a]	±Std. Dev.	Mean[a]	±Std. Dev.
RP/BR	Aluminum	13.90	3.96	16.15	2.62	14.35	2.47	18.50	2.26	17.25	5.73
	Arsenic[b]	0.05[c]	0.00	0.07	0.03	0.30	0.00	0.65	0.07	2.55	0.49
	Chromium	3.25	3.46	1.00	0.57	1.30	0.71	1.55	0.07	2.40	1.13
	Lead	7.15	0.21	6.70	2.55	6.70	2.97	10.15	1.06	9.60	1.84
	Molybdenum	5.20	6.65	10.85	4.03	7.70	0.57	10.05	1.48	7.00	6.36
	Phosphorus[b]	2428.20	322.72	3312.95	1093.12	4057.05	204.42	7475.65	117.59	15287.50	2245.06
WP	Aluminum[d]	18.25	2.47	18.45	0.64	26.00	4.10	19.75	4.03	28.30	3.39
	Arsenic[e]	0.13	0.11	0.13	0.11	0.15	0.07	0.15	0.07	0.90	0.28
	Chromium	0.95	0.21	1.85	1.48	1.70	0.57	1.25	0.92	2.15	1.20
	Lead[d]	7.75	0.35	9.00	1.56	14.20	1.13	10.00	4.95	15.70	0.99
	Molybdenum	9.25	1.20	10.30	0.71	11.15	1.06	9.15	1.77	11.25	1.20
	Phosphorus[f]	2118.80	176.78	2654.40	128.13	5326.90	2194.72	4569.40	1130.52	12852.00	1162.48

[a] N = two carts per target dose; biomass was composited for all species and all three microcosms per cart before chemical analysis.
[b] Statistically significant linear dose-response, $p < 0.001$.
[c] Detection limit has been set at 0.05 μg/g; values less than that have been set to 0.05.
[d] Statistically significant linear dose-response, $p < 0.05$.
[e] Statistically significant quadratic dose-response, $p < 0.001$.
[f] Statistically significant linear dose-response, $p < 0.0001$.

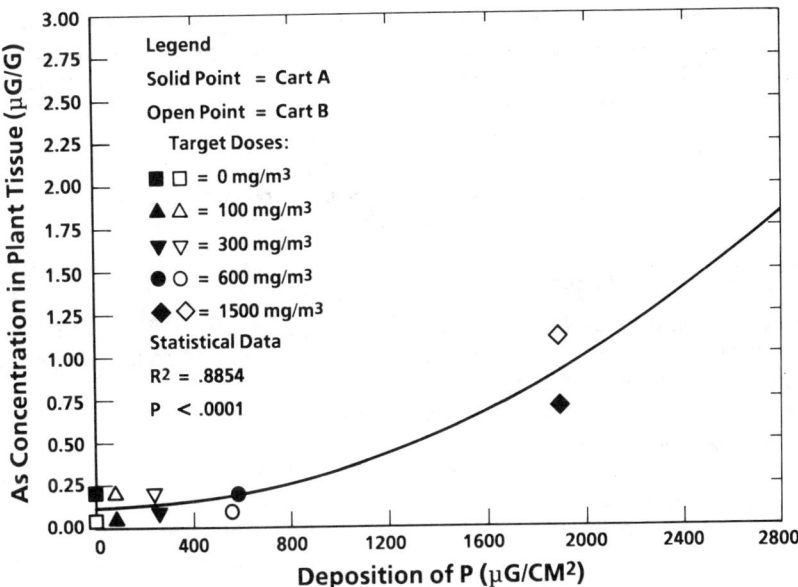

FIG. 4.—*Effect of exposure to WP smoke on arsenic concentrations in plant tissue: Observed values and fitted quadratic regression curve.*

level of WP smoke may be due to the beneficial effect of As on wheat. Liebig [30] reports that wheat yields were improved after application of calcium arsenate.

Uptake of As and Pb does not pose a potential short-term threat to even the most significant grazing animals for which acute toxicity data are available [7]. The highest mean As concentration (2.55 ppm) in plants exposed to either smoke is still below the level (>3.4 ppm) considered toxic to sheep. Similarly, the highest mean Pb concentration (15.7 ppm) in plants exposed to either smoke is well below the level (>80 ppm) considered toxic to horses.

The improved yield in sweetclover biomass at intermediate treatment levels (Table 3 and Fig. 3) is not due to correction of a P deficiency. Plant tissue concentrations for P given in Table 5; are well above the levels indicative of a P deficiency [31]. In addition, half-strength Hoagland's nutrient solution was added based on fertilizer recommendations from soil analysis.

Conclusions and Recommendations

Utility of Microcosm Technique

The terrestrial microcosm and static exposure chamber system used in this study produced statistically-significant dose-response curves for three types of ecological effects with two, low-acute-toxicity aerosols. Thus the system appears to be appropriate for evaluating the static, long-term, particle-deposition effects of other aerosols. Although this test involved only 6 months of plant growth in the microcosms, which ended in 8 weeks of smoke exposure followed by 4 weeks of post-exposure monitoring, another ecological effects test with the same type of microcosm produced field-validated results for two years [10,15].

The terrestrial microcosm technique is particularly relevant to hazard assessment. Firstly, this laboratory technique has been shown to accurately predict effects recorded from equivalent treatment of two waste materials applied in the field [14,15]. Secondly, it consists of numerous species of plants and soil organisms in an intact soil column and thus duplicates field-level physical, chemical, and biological processes more accurately than single-species studies. Compensatory mechanisms that occur in terrestrial ecosystems occur naturally in the microcosm. Thirdly, the terrestrial microcosm design used in this study has been shown to accurately predict effects over periods as long as two years [15], and thus can be continued for long-term chronic studies if preliminary data suggest a potential problem.

Use of Expanded Range-Finding Test

The expanded range-finding type of microcosm test produces input for hazard assessment at a considerable saving in cost and time compared to the more traditional range-finding plus definitive test strategy used in other studies by these authors [14,15]. The plant stress ethylene test essentially replaces a microcosm range-finding test at an estimated 20% time and cost savings. The expanded microcosm range-finding test combines aspects of both the range-finding and definitive tests and offers an estimated 50% savings over the time and cost required for a full-growing season definitive microcosm test. Our results corroborate those in Brusick and Young [20], where the use of expanded range-finding tests was recommended for a large number of different static acute toxicity tests as offering considerable savings in time and cost without compromising the quality and utility of the resulting data for use in hazard assessments.

Potential Ecological Effects and Implications for Smoke Use

The ecosystem-level effects of 16 semiweekly static exposures of intact soil-core microcosms to RP/BR or WP smokes were determined to be insignificant at expected field concentrations by monitoring nutrient loss, biomass yield, and element uptake. Minor effects were noted in these three parameters at the highest (1500 mg/m^3) treatment level, but this concentration is far above expected field concentrations (50 to 400 mg/m^3). Minor ecosystem-level effects due to exposure to the highest RP/BR smoke treatment included Ca loss in leachate, decline in sweetclover biomass, and increased uptake of As to potentially toxic levels. For WP smoke, minor ecosystem effects at the highest treatment level included increased wheat biomass, decreased sweetclover biomass, and increased uptake of As to potentially toxic levels. The increase in As uptake may have caused the decline in sweetclover biomass [30] as well as indirectly causing loss of Ca in leachate [26,33,34].

Since typical field concentrations are less than 600 mg/m^3 and the number of repeat exposures at a single training ground are less than the 16 used in this test for both smokes, no impacts on terrestrial ecosystems with plant species and soil type similar to the ones used in this study are expected from typical use of these smokes. It should be cautioned that the static exposures used in this study were designed to mimic the short-term passage of peak smoke concentrations over vegetation, under very low wind. Continuous smoke generation, where vegetation is exposed to peak smoke concentrations lasting for more than 10 min, or smoke impacts under windy conditions were not evaluated.

The results of the study suggest that minor negative effects, although unlikely, could occur with sensitive plant species or soil types. If the single, acute smoke concentration exceeds 20 000 mg/m^3, sensitive plant species such as sweetclover may be directly affected. If the chronic, repeated smoke concentrations reach or exceed 1500 mg/m^3 and/or the frequency of exposures exceeds at least double the number (16 exposures in 8 weeks) used in this study, negative ecosystem effects may include increased Ca loss, decreased biomass yield, or increased plant uptake of potentially toxic elements. If high smoke concentrations and exposure frequencies are used on training grounds with noncalcareous, low-cation-exchange-capacity, circumneutral soils, the resulting drop to an acid soil pH will make potentially toxic elements more available to plants, especially Al. The soils considered potentially sensitive to acid precipitation (and thus to acidic aerosols of WP or RP/BR smoke) have been mapped by acid rain research [35,36].

Acknowledgements

The financial support of the USAMRDC under Contract DAMD17-84-C-4001 for conducting this research is greatly appreciated. The close cooperation of Mr. Jesse J. Barkley, Jr., and CPT (P) Gary M. Bratt of USAMBRDL is gratefully acknowledged. The assistance of the following Battelle technical staff on this research also deserves special thanks: Drs. Michael R. Kuhlman, Danny R. Jackson, and Paul I. Feder, and Messrs. Vladimir Kogan and Dean P. Margeson. Special thanks are due technicians Thomas C. Zwick and G. Kelly O'Brien, who assumed the lion's share of the plant maintenance and microcosm exposure activities.

References

[1] Ballou, J. E., "Chemical Characterization and Toxicologic Evaluation of Airborne Mixtures," NTIS Report No. ADA102678, Contract No. DAMD17-79-C-9160, U.S. Army Medical Research and Development Command, Fort Detrick, Frederick, Md., 1981.
[2] Wasti, K., Abaidoo, K. J. R., Villaume, J. E., and Craig, P. N., "A Literature Review: Problem Definition Studies on Selected Toxic Chemicals, Vol. 2 of 8: Occupational Health and Safety Aspects of Phosphorus Smoke Compounds," NTIS Report No. ADA056019, Contract No. DAMD17-77-C-7020, U.S. Army Medical Research and Development Command, Fort Detrick, Frederick, Md., April 1978.
[3] Katz, S., Snelson, A., Butler, R., Bock, W., Rajendran, N., and Relwani, S., "Physical and Chemical Characterization of Military Smokes: Part III—White Phosphorus-Felt Smokes," NTIS Report No. ADA115657, Contract No. DAMD17-78-C-8085, U.S. Army Medical Research and Development Command, Fort Detrick, Frederick, Md., May 1981.
[4] Burton, F. G., Clark, M. L., Miller, R. A., and Schirmer, R. E., *American Industrial Hygiene Association Journal*, Vol. 43, 1982, pp. 767–772.
[5] Tolle, D. A., Arthur, M. F., Chesson, J., Duke, K. M., Jackson, D. R., Kogan, V., Kuhlman, M. R., and Margeson, D. P., "Terrestrial Microcosm Evaluation of Two Army Smoke-Producing Compounds," NTIS Report No. ADA190797, Final Report on Contract DAMD17-84-C-4001 to U.S. Army Medical Research and Development Command, Fort Detrick, Md., by Battelle Columbus Division, Columbus, Ohio, Jan. 1988.
[6] Van Voris, P., Cataldo, D. A., Ligotke, M. W., Garland, T. R., McFadden, K. M., Fredrickson, J. K., Li, S. W., Bean, R. M., Thomas, B. L., and Carlile, D. W., "Evaluate and Characterize Mechanisms Controlling Transport, Fate and Effects of Army Smokes in the Aerosol Wind Tunnel," NTIS Report No. ADA191109, Final Report on Project Order No. 84PP4819 to U.S. Army Medical Research and Development Command, Fort Detrick, Md., by Pacific Northwest Laboratory, Richland, Wash., Oct. 1987.
[7] Dvorak, A. J. and Lewis B. G., Eds., "Impacts of Coal-Fired Power Plants on Fish, Wildlife, and Their Habitats," FWS/OBS-78/29, U.S. Department of the Interior, Fish and Wildlife Service, Office of Biological Services, Washington, D.C., March 1978.
[8] Gillett, J. W. and Witt, J. M., Eds., "Terrestrial Microcosms: The Proceedings of the Workshop on Terrestrial Microcosms," NSF/RA 79-0027, National Science Foundation, Washington, D.C., 1979.
[9] Hammons, A. S., *Methods for Ecological Toxicity: A Critical Review of Laboratory Multispecies Tests*, Ann Arbor Science Publishers, Ann Arbor, Mich., 1981.
[10] Tolle, D. A., Arthur, M. F., Chesson, J., and Van Voris, P., *Environmental Toxicology and Chemistry*, Vol. 4, 1985, pp. 501–509.
[11] Tolle, D. A., Van Voris, P., Arthur, M. F., Morris, J. P., and Larson, M., *Bulletin of the Ecological Society of America*, Vol. 62, 1981, pp. 141–142.
[12] Tolle, D. A., Arthur, M. F., and Van Voris, P., *The Science of the Total Environment*, Vol. 31, 1983, pp. 243–261.

[13] Van Voris, P., Tolle, D. A., and Arthur, M. F. in *Proceedings*, First International Waste Recycling Symposium, Clean Japan Center, Tokyo, Japan, 1981, pp. 389–404.
[14] Van Voris, P., Arthur, M. F., and Tolle, D. A., "Evaluation of Terrestrial Microcosms for Assessing Ecological Effects of Utility Wastes," NTIS Report No. DE82903730, EPRI EA-2364, Electric Power Research Institute, Palo Alto, Calif., April 1982.
[15] Van Voris, P., Tolle, D. A., Arthur, M. F., Chesson, J., and Zwick, T. C., "Development and Validation of a Terrestrial Microcosm Test System for Assessing Ecological Effects of Utility Wastes," NTIS Report No. DE85001913 (Final Report), EPRI EA-3672, Electric Power Research Institute, Palo Alto, Calif., Aug. 1984.
[16] Zwick, T. C., Arthur, M. F., Tolle, D. A., and Van Voris, P., *Plant and Soil*, Vol. 77, 1984, pp. 395–399.
[17] Arthur, M. F., Zwick, T. C., Tolle, D. A., and Van Voris, P., *Water, Air, and Soil Pollution*, Vol. 22, 1984, pp. 209–216.
[18] U.S. Environmental Protection Agency, "Guideline 797.3775, Soil-Core Microcosm Test," *Federal Register*, Vol. 52, 1987, pp. 36363–36371.
[19] ASTM Guide for Conducting a Terrestrial Soil-Core Microcosm Test (E 1197), in *1988 Annual Book of ASTM Standards*, Philadelphia, Vol. 11.04, pp. 743–755.
[20] Brusick, D. J. and Young, R. R., "IERL-RTP Procedures Manual: Level 1 Environmental Assessment Biological Tests," EPA-600/8-81-024, U.S. Environmental Protection Agency, Research Triangle Park, N.C., Oct. 1981.
[21] Greenburg, A. E., Conners, J. J., and Jenkins, D., Eds., "Standard Methods for the Examination of Water and Wastewater," 15th ed., APHA-AWWA-WPCF, American Public Health Assn., Washington, D.C., 1981.
[22] U.S. Environmental Protection Agency, "Test Methods for Evaluating Solid Waste: Vol. 1C—Laboratory Manual Physical/Chemical Methods," 3rd ed., SW-846, Office of Solid Waste and Emergency Response, U.S. EPA, Washington, D.C., 1986.
[23] Rand, M. D., Greenburg, A. E., and Taras, M. J., Eds., "Standard Methods for the Examination of Water and Wastewater," 14th ed., APHA-AWWA-WPCF, American Public Health Assn., Washington, D.C., 1976.
[24] Hurlbert, S. H., *Ecological Monographs*, Vol. 54, 1984, pp. 187–211.
[25] Likens, G. E., Bormann, F. H., Johnson, N. M., Fisher, D. W., and Pierce, R. S., *Ecological Monographs*, Vol. 40, 1970, pp. 23–47.
[26] Van Voris, P., O'Neill, R. V., Emanuel, W. R., and Shugart, H. H., Jr., *Ecology*, Vol. 61, 1980, pp. 1352–1360.
[27] Aumus, B. S., Dodson, G. J., and Jackson, D. R., *Water, Air, and Soil Pollution*, Vol 10, 1978, pp. 19–26.
[28] Jackson, D. R., Washburne, C. D., and Ausmus, B. S., *Water, Air, and Soil Pollution*, Vol. 8, 1977, pp. 279–284.
[29] Foth, H. D., *Fundamentals of Soil Science*, 6th ed., Wiley, New York, 1978.
[30] Liebig, G. F. in *Diagnostic Criteria for Plants and Soils*, H. D. Chapman, Ed., Quality Printing Company, Abilene, Tex., 1965, Chapter 2, pp. 13–23.
[31] Chapman, H. D., Ed., *Diagnostic Criteria for Plants and Soils*, Quality Printing Company, Abilene, Tex., 1965.
[32] Allaway, W. H., *Advances in Agronomy*, Vol. 20, 1968, pp. 235–274.
[33] Jackson, D. R., Selvidge, W. J., and Ausmus, B. S., *Water, Air, and Soil Pollution*, Vol. 10, 1978, pp. 13–18.
[34] Jackson, D. R., Ausmus, B. S., and Levin, M., *Water, Air, and Soil Pollution*, Vol. 11, 1979, pp. 13–21.
[35] McFee, W. W., "Sensitivity of Soil Regions to Acid Precipitation," EPA-600/3-80-013, Environmental Research Laboratory, Office of Research and Development, U.S. Environmental Protection Agency, Corvallis, Ore., Jan. 1980.
[36] McFee, W. W., *Environmental and Experimental Botany*, Vol. 23, 1983, pp. 203–210.

Curtis J. Richardson,[1] *Thomas W. Sasek,*[1] *and Richard T. Di Giulio*[1]

Use of Physiological and Biochemical Markers for Assessing Air Pollution Stress in Trees

REFERENCE: Richardson, C. J., Sasek, T. W., and Di Giulio, R. T., "**Use of Physiological and Biochemical Markers for Assessing Air Pollution Stress in Trees,**" *Plants for Toxicity Assessment, ASTM STP 1091,* W. Wang, J. W. Gorsuch, and W. R. Lower, Eds., American Society for Testing and Materials, Philadelphia, 1990, pp. 143–155.

ABSTRACT: Air pollutants such as O_3, NO_x, SO_2, and H_2O_2 are powerful oxidants that can generate extremely reactive oxygen free radicals that may cause enzyme breakdown, membrane damage, and DNA alterations, all resulting in reduced growth. In this study, specific gas exchange measurements were used diagnostically as indicators of stress and as a means of separating stomatal from biochemical effects on photosynthesis. Similarly, biochemical antioxidants and oxidant stress indicators were hypothesized to be useful as early biomarkers of oxidant stress in trees.

Loblolly pine (*Pinus taeda* L.) seedlings were exposed to a range of ozone treatments in open-top chambers at two study sites in North Carolina. Treatments consisted of charcoal-filtered (CF) air and proportional additions of ozone to non-filtered air in relation to the ambient ozone concentration. Diagnostic gas exchange measurements included the response of photosynthesis and stomatal conductance to irradiance and to intercellular CO_2 concentration (C_i). Biochemical measurements included the content of malondialdehyde (MDA), an indicator of lipid peroxidation, and the activities of the enzymatic antioxidants superoxide dismutase (SOD) and peroxidase (Px). These biochemical characteristics were also related to concurrent measurements of light-saturated photosynthesis.

In loblolly pine, elevated ozone (3.0× ambient ozone) decreased photosynthetic capacity under light-saturating conditions by up to 40% and decreased the apparent quantum yield of photosynthesis by up to 54% compared to the charcoal-filtered control. The same elevated ozone concentrations also reduced the initial slope of the photosynthesis versus C_i relationship by almost 34% and reduced the CO_2-saturated rate of photosynthesis by 28%. However, stomatal resistance reduced potential rates of photosynthesis by approximately 34% in both the elevated ozone treatment and the charcoal-filtered control. Thus reduced rates of photosynthesis at elevated ozone concentrations were due almost entirely to internal biochemical processes, not to stomatal effects. These biochemical effects may include light-harvesting and biochemical efficiencies of the photosystems, the activity of RuBP carboxylase, the regeneration rate of RuBP, and the electron transport capacity.

Malondialdehyde levels in loblolly seedlings exposed to 3.0× ambient ozone were twice those of pines exposed to charcoal-filtered air, indicating a significant increase in lipid peroxidation. At the same time, the activities of the antioxidants SOD and Px were higher in pines exposed to intermediate levels of ozone (1.5× ambient ozone), and more than twice the controls in the 3.0× ambient ozone treatment.

Our results suggest that diagnostic gas exchange techniques can be used as a sensitive indicator for separating stomatal from biochemical control of photosynthesis in trees exposed to oxidizing air pollutants. Additionally, biochemical antioxidants, such as Px and SOD, and oxidant stress indicators, such as MDA, may be useful as early biomarkers of oxidant stress.

KEY WORDS: loblolly pine, *Pinus taeda,* ozone, photosynthesis, conductance, free radicals, antioxidants, malondialdehyde, superoxide dismutase, peroxidase, biomarkers

[1] School of Forestry and Environmental Studies, Duke University, Durham, NC 27706.

Air pollutants including O_3, NO_x, SO_2, H_2O_2, and acid precipitation have been linked to forest decline in the United States and Europe [1–3]. Nearly a dozen hypotheses have been proposed to explain the inconsistency of damage symptoms that have been reported for forests stressed by air pollutants [4–6]. For key processes such as photosynthesis and metabolism, sensitive and specific physiological and biochemical bioassays are needed to assess the effects of air pollutants. In this paper, we present a two-phase approach to assessing air pollution damage. The approach is based on (1) diagnostic gas exchange techniques and (2) biochemical assays for oxidant stress (oxidant biomarkers).

Diagnostic Gas Exchange

Short-term or instantaneous measurements of photosynthesis and stomatal conductance are often used as an index of the physiological status of plants. Rates obtained from these studies are useful for comparing major physiological differences between trees or branches but do not allow for an understanding of mechanisms controlling photosynthetic capacity and water flux, especially under pollution stress conditions [7]. In addition, photosynthesis and transpiration rates may vary greatly during the day, seasonally, and even from branch to branch. In contrast, measurements of the response of photosynthesis and transpiration to changes in light, temperature, or CO_2 can be used to diagnose interactions between biological processes and environmental factors, providing clues to more long-term biochemical changes within the plant in response to stress [8]. By studying these response parameters diagnostically over a range of environmental conditions, the effects of temporal and micrometeorological variability are reduced.

Light Response Curves

The relationship between irradiance and photosynthesis can be used diagnostically to study the effects of stress on photosynthetic capacity and photochemical efficiency. At low irradiances, net fixation of CO_2 is linearly dependent on irradiance because the speed of the photochemical reactions limits photosynthesis. The initial slope of the linear portion of the curve is termed the apparent quantum yield [9]. Changes in the apparent quantum yield of photosynthesis are correlated to changes in the light harvesting system and to the efficiency of the photochemical reactions. At high irradiances, net fixation of CO_2 is limited by enzymatic reactions of photosynthesis and the rate of supply of CO_2. Changes in photosynthetic capacity at light saturation due to stress can be correlated to these biochemical processes.

The quantum yield of photosynthesis and the photosynthetic capacity, as measures of photosynthetic efficiency, depend primarily on long-term adaptation to prevailing environmental conditions, especially irradiance. For example, the development of "sun" and "shade" leaves results in morphological, anatomical, and biochemical changes of the entire photosynthetic apparatus. Chronic or acute environmental stressors can also affect the quantum yield of photosynthesis and photosynthetic capacity. Damage or inactivation of chlorophyll or the accessory pigments may reduce light capture. Similarly, membrane damage can affect the photosystems directly (e.g., by disrupting photosystem structure) or affect enzymatic and biochemical processes that regulate photosynthesis (e.g., by reducing rates of NADPH transport).

Carbon Dioxide Response Curves

Photosynthetic fixation of carbon depends on the concentration of carbon dioxide in the atmosphere. The relationship between net photosynthesis and CO_2 concentration therefore reflects both (1) the reactions of carbon fixation and (2) the extent of stomatal and biochemical limitations on photosynthesis.

Under normal conditions, the greatest resistance for the diffusion of CO_2 along the pathway from the atmosphere to the site of fixation is through the stomates [8]. Thus, the internal, intercellular CO_2 concentration will be less than the external, ambient CO_2 concentration. If stomatal conductance changes at different CO_2 concentrations, the relationship between intercellular CO_2 and ambient CO_2 may not be constant. To eliminate the confounding effects of stomatal conductance, the photosynthetic rate can be expressed in relation to the intercellular concentration of CO_2.

The intercellular concentration of CO_2 (C_i) is estimated from the external CO_2 concentration (C_a), the experimentally measured rate of photosynthesis (A), and the stomatal conductance to CO_2 (G_c), which is determined from the measured rate of transpiration (E). However, as water evaporates from the leaf, some of the CO_2 diffusing into the leaf is carried back out by mass flow. The calculation described by von Caemmerer and Farquhar [8] (Eq 1) relates photosynthesis to the CO_2 gradient and the mass flow loss of CO_2:

$$A = G_c(C_a - C_i) - \frac{E(C_i + C_a)}{2} \tag{1}$$

This equation can be rearranged to calculate C_i:

$$C_i = \frac{\left(G_c - \frac{E}{2}\right)C_a - A}{\left(G_c + \frac{E}{2}\right)} \tag{2}$$

Although the intercellular concentration of CO_2 is a calculated value, research has shown that the equation accurately predicts measured values [10].

Stomatal Limitations to Photosynthesis

Stomatal limitations to photosynthesis have been analyzed by Farquhar and Sharkey [11]. Stomatal limitations are estimated using the experimentally determined relationship between A and C_i. At a given external, ambient concentration of CO_2 (C_a), A and C_i are known. However, if stomatal conductance is assumed to be infinite, C_i would equal C_a (Fig. 1a). Therefore the difference between the potential rate of photosynthesis at C_a and the actual rate of photosynthesis at C_i is due to stomatal limitations. This reduction can be expressed as a percentage decrease from the potential photosynthetic rate; it is an index of the amount of stomatal control over photosynthesis.

Biochemical Limitations to Photosynthesis

The shape of the A versus C_i curve is related to the biochemistry of photosynthesis. These relationships to biochemistry have been developed by von Caemmerer and Farquhar [8] and are based on considerable experimental evidence and validation. The following discussion highlights only the diagnostic uses of this type of analysis.

At low C_i, the net fixation of CO_2 is determined by the kinetics of the enzyme ribulose bisphosphate carboxylase/oxygenase (RuBPcase), since its substrate RuBP is present in saturating amounts. Therefore the initial slope of the CO_2 response curve is linearly related to

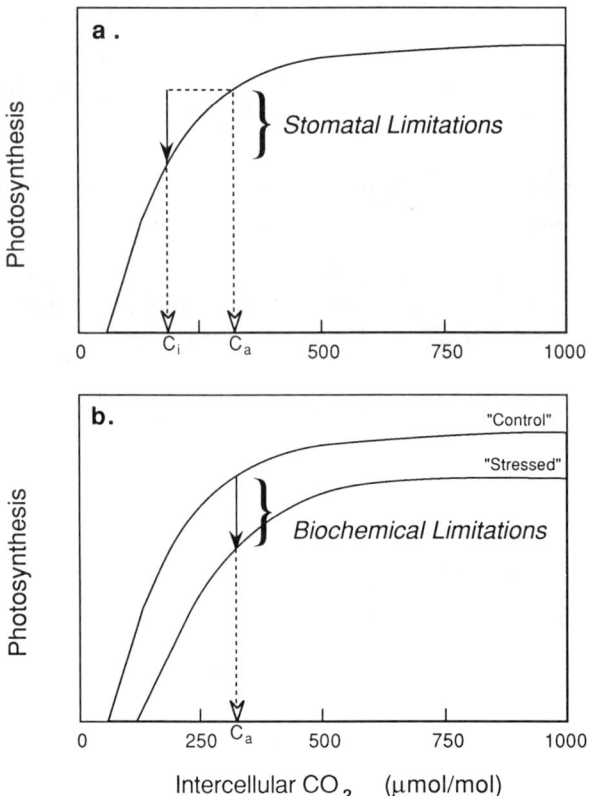

FIG. 1—*Theoretical response of photosynthesis to intercellular CO_2 concentration.* (a) *Stomatal limitations to photosynthesis are calculated from the ambient CO_2 concentration (C_a) to the experimentally determined intercellular CO_2 concentration (C_i).* (b) *Biochemical limitations to photosynthesis are calculated as the difference between the curves at a specific CO_2 concentration, such as C_a.*

CO_2. In saturating light, however, the initial slope is affected by O_2 concentration and temperature. The initial slope is often termed the carboxylation efficiency. If the maximum activity of RuBPcase is assumed to be proportional to the amount of RuBPcase present, then the initial slope can be used as an index of the size of the RuBPcase pool.

At high concentrations of CO_2, the RuBP-saturated kinetics of RuBPcase alone cannot explain the rate of CO_2 fixation. At high C_i concentrations, the rate at which RuBP is regenerated is limiting. The regeneration of RuBP depends on the regeneration of ATP and NADPH and therefore depends on the electron transport capacity. Thus the saturated rate of net photosynthesis depends on irradiance and temperature but not on the concentration of CO_2. However, net photosynthesis increases slightly with increasingly high CO_2 concentration as the $CO_2:O_2$ ratio increases.

By comparing the A versus C_i curves from plants exposed to different pollutant treatments, it is possible to suggest biochemical mechanisms of response to the toxicants. It is also possible to determine the extent of biochemical changes caused by the pollutants. However, before calculating biochemical limitations due to pollutant exposure, it is important to consider differences in stomatal limitations, because stomates are often affected by experimental treatments. The degree of stomatal resistance does not change the relationship between A and C_i; it only affects the position on the curve. However, if the relationship between A and C_i differs for control and stressed plants, non-stomatal factors have affected one or more biochemical processes of photosynthesis (Fig. 1b). Stomatal limitations should then be calculated from the specific A versus C_i curve for each treatment before analyzing the curves for biochemical effects. The relative differences between A versus C_i curves may not be the same at all values of C_i. These differences may provide additional kinds of diagnostic information. However, in most cases, it may be most appropriate to calculate the biochemical limitations at the ambient CO_2 conditions under which the plant grows.

Oxidant Biomarkers

Photochemical oxidants such as SO_2, NO_x, and O_3 can damage plants and animals via the generation of free radicals [12–15]. A free radical is any atom, group of atoms, or molecule that has one or more unpaired electrons occupying the outer orbital [16]. Free radicals can be produced by oxidation or reduction reactions, or by ionizing radiation. Dioxygen reduction occurs commonly: the intermediates of this reduction (oxyradicals) are the superoxide anion radical ($O_2^- \cdot$), hydrogen peroxide (H_2O_2), and the extremely reactive hydroxyl radical ($OH \cdot$) (Fig. 2). While not a true free radical, H_2O_2 is generally grouped with oxyradicals due to its reactivity and role in generating $OH \cdot$. The hydroxyl radical can extensively damage a cell by oxidizing organic molecules, including enzymes and membrane lipids; it can also damage DNA [15,16]. The primary defense against oxyradicals, the antioxidant defense system, includes enzymatic and non-enzymatic components. Key antioxidant enzymes are superoxide dismutase (SOD), peroxidases (P_x), and catalase (Fig. 2). Non-enzymatic components include α-tocopherol (Vitamin E), β-carotenes, glutathione, and ascorbic acids [6,14]. Glutathione and ascorbate also play important roles in driving the activities of antioxidant peroxidases. Collectively, these enzymes and compounds are known as antioxidants. We hypothesize that an increase in the production or activity of these antioxidants by some genotypes within a population would result in increased biochemical resistance to air pollutants. It has also been suggested that changes in the activities or concentrations of these antioxidants could also be used as biomarkers of oxidant exposure in trees [17–21]. A key biochemical toxicity that occurs when antioxidant defenses are overwhelmed is lipid peroxidation [13]; its quantification may provide a useful marker for oxidative stress due to air pollutants. Our studies address these possibilities.

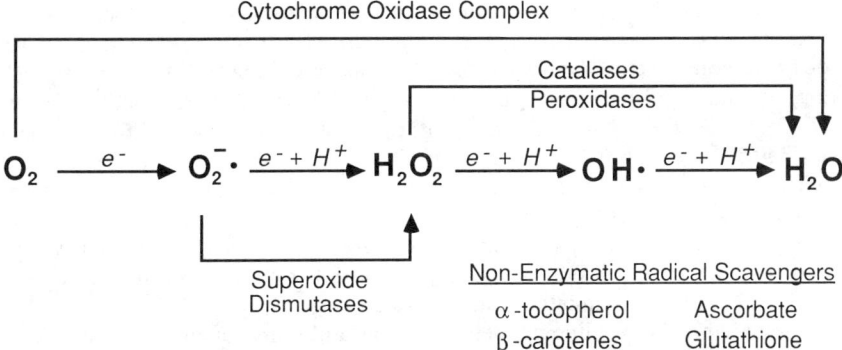

FIG. 2—*Oxygen radical generation and antioxidant defense mechanisms.*

Materials and Methods

Seedlings of loblolly pine (*Pinus taeda* L.) were examined from an acid rain and ozone exposure study at the Duke Forest Southern Commercial Forest Research Cooperative study site near Durham, North Carolina [22] and a three-year ozone exposure study on young loblolly pine near Raleigh, North Carolina [7]. Open-top field chambers [23], 3 m in diameter and 2.5 m tall, were used for exposures. Chambers with charcoal-filtered air (CF) removed approximately 50% of ambient ozone. Non-filtered (NF) chambers paralleled ambient ozone concentration. Ambient ozone concentration was continually monitored and additional ozone was proportionally dispensed to non-filtered air in the elevated treatment chambers by a computer-controlled system [24] to achieve target concentrations of $1.5\times$ or $3.0\times$ ambient ozone at the Duke Forest site and 2.0 times ambient ozone at the Raleigh site. Elevated ozone exposures occurred from 8:00 am to 8:00 pm (EST) except during precipitation events (natural or simulated) or when ambient ozone concentrations fell below 20 ppb. Ozone fumigation began 27 March and ended 6 December 1987 at the Duke site, and began 8 April and ended 1 October 1987 at the Raleigh site. The trees at the Raleigh site were also fumigated in 1985 and 1986 [7]. The highest ambient daytime peak concentrations of ozone during the summer of 1987 were 125 and 120 ppb at the Duke and Raleigh sites, respectively, but averaged 48 and 50 ppb (12 h mean), respectively, over the growing season. The trees sampled for the biochemical analyses were in a subgroup of chambers at the Duke site in which elevated ozone exposures occurred for 24 h per day. There were two replicate chambers of each treatment at both sites. At the Duke site, replicates were blocked to account for site variation.

The diagnostic gas exchange measurements were made with a portable null-balancing, infrared gas analysis system (PACsys 9900, Data Design Group, La Jolla, California; Software ver. 2.0) with a controlled-environment cuvette. The system controls temperature, relative humidity and CO_2 concentration within the cuvette. Irradiance was reduced by placing layers of neutral-density screening above the cuvette. During comparable measurements, similar environmental conditions were maintained. Typical conditions were 30°C, 350 ppm CO_2, and a 1.5 kPa vapor pressure gradient (leaf to air).

Loblolly pine produces several flushes of growth per year and has fascicles that normally function for two growing seasons. Gas exchange measurements in this study were conducted in late summer 1987 on fascicles from the first flush produced in 1987. Three attached fascicles (nine needles) were placed in the cuvette per measurement. Total fascicle surface area was estimated from a regression equation relating surface area to length [7], following a method modified from Johnson [25].

Bimonthly harvests of fascicles were conducted at approximately mid-month during the growing season. At each harvest, at least six seedlings per chamber were examined for *in situ* physiological responses before being harvested for biochemical assays. Immediately before harvesting, simultaneous measurements of photosynthesis and stomatal conductance were measured *in situ* with a portable photosynthesis system (LI-6200, LI-COR, Lincoln, Nebraska; Software ver. 1.03) with an uncontrolled environment cuvette (0.25 L volume). Two fascicles (six needles) were placed in the cuvette and each measurement lasted approximately 20 to 60 s. The system was controlled to maintain relatively constant vapor pressure ($\pm 5\%$) during the measurement, although temperature increased slightly ($<2°C$). Measurements were only made when irradiance was high enough to saturate photosynthesis (>1000 μmol m^{-2} s^{-1} PPFD [photosynthetic photon flux density]).

Needles were harvested by clipping them into vials and immediately immersing them in liquid nitrogen. The samples were stored at $-70°C$ at the Ecotoxicology Laboratory at Duke until analysis. Activities of superoxide dismutase were determined using the methods of McCord and Fridovich [26], Asada et al. [27], and Tandy et al. [21]. Peroxidase was analyzed using the methods of Putter and Becker [28]. Lipid peroxidation was indexed using the malondialdehyde (MDA) method described by Uchiyama and Mihara [29].

Results and Discussion

Diagnostic Gas Exchange

The response of 1987 first-flush fascicles to light was strongly affected by ozone. After approximately 150 days of exposure, photosynthetic capacity of the fascicles was reduced by 21% at 2.0× ambient ozone relative to the charcoal-filtered treatments at the Raleigh site, and by 40% at 3.0× ambient ozone at the Duke site (Fig. 3). The apparent quantum yield

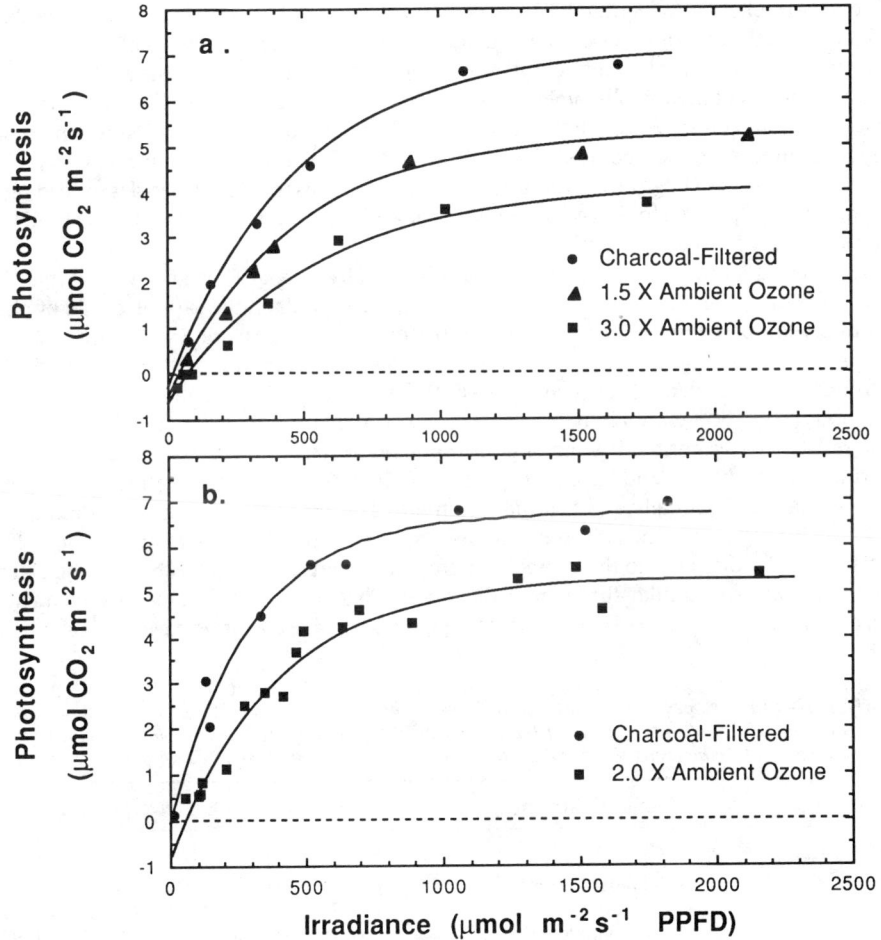

FIG. 3—*Response of photosynthesis of the 1987 first-flush needles to irradiance level in loblolly pine seedlings at the Duke Forest site (a) or at the Raleigh site (b) under various ozone treatments. All measurements were made at 33°C and 50% relative humidity. Results shown are combined from three trees within each treatment. The curves were fit by nonlinear regression with the function:*

$$Y = Y_{max} + (Y_{min} - Y_{max})e^{(-k \cdot Light)}$$

of photosynthesis at 2.0× ambient ozone was reduced by 39% relative to the charcoal-filtered treatment, and was reduced by 54% at 3.0× ambient ozone (Table 1).

The differences between elevated ozone exposures and charcoal-filtered treatments suggested that there may have been significant changes in biochemical efficiency and in the total light-harvesting efficiency of the pine photosystem. In poplar trees (*Populus deltoides* × *trichocarpa*), chronic exposure to ozone similarly reduced photosynthetic capacity and quantum yield compared to charcoal-filtered controls [30]. Reductions in apparent quantum yield of pine suggest that ozone can affect the photochemical reactions of the photosystems. Reduced photosynthetic capacity could result from higher rates of dark respiration, accelerated physiological aging of the leaves, or from direct biochemical effects [30]. Reduced photochemical efficiency can also be correlated to reduced concentrations of chlorophyll, though it is not possible to state cause-and-effect relationships [30].

Ozone is unlikely to penetrate past the cell plasmalemma before reacting, but secondary products and the consequences of ozone oxidations can still result in lower chlorophyll concentrations [31]. Chlorophyll and carotenoid concentrations per unit needle surface area in the oldest flushes of the loblolly pines at the Raleigh site were 20% less in the 2.0× ambient ozone treatment compared to CF [7]. Enzymatic degradation of chlorophylls may be more likely than direct oxidation because chlorophyll content does not decrease immediately after exposure to ozone [32]. However, chlorophyll concentrations in loblolly needles exceed rate-limiting levels [7]. Therefore decreased photosynthesis should not be directly attributed to decreased chlorophyll concentrations [30].

The A versus C_i curve from the three year old loblolly pine at Raleigh revealed that the 2.0× ambient ozone treatment reduced both the initial slope and the saturated rate of photosynthesis compared to the charcoal-filtered treatment (Fig. 4). An analysis of stomatal limitations showed that stomates reduced photosynthesis by approximately 35% in both the control and treated plants (Fig. 4 and Table 2). Teskey et al. [33] found similar stomatal limitations (20 to 30%) in loblolly pine under ambient environmental conditions. Thus stomatal resistance was not the key factor controlling photosynthesis in loblolly pine under the experimental conditions, and long-term exposure (150 d) to elevated concentrations of ozone (2.0×) did not differentially affect stomatal activity. The analysis of biochemical limitations showed that long-term exposure to ozone decreased photosynthesis by approximately 28% at ambient CO_2 compared to the charcoal-filtered treatment (Fig. 4 and Table 2). The extent of the reduction was similar throughout the curve. Changes in the gas exchange characteristics of loblolly pine due to ozone exposure suggested possible effects on several biochemical

TABLE 1—*Initial slopes of the relationship between photosynthesis and light or intercellular CO_2 concentration for loblolly pine grown at the Duke Forest site and the Raleigh site in different ozone exposure treatments. The regression for each treatment is derived from combined data from three trees.*

	Ozone Treatment	Initial Slope	
LIGHT CURVE:			
Duke Forest	Charcoal-Filtered	0.0100 ± 0.0019	$R^2 = 0.961$
	1.5× Ambient	0.0078 ± 0.0005	$R^2 = 0.996$
	3.0× Ambient	0.0046 ± 0.0006	$R^2 = 0.970$
Raleigh	Charcoal-Filtered	0.0135 ± 0.0008	$R^2 = 0.993$
	2.0× Ambient	0.0082 ± 0.0007	$R^2 = 0.952$
INTERCELLULAR CO_2 CURVE:			
Raleigh	Charcoal-Filtered	0.0350 ± 0.0005	$R^2 = 0.998$
	2.0× Ambient	0.0230 ± 0.0009	$R^2 = 0.980$

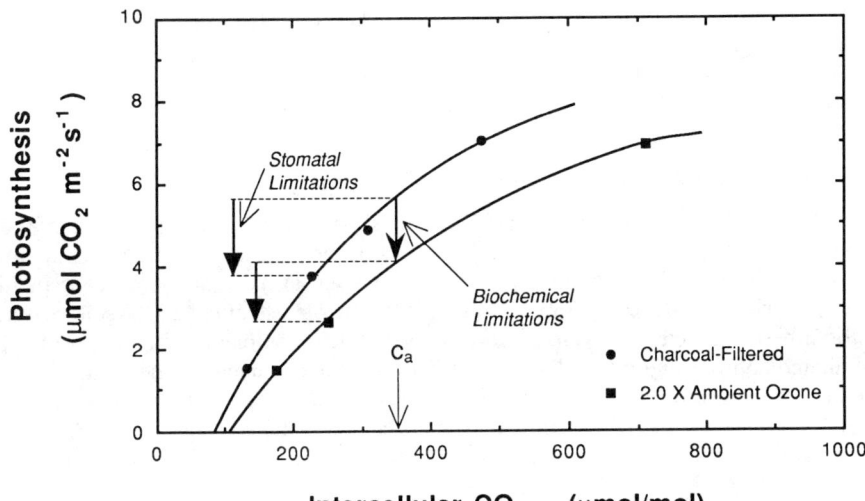

FIG. 4—*Response of photosynthesis of the 1987 first-flush needles to intercellular CO_2 concentration in three-year-old loblolly pine seedlings grown in charcoal-filtered air or 2.0× ambient ozone at the Raleigh site. Measurements were made at 33°C and 50% relative humidity. The decreases due to stomatal resistance or biochemical effects are indicated at 350 ppm ambient CO_2 concentration (C_a). The curves were fit by nonlinear regression with the function:*

$$Y = Y_{max} + (Y_{min} - Y_{max})e^{(-k*Light)}$$

processes. A decrease in initial slope, correlated to the carboxylation activity, suggested that ozone exposure may have decreased the amount of RuBPcase, as reported for other species [*30,34*]. The decrease in the CO_2-saturated rate of photosynthesis in loblolly pine at 2.0× ambient ozone also suggested a decrease in the rate of RuBP regeneration, but this speculation has not yet been verified.

Direct and indirect effects of ozone exposure have also been hypothesized to alter the chemical structure of chlorophylls and other pigments or to disrupt their organization,

TABLE 2—*Calculation of the stomatal and biochemical limitations to photosynthesis for three year old loblolly pines grown in charcoal-filtered or 2.0× ambient ozone. All measurements were made at 30°C and 50% relative humidity.*

	Ozone Treatment	
	Charcoal-Filtered	2.0× Ambient
Intercellular CO_2	228 ppm	251 ppm
Ambient CO_2	350 ppm	350 ppm
Photosynthesis at C_i	3.8 μmol CO_2 m^{-2} s^{-1}	2.7
Photosynthesis at C_a	5.7	4.1
STOMATAL LIMITATION:		
$\frac{Psyn(C_a) - Psyn(C_i)}{Psyn(C_a)}$	33.4%	34.9%
BIOCHEMICAL LIMITATION:		
$\frac{Psyn(CF) - Psyn(2.0 \times O_3)}{Psyn(CF)}$		28.1%

thereby inhibiting the transfer of electrons from NADPH to the electron transport system [*7,35*]. Although ozone primarily affects the electron transport system, photophosphorylation rates can also be inhibited as a consequence of the decreased electron transport capacity [*17,35*].

Photosynthetic Response and Biochemical Oxidants

First-flush loblolly pine needles in 1987 did not show a significant response after 90 days of exposure at 1.5× ambient ozone in the 24 h/d ozone treatment (from March to June); photosynthesis was reduced only 11% (Fig. 5). However, at 3.0× ambient ozone in June, photosynthesis was reduced by 36% in June (Fig. 5). In August, after 150 days of treatment, photosynthesis was reduced by an average of 29% at 1.5× ambient ozone relative to charcoal-filtered controls and was reduced by 67% in the 3.0× ambient ozone treatment. After 210 days of exposure (October), the 1.5× ambient ozone treatment seedlings had a 20% reduction in photosynthetic rate compared to the charcoal-filtered controls. The 3.0× ambient ozone treatment could not be measured because the 1987 first-flush fascicles had prematurely senesced.

Biochemical analysis of June needle tissue collected concurrently with gas exchange measurements showed that malondialdehyde (MDA) did not increase above controls at 1.5× ambient ozone (Fig. 6), while photosynthesis decreased 11% (Fig. 5). Photosynthesis dropped an average of 36% (Fig. 5) at 3.0× ambient ozone, while MDA increased nearly 100%.

A seasonal comparison of photosynthetic rates, SOD, and Px responses revealed that the highest activities of the antioxidant enzymes were found in the 3.0× ambient ozone treatments in June and August, when photosynthesis was reduced by 36% and 67%, respectively (Fig. 7). SOD was almost three times higher than controls in June and 165% in August, while Px was nearly two times higher than controls in both months. In the 1.5× ambient ozone treatment in June, SOD and Px increased by 26 and 46% over controls while photosynthesis decreased by 11%. These data suggest that antioxidant levels may be used as indicators of oxidative stress in trees early in the growing season. In August, at the 1.5× ambient ozone, photosynthesis decreased by 29% and SOD and Px increased by 19 and 17%, respectively.

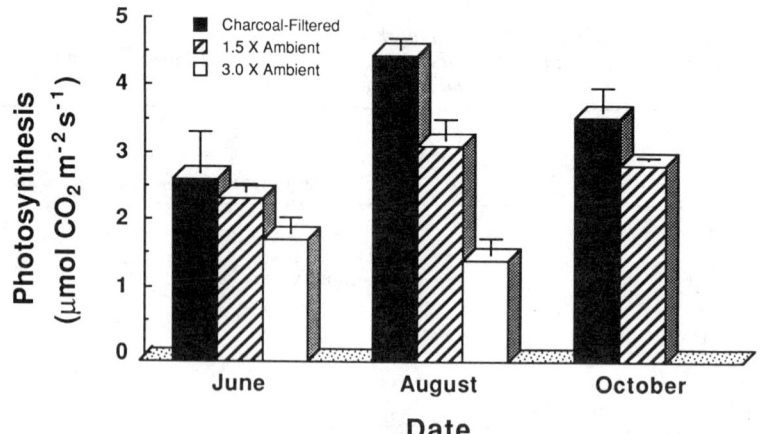

FIG. 5—*Photosynthetic response (mean ± 1 SE; n = 2 for each treatment; n = 27 in each of the two replicate chambers) of loblolly pine exposed for 24 h/d to charcoal-filtered air (control), 1.5× ambient ozone or 3.0× ambient ozone at the Duke Forest site.*

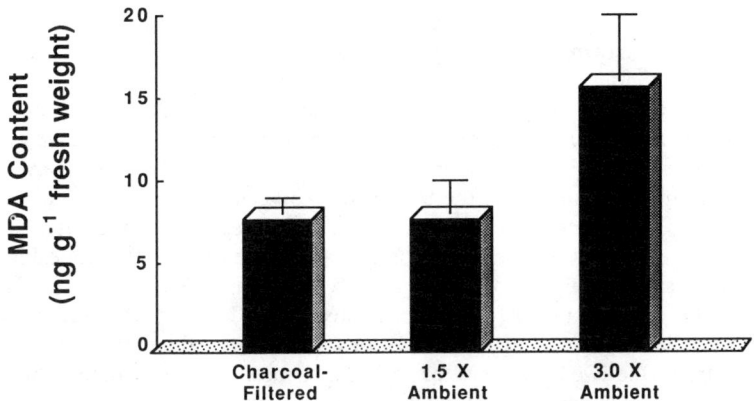

FIG. 6—*Malondialdehyde (MDA) content (mean ± 1 SE; n = 6 in each treatment) in loblolly pine seedlings exposed for 90 d to charcoal-filtered air (control), 1.5× ambient ozone or 3.0× ambient ozone. Data are for June 1987 for tree seedlings from the Duke site.*

By fall, SOD and Px had both increased over early summer values, and the treated plants had antioxidant values that were nearly 50% higher than controls; photosynthesis had decreased by 20% in the exposed plants (Fig. 7).

Other researchers have also noted increases in antioxidants in trees exposed to air pollutants, but did not determine photosynthetic characteristics [18–21]. For example, activity of

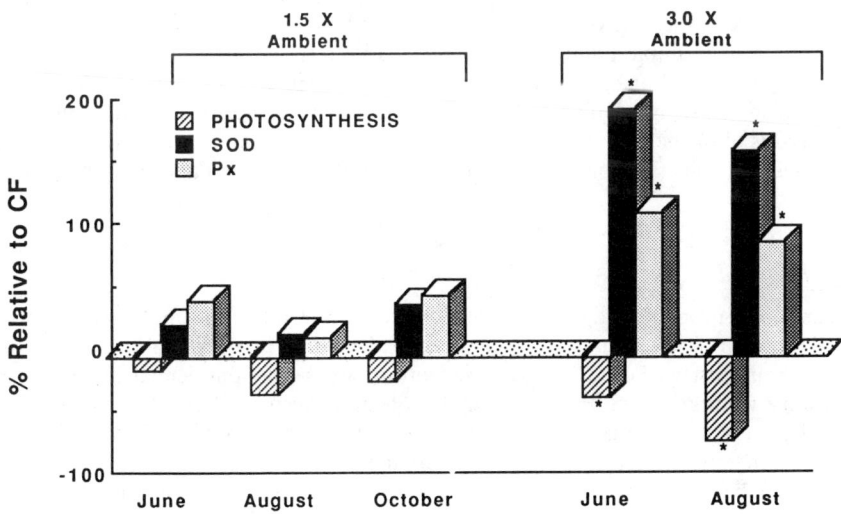

FIG. 7—*Seasonal changes in photosynthesis and the activity of the antioxidants superoxide dismutase (SOD) and peroxidase (Px) in loblolly pine seedlings exposed 24 h/d for 90 d to charcoal-filtered air (control), 1.5× ambient ozone or 3.0× ambient ozone at the Duke site. Asterisks (*) indicate statistically significant differences compared to CF ($\alpha = 0.05$).*

Px, SOD, and catalase in needles of *Pinus sylvestris* (scotch pine) after 240 h of exposure to 0.3 ppm of SO_2 was shown to depend on needle age and reached a maximum in November [18]. In addition, there was no visible needle injury or necrosis but significant increases in Px and SOD activities in SO_2-exposed plants versus controls. Catalase did not show an increase over controls. The activity of SOD in the intercellular fluid of *Picea abies* (Norway spruce) also increased significantly, after 30 days of 7 h exposures with 300 $\mu g\ m^{-3}$ of ozone [19]. The 30-d fumigation doubled the SOD activity in cell material relative to a 2-d treatment. A decrease in the ascorbic acid content of the same needles was attributed to the scavenging of toxic free radicals by ascorbate-specific peroxidase [19]. Glutathione and Vitamin E also increased significantly in both *Abies alba* (white fir) and *Picea abies* after exposure to O_3, SO_2 and SO_2+O_3 treatments [20]. Trees treated with SO_2 had a greater increase of antioxidants in needles than those exposed to O_3 or charcoal-filtered air. In addition, the combined exposure to the two gases resulted in higher concentrations of both Vitamin E and total glutathione than exposure to ozone or filtered air alone.

Our data showed that antioxidants and MDA increased significantly over controls while photosynthesis decreased. To what extent increased levels of antioxidants may ameliorate the effects of oxidant pollutants on photosynthesis and other physiological processes is unclear. However, antioxidants do seem to be useful biomarkers for indexing oxidant air pollutant stress in trees. Note that the highest levels of SOD and Px were found in August, a period of highest photosynthesis in control plants and lowest exchange rates in the treated plants. There may be important ramifications of increased antioxidant activities on carbohydrate allocation, nutrient use, growth, etc.

Future Directions

The possibility of tracking oxidant stress from diagnostic gas exchange and oxidant-mediated biochemical markers is encouraging. However, much more experimental work is needed to establish mechanisms of action, molecular bases of antioxidant increases, seasonal effects, effects of multiple pollutants, as well as dose response relationships for individual species and ecotypes. Further work is also needed to adequately test the hypothesis that air pollutant oxidants significantly increase antioxidants *in vivo*. Future research should employ combinations of ozone with other free radical generating pollutants, such as SO_2, NO_2, and H_2O_2 to test the effects of complex gas mixtures on photosynthesis and other physiological processes.

Acknowledgments

Major support for this project was from the National Council of the Paper Industry for Air and Stream Improvement, Inc. (NCASI, 260 Madison Ave., New York, NY 10016) and from the Southeastern Forest Experiment Station, Southern Commercial Forest Research Cooperative within the U.S. Environmental Protection Agency—U.S. Department of Agriculture Forest Service Forest Response Program in cooperation with NCASI. The Forest Response Program is part of the National Acid Precipitation Assessment Program. This paper has not been subject to EPA or Forest Service peer review and should not be construed to represent the policies of either Agency. David A. Keen, Mark J. Pistrang, Edward A. Fendick, and Norman E. Tandy assisted in measurements, analyses, and data analysis. Many thanks to Bill Winner for input into the design of the diagnostic gas exchange approach used in this study. The trees used in the Raleigh study were part of a larger cooperative investigation of the USDA-ARS and North Carolina State University by Steven R. Shafer and Allen S. Heagle. We thank them for providing assistance at the site and for reporting ozone exposure data.

References

[1] Pye, J. M., *Journal of Environmental Quality*, Vol. 17, 1988, pp. 347–360.
[2] Woodman, J. N. and Cowling, E. B., *Environmental Science and Technology*, Vol. 21, 1987, pp. 120–126.
[3] Reich, P. B. and Amundson, R. G., *Science*, Vol. 230, 1985, pp. 566–570.
[4] Mohnen, V. A., *Scientific American*, Vol. 259, 1988, pp. 30–38.
[5] Wellburn, A. R., in *Methods for Assessing the Effects of Mixtures of Chemicals, SCOPE 1987*, V. B. Vouk, G. C. Butler, A. C. Upton, D. V. Parke, and S. C. Asher, Eds., pp. 813–829.
[6] Di Giulio, R. T. and Richardson, C. J., Annual Report to Spruce/Fir Cooperative, Broomall, Penna., 1987.
[7] Sasek, T. W. and Richardson, C. J., *Forest Science*, Vol. 35, 1989, pp. 745–755.
[8] Von Caemmerer, S. and Farquhar, G. D., *Planta*, Vol. 153, 1981, pp. 376–387.
[9] Milthorpe, F. L. and Moorby, J., *An Introduction to Crop Physiology*, Cambridge University Press, 1979, p. 98.
[10] Sharkey, T. D., Katsu, I., Farquhar, G. D., and Cowan, I. R., *Plant Physiology*, Vol. 69, 1982, pp. 657–659.
[11] Farquhar, G. D. and Sharkey, T. D., *Annual Review of Plant Physiology*, Vol. 33, 1982, pp. 317–345.
[12] Thomas, C. E. and Aust, S. D., *Annals of Emergency Medicine*, Vol. 15, 1986, pp. 1075–1983.
[13] Kappus, H., *Archives Toxicology*, Vol. 60, 1987, pp. 144–149.
[14] Richardson, C. J., Di Giulio, R. T., and Tandy, N. E. in *Biological Markers of Air-Pollution Stress and Damage in Forests*, R. B. Smythe and D. Polincansky, Eds., National Academy Press, 1989, pp. 251–260.
[15] Rabinowitch, H. D. and Fridovich, I., *Photochemistry and Photobiology*, Vol. 37, 1983, pp. 679–690.
[16] Del Maestro, R. F., *Acta Physiologica Scandinavica*, Vol. 492, 1980, pp. 153–168.
[17] Alscher, R. G. and Amthor, J. S. in *Air Pollution and Plant Metabolism*, S. Shulte-Hostede, N. M. Darrall, L. W. Blank, and A. R. Wellburn, Eds., Elsevier Applied Science, New York, 1988, pp. 94–115.
[18] Schultz, H., *Biochemie und Pysiologie der Pflanzen*, Vol. 81, 1986, pp. 241–256.
[19] Castillo, F. J., Miller, P. R., and Greppin, H., *Experientia*, Vol. 43, 1987, pp. 111–115.
[20] Mehlhorn, H., Seaufart, G., Schmidt, A., and Kurnert, K. J., *Plant Physiology*, Vol. 82, 1986, pp. 336–338.
[21] Tandy, N. E., Di Giulio, R. T., and Richardson, C. J., *Plant Physiology*, Vol. 90, 1988, pp. 742–748.
[22] Richardson, C. J., Sasek, T. W., Keen, D. E., Fendick, E. A., and Di Giulio, R. T., Annual Report to the Southern Commercial Forestry Research Cooperative, Raleigh, N.C., 1988.
[23] Heagle, A. S., Body, D. E., and Heck, W. W., *Journal of Environmental Quality*, Vol. 2, 1973, p. 365.
[24] Heagle, A. S., Philbeck, R. B., Rogers, H. H., and Letchworth, M. B., *Phytopathology*, Vol. 68, 1979, p. 15.
[25] Johnson, J. D., *Forest Science*, Vol. 30, 1984, pp. 913–921.
[26] McCord, J. M. and Fridovich, I., *Journal of Biological Chemistry*, Vol. 244, 1969, pp. 6049–6055.
[27] Asada, K., Takahashi, M. and Nagate, M., *Agricultural Biological Chemistry*, Vol. 38, 1974, pp. 471–473.
[28] Putter, J. and Becker, R., *Methods of Enzymatic Analysis*, H. U. Bergmeyer, Ed., Verlag Chamie, Deerfield Beach, Fla., 1983.
[29] Uchiyama, M. and Mihara, M., *Analytical Biochemistry*, Vol. 86, 1978, pp. 271–278.
[30] Reich, P. B., *Plant Physiology*, Vol. 73, 1983, pp. 291–296.
[31] Heath, R. L., *Annual Review of Plant Physiology*, Vol. 31, 1980, pp. 395–431.
[32] Craker, L. E. and Starbuck, J. S., *Canadian Journal of Plant Science*, Vol. 52, 1972, pp. 589–597.
[33] Teskey, R. D., Fites, J. A., Samuelson, L. J., and Bongarten, B. C., *Tree Physiology*, Vol. 3, 1986, pp. 41–61.
[34] Pell, E. J. and Pearson, N. S., *Plant Physiology*, Vol. 73, 1983, pp. 185–187.
[35] Coulson, C. and Heath, R. L., *Plant Physiology*, Vol. 53, 1974, pp. 32–38.

K. W. Kromroy,[1] M. F. Olson,[1] D. F. Grigal,[2] P. S. Teng,[3] D. R. French,[1] and G. H. Amundson[1]

A Bioindicator System Assessing Air Quality Within Minnesota

REFERENCE: Kromroy, K. W., Olson, M. F., Grigal, D. F., Teng, P. S., French, D. R., and Amundson, G. H., "**A Bioindicator System Assessing Air Quality Within Minnesota,**" *Plants for Toxicity Assessment, ASTM STP 1091*, W. Wang, J. W. Gorsuch, and W. R. Lower, Eds., American Society for Testing and Materials, Philadelphia, 1990, pp. 156-169.

ABSTRACT: The Minnesota Bioindicator Study is a long-term field research project. It has three main objectives: (1) to develop a biological system to indicate the spatial and temporal distribution of air pollutants; (2) to determine the effects of air pollutants on important agronomic and forest species; and (3) to archive biological and environmental data, including samples of soil and plant material for future elemental analyses. Ten field plots were established throughout the state and planted with a variety of commercial and non-commercial plant species. Plants were grown in one or more of three different soil regimes. Plant growth and symptom data were collected during the growing seasons of 1983-1987. Air pollutant and weather data were obtained from local and national agencies. No symptoms of sulfur dioxide injury were confirmed, but symptoms of ozone injury were observed in all five years on milkweed, soybean, and/or potato. Preliminary analyses do not indicate statewide effects of ozone on alfalfa yield. About 2000 samples of soil and plant material are stored, and 1.75 million data points have been accumulated. Data from this study are to provide a baseline for documenting long-term changes in the quality of Minnesota's environment.

KEY WORDS: air pollutants, ozone, sulfur dioxide, crops, yield loss

Air pollutants affect plants in many different ways. Research has been conducted on air pollutant-induced changes in plant biochemistry, physiology, morphology, and productivity. Assessment of air pollutant effects has ranged from the cellular level to ecosystems and entire countries [1,2]. Many studies have been conducted in recent years to establish relationships between various pollutants and yield loss of crop species, and work continues on the response of tree species to air pollutants [3-5].

In addition to pollutant type and dose, many other factors affect the plant response to air pollutants [6]. Air pollutant-induced stress may also affect plant susceptibility to other abiotic and biotic stress factors [7]. Because of their ability to integrate biological, cultural, and environmental factors, plants have long been valued as indicators of the quality of the environment in which they live. Bioindicator systems, systems that use cultivated plants as indicators of environmental quality, have been implemented in several countries and have been previously reviewed [8-10].

One of the most extensive bioindicator systems has been developed in The Netherlands

---

[1] Department of Plant Pathology, University of Minnesota, St. Paul, MN 55108.
[2] Department of Soil Science, University of Minnesota, St. Paul, MN 55108.
[3] Department of Plant Pathology, International Rice Research Institute (IRRI), P.O. Box 933, Manilla, The Philippines.

[11,12]. In this system, a variety of plant species are included and several parameters are assessed to indicate the presence and effects of different air pollutants. After several years of operation of this national network, Posthumus concludes that, although biological systems cannot replace physical-chemical monitoring of air pollutants, there are advantages to their use. These systems provide a method to assess the effects of air pollutants on living organisms in conjunction with other kinds of stress, and may function as early warning systems which may instigate action to prevent more disastrous air pollutant effects [12].

The Minnesota Bioindicator Study is a long-term field research project. The project was designed to collect information on air quality in Minnesota by measuring the effects of air pollutants on various commercial and non-commercial plant species. It includes both bioindicator and dose-response (regression linking pollutant dose to plant growth) components. In this study, air pollutant effects on plants were evaluated in several ways, including the assessment of foliar symptoms, biomass production during the growing season, final season yield, and elemental analyses.

The objectives of the Minnesota project are threefold:

1. To develop a biological system that can be used to indicate the temporal and spatial distribution of phytotoxic concentrations of air pollutants, primarily ozone (O_3) and sulfur dioxide (SO_2) in the state: *Indicator Objective.*
2. To determine the effects of air pollutants, primarily O_3 and SO_2, on the growth and productivity of certain agronomic and forest species: *Yield Objective.*
3. To archive biological and environmental data, including samples of soil and plant material that in the future will be subjected to elemental analyses: *Accumulator/Data Archive Objective.*

Information obtained from the Bioindicator Study provides a baseline for documenting changes in the quality of Minnesota's environment. Data collected on plant growth, yield, elemental concentrations, and foliar symptomatology will provide evidence of short-term damage and long-term changes in measured parameters. If these changes can be related to changes in air pollutant concentrations, this information will provide a basis for evaluating existing and proposed air quality standards.

The purpose of this paper is to describe the implementation and general methodology of the Minnesota system, and to present a summary of results of preliminary analyses for the Indicator and Yield Objectives. Because of the geographic scope of this study, the number of objectives, and the complexity of field sampling, the amount and variety of data collected far exceeded the original estimates, and statistical analysis of the data has lagged behind data collection. Also, project experience indicates that data collection tends to be more immediately demanding of resources than data analysis. Current emphasis is on completing analysis of all the data, and results will be available as they are completed.

General Materials and Methods

Plot Installation and Design

Seven field plots were established in 1982, with three plots added in 1985. Plots were located to represent as many different regions of the state as possible (Fig. 1). Other factors important in selecting general plot locations included proximity to air pollution sources, access to public lands, and availability of physically monitored air quality data. Specific plot locations were also based on landform, soil type, and vegetation present. Each plot consists

FIG. 1—*Location of bioindicator plots.*

of a 15.25 by 15.25 m cleared area, enclosed with a chain-linked fence. In forested areas, this is surrounded by a perimeter of cleared land, 15.25 m wide.

Plant species used to evaluate air quality were selected because of their sensitivity to O_3 and/or SO_2 or their reported efficiency of accumulation of metals from the atmosphere. This information was based on controlled exposure studies early in the project and on information available in the literature [8,13–16]. A variety of species were planted in each plot: alfalfa (*Medicago sativa* L.), soybean (*Glycine max* [L.] Merr), potato (*Solanum tuberosum* L.), bachelor buttons (*Centaurea cyanus* L.), white pine (*Pinus strobus* L.), common ragweed (*Ambrosia artemisiifolia* L.), downy bromegrass (*Bromus tectorum* L.), raspberry (*Rubus idaeus* L.), common milkweed (*Asclepias syriaca* L.), and perennial ryegrass (*Lolium perenne* L.). Ozone-sensitive tobacco cultivar Bel-W3 and ozone-resistant tobacco cultivar Bel-B were grown for two years, but their use was discontinued due to difficulties in maintaining healthy plants.

For each of the agronomic species, two cultivars were used, one that was reported to be more sensitive and the other more resistant to O_3, based on foliar symptomatology [6,17,18]. These cultivars, commonly grown in Minnesota, include Vernal and Iroquois alfalfa, Corsoy and Swift soybeans, and Norland and Kennebec potatoes. With the exception of the bachelor buttons, cultivars Jubilee Gem and Polka Dot [8], information describing differential sensitivity to SO_2 was not available when cultivars were selected.

Plants of all species were grown in the native soil (subsequently referred to as the Native Soil Rows [NSR]). In order to have some measure of the effect of the varying soils at the different plots and to enhance growth comparisons between plots, some plants were also grown in a standard replaced soil (SRS). The SRS is a fertile, loamy topsoil obtained from a reserved area on a University of Minnesota Experiment Station (Rosemount). This soil is contained in open-ended sections of polyvinyl chloride (PVC) pipe 0.61 m in diameter by 0.91 m long, with a wall 6 mm thick, set 0.85 m in the ground.

A third regime was the Standard Soil Trays (SST). These are transportable plastic trays (45.7 by 30.5 by 15.2 cm) equipped with a filter candle watering system and filled with the standard soil, designed after those of Posthumus [19]. The trays were used as part of the Indicator Objective in order to have plants growing in a regime where the soil was standardized and where the watering system would provide more uniform soil moisture. Also, by

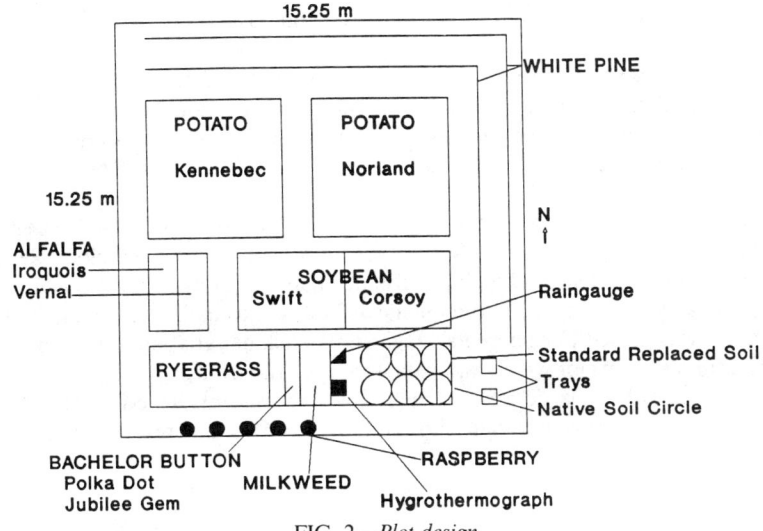

FIG. 2—*Plot design.*

periodic replacement with trays of young plants, this regime provided plants that were of a uniform, and ideally sensitive, age at all the plots throughout the growing season.

The layout of the entire plot was designed to accommodate the variety of objectives, species, and types of data to be collected (Fig. 2).

Data Collection

From mid-May through mid-June the plots were planted, and from mid-September through mid-October they were harvested. The climatic range of the plots allowed both planting and harvesting to be done sequentially across the state. During the growing season, plots were visited weekly for data collection and maintenance work. University project members and local cooperators collected data.

Many kinds of plant data were collected and are described in more detail below for each of the three objectives. Sample sizes varied, depending on the objective. A variety of quality control procedures were implemented, appropriate to specific aspects of field data collection, sample and data processing, and elemental analyses. Field data have been collected for five growing seasons, 1983–1987.

The Minnesota Pollution Control Agency (MPCA) provided hourly O_3 and SO_2 data for all monitoring stations in Minnesota for 1983–1987. Ozone data for 1987 for the Voyageurs plot were provided by the National Park Service. Ozone concentrations were measured with an ultraviolet photometry monitor.

There were five to seven O_3 monitors in operation each year and, in any single year, one to four sites were considered to be either collocated with O_3 monitors or to be proximate so that data were directly applicable. For collocated plots (Campus, Ely, Hastings, and Voyageurs), the distance between plot and monitor ranged from immediately adjacent to the plot, to about 5 km west of the plot. Ozone data for a fifth plot, Forestville, were obtained from a monitor located about 45 km north-northeast of the plot. These data were considered acceptable for use in current analyses for the Yield Objective, based on statistical analysis of O_3 data from this monitor, in conjunction with data from other nearby monitors, which indicated little regional differences in O_3 concentrations using averages of 24 h or longer [20,21].

Sulfur dioxide concentrations were measured with pulsed fluorescence, flame photometric, coulemetric, and ultraviolet stimulated fluorescence monitors (concentrations from the different monitors are equivalent as they are measured). Sulfur dioxide data were available from about 24 sites during each year. The data were obtained on magnetic tape in Storage and Retrieval of Aerometric Data (SAROAD) format.

Temperature and rainfall data were obtained from the University Soil Science Department and the Minnesota Department of Natural Resources (DNR) Climatology Service for weather stations located nearest the plots. Distance between plots and stations from which temperature data were used ranged from on-site to 21 km. Extrapolation of temperature data to this distance was judged to be acceptable, given the lack of variation in terrain.[4] Distance between plots and stations for which rain data were used ranged from on-site to 14 km. Rain was also collected with a plastic gage in each plot and amounts were recorded weekly. A comparison of the two sets of rain data showed few significant differences.

Indicator Objective

The Indicator Objective of the Minnesota Bioindicator Project was to design a biological system that could be used to index air quality, with a focus on the gaseous pollutants O_3 and SO_2. As previously summarized [8–10], other bioindicator systems have used a variety of species, including tobacco, morning glory, spinach, milkweed and petunia as O_3 indicators; alfalfa, buckwheat, and honeysuckle are a few of the species that have been used as SO_2 indicators.

Materials and Methods

In addition to the SO_2 indicators of bachelor buttons, downy bromegrass, and common ragweed that were selected as a result of controlled exposures [8,14], alfalfa, soybean, and perennial ryegrass were also regularly evaluated for symptoms of SO_2-type injury. Raspberry (cultivar September) was added to the plots in 1984.

Common milkweed was the original O_3 indicator species, but plants of all species and cultivars were routinely evaluated for symptoms of O_3 injury. Indicator plants were grown in the NSR, the SRS, and/or in the SST.

Data Collection

During the first three years of data collection, weekly evaluations for symptoms of disease, insects, air pollution injury, and other types of biotic and abiotic stress were carried out on individual plants; these were termed *specific evaluations*. For each plant, all observed symptoms were coded using a standardized set of symptom codes, the percent area of the whole plant with each symptom was estimated, and the tissue ages which were affected were recorded. For alfalfa, soybean, and potato, the evaluated plants were the same plants that were sampled for biomass measurements for the Yield Objective. Beginning in 1986, *general evaluations* of symptoms for the entire stand were implemented and alternated with specific evaluations every second week. In the general evaluations, the entire stand of each species, or cultivar, was evaluated as if it were one plant, using the same set of symptom codes.

---

[4] Personal communication, Mark Seeley, Agricultural Climatologist, Department of Soil Science, University of Minnesota, St. Paul, Minnesota.

TABLE 1—*Bioindicator plots in which O_3 injury was observed (1983–87).*

Plot	1983	1984	1985	1986	1987
Ely	X	X		X	
Deer River				X	
Hastings	X	X	X	X	X
Big Lake	X		X	X	X
Lake Maria	X		X	X	X
Lamberton	X	X	X	X	
Voyageurs	X		X	X	
Agassiz	N	N	X	X	
Forestville	N	N			X
Campus	N	N		X	X
Other[a]	X		X	X	X

[a] Observations of O_3 injury on species growing outside Bioindicator plots (milkweed, soybean, grape). Symptoms have occurred on soybean cvs. Corsoy and Swift; potato cvs. Norland and Kennebec; and on common milkweed.
N = plot not established yet.

Manuals that included symptom codes and descriptions, leaf area diagrams, photographs, and pressed plant examples of air pollution and other common damage on the indicators were provided to Project staff and cooperators. Project staff spent time with each cooperator to calibrate their evaluations and estimates. Cooperators mailed their data sheets to the University of Minnesota each week.

Results and Discussion

The results presented here provide examples of the type of information collected in the Minnesota study. At this time, quantitative analyses of all the data for the Indicator Objective are not complete, but some major trends have been identified.

Symptoms that could definitely be attributed to SO_2 were not observed, and the concentrations of SO_2 monitored in Minnesota during the five-year study period were lower than those required to cause foliar injury on the indicator species [8,14,22]. Some of the symptoms that can be caused by SO_2, such as marginal and tip necrosis, chlorosis, and premature senescence, can also be caused by other biotic and abiotic factors, which can make it difficult to diagnose and confirm SO_2 in the field using only foliar symptoms [23,24].

Ozone injury was observed for all five years of field data collection at the Bioindicator plots (Table 1). Symptoms were reported on all cultivars of soybean and potato, and on common milkweed. Interveinal pigmented and/or necrotic flecking was the most commonly observed injury. Symptoms were more frequent and severe on potato cv. Norland than on cv. Kennebec. The difference in sensitivity between the two soybean cultivars was not as distinct. Injury was also observed once on alfalfa, and on several occasions symptoms were observed on other species outside the plots.

The incidence and severity of O_3 injury varied by location and year. Using the more sensitive potato cultivar, Norland, Fig. 3 shows examples of the differences in the incidence of O_3 injury at two plots, Hastings and Ely, for the periods of plant exposure (emergence to harvest) in the 1985–1987 growing seasons. Figure 4 shows the number of days during each period on which the 1-h average concentration exceeded 0.08 ppm.[5] No O_3 injury was

---

[5] 1 ppm O_3 = 1996.5973 $\mu g/m^3$ at standard pressure and 20°C.

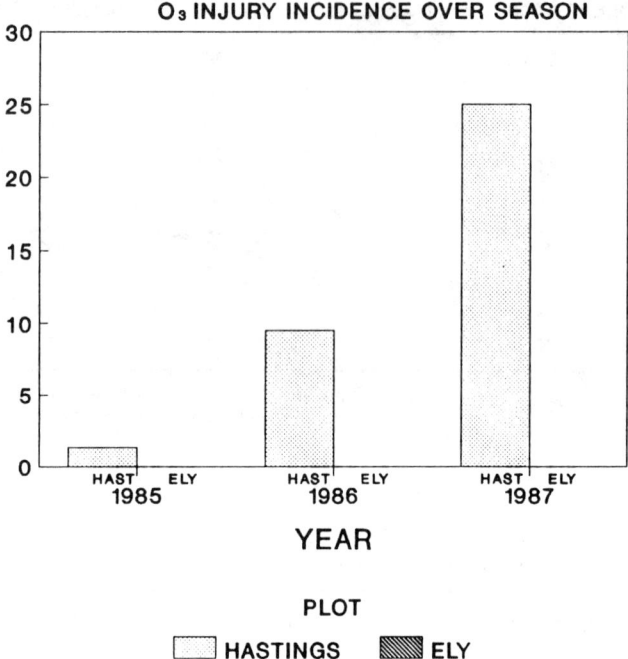

FIG. 3—*Incidence of O_3 injury on potato cultivar Norland at the Hastings and Ely plots (1985–1987).*

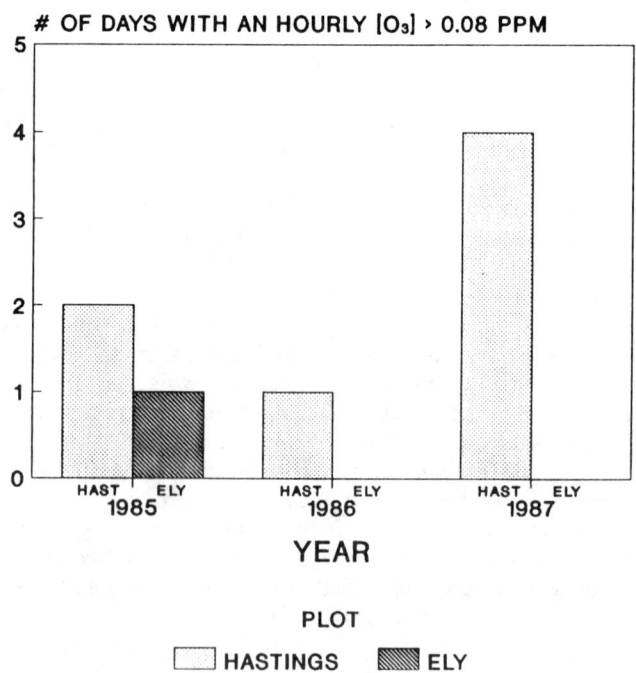

FIG. 4—*Numbers of days on which hourly O_3 concentrations exceeded 0.08 ppm at the Hastings and Ely plots (1985–1987).*

observed on Norland potatoes at the Ely plot, and the available data show that during these three growing seasons, there was only one day on which the hourly O_3 concentration exceeded 0.08 ppm. At the Hastings plot, the numbers of days with hourly concentrations greater than 0.08 ppm was 2, 1, and 4 for the 1985, 1986, and 1987 growing seasons, respectively. The incidence of Norland plants on which O_3 injury was observed at this plot was approximately 1, 10, and 25% of the total numbers of plants evaluated (75, 42, and 24 plants) for 1985, 1986, and 1987, respectively. The total number of plants evaluated was reduced from 1985 to 1987, based on methodology refinements which determined that equivalent information could be obtained with reduced numbers of evaluations (i.e., using general stand evaluations etc.). The diurnal curves for O_3 concentrations at the Hastings and Ely sites were typical for urban and non-urban sites, respectively, with much flatter curves, but similar overall averages at the Ely site compared to the Hastings site [22,25].

Using Norland cultivar potato, Fig. 5 further illustrates the spatial and temporal variation in the occurrence of O_3 injury at all ten plots for the 1985–1987 growing seasons.

Based on data from all monitors in the state, O_3 averages were higher in 1985 than in 1986 [22], but the incidence of O_3 injury across all plots was higher in 1986 than in 1985. This may have been due to the timing and concentrations of the peak values in 1986. For the five monitoring sites that were in operation in both years, peak 1-h concentrations in June, July, and August 1986 were 0.005 to 0.05 ppm higher than the peaks for those months in 1985 at several of the monitoring sites. Maximum 1-h concentrations in 1986 occurred on 18 July at three of the monitoring sites; the majority of ozone injury was observed after that, beginning about 21 July.

During the five-year study period, 1983–1987, the highest 1-h O_3 concentration monitored in Minnesota was 0.148 ppm, recorded in 1984 at a site east of the St. Paul/Minneapolis metropolitan area. Only one violation of the 0.12 ppm Federal Air Quality Standard for O_3 occurred during the five years, and there were two years where 0.12 ppm was not monitored at any of the sites. These data indicate that O_3 concentrations in Minnesota are generally low compared to many areas [26] in the United States but, based on results of the Indicator Objective, are high enough to visibly injure vegetation.

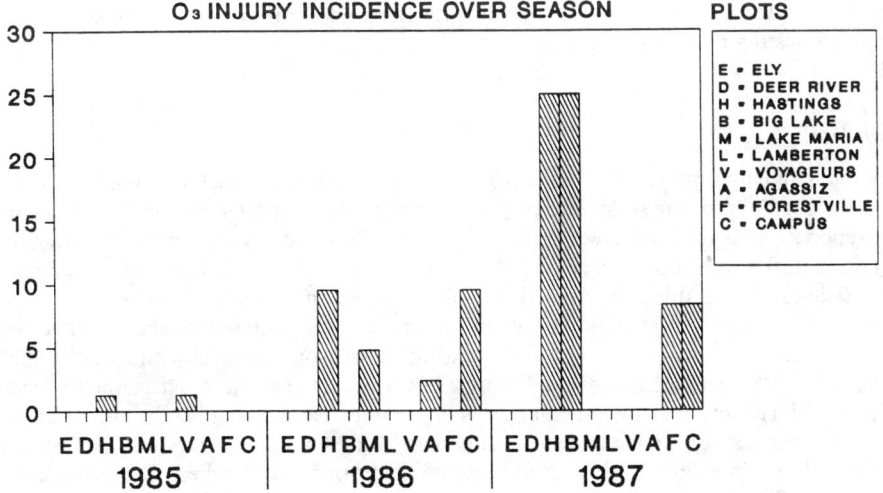

FIG. 5—*Incidence of O_3 injury on potato cultivar Norland at all plots (1985–1987).*

Summary

The absence of injury on the SO_2 indicator species is consistent with the generally low concentrations of that pollutant which were monitored. The Bioindicator System successfully indicated O_3 concentrations that should be of concern. Injury was reported on several occasions throughout the five years of study on a variety of species and cultivars, including the important agronomic crops, soybeans, and potatoes. Results of analyses currently in progress should indicate whether or not O_3 is significantly affecting yield of these crops.

Yield Objective

The goal of this component was to assess the effect of air pollutants on growth and yield of certain agronomic and forest species in relation to weather, diseases, and insects as they occur in the field. The objective was to evaluate these relationships during the growing season and at the end of the season. Although there have been many studies in controlled environments, there have only been a few studies similar to this in which effects of air pollutants on the growth of plants have been so intensively studied in an uncontrolled, field environment [27-29].

Materials and Methods

Alfalfa, soybean, potato, and white pine were selected for this component because of their reported sensitivity to O_3 and/or SO_2, and because of their importance to Minnesota's economy. Alfalfa, soybean, and potato were grown in the NSR areas, and alfalfa and soybean were also planted in the SRS. Two-year old white pine seedlings were introduced in 1982 into the native soil and were replaced as necessary through 1987.

The crop species were managed according to accepted grower practices. Fertilization of the NSR area that was planted to agronomic crops was begun in 1985 to maintain adequate yields. Soil sampling was conducted regularly in the native soil to determine whether fertilizer moved laterally into other areas of the plots. Soybeans and potatoes were planted and harvested annually, and alfalfa was replanted every second year. Depending on its growth, alfalfa was harvested once or more in mid-season in addition to a final harvest. Standard alfalfa harvesting practices were used.

Plant Data Collection

Individual plant samples of the three agronomic crops were collected from the NSR several times during the growing season (sequential samples). At each sampling time, growth stage was reported, and in some cases plant height was measured. Symptoms were assessed as described in the Indicator Objective. After collection, samples were separated into components (leaves, shoots, tubers, pods), dried, and weighed using specific protocols.

In the SRS, growth stage, plant height, and symptoms were assessed regularly, but destructive sampling was only carried out at the end of the season, except for mid-season alfalfa harvests. In 1985, leaf areas of the agronomic crops were measured to determine leaf/shoot ratios and the effects of environmental stress on biomass partitioning. At the end of the season, stand counts were made and the entire planted area of each species was harvested for the overall season yields. All harvested samples were dried and weighed. Growth measurements, including trunk diameter and height, were made on the white pine once or twice per season.

Statistical Analysis

Alfalfa was selected as the first species for statistical analysis, because it was the most complete in terms of sample processing and data entry at the time analyses began. The entire aboveground plant material is the actual yield for this species; therefore a direct assessment can be made of the relationship between pollutant concentrations and yield. Early studies establishing the sensitivity of alfalfa to O_3 have been confirmed by other studies in recent years [30–34], and the analysis focused on O_3. Although yield reductions have also been reported as a result of exposure to SO_2, the concentrations required are generally higher than those monitored in Minnesota [22,28,31,35–37].

A subset of data from selected plots and years was used in the analyses presented here. Data used in these first analyses were included based on the availability of physically monitored O_3 data and the need to have a combination of plots and years that offered the opportunity for meaningful comparisons at this stage of the analysis. The plant data consisted of dry weight and symptom data from the sequential samplings at the Campus, Hastings, and Ely plots for 1985 and 1986, and at the Forestville plot for 1985. Those data were combined with the O_3 and weather data obtained from local agencies, as previously discussed. Data were analyzed in Statistical Analysis System (SAS) using the General Linear Models procedure. A variety of statistics to express ozone, temperature, rainfall, and time were computed and evaluated, and several different models were developed and compared.

Data from other plots and years will be analyzed based on results of current analyses. Methods such as kriging and/or spatial interpolation may be used to estimate O_3 concentrations for sites not collocated with monitors.

Results and Discussion

Approximately 50 different models were reviewed. Models selected for final consideration included variables that were biologically meaningful, were relatively simple, and produced reasonable R^2 values.

The final model selected for application to the individual treatments (a single cultivar grown at a single plot in a single year) was

STEM DRY WT = HDAYS CUMRAIN MEANTEMP PROD MEAN12

where

- STEM DRY WT = dry weight of the aboveground portion of sampled plant, based on average of 3 to 6 plants per sample date, 6 to 9 sequential sampling dates per season;
- HDAYS = the number of days between harvests (not sequential samplings);
- CUMRAIN = accumulated daily rainfall for the interval defined by HDAYS;
- MEANTEMP = the average daily temperature for the interval defined by HDAYS;
- PROD = an index of foliar symptoms [38]; and
- MEAN12 = the 12-h O_3 average, from 0800 to 2000, for the interval defined by HDAYS.

Table 2 summarizes the data for the plots and years in these analyses as cumulative season totals or averages for each variable.

A model similar to the above model, but including interaction variables for plot/year/cultivar was applied to the combined data set. Results of the model applied to the combined data set showed that the relationships between alfalfa yield and O_3 were different for the different plots, years, and cultivars, with no significant overall effect of ozone. The significance of rain, temperature, and foliar symptoms, as expressed by the variables in the equation, also varied for different plots, years, and cultivars.

TABLE 2—*Overall season averages and/or totals of alfalfa and environmental variables for selected plots and years.*[a]

Site	Year	CV	STEMDRY	VAR	Days	Ozone	Rain	Temp	Prod
Ely	1985	IRO	0.043	0.0008	109	0.039	14.02	61.3	0.973
Ely	1985	VER	0.058	0.0008	109	0.039	14.02	61.3	0.970
Hastings	1985	IRO	0.139	0.0064	106	0.044	8.39	67.4	0.950
Hastings	1985	VER	0.111	0.0042	106	0.044	8.39	67.4	0.962
Forestville	1985	IRO	0.288	0.0208	114	0.035	10.01	65.8	0.878
Forestville	1985	VER	0.269	0.0210	114	0.035	10.01	65.8	0.867
Campus	1985	IRO	0.152	0.0049	94	0.025	13.19	68.4	0.917
Campus	1985	VER	0.209	0.0113	94	0.025	13.19	68.4	0.904
Ely	1986	IRO	0.121	0.0132	127	0.033	17.87	61.6	0.958
Ely	1986	VER	0.082	0.0019	127	0.033	17.87	61.6	0.909
Hastings	1986	IRO	0.278	0.0405	172	0.034	24.30	62.3	0.780
Hastings	1986	VER	0.269	0.0865	172	0.034	24.30	62.3	0.761
Campus	1986	IRO	0.391	0.1131	163	0.031	25.09	63.6	0.884
Campus	1986	VER	0.344	0.0813	152	0.032	24.20	63.5	0.934

[a] CV = alfalfa cultivars Iroquois (IRO) and Vernal (VER).
STEMDRY = average alfalfa stem dry weight (aboveground plant material) for sequential samples, in grams.
VAR = variance of alfalfa stem dry weight.
DAYS = total number of days in defined growing season.
OZONE = average O_3 concentration for 0800 to 2000 hours over the growing season, in ppm (1 ppm O_3 = 1996.5973 $\mu g/m^3$ at standard pressure and 20°C).
RAIN = total rainfall over growing season, in inches (1 in. = 2.54 cm).
TEMP = average temperature over growing season, in °F (°F = ⅘°C + 32).
PROD = an index of foliar symptoms [*38*]; a proportion.

Results of the above model applied separately to each plot/year/cultivar indicated that O_3 may have had a significant effect on alfalfa yield of both cultivars at the Hastings plot in 1985. Although the relationships between yield and O_3 were statistically significant, these are preliminary results, and their biological meaning requires further interpretation.

Hastings, 1985, had a 12-h growing-season O_3 average of 0.044 ppm, which was the highest of all the plots and years in the analysis. This is at the low end of the range of O_3 concentrations commonly reported in other studies, although significant effects on yield have been reported at this level [*34*]. Season 12-h average O_3 concentrations for the other plots and years ranged from 0.025 to 0.039 ppm. Lack of statistically significant negative relationships for O_3 and yield for these other plot/years may be related to these lower concentrations. More detailed results of these analyses have been previously reported [*22*].

Summary

Results of the statistical analyses completed to date did not identify a statewide effect of O_3 on alfalfa yield, but they did indicate that yield reductions may be occurring at higher O_3 concentrations, such as those found at the Hastings plot in 1985. More detailed results of yield analyses will be published for alfalfa, soybean, and potato as they are completed.

Accumulator/Data Archive Objective

This Objective had two components. One component was to use biological material as "accumulators" that could be analyzed for concentrations of various elements. Regular collection and storage of samples of plant tissue and soils from each year provide an archive for later elemental analyses to detect changes over time. It was understood from the initiation of the project that resources would not be available for analyses of all the samples, and so procedures for organizing and safely storing these samples were developed. Samples were collected at all ten field plots.

The second component of this objective was to archive all the plant and environmental data that were collected or acquired throughout the field study. The data were to be safely stored and accessible to the scientific community.

Materials and Methods

Soil—As part of the original plot characterizations, soil samples were collected from each soil horizon at each plot down to a depth of 1.5 m. A composite sample was also taken from the upper 15 to 20 cm of the soil. During the five years of data collection, soil samples were also collected at least once per year at each plot. Soil cores were taken from the corners and center of the plot and from the SRS. Elemental analysis has been performed on subsamples of the material collected from the upper horizons during the initial characterization and from the standard soil. Analysis was for over 30 metals and metallic earths, ranging from beryllium to uranium on the periodic table. Most of those analyses were by inductively-coupled plasma spectroscopy. In addition, nutritionally important anions including nitrate, phosphate, and sulfate were analyzed colorimetrically. Analysis of all samples has not yet been completed; statistical analyses are awaiting that completion. Samples that have not been analyzed are currently in cold storage at 4°C.

Plants—The use of vascular plants to evaluate the deposition and accumulation of certain elements in terrestrial systems has been reviewed [8,16,39]. Alfalfa cultivars and perennial ryegrass cv. Delray were used as accumulators. These species were grown in the NSR and the SRS to provide information on the contribution of the soil to elemental loading. Bulk samples of the alfalfa and ryegrass were taken from the NSR and the SRS one to three times during the growing season. These samples were dried at 60°C for two weeks, weighed using a Sartorius balance, then ground with a stainless steel Wiley mill. The ground samples were placed in sterile plastic bags and stored at room temperature.

Samples collected during the 1983 growing season have been analyzed for the same elements as were the soils. Approximately 1500 alfalfa samples and 300 ryegrass samples are currently stored.

Data Archive—The Project has collected and accumulated approximately one and three-quarter million data points. About 95% of these data have been entered into a computer system–mainframe and/or personal computer (PC). Approximately two years ago, a decision was made to move the data and analysis from the mainframe to a PC. This has proved to be less expensive and just as flexible. Data that were stored on magnetic tape on the mainframe system are downloaded to the PC as needed. All the data are inventoried and accessible, and all the data that have been used in statistical analyses are in a microcomputer relational data base management system.

Summary

As of January 1989, approximately 150 samples have been analyzed for elemental concentrations, but in-depth statistical analyses and interpretations must be completed in order to select additional samples for analyses. In addition, 40 bags of soil samples, 1500 ground alfalfa samples, and 300 ground ryegrass samples are in storage. The data base developed from this project is extensive in terms of the diversity and consistency of the data and the geographic range and number of years over which they were collected. The significance of these samples and data lies in their ability to provide information on numerous biological and environmental issues, in keeping with the original overall objective of establishing a baseline for evaluating long-term changes in the environment of Minnesota.

Acknowledgments

This research has been supported by the Minnesota Environmental Quality Board, State Planning Agency. Additional support was provided by the University of Minnesota Department of Plant Pathology, the National Park Service, and Voyageurs National Park. The authors would like to thank Dr. David Lang for his key role in the initiation and development of this project, and Dr. John Laurence for his ongoing support and assistance. The authors also thank Sean Harrington for his consistent technical support and Ms. Katherine O'Connell for her statistical advising. The plot cooperators who collected data and maintained the outstate plots, and the many persons involved in the management of the land on which the plots are located, are gratefully acknowledged.

References

[1] Treshow, M., Ed., *Air Pollution and Plant Life*, Wiley, Chichester, 1984, 486 pp.
[2] Legge, A. H. and Krupa, S. V., *Air Pollutants and Their Effect on the Terrestrial Ecosystem*, Wiley, New York, 1986, 662 pp.
[3] Heck, W. W., Taylor, O. C., and Tingey, D. T., Eds., *Assessment of Crop Loss from Air Pollutants*, Elsevier Applied Science, London, 1987, 552 pp.
[4] Dempster, J. P., and Manning, W. J., Eds., Mellanby, K., Chairman, *Environmental Pollution*, Vol. 53, Nos. 1-4, 1988, 478 pp.
[5] U.S. Environmental Protection Agency, *Air Quality Criteria for Ozone and Other Photochemical Oxidants*, Vol. 3, EPA/600/8-84/020CF, Research Triangle Park, N.C., Aug. 1986, 417 pp.
[6] Kromroy, K. W., *Comparative Evaluation of the Short Term Response of Six Soybean Cultivars to Ozone and Sulfur Dioxide, Singly and in Combinations, Under Controlled Environment*, M.S. thesis, University of Minnesota, St. Paul, 1982, 129 pp.
[7] Huttunen, S. in *Air Pollution and Plant Life*, M. Treshow, Ed., Wiley, Chichester, 1984, Chapter 14, pp. 321-356.
[8] Laurence, J. C. and Greitner, C. S. in *Development of a Biological Air Quality Indexing System*, Minnesota Environmental Quality Board Report SEO184B, St. Paul, Minn., Jan. 1984, 380 pp.
[9] Steubing, L. and Jager, H. J., *Monitoring of Air Pollutants by Plants*, Dr. W. Junk, The Hague, The Netherlands, 1982, 162 pp.
[10] Manning, W. J. and Feder, W. A., *Biomonitoring Air Pollutants with Plants*, Applied Science, London, 1980, 142 pp.
[11] Posthumus, A. C., "Environment and the Quality of Life: Elaboration of a Communitive Methodology for the Biological Surveillance of the Air Quality by the Evaluation of the Effects on Plants," Final Report to the Environment and Consumer Protection Service, Commission of the European Communities, 1980, 40 pp.
[12] Posthumus, A. C. in *Air Pollution and Plant Life*, M. Treshow, Ed., Wiley, Chichester, 1984, Chapter 5, pp. 73-96.
[13] Benson, F. J., Krupa, S. V., Teng, P. S., Welsch, D. E., Chen, C., and Kromroy, K., *Economic Assessment of Air Pollution Damage to Agricultural and Silvicultural Crops in Minnesota*, St. Paul, Minn., March 1982, 270 pp.
[14] Laurence, J. A., Reynolds, K. L., and Greitner, C. S., *Environmental Pollution (Series A)*, Vol. 37, 1985, pp. 43-52.
[15] Duchelle, S. E. and Skelly, J. M., *Plant Disease*, Vol. 65, 1981, pp. 661-663.
[16] Martin, M. H. and Coughtrey, P. J., *Biological Monitoring of Heavy Metal Pollution*, Applied Science, London, 1982, 475 pp.
[17] Howell, R. K., Devine, T. E., and Hanson, C. H., *Crop Science*, Vol. 11, 1971, pp. 114-115.
[18] Mosley, A. R., Rowe, R. C., and Weidensaul, T. C., *American Potato Journal*, Vol. 55, 1978, pp. 147-153.
[19] Posthumus, A. C. in *Proceedings*, Kuopie Meeting on Plant Damage Caused Air Pollution, Sci. Pap. Symp., 1976, pp. 115-120.

[20] Hendrickson, R. C., *Rural Ozone in Southern Minnesota,* Report 1, Minnesota Environmental Quality Board, St. Paul, Minn., Feb. 1985, 13 pp.
[21] Hendrickson, R. C., *Ozone Data and the Minnesota Bioindicator Study,* Report 2, Minnesota Environmental Quality Board, St. Paul, Minn., July 1985, 33 pp.
[22] Kromroy, K. W., Grigal, D. F., Olson, M. F., French, D. R., Amundson, G. H., and Harrington, S., *A Biological System for Indexing Air Quality and Assessing Effects on Vegetation: (Minnesota Bioindicator Study): Five Year, Midterm Report,* Report to the Minnesota Environmental Quality Board, St. Paul, Minn., Feb. 1989, 371 pp.
[23] Treshow, M. in *Air Pollution and Plant Life,* M. Treshow, Ed., Wiley, Chichester, 1984, Chapter 6, pp. 97–112.
[24] Guderian, R. and van Haut, H., *Staub-Reinhalt. Luft,* Vol. 30, No. 1, 1970, pp. 22–35.
[25] Tilton, B. E. and Meeks, S. A. in *Atmospheric Ozone Research and Its Policy Implications,* T. Schneider, S. D. Lee, G. J. R. Wolters, L. D. Grant, Eds., Elsevier, Amsterdam, 1989, pp. 177–194.
[26] Walker, H. M., *Journal of the Air Pollution Control Association,* Vol. 35, 1985, pp. 903–912.
[27] Ashmore, M. R., Bell, N. B., and Mimmack, A., *Environmental Pollution,* Vol. 53, 1988, pp. 99–121.
[28] Guderian, R. and Stratmann, H. C., "Field Experiments for Determining Effects of Sulfur Dioxide on Vegetation: Part III—Threshold Values of Harmful SO_2 Emissions for Fruit and Forest Trees and for Agricultural and Garden Plant Species," Research Reports by the State of North Rhine-Westphalia, No. 1920, West German Press, Cologne, 1968. Translated from German by Leo Kanner Association, 1971, 157 pp.
[29] Krupa, S. V. and Manning, W. J., *Environmental Pollution,* Vol. 50, 1988, pp. 101–137.
[30] Oshima, R. J., Poe, M. P., Braegelmann, P. K., Baldwin, D. W., and Van Way, V., *Journal of the Air Pollution Control Association,* Vol. 26, No. 9, 1976, pp. 861–865.
[31] Tingey, D. T. and Reinert, R. A., *Environmental Pollution,* Vol. 9, 1975, pp. 117–125.
[32] Cooley, D. R. and Manning, W. J., *Environmental Pollution,* Vol. 49, 1988, pp. 19–36.
[33] Takemoto, B. K., Hutton, W. J., and Olszyk, D. M., *Environmental Pollution,* Vol. 54, 1988, pp. 97–107.
[34] Temple, P. J., Benoit, L. F., Lennox, C. A., Reagan, C. A., and Taylor, O. C., *Journal of Environmental Quality,* Vol. 17, No. 1, 1988, pp. 108–113.
[35] Lorenzini, G., Mimack, A., and Ashmore, M. R., *Rivista di Patologia Vegetale,* S. IV, 21, 1985, pp. 13–27.
[36] Stevens, T. H. and Hazelton, T. W., *New Mexico Agricultural Experiment Station Bulletin,* No. 647, 1976, 22 pp.
[37] Booth, J. A., Thorneberry, G. O., and Lujan, M., *New Mexico Agricultural Experiment Station Bulletin,* No. 645, 1976, 26 pp.
[38] Johnson, K. B., Teng, P. S. and Radcliffe, E. B., *Zeitschrift für Pflanzenkrankheiten und Pflanzenschutz (Journal of Plant Diseases and Protection),* Vol. 94, 1987, pp. 22–33.
[39] Teng, P. S. and Kromroy, K. W., *A Biological System for Indexing Air Quality and Assessing Vegetation Effects,* Annual Report to the Minnesota Environmental Quality Board, St. Paul, Minn., Dec. 1984, 173 pp.

Chen Guangrong,[1] Jin Bo,[1] Li Ming,[1] and Weng Xingguo[1]

The *Vicia* Leaf Tip Cell-Micronucleus Bioassay and Air Pollution Monitoring

REFERENCE: Chen Guangrong, Jin Bo, Li Ming, and Weng Xingguo, "The *Vicia* Leaf Tip Cell-Micronucleus Bioassay and Air Pollution Monitoring," *Plants for Toxicity Assessment, ASTM STP 1091*, W. Wang, J. W. Gorsuch, and W. R. Lower, Eds., American Society for Testing and Materials, Philadelphia, 1990, pp. 170–174.

ABSTRACT: We study the *Vicia faba* leaf tip cell-micronucleus test (*Vicia*-LTMT) to expand the monitoring scope of the *Vicia* plant. The paper reports positive findings in detecting poisonous gas (xylene, hydrogen sulfide) by the utilization of *Vicia*-LTMT. The *in situ* monitoring of air pollution in Huangshi, an industrial area, was conducted at four locations, where seedlings were exposed for 4 h. The micronucleus (MCN) frequencies of the *V. faba* leaf tip cells in each location were obviously higher than the control ($F > F_{0.01}$, $P < 0.01$) by statistical analysis. The results show various degrees of pollution in these industrial areas.

KEY WORDS: *Vicia faba*, micronucleus test, bioassay, air pollution monitoring, xylene, hydrogen sulfide

Micronuclei in fast dividing cells are indicators of chromosome breakages induced by environmental agents [1]. Of the number of micronucleus (MCN) bioassays, few are useful for conducting on-site monitoring. Those suitable for on-site monitoring are mostly of eukaryotic plant types; for example, the *Tradescantia*-MCN test [2,3] has a relatively large database. *Vicia faba*, a classical cytogenetic material, is well-known for its chromosome aberration and MCN test by using its secondary root tip for clastogens [4]. Since 1983, we have applied *V. faba* to water pollution testing by using its primary root tip [5–7]. This biological monitoring technique has been examined and approved by the Chinese National Environmental Protection Agency and included in "Biological Monitoring Technique Standards" (of water environment). To expand the monitoring scope of the *Vicia* plant, the current study was designed to develop a standard protocol for the *Vicia* leaf tip cell-MCN bioassay [8]. Thereby, a single plant material can be used for both water and air pollution monitoring.

Materials and Methods

V. faba var. Song Zhi Qing Pi Du (SZQPD), a special variety cultivated in an experimental plot for two years to establish its genetic homogeneity, was selected for this investigation. The seeds were soaked in a 25°C incubator for 24 h and then germinated on a piece of cheesecloth in a moist box [5]. The germinated seedlings were transferred into a sand beaker when the roots were about 2 cm long. About 5 to 6 days later, the seedlings with 2 to 3 true leaves were maintained in a beaker filled with sand (⅚ full) and used to conduct genotoxicity testing in laboratory and pollution monitoring experiments. Usually, 6 to 7 seedlings per beaker were used for exposure (Fig. 1). The xylene fumes evaporated from the liquid at doses of 20,

---

[1] Department of Biology, Central China Normal University, Wuhan, People's Republic of China.

FIG. 1—*Seedlings for* in situ *monitoring.*

100, and 500 mg/m^3 were used to treat the leaf tips in an air-tight bell jar. The hydrogen sulfide gas generated from a chemical reaction ($Na_2S + 2HCl \rightarrow 2NaCl + H_2S$) was used to treat the leaf tip under a similar air-tight bell jar.

The *in situ* monitoring was conducted at four different functional divisions in Huangshi City. We used a chemical monitoring method at locations set up by the Health and Anti-Epidemic Station of Huangshi. All seedlings were exposed at these areas for 4 h. The control groups of xylene- and hydrogen sulfide-treated groups were maintained in a clean air bell jar for the same duration as that of the treated groups.

Since our university is far from Huangshi City, two control groups were maintained for the air pollution monitoring experiment. One was an "on the road" control group that accompanied the plant samples to and from the sites; the other was kept on Guizhi Hill (a clean air area). The seedlings were washed in tap water and recovered for 20 h. The exposed leaves and their stems were fixed in aceto-alcohol (1:3) for 24 h and washed in distilled water. These fixed materials were hydrolyzed in concentrated HCl and 95% ethanol (1:1) at room temperature for 2 to 3 min and washed thoroughly in distilled water before storage. Temporary slides were prepared by staining the dispersed cells from the 2 mm tips of the leaf with carbonfuchsin on a clean slide and squashed under a coverglass. The MCN frequencies (Fig. 2) were scored from each of the five slides (1000 cells per slide) of an experimental group. The mean frequencies and standard errors were derived from 5000 cells of each group. Student's *t*-test or *F*-test was applied to determine the significance of the difference between treated and control MCN frequencies at the 0.05 probability level.

Results and Discussion

The results of the *Vicia*-LTM (leaf tip micronucleus) tests on the two gaseous chemical agents and the *in situ* monitoring of industrial sites of Huangshi City show the validity of the bioassay. Test results of xylene at three concentrations are given in Table 1.

Degrassi et al. [4] reported a good dose-effect relationship for MCN of *Vicia* root tip cells.

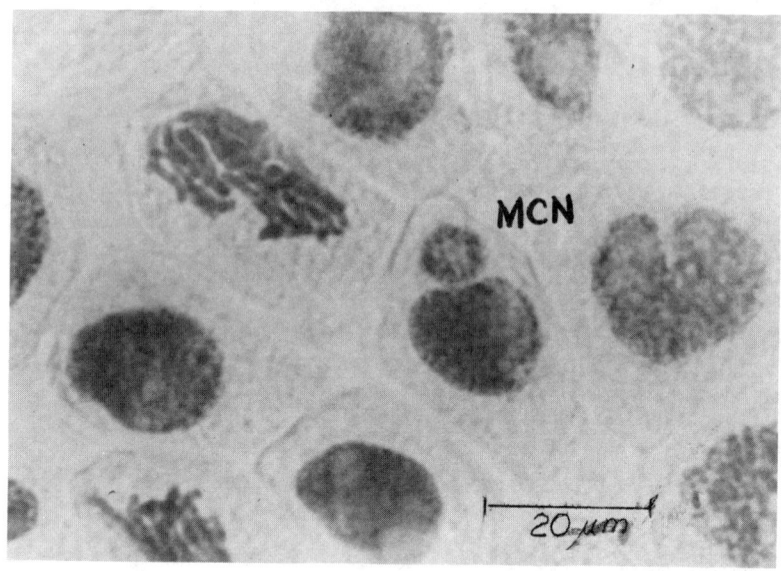

FIG. 2—*Photomicrograph of micronucleus induced by xylene in the* Vicia *leaf tip cells.*

In 1989, Chen et al. [9] observed that a good direct correlation exists between the leaf tip cells and root tip cells in the MCN frequency.

The study reconfirmed that the MCN frequencies of the three dosage-treated groups were markedly different compared with that of the clean air control group. Positive responses were detected at every dose level. The MCN frequencies through the range of xylene concentrations exhibited a dose-related increase.

Results of tests on hydrogen sulfide at two dosage levels are shown in Table 2. Xylene (100 mg/m^3) and H$_2$S (10 mg/m^3) in the air of the workshops (China's permitted discharge standards) were detected by MCN tests. Results showed that air in the workshops was markedly different from the control (clean air) for the micronucleus effect. This indicated that an MCN test in *V. faba* leaf tip cells was the more suitable.

Positive results in this study were similar to that of inhaled aerosols in mice by MCN tests [10] and demonstrated that, by using MCN tests, the cytogenetic effects of the seedlings of *V. faba* exposed to poisonous air might be used for genotoxicity estimations.

Currently, few bioassays are capable of detecting the genotoxicity of air pollutants on site, and almost all the *in situ* biological monitors are of the plant types. Among them, the *Tra-*

TABLE 1—*Frequencies of micronuclei induced by xylene fumes in the chamber (t-test).*

Experimental Groups	Dosages, mg/m^3	MCN/1000 Cells		Significant P
		Means	S.E.	
T-1	20	9.14	0.92	$P < 0.05$
T-2	100[a]	12.80	1.00	$P < 0.05$
T-3	500	14.33	0.03	$P < 0.01$
C-1	clean air	4.66	0.22	

[a] Permissible discharge standards in People's Republic of China.

TABLE 2—*Frequencies of micronuclei induced by H_2S in the chamber (t-test).*

Experimental Groups	Dosages, mg/m^3	MCN/1000 Cells		Significant P
		Means	S.E.	
T-1	10[a]	11.71	0.86	$P < 0.05$
T-2	100	14.42	0.54	$P < 0.01$
C-1	clean air	4.42	0.12	

[a] Permissible discharge standards in People's Republic of China.

descantia-MCN test has been extensively used for on-site monitoring [2,3]. Results of the *in situ* monitoring of polluted air at four industrial locations in Huangshi City are given in Table 3.

For the statistical analysis, the values (MCN%) in effects of MCN in each area of air pollution, either minus the controls in "on-the-road" or minus the values in the controls of the clean air (Guizhi Hill), were highly significant (F-test, $F \gg F_{0.01}$, $P \ll 0.01$). The results showed different degrees of pollution in each area of air in Huangshi City. Those in the Red Flag Steel Factory and Huangshi Station were higher than in the controls, because Huangshi is moderately industrial, containing steel factories, thermal power plants, coal pits, etc., thus having greater air pollution. We consulted data on chemical monitoring from the Health and Anti-Epidemic Station of Huangshi City. According to the highest allowable concentration of poisonous substances in the atmosphere of residential districts, the daily average concentrations of sulfur dioxide, dust float, and benzophrene in these areas all exceeded the highest allowable concentration.

Clive et al. [11] reported a correlation between eight kinds of chemical mutagen on plant, mammal, *Drosophila melanogaster,* and bacteria. They identified plant detection effects with mammals. Therefore micronucleus tests on *V. faba* plants may reflect the toxicity effects of pollutants on mammals. However, *V. faba* as a testing material for biological monitoring is more manageable and cheaper than other plants and animals.

The MCN tests in *V. faba* leaf tip cells may be widely used as a rapid *in vivo* method for detecting the cytogenetic damage of environmental pollution.

Acknowledgments

We are grateful to Dr. Te-Hsiu Ma for his assistance in the preparation of the manuscript and the support of the National Natural Science Foundation of China.

TABLE 3—*Frequencies of micronuclei induced by air pollutants at four different sites in Huangshi City.*

Experimental Groups	Monitoring Site	Duration, h	MCN/1000 Cells		Signif. 0.01[a]	Remarks
			Mean	S.E.		
L-1	Customs Hill Hotel	4	4.16	0.57	+	...
L-2	Cotton Hill	4	4.14	0.77	+	...
L-3	Huangshi Station	4	5.60	0.52	+	...
L-4	Red Flag Steel Mill	4	10.32	1.06	+	...
C-R	...	...	3.48	0.96	...	on-the-road
C-1	Guizhi Hill	4	2.54	0.53	...	clean air

[a] P-value of F-test.

References

[1] Ma, T. H., Sparrow, A. H., Schairer, L. A., and Nauman, A. F., "Effect of 1,2-Dibromoethane (DBE) on Meiotic Chromosomes of *Tradescantia*," *Mutation Research*, Vol. 58, 1978, pp. 251–258.

[2] Ma, T. H., Anderson, V., and Ahmed, I., "Environmental Clastogene Detected by Meiotic Pollen Mother Cell of *Tradescantia*," in *Genotoxic Effects of Airborne Agents*, R. R. Tice, D. L. Costa, and K. M. Schaich, Eds., Plenum Publishing Corp., New York, 1982, pp. 141–157.

[3] Fang, T., "A Report on the Studies of Effects of Environmental Pollution on Chromosomes: A Sino-American Collaborative Research Project, II. *Tradescantia*-Micronucleus Bioassay on Environmental Mutagens in the Air and Water Samples from Some Industrial Areas of Qingdao, PRC, and the Pesticide-DDV," *Journal of the Shandong College of Oceanology*, Vol. 11, 1981, pp. 9–11.

[4] Degrassi, F. and Pizzoni, M., "Micronucleus Test in *Vicia Faba* Root Tips to Detect Mutagen Damage in Freshwater Pollution," *Mutation Research*, Vol. 97, 1982, pp. 19–33.

[5] Chen, G., Jin, B., and Li, M., "The Utilization of Micronucleus Tests in *Vicia faba* Root Tips to Detect Pesticide and Mutagen," *Journal of the Central China Teachers College (Natural Science Edition)*, No. 4, 1983, pp. 69–75.

[6] Chen, G., Jin, B., and Li, M., "A Preliminary Study on the Utilization of Micronucleus Test Technique in *Vicia Faba* Root Tips to Detect the Pollution of Qing Shan Lake," *Journal of the Chinese Society of Environment Science*, Vol. 5, No. 4, 1985, pp. 2–7.

[7] Chen, G., Jin, B., and Li, M., "Utilization of Pollution Index in Micronucleus Test Technique in Detection of Water Pollution," *Journal of the Chinese Society of Environment Science*, Vol. 6, No. 2, 1986, pp. 60–63.

[8] Chen, G., Jin, B., and Li, M., "Utilization of Micronucleus Test in *Vicia Faba* Leaf Cells to Detect Air Pollution," *Journal of the Chinese Society of Environment Science*, Vol. 7, No. 1, 1987, p. 48.

[9] Chen, G., Jin, B., and Li, M., "A Study on Micronucleus Effect of Leaf Tip and Root Tip Cells and Correlation between Them in *Vicia Faba*," *Journal of the Central China Normal University (Nat. Sci)*, Vol. 23, No. 1, 1989, pp. 82–86.

[10] Odagiri, Y., Adachi, S., Katayama, H., and Takemoto, K., "Detection of the Cytogenetics Effect of Inhaled Aerosols by the Micronucleus Test," *Mutat. Res.*, Vol. 170, 1986, pp. 79–83.

[11] Clive, D. "Comparative Mutagenesis: An Overview," in *Proceedings, Comparative Chemical Mutagenesis Workshop*, F. J. de Serres, Ed., NIEHS, Research Triangle Park, N.C., 1978.

General Phytotoxicology

Greg Linder,[1] Joseph C. Greene,[2] Hilman Ratsch,[2] Julius Nwosu,[1] Sheila Smith,[1] and David Wilborn[1]

Seed Germination and Root Elongation Toxicity Tests in Hazardous Waste Site Evaluation: Methods Development and Applications

REFERENCE: Linder, G., Greene, J. C., Ratsch, H., Nwosu, J., Smith, S., and Wilborn, D., "Seed Germination and Root Elongation Toxicity Tests in Hazardous Waste Site Evaluation: Methods Development and Applications," *Plants for Toxicity Assessment, ASTM STP 1091*, W. Wang, J. W. Gorsuch, and W. R. Lower, Eds., American Society for Testing and Materials, Philadelphia, 1990, pp. 177–187.

ABSTRACT: Techniques modified from methods originally developed in the plant and weed science disciplines have yielded short-term tests that assess toxic chemical effects on plants. The seed germination and root elongation bioassays are laboratory toxicity tests that evolved to assess the direct and indirect soil toxicity of hazardous waste sites. These assessment tools are part of a collection of single-species toxicity tests that evaluate toxicity endpoints (seed germination and root elongation) pertinent to ecological assessments for terrestrial hazardous waste sites. Seed germination tests measure toxicity associated with soils directly, while root elongation tests consider the indirect effects of water-soluble constituents that may be present in site samples. In the seed germination toxicity test, site soil is mixed with a reference soil to yield a logarithmic series of exposure concentrations into which test seeds are planted. Germination is evaluated after a five-day exposure, and effective concentrations associated with a 50% reduction in seed germination are calculated. Contrasted to this direct test of soil toxicity, the root elongation test evaluates soil eluates that are prepared from site samples and contain water-soluble soil constituents potentially available to plants on-site and off-site. For the root elongation test, seeds are placed onto moistened filter paper that lines petri dish exposure chambers. Then, the exposure chambers are covered and incubated in complete darkness for five days; inhibition of root elongation is calculated as an EC_{50} (exposure concentration that yields a 50% reduction in root length relative to controls) upon termination of the test. By using a variety of plant species and developing a comparative toxicity database, both seed germination and root elongation toxicity tests may be applied on a site-specific basis and contribute to the toxicity assessment required as part of an ecological assessment for a hazardous waste site.

KEY WORDS: seed germination, root elongation, terrestrial toxicity testing, ecological assessment, toxicity assessment, hazardous waste site remediation

Plants are recognized as critical elements of terrestrial ecosystems and must be considered as part of any ecological hazard assessment [1,2]. Ecological hazard assessment requires interrelated evaluations of toxicity and exposure. While assessment methods are less well developed for terrestrial sites [3], the evaluation of ecological effects in these systems requires methods not unlike those frequently outlined for hazard evaluations for human health or

---
[1] NSI Technology Services Corporation, CERL, Corvallis, OR 97333.
[2] U.S. Environmental Protection Agency, Environmental Research Laboratory, Corvallis, OR 97333.

aquatic toxicity [4]. For example, hazardous waste sites potentially exert adverse effects on terrestrial ecosystems, and plant populations and communities may reflect those adverse effects. Plants may be evaluated in both laboratory and field [5,6] and contribute to hazardous waste site evaluations. Within laboratory settings, plant toxicity assessment includes methods that assess critical developmental stages in plant life history [7] and yield information pertinent to the evaluation of ecological effects in the field [2]. The seed germination and root elongation assays are both derived from tests traditionally used in plant science, as well as the fertilizer and herbicide evaluation [e.g., 8-14]. In this paper we will review these two phytotoxicity test methods, consider the development of the root elongation toxicity test and its miniaturization to yield a cost-effective tool for toxicity assessment, and illustrate how the seed germination test may contribute to the toxicity assessment process for a waste site.

Test Matrices for Phytotoxicity Assessments

Within laboratory settings, site soils must be managed to maintain their integrity so that they remain representative of the site being evaluated [2]. At the same time, site samples must be amenable to laboratory toxicity test conditions and conform to the specifications of standard test methods. Samples obtained from hazardous waste sites frequently represent heterogeneous mixtures of anthropogenic chemicals and chemicals that occur naturally in the substrate (e.g., clays, silts, and sands in varying proportions) [15-17]. Field sampling of site soils may be considered the most critical step in soil toxicity assessment, particularly if those assessments are derived from toxicity testing completed at off-site laboratories [2]. Furthermore, to establish linkages between chemicals present on the site and biological effects noted in the toxicity assessment, soil samples collected for plant toxicity testing should be compatible with sample requirements for soil chemical analyses.

Critical Life Stage Toxicity Assessment Methods Using Plants

Seed Germination (120-h) Toxicity Test

The seed germination test yields toxicity information that reflects the potential hazard directly associated with soil [2]. Seed germination represents a critical stage in the developmental biology of plants [18], and the integrative responses associated with exposures of seeds to site-soil contaminants may yield toxicity information pertinent to site evaluation and characterization. The seed germination test summarized here [7] represents a modification of a method developed by Thomas and Cline [19].

Site Soil Preparation and Toxicity Testing

Toxicity tests like the seed germination bioassay are performed directly upon site soils and require site soil samples being mixed with artificial soil to derive test concentrations. Depending upon site-specific requirements, exposure series may be adjusted for pH and hydration and cover a logarithmically spaced series of test concentrations. Test soils are loaded into petri dish bottoms to yield a range of exposure concentrations (e.g., 80, 40, 20, 10, 5, and 0% site sample mixed with artificial soil). Forty seeds (seeds of commercially important species are commonly tested) are planted onto each replicate test soil, then covered with 16-mesh silica sand and irrigated to a nominal 85% water holding capacity. The loaded petri dish is incubated in a sealed Ziploc bag, with as much air space as possible left inside the closed system (Fig. 1, after Thomas and Cline [19]). Incubation occurs in a growth chamber for 120 h (24 ± 2°C); the first 48 h are completed in total darkness and the balance of the incubation occurs at 16:8 light:dark. During the last 72 h the photophase characteristically

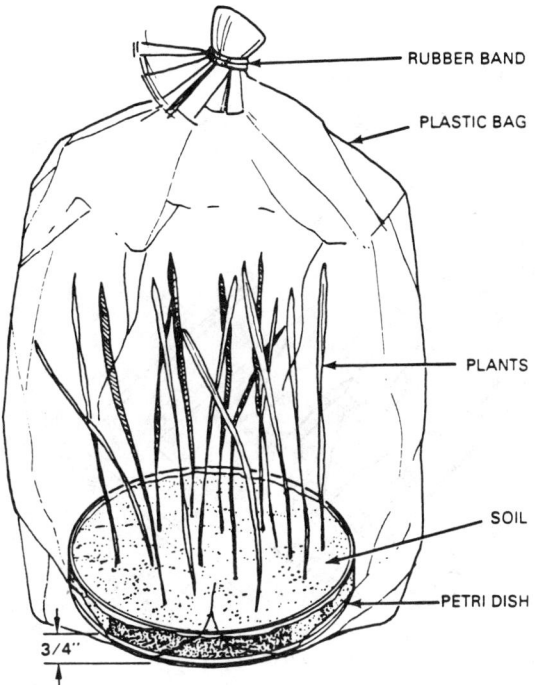

FIG. 1—*After Thomas and Cline* [19].

occurs at 4304 ± 430 lx. After 120 h the number of seeds that have germinated are counted and toxicity estimates (LC_{50}s) are calculated using the trimmed Spearman-Karber method [20] or other appropriate statistics of bioassay [e.g., 21–24].

Root Elongation (120-h) Toxicity Test

Unlike the Neubauer seed germination toxicity test, the short-term (120-h) root elongation bioassay evaluates only the water-soluble constituents of a site soil eluate and is completed without site soils being included in the test system [2]. Depending upon the water solubility of soil contaminants, root elongation is generally inhibited at lower toxicant concentrations than is seed germination [7,25–28]; hence root elongation may be a more sensitive indicator of biological effects. As with the seed germination toxicity test, root elongation tests may be performed with any number of species which represent economically important plants that are readily available, germinate, and grow rapidly [29].

Soil Eluate Preparation and Toxicity Testing

Indirect tests which evaluate the mobility of chemical constituents characteristic of hazardous waste site complex mixtures may be completed with eluates prepared from untreated site soils. Exposures in these indirect toxicity bioassays reflect percent eluate mixed with dilution water and are performed using a logarithmically spaced series of concentrations capable of yielding the desired toxicity measures [7]. The test is performed using graded seeds which are placed in filter paper-lined petri dish exposure chambers. In addition to deionized water controls, a logarithmic series of test concentrations (volume:volume percent) is prepared and

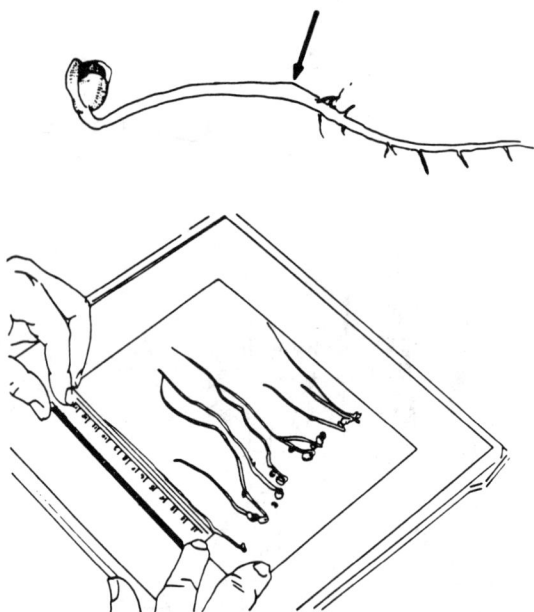

FIG. 2—*After Ratsch [32]. Arrow indicates transition point between hypocotyl and root used in measuring root length (transition point to root tip).*

added to the filter paper-lined Petri dish bottoms. Then, test seeds are spaced equidistantly on the saturated filter paper. The covered petri dishes are incubated in a darkened, humid container for 120 h at $24 \pm 2°C$. At the end of 120 h, root length is measured. While morphological changes in roots may suggest potential toxic effects associated with exposure, the measurement endpoint in these tests is linear growth. Root elongation is measured along the primary growth axis (manually [see Fig. 2] or with planimeters or root area meters), and effective concentrations which yield 50% inhibition in root elongation (EC_{50}s) are calculated according to statistical procedures outlined in Stephan [30] and Stephan and Rogers [31].

Root Elongation Toxicity Test Development and Validation

The root elongation toxicity test outlined above exemplifies the development of tools applicable to toxicity assessment. As a modification of the method used by Vandecaveye [14], the development of the root elongation toxicity test required comparisons to previously completed interlaboratory tests as an indirect evaluation of the method's ruggedness. Ratsch [32] in an interlaboratory test followed a "larger-scale" procedure originally developed by industry [33] as part of its compliance with the Toxic Substances Control Act (TSCA). In the "larger-scale" procedure, tested seeds were "planted" on glass plates that supported chromatography filter paper substrates saturated with test solution. These glass plates were placed into molded glass incubation chambers containing 1 L of test solution, which assured that the bottom 2 cm of the filter paper-covered plate was immersed in the test solution (Fig. 3, after Ratsch [32]). Incubation conditions and measurements of root growth (root lengths measured from the transition point between the hypocotyl and root to the tip of the root; see Fig. 2) were similar to those in the petri dish method summarized above.

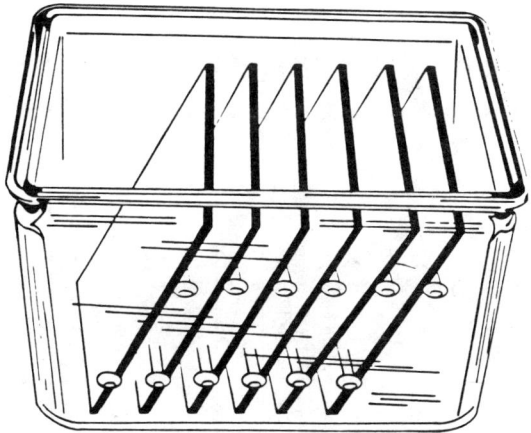

FIG. 3—*After Ratsch* [32].

The "scaled-down" petri dish technique reduces test solution volumes and also reduces the cost associated with test materials and their disposal upon completion of the test. A comparison between the "larger-scale" test system [32] and the "scaled-down" petri dish modification [7] illustrates the comparability in root elongation test results generated by each method for a selected group of chemicals. Table 1 summarizes pertinent toxicity endpoint comparisons. On the basis of effective concentrations that yield 50% reductions in root elongation (EC_{50}s), the two methods presented no statistically significant differences among chemical rankings or among species responses to the test chemicals (Mann-Whitney test, $p = 0.05$ [34]). The application of the modified test to phytotoxicity assessment then finds support from these comparative analyses; the "scaled-down" petri dish modification yields results not unlike those generated by the "larger-scale" method used by Ratsch [32] earlier. For example, each method suggested a similar ranking of species sensitivities and presented a highly correlated ranking (Spearman rank correlation, $p = 0.05$ [34]) of test chemicals when these toxicity tests were used to compare inhibition of root elongation (Table 1). While quantitative differences in their measurements of root elongation were apparent (e.g., EC_{50} estimates derived from the "scaled-down" method were generally higher than those derived from the "larger-scale" method; see Table 1), those differences were not surprising in view of each method's unique exposure system. Furthermore, the measurements of root elongation in the modified, "scaled-down" technique reflected test species variability, which was not inconsistent with the findings of Ratsch [32] in his initial interlaboratory comparison completed with the "larger-scale" test system.

Seed Germination Tests and Waste Site Toxicity Assessments

As illustrated by the root elongation toxicity test, cost-effectiveness may drive toxicity test development and modification, particularly when phytotoxicity assessments at hazardous waste sites are being developed. Similarly, ecological assessments for hazardous waste sites may encourage additional modifications and applications for phytotoxicity tests throughout site characterization and evaluation [5]. For example, toxicity estimates for a site soil may be derived from either seed germination or root elongation tests and may be used to improve site evaluations. Toxicity-based criteria then may be integrated into management decisions

TABLE 1—*Comparison of root elongation toxicity test endpoints. (Ratsch [32] and Greene et al. [7] methods were used to evaluate biological effects of reference toxicants.)*

Toxicant[a]	Clover[b]		Lettuce[b]		Wheat[b]		Root Elongation "Scaled-Down" > "Larger-Scale"
	"Scaled-Down"	"Larger-Scale"	"Scaled-Down"	"Larger-Scale"	"Scaled-Down"	"Larger-Scale"	
$CdCl_2$	29.3 (4.7)	18.4 (4.1)	10.5 (1.4)	8.6 (1.9)	35.9 (7.2)	99.7 (17.9)	4 > 1
$AgNO_3$	131.2 (16.6)	124.7 (35.9)	21.9 (3.0)	3.3 (1.0)	87.4 (23.5)	167.7 (28.0)	4 > > 1
NaF	277.6 (27.7)	311.6 (29.3)	467.8 (41.3)	496.7 (34.4)	501.9 (79.3)	238.1 (13.1)	3 > 2
MA	32.1 (8.4)	18.9 (2.3)	36.6 (4.5)	20.3 (3.6)	405.2 (143.1)	681.3 (204.5)	4 > > 1
2,4-D	0.16 (0.11)	0.053 (0.014)	0.019 (0.005)	0.043 (0.009)	0.34 (0.068)	3.75 (1.31)	3 > 2
Monuron	155.0 (18.4)	64.6 (11.5)	54.6 (9.5)	83.4 (12.4)	40.8 (10.1)	179.5 (19.2)	3 > 2

Toxicant[a]	Cucumber[b]		Radish[b]		Rank Score (±1 S.D.)		
	"Scaled-Down"	"Larger-Scale"	"Scaled-Down"	"Larger-Scale"	"Scaled-Down"	"Larger-Scale"	
$CdCl_2$	41.5 (5.5)	19.7 (3.5)	153.0 (24.6)	18.9 (3.5)	2.4 (0.9)	2.2 (0.4)	
$AgNO_3$	105.5 (10.9)	27.4 (9.1)	275.2 (37.1)	73.9 (13.6)	3.8 (0.8)	3.4 (1.1)	
NaF	227.9 (38.6)	166.4 (10.1)	412.5 (29.6)	354.3 (34.5)	6.0 (NC)[c]	5.8 (0.4)	
MA	146.3 (23.8)	92.9 (12.3)	87.1 (47.6)	28.7 (6.2)	3.8 (1.3)	4.2 (1.3)	
2,4-D	0.074 (0.010)	0.045 (0.019)	0.065 (0.010)	0.038 (0.006)	1.0 (NC)	1.0 (NC)	
Monuron	136.2 (19.7)	89.9 (17.0)	151.4 (16.8)	99.6 (19.0)	4.0 (1.0)	4.4 (0.5)	

[a] Toxicant concentrations in mg/L (as [Cd]; [Ag]; [F]; [methyl arsonate]; [2,4-D]; [Monuron]; MA = methanearsonic acid.
[b] EC_{50} expressed as toxicant concentration associated with 50% reduction in root length after 120 h.
[c] NC = not calculable.

regarding the disposition of the site [6]. Such toxicity-based guidelines potentially complement hazard assessment strategies formerly reliant chiefly upon site sample chemical analyses; for most hazardous waste sites, toxicity-based criteria have become increasingly important owing to the complex chemical mixtures characteristic of these sites. Phytotoxicity assessment tools may also suggest the efficacy of proposed remediation methods for site cleanup. Within the context of site characterization and evaluation, recent studies [35] illustrate the role that seed germination tests may contribute to the process.

In 1982, the Bunker Hill Superfund site located in the Silver Valley of Idaho was listed as a priority site for cleanup by the U.S. Environmental Protection Agency under the authority of the Comprehensive Environmental Response, Compensation, and Liability Act of 1980 (CERCLA). Being a mine tailings waste site, Bunker Hill presented high concentrations of heavy metals in site soils, and the chief concerns were contaminants like lead, cadmium, and zinc [35]; preliminary chemical analysis determined that these chemical constituents were grossly elevated in the contaminant mixture present in the site soil (Table 2).

As a result of a preliminary field survey that identified altered vegetative characteristics at the site, both laboratory and greenhouse phytotoxicity assessments were included as part of the site evaluation and were completed with site soil samples. Preliminary phytotoxicity assessments completed in the laboratory included the modified Neubauer seed germination toxicity test. With both site samples and site samples mixed with proposed soil amendments (zeolites were identified as additives potentially capable of sequestering heavy metal contaminants), preliminary screening tests adapted from the seed germination toxicity test [7] were completed in conjunction with greenhouse exposures. A summary of these laboratory-generated phytotoxicity measurements is included in Table 3, which also includes brief summaries of the greenhouse potted-plant toxicity assessments that were completed with the same site samples [35].

From the greenhouse experiments completed by Krawczyk et al. [35], germination was variable for seeds (bush bean, *Phaseolus vulgaris* L. var. Blue Lake 290 and tall fescue, *Festuca arundinacea* Shreb) exposed to test soils, and quantitative comparisons were not possible because of variable environmental conditions at the soil surface. The complementary phytotoxicity assessment using lettuce seed in the modified Neubauer seed germination toxicity test, however, indicated a zeolite concentration-dependent reduction in test seed germination, which suggested that critical early developmental effects would remain, even in amended site soils. Indeed, as suggested by the biomass data in Table 3, Krawczyk et al. [35] found that bean and fescue showed decreased yields in amended soils and that site soils amended with zeolites were no less toxic to plants than the unamended soil. As evident from these results, the two assessment tools were complementary and suggested that the proposed amendments may not significantly diminish the phytotoxicity associated with the Bunker

TABLE 2—*Bulk chemical analysis of major and minor constituents of unamended Bunker Hill soil (after Krawczyk et al. [35]).*

Major Constituents (Percent Concentrations)		Minor Constituents (mg/kg)	
Fe	14.1 ± 1.3	Cu	483 ± 42
Mn	1.4 ± 0.1	Cd	65 ± 7
Pb	2.2 ± 0.2	P	314 ± 27
S	0.8 ± 0.1	Na	256 ± 109
Zn	0.9 ± 0.1	Sr	8.5 ± 0.6

[a] Analyses completed using U.S. Environmental Protection Agency Methods 3050 and 6010; mean (±S.D.) of triplicate determinations.

TABLE 3.—*Summary data sheet for toxicity assessments on Bunker Hill site samples completed using modified Neubauer toxicity test and greenhouse potted plant exposures.*

Site Sample	Seed Germination[a]	Total Biomass[b]	
		Bean	Fescue
Zeolite, Bunker Hill (unamended)	95 ± 5	1.0	0.8
Zeolite, Bunker Hill (limestone amendment)	93 ± 4	4.0	0.4
Zeolite, Bunker Hill (XY 5,[e] 4% amendment)	90 ± 11	0.6	1.1
Zeolite, Bunker Hill (XY 5, 12% amendment)	77 ± 2	1.2	0.7
Zeolite, Bunker Hill (XY 5, 20% amendment)	40 ± 4	1.3	1.2
Zeolite, Bunker Hill (CH 20,[f] 4% amendment	90 ± 2	1.4	0.7
Zeolite, Bunker Hill (CH 20, 12% amendment)	73 ± 10	0.8	0.8
Zeolite, Bunker Hill (CH 20, 20% amendment)	62 ± 4	1.1	0.7
Control Soils:			
Hyslop control	not tested	12.0	11.0
Germination control[c]	95 ± 5[d]	26.0	18.0
Greenhouse control	not tested	32.0	16.0

[a] Mean (±S.D.) percent lettuce seed germination at 100% site sample; $n = 3$.
[b] Modified from Krawczyk et al. [35]; mean dry weight/plant after 60 days growth.
[c] Germination control consists of 70% sand, 20% kaolin, 10% screened peat, and 0.42% $CaCO_3$.
[d] Data base developed in support of Greene et al. [7] using Thomas and Cline method [19] with sand as control matrix.
[e] XY 5 = percent composition on dry weight basis: SiO_2, 73.8; Al_2O_3, 15.0; TiO_2, 0.29; Fe_2O_3, 2.58; CaO, 3.23; MgO, 1.27; Na_2O, 1.60; K_2O, 1.91.
[f] CH 20 = percent composition on dry weight basis: SiO_2, 75.02; Al_2O_3, 13.27; TiO_2, 0.50; Fe_2O_3, 3.59; CaO, 0.93; MgO, 0.86; Na_2O, 1.05; K_2O, 4.78.

Hill soils [35]. Additionally, these toxicity assessments corroborated the preliminary biological survey completed at the site, and suggested that additional phytotoxicity assessment work may be a valuable contribution to future ecological evaluations for the site.

Phytotoxicity Tests and Ecological Assessments at Waste Sites

Within the ecological assessment process, both the seed germination and root elongation phytotoxicity tests may contribute to hazardous waste site evaluations [6,36,37]. As previously suggested [5], toxicity assessments may be valued components in the site characterization, evaluation, remediation, or monitoring process, and, if considered alone, the seed germination and root elongation tests may directly contribute to an assessment of the phytotoxicity associated with hazardous waste sites. But, if regarded within an integrated laboratory and field assessment strategy, their contributions as measurements of toxicity may suggest the possible ecological consequences potentially associated with exposures in the field and more firmly establish linkages between site-specific chemical constituents and the biological responses characteristic of the site. Thomas et al. [38] suggested, for example, that laboratory toxicity assessments may contribute to waste site evaluations. By applying spatial statistic techniques such as kriging to endpoints derived from the lettuce seed germination test, Thomas et al. [38] "mapped" toxicity data and suggested that such methods could potentially contribute to site cleanup recommendations. Similarly, the brief consideration of the integrated laboratory-greenhouse evaluations completed for the Bunker Hill site suggested that these more controlled laboratory exposure systems may complement field surveys and contribute to site remediation decisions. As illustrated here, the potential role of laboratory methods in site remediation may provide pertinent information regarding the potential effectiveness of various remedial actions.

Considering the demands placed upon phytotoxicity test systems, methods development has been responsive to a variety of problems, and limitations in sample volumes have frequently been a consideration for hazardous waste site toxicity assessments. As suggested by the development of the root elongation test outlined here, method modifications and refinements allow reduced site sample volumes for testing yet yield toxicity endpoint estimates not unlike those obtained from techniques that require larger sample volumes. The root elongation method summarized here correlates well with other test methods used in phytotoxicity assessments [*39–43*]; these phytotoxicity test methods differ with respect to quantitative estimates of their common toxicity endpoint, root elongation, but as assessment tools, they yield quite similar rankings for chemicals tested to date [*39*]. The root elongation test as well as the modified Neubauer seed germination test minimize the test sample volumes required, but the database currently available suggests the toxicity endpoints relevant to site evaluation are not unduly compromised. The root elongation and seed germination methods summarized here, however, do reflect work that has been completed primarily with reference toxicants in single compound exposures. Work with defined chemical mixtures and complex chemical mixtures from waste sites is needed in order to describe fully the limitations and robust characteristics of these phytotoxicity test systems.

Conclusions

Phytotoxicity assessments may play a critical role in the evaluation of adverse ecological effects at hazardous waste sites, and plant toxicity tests that are amenable to the task have been developed or modified in response to this increasing demand. Seed germination is a critical stage in plant life, as is root elongation, and both biological events have been evaluated as toxicity endpoints relevant to phytotoxicity assessment. As derivatives from plant and weed science, the seed germination and root elongation tests have both undergone modifications that have increased their applicability to toxicity assessment. Reductions to minimize testing volumes have been central to methods development and have yielded techniques that measure the toxicity of soils or soil sample derivatives such as eluates and leachates, while minimizing the amount of material required for testing and subsequent disposal. Similarly, plant toxicity assessment methods have gained increased application in waste site evaluations and have been used during all phases of site assessment from site characterization through site remediation and monitoring. *In situ* methods that complement these techniques for laboratory phytotoxicity assessment are currently being developed, and the integration of field assessment tools with these techniques will enhance their contribution to ecological assessments for hazardous waste sites.

Acknowledgments

Our paper represents the combined efforts of the co-authors as well as numerous workers associated with the Environmental Research Laboratory, Corvallis, Oregon. We gratefully acknowledge the contributions of Jack Cline and John Thomas, Battelle Pacific Northwest Laboratory; Daniel Krawczyk, Claudia Wise, Lawrence Kapustka, Spencer Peterson, Harold V. Kibby, US EPA/CERL; Michael Bollman, Cathy Bartels, Merline Robideaux, NSI Technology Services Corporation; John Fletcher, University of Oklahoma. We thank Safa Shirazi and Tom Pfleeger for preliminary reviews of the paper.

References

[1] Kapustka, L., "Vegetative Assessment," *Ecological Assessments of Hazardous Waste Sites: A Field and Laboratory Reference Document,* W. Warren-Hicks, B. Parkhurst, and S. Baker, Jr., Eds., U.S. Environmental Protection Agency, Corvallis Environmental Research Laboratory, Corvallis, Ore., 1989, pp. 8-40 to 8-57.

[2] Linder, G. and McBee, K., "Terrestrial Toxicity Tests," in *Ecological Assessments of Hazardous Waste Sites: a Field and Laboratory Reference Document,* W. Warren-Hicks, B. Parkhurst, and S. Baker, Jr., Eds., U.S. Environmental Protection Agency, Corvallis Environmental Research Laboratory, Corvallis, Ore., 1989, pp. 6-27 to 6-43.

[3] Fava, J. A., Adams, W. J., Larson, R. J., Dickson, G. W., Dickson, K. L., and Bishop, W. E., "Research Priorities in Environmental Risk Assessment," Workshop Report, Society of Environmental Toxicology and Chemistry, Rockville, Md., 1987.

[4] *Analyzing the Hazard Evaluation Process,* K. L. Dickson, A. W. Maki, and J. Cairns, Jr., Eds., American Fisheries Society (Water Quality Section), Bethesda, Md., 1979.

[5] Athey, L. A., Thomas, J. M., Skalski, J. R., and Miller, W. E., "Role of Acute Toxicity Bioassays in the Remedial Action Process at Hazardous Waste Sites," EPA/600/8-87/044, U.S. Environmental Protection Agency, Environmental Research Laboratory, Corvallis, Ore., 1987.

[6] *Ecological Assessments of Hazardous Waste Sites: a Field and Laboratory Reference Document,* W. Warren-Hicks, B. Parkhurst, and S. Baker, Jr., Eds., U.S. Environmental Protection Agency, Corvallis Environmental Research Laboratory, Corvallis, Ore., 1989.

[7] Greene, J. C., Bartels, C. L., Warren-Hicks, W. J., Parkhurst, B. R., Linder, G. L., Peterson, S. A., and Miller, W. E., *Protocols for Short-Term Toxicity Screening of Hazardous Waste Sites,* U.S. Environmental Protection Agency, Corvallis, Ore., 1988.

[8] Horowitz, M., "A Rapid Bioassay for PEBC and Its Application in Volatilization and Absorption Studies," *Weed Research,* Vol. 6, 1966, pp. 22-36.

[9] Myhill, R. R. and Konzak, C. F., "A New Technique for Culturing and Measuring Barley Seedlings," *Crop Science,* Vol. 7, 1967, pp. 275-276.

[10] Parker, C., "The Importance of Shoot Entry in the Action of Herbicides Applied to the Soil," *Weeds,* Vol. 14, 1966, pp. 117-121.

[11] Ready, D. and Grant, V. Q., "A Rapid Sensitive Method for Determination of Low Concentrations of 2,4 Dichloro-Phenoxyacetic Acid in Aqueous Solution," *Botanical Gazette,* Vol. 108, 1947, pp. 39-44.

[12] Santelmann, P. W., "Herbicide Bioassay," in *Research Methods in Weed Science,* Weed Science Society, Champaign, Ill., pp. 91-101.

[13] Swanson, C. P., "A Simple Bioassay Method for the Determination of Low Concentrations of 2,4 Dichloro-Phenoxyacetic Acid in Aqueous Solutions," *Botanical Gazette,* Vol. 107, 1946, pp. 507-509.

[14] Vandecaveye, S. C., "Biological Methods of Determining Nutrients of Soils," *Diagnostic Techniques for Soil and Crops,* H. B. Kitchen, Ed., American Potash Institute, Washington, D.C., 1948.

[15] Bohn, H. L., McNeal, B. L., and O'Connor, G. A., *Soil Chemistry,* Wiley, New York, 1979.

[16] Brady, N. C., *The Nature and Properties of Soils,* Macmillan, New York, 1974.

[17] Morrill, L. G., Mahilum, B. C., and Mohiuddin, S. H., *Organic Compounds in Soil: Sorption, Degradation, and Persistence,* Ann Arbor Science Publishers, Ann Arbor, Mich., 1982.

[18] Mayer, A. M. and Poljokoff-Mayer, A., *The Germination of Seeds,* 3rd ed., Pergamon Press, Oxford, England, 1982.

[19] Thomas, J. M. and Cline, J. E., "Modification of the Neubauer Technique to Assess Toxicity of Hazardous Chemicals in Soils," *Environmental Toxicology and Chemistry,* Vol. 4, 1985, pp. 201-207.

[20] Hamilton, M. A., Russo, R., and Thurston, R. V., "Trimmed Spearman-Karber Method for Estimating Median Lethal Concentrations in Toxicity Bioassays," *Environmental Science and Technology,* Vol. 11, 1977, pp. 714-719, (Correction: Vol. 12, p. 417).

[21] Finney, D. J., *Statistical Method in Biological Assay,* 3rd ed., Charles Griffin and Company, London, 1978.

[22] Thompson, W. R., "Use of Moving Averages and Interpolation to Estimate Median-Effective Dose," *Bacteriological Reviews,* Vol. 11, 1947, pp. 115-145.

[23] Thompson, W. R. and Weil, C., "On the Construction of Tables for Moving-Average Interpolation," *Biometrics,* Vol. 8, 1952, pp. 51-54.

[24] Weil, C. S., "Tables for Convenient Calculation of Median-Effective Dose (LD_{50} or ED_{50}) and Instructions in Their Use," *Biometrics,* Vol. 8, 1952, pp. 249-263.

[25] Luessem, H. and Rahman, A., "Root Length Test with Garden Cress-Simple Ecotoxicological Test," *Vom Wasser,* Vol. 54, 1980, pp. 29-35.

[26] Wang, W., "The Use of Plant Seeds in Toxicity Tests of Phenolic Compounds," *Environment International*, Vol. 11, 1985, pp. 49–55.
[27] Wang, W., "Comparative Toxicology of Phenolic Compounds Using Root Elongation Method," *Environmental Toxicology and Chemistry*, Vol. 5, 1986, pp. 891–896.
[28] Wang, W., "Root Elongation Method for Toxicity Testing of Organic and Inorganic Pollutants," *Environmental Toxicology and Chemistry*, Vol. 6, 1987, pp. 409–414.
[29] Fletcher, J. S., Muhitch, M. J., Vann, D. R., McFarlane, J. C., and Benenati, F. E., "Review: PHYTOTOX Database Evaluation of Surrogate Plant Species Recommended by the U.S. Environmental Protection Agency and the Organization of Economic Cooperation and Development," *Environmental Toxicology and Chemistry*, Vol. 4, 1985, pp. 523–532.
[30] Stephan, C. E., "Methods for Calculating an LC_{50}," in *Aquatic Toxicology and Hazard Evaluation, ASTM STP 634*, F. L. Mayer and M. L. Hamelink, Eds., American Society for Testing and Materials, Philadelphia, 1977, pp. 65–84.
[31] Stephan, C. E. and Rogers, J. W., "Advantages of Using Regression Analysis to Calculate Results of Chronic Toxicity Tests," in *Aquatic Toxicology and Hazard Assessment: Eighth Symposium, ASTM STP 891*, R. C. Bohner and D. J. Hansen, Eds., American Society for Testing and Materials, Philadelphia, 1985, pp. 328–338.
[32] Ratsch, H., "Interlaboratory Root Elongation Testing of Toxic Substances on Selected Plant Species," NTIS, PB 83-226 126, 1983.
[33] Brusick, D. J. and Young, R. R., "Root Elongation Test," *IERL-RTP Procedures Manual: Level 1 Environmental Assessment-Biological Tests*, EPA-600/8-81-024, U.S. Environmental Protection Agency, Washington, D.C., 1981.
[34] Zar, J. H., *Biostatistical Analysis*, Prentice-Hall, Englewood Cliffs, N.J., 1974.
[35] Krawczyk, D. F., Fletcher, S. J., Wise, C. M., and Robideaux, M. L., "Toxicity of a Metal-Contaminated Superfund Site with and without Zeolite Amendment to Plants," Corvallis Environmental Research Laboratory, U.S. Environmental Research Laboratory, Corvallis, Ore., manuscript in review.
[36] Thomas, J. M., Cline, J. F., Cushing, C. E., McShane, M. C., Rogers, J. E., Rogers, L. E., Simpson, J. C., and Skalski, J. R., *Field Evaluation of Hazardous Waste Site Bioassessment Protocols*, Vol. 1, PNL-4614, Pacific Northwest Laboratory, Richland, Wash., 1983.
[37] Thomas, J. M., Cline, J. F., Gano, K. A., McShane, M. C., Rogers, J. E., Rogers, L. E., Simpson, J. C., and Skalski, J. R., *Field Evaluation of Hazardous Waste Site Bioassessment Protocols*, Vol. 2, PNL-4614 (Vol. 2), Pacific Northwest Laboratory, Richland, Wash., 1984.
[38] Thomas, J. M., Skalski, J. R., Cline, J. F., McShane, M. C., Simpson, J. D., Miller, W. E., Peterson, S. A., Callahan, C. A., and Greene, J. C., "Characterization of Chemical Waste Site Contamination and Determination of Its Extent Using Bioassays," *Environmental Toxicology and Chemistry*, Vol. 5, 1986, pp. 487–501.
[39] Ratsch, H. and Johndro, D., "Comparative Toxicity of Six Test Chemicals to Lettuce Using Two Root Elongation Test Methods," *Environmental Monitoring Assessment*, Vol. 6, 1986, pp. 267–276.
[40] U.S. Food and Drug Administration, *Seed Germination and Root Elongation*, FDA Environmental Assessment Technical Guide No. 4.06, Center for Food Safety and Applied Nutrition and Center for Veterinary Medicine, U.S. Department of Health and Human Services, Washington, D.C., 1987.
[41] Organization for Economic Cooperation and Development, *Terrestrial Plants: Growth Test*, OECD Guideline for Testing of Chemicals, No. 208, Paris, 1984.
[42] U.S. Environmental Protection Agency, *Seed Germination/Root Elongation Toxicity Test*, EG-12, Office of Toxic Substances, Office of Pesticides and Toxic Substances, Washington, D.C., 1982, 22 pp.
[43] Food and Drug Administration, *Seed Germination*, Environmental Assessment Technical Guide No. 11.06 (Draft), Center for Food Safety and Applied Nutrition, Center for Veterinary Medicine, U.S. Department of Health and Human Services, Washington, D.C., 1984.

D. L. Sirois[1]

Survey of Rocky Mountain Arsenal for Phytotoxic Substances

REFERENCE: Sirois, D. L., "**Survey of Rocky Mountain Arsenal for Phytotoxic Substances,**" *Plants for Toxicity Assessment, ASTM STP 1091,* W. Wang, J. W. Gorsuch, and W. R. Lower, Eds., American Society for Testing and Materials, Philadelphia, 1990, pp. 188–197.

ABSTRACT: For some years, beginning in 1941, military and industrial wastes were disposed of at the Rocky Mountain Arsenal (RMA) located near Denver, Colorado. A comprehensive study of RMA supported by the United States Army was conducted several years ago. Efforts were made to identify suspected toxic substances within the confines of or possibly migrating from RMA. For this study, bioassays were made in the glasshouse on soil samples taken from RMA. The tests were patterned after tests routinely used to identify herbicidal activity. Inhibition of seed germination or retardation of growth and development of seedlings grown in the soils were major criteria for indicating the presence of substances toxic to plants. In addition, observations were made of chlorosis, foliar necrosis, formative effects, and other symptoms indicative of phytotoxicity. Test results showed evidence of toxicity on plants grown in soil samples from Section 36 in the center of RMA. The experimental approach proved appropriate. Influences due to test species, seasonal variation of the greenhouse environment, and data analysis were evaluated.

KEY WORDS: organic compounds, industrial wastes, contaminants, bioassay, germination, plant growth, toxicity

The study of herbicides has provided a large body of knowledge on how chemicals affect plant life. Indeed, procedures used to discover new herbicides are applicable to assessing the presence of phytotoxic substances in the environment. Tests [1] conducted at Boyce Thompson Institute to determine the presence of phytotoxins in the soil at the Rocky Mountain Arsenal (RMA) provide a valuable case study.

Military and industrial wastes were disposed of at RMA beginning in 1941 [2]. In 1976 soils from several sections of RMA were sampled and bioassayed as part of a study conducted by the U.S. Army. To focus on methodology and not the distribution of contamination at RMA, only data from Section 36 is considered for the present discussion. Section 36 was extensively sampled, with 120 soil samples tested. This section is centrally located at RMA [2,3], where contaminants were known to be present [2]. Also, soil samples from this section were among the few from RMA in which plant bioassays showed evidence of phytotoxins [2]. Other studies [3] also have demonstrated the presence of phytotoxic substances in Section 36. Thus the results of bioassays of soils from Section 36 were used to demonstrate the utility of the test system.

Procedures

Herbicide screening procedures used for many years at Boyce Thompson Institute were adapted for RMA soil sample testing. The tests involved the germination of seeds and growth

[1] Boyce Thompson Institute for Plant Research at Cornell University, Ithaca, NY 14853.

of selected indicator plants in soil samples taken from different depths at various locations at RMA. Representative monocotylenous and dicotyledenous plant species were included. These tests detect the inhibition or retardation of the emergence of seedlings and seedling growth. Interference with chlorophyll development, necrosis of foliage, and death of the plants also were detected.

Soil Sampling Program

The following description of sampling program for the Rocky Mountain Arsenal was excerpted from an unpublished report [2]. A grid pattern was used for locating sampling sites. Grid spacing varied for the various sections of RMA according to the intensity of sampling of the particular sections. In Section 36, grid spacing was 134.1 m (440 ft) (Fig. 1). A wider grid spacing was used in less intensively studied areas.

Cores were drilled at designated sites to 213 cm (7 ft), groundwater or bedrock. Each core was divided into segments of 0 to 61 cm (0 to 2 ft), 61 to 213 cm (2 to 7 ft), and subsequent 152 cm (5 ft) segments to the end of the core.

Each sample as collected was labeled with a unique eleven digit identification code of the form 00-0000-0000-0. In the identification system, the first two digits specify the section at

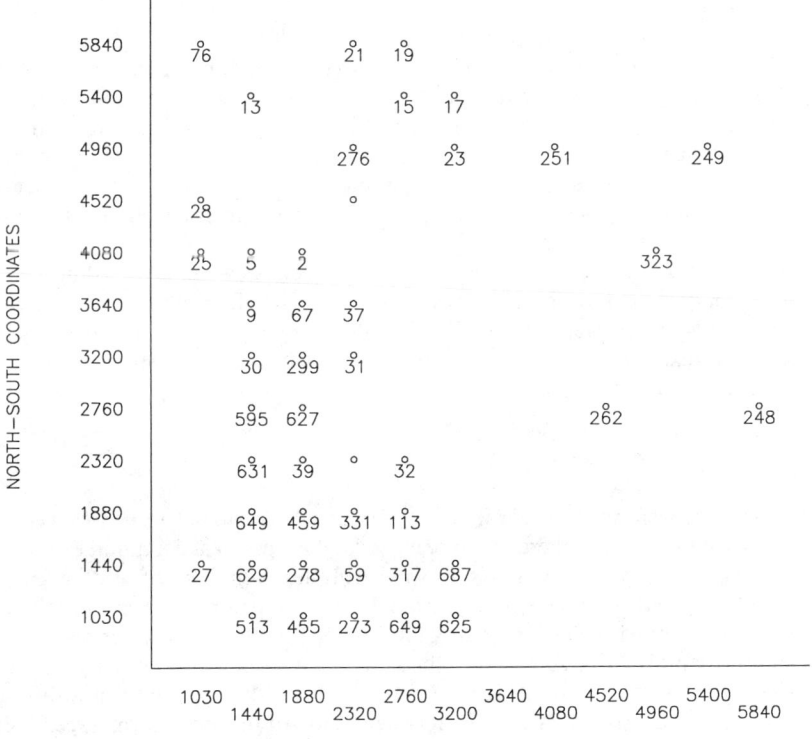

FIG. 1—*Distribution of sites sampled within Section 36 of Rocky Mountain Arsenal. Numerals indicate sample number assigned to soil samples, from the 0 to 61 cm depth, tested in bioassays at Boyce Thompson Institute.*

RMA, the next four digits are the number of feet east of the section's southwest corner + 1000. The following four digits are the number of feet north of the section's southwest corner + 1000. The last digit is the designator of sample depth. If the depth designator is 1, the 0 to 61 cm (0 to 2 ft) sampling depth is indicated. For samples deeper in the profile, the depth designator is increased by one for each 152 cm (5 ft) increment. Thus 36-2320-3640-2 designates a soil sample collected from Section 36, 402.3 m (1320 ft) east of the SW corner, 804.7 m (2640 ft) north of the SW corner, and at a depth of 61 to 213 cm (2 to 7 ft).

Soil segments from the cores were air dried and passed through a No. 4 screen. This screen removed large stones and facilitated mixing. The samples were then passed through a chambered sample splitter to ensure homogeneity. Subsamples of approximately 1000 g were shipped to Boyce Thompson Institute in capped 2 L jars.

Primary Tests

Indicator plants selected for this study were sugar beet, *Beta vulgaris* L., cv. Great Western; mustard, *Brassica nigra* L., cv. Florida broadleaf; perennial ryegrass, *Lolium perenne* L., cv. NK-100; and foxtail millet, *Setaria italica* L. Beaouv cv. Golden. These species, at one time or another, have been used in routine herbicide screening tests at Boyce Thompson Institute. These representative broadleaf and grass species are of intermediate sensitivity to phytotoxic substances. Sugar beet and perennial ryegrass are examples of cultivated plants that might be grown at RMA.

Soil samples from RMA were placed in fiber trays (19 by 13 by 7.6 cm). The soil was firmly pressed and then marked with four rows. Seeds of perennial ryegrass, foxtail millet, mustard, and sugar beet were sown in separate rows. A layer of the sample soil approximately 1 cm deep was used to cover the seeds. Trays filled with fertile compost, sown with seeds of the indicator plants, served as controls. Before planting all seeds were treated with thiram, a soil fungicide. The trays were sprinkled to optimum moisture level for seed germination and placed on a glasshouse bench. Glasshouse temperature was maintained at a minimum of 15°C.

At weekly intervals the emergence and growth of seedlings were rated. Additional observations were made on chlorosis, foliage necrosis, formative effects, and other abnormalities. After three weeks, plants were measured for median shoot growth and removed from the soil. Roots were examined for necrosis and any abnormality, such as stubby growth, and maximum root lengths recorded.

Secondary Tests

Secondary tests were made when it was not clear whether plant responses were due to phytotoxic substances or to inherently difficult soils. Poor physical condition or low fertility were suspected with some soil samples. In some samples, inhibition of seedling emergence or suboptimum growth may have been attributable to soil factors. To minimize these problems, secondary testing involved some modifications of the primary test procedure.

Problem soil samples were seeded with the same plant species as in primary testing. After the fungicide-treated seeds were sown, they were covered with a thin layer of washed sand to prevent crusting of the soil surface. This treatment allowed for more normal seedling emergence. In addition, after the seeds were covered, the trays were watered with a balanced fertilizer to correct possible nutrient deficiencies. All trays were placed in the glasshouse and phytotoxicity assessed in the same manner as in the primary tests.

Results

Primary Tests

It is clear from the emergence and growth ratings presented in Table 1 that problem soils were encountered near the surface in the western half of RMA Section 36. Seeds sown in soil samples from Site 36-2320-1440 and from Site 36-2320-1880 in the southwest quadrant of this section showed especially poor emergence. Soil samples from sites further north also appeared to have problems. In particular, seedling emergence appeared severely inhibited in soils from Sites 36-1880-2760, 36-1440-3640, 36-2320-3640, and 36-1880-4080. Low growth ratings recorded for plants grown in soil from these sites are, largely, a reflection of poor seedling emergence. Seedling emergence also appeared inhibited, although to a lesser degree, in samples from immediately adjacent sites. Likewise, low growth ratings were observed in plants grown on soils from the adjacent sites. Thus the test results indicated the existence of phytotoxins or other soil problems in these areas.

The variation of tests results with depth at six selected sites in Section 36 is evident in the data presented in Table 2. The poor emergence and growth ratings of indicator plants grown in soil samples from beneath the surface at Sites 36-2320-1440 and 36-2320-1880 suggest that the contamination or poor soil conditions found at the surface extends to lower depths at these locations. However, this is less evident at Sites 36-1440-3640, 36-2320-3640, 36-1880-2760, and 36-1880-4080. Nevertheless, these results help to impart confidence in the ability of the test system to identify problem sites.

Growth of indicator plants was influenced by fluctuation of the glasshouse environment. This influence is evident in Fig. 2, which shows variation in shoot height observed with control plants grown in compost for each test. With each of the indicator plants there are apparent seasonal differences. During midsummer, growth was more pronounced than in the winter, spring, or late summer. This largely was attributable to the lack of air conditioning in the glasshouse used in the tests. Maximum glasshouse temperatures tended to be high in summer. Growth differences between the two duplicate controls in each test also was apparent.

Normalization of the growth measurements helped to smooth out growth differences among the tests. To normalize the growth data shoot and root measurements of the indicator plants grown in the 120 soil samples from Section 36 were expressed as a percent of the controls. From the normalized data, 26 soil samples were identified where root length or shoot length of two or more indicator species fall below the lower quartile. The normalized growth measurements of the plants grown in these 26 soil samples are shown in Table 3. The data for these samples are listed in ascending order according to the north-south coordinates of the sampling sites.

It is not surprising that severe growth inhibition was recorded with Samples 46 and 59 from Site 36-2320-1440 and with Samples 260 and 331 from Site 36-2320-1880. These samples were identified as possibly being contaminated by the emergence rating and growth rating data. Also not surprising is that Sample 627 from Site 36-1880-2760, Sample 9 from Site 36-1440-3640, Sample 37 from Site 36-2320-3640, and Sample 2 from Site 36-1880-4080 appear on the list. Growth ratings and emergence ratings of plants grown in these samples also were low.

Low growth ratings, generally, were confirmed by the reduced shoot and root length of the test plants. Some overlap between poor seedling emergence and growth was apparent with some soil samples. Indeed, data presented in Table 4 show positive test results with approximately the same number of soil samples using the various test criteria. With few exceptions, the same soil samples were identified with each of the criteria.

TABLE 1—*Emergence and growth ratings of indicator plants after 3 weeks' growth in soil samples from a depth of 0 to 61 cm at sites in Section 36 of the Rocky Mountain Arsenal.*

Sample	Site Designator	Emergence Rating[a]				Growth Rating[b]			
		Sug	Msd	Fox	Rye	Sug	Msd	Fox	Rye
27	36-1030-1440-1	10	10	10	10	6	6	6	8
25	36-1030-4080-1	10	10	10	10	4	4	5^y	6
28	36-1030-4520-1	10	10	10	10	6	7^y	7^y	8
76	36-1030-5840-1	10	10	10	10	6	8	10^y	9
513	36-1440-1030-1	10	10	5	10	8	9	4^s	5^s
629	36-1440-1440-1	10	10	8	10	10	6	8	8
649	36-1440-1880-1	10	10	10	10	10	10	10	10
631	36-1440-2320-1	10	10	10	10	10	7	9	10
595	36-1440-2760-1	10	10	10	10	8	8^n	8	9
30	36-1440-3200-1	10	10	8	10	5^y	6^y	8^y	7
9	36-1440-3640-1	9	0	0	1	3	0	0	1
5	36-1440-4080-1	10	10	10	10	10	9	8	9
13	36-1440-5400-1	10	10	10	10	6^y	7^y	10^y	9
455	36-1880-1030-1	10	10	10	10	6	8	8	8
278	36-1880-1440-1	10	10	3	2	4^s	5^s	2^s	1^s
459	36-1880-1880-1	10	10	10	9	9	9	8	6^s
39	36-1880-2320-1	10	10	10	10	6	6	5^y	7
627	36-1880-2760-1	8	2	1	2	3	1	1	1
299	36-1880-3200-1	10	5	5	9	9	3^s	5^s	8
67	36-1880-3640-1	10	8	4	5	4	3^s	2^n	3^s
2	36-1880-4080-1	0	0	0	5	0	0	0	5
273	36-2320-1030-1	10	10	3	6	9	9	4^s	3^s
59	36-2320-1440-1	1	0	0	0	1	0	0	0
331	36-2320-1880-1	0	0	0	0	0	0	0	0
31	36-2320-3200-1	9	9	dead	3	2^n	2^c	dead	3
37	36-2320-3640-1	1	1	1^n	9	1	1	2^n	6
276	36-2320-4960-1	5	8	8	8	3^s	$4^{s,y}$	4	5
21	36-2320-5840-1	10	10	10	10	8	9^y	9^y	10
619	36-2760-1030-1	10	10	10	10	9	9	8	10
317	36-2760-1440-1	10	10	10	10	7	8	7^y	8
113	36-2760-1880-1	7	9	1	4	5^s	7	1^s	3^s
32	36-2760-2320-1	10	10	3	10	7	6	5	6
15	36-2760-5400-1	10	10	10	10	2	6	8^y	8
19	36-2760-5840-1	9	10	10	10	6^y	6	9^y	8
625	36-3200-1030-1	10	10	10	10	10	10	10	10
687	36-3200-1440-1	10	10	3	10	8	8	3	9
23	36-3200-4960-1	10	10	10	10	10	9	10^y	10
17	36-3200-5400-1	10	10	10	10	10	10	10	10
251	36-4080-4960-1	10	10	10	10	10	10	10	10
262	36-4520-2760-1	10	10	10	10	10	10	10	10
323	36-4960-4080-1	10	10	10	10	10	10	10	10
249	36-5400-4960-1	10	10	10	10	9	9	10	10
243	36-5840-2760-1	10	10	10	10	10	10	10	10

[a] Emergence rating key: Emergence of seedlings rated on a scale of 0 – 10, where 0 = no emergence and 10 = emergence comparable to controls.
[b] Growth rating key: Growth rated on a scale of 0 – 10, where 0 = no growth and 10 = growth comparable to controls. Superscripts indicate as follows: n = necrosis, s = stunting, y = yellowing, and c = severe chlorosis.

TABLE 2—*Variation of emergence and growth ratings of indicator plants after 3 weeks' growth in soil samples from different depths at selected sites in Section 36 of the Rocky Mountain Arsenal.*

Sample	Site Designator	Depth, cm	Emergence Rating[a]				Growth Rating[b]			
			Sug	Msd	Fox	Rye	Sug	Msd	Fox	Rye
59	36-2320-1440-1	0–61	1	0	0	0	1	0	0	0
46	36-2320-1440-2	61–213	5	2	2	2	2	1^c	1	1^c
331	36-2320-1880-1	0–61	0	0	0	0	0	0	0	0
260	36-2320-1880-2	61–213	0	0	0	0	0	0	0	0
311	36-2320-1880-3	213–370	10	10	5	5	4^s	5^s	1^s	3^s
258	36-2320-1880-4	370–540	10	10	8	9	5	7	3^s	4^s
627	36-1880-2760-1	0–61	8	2	1	2	3	1	1	1
613	36-1880-2760-2	61–213	10	10	10	8	8	7	5	5
9	36-1440-3640-1	0–61	9	0	0	1	3	0	0	1
7	36-1440-3640-2	61–213	10	10	10	10	10	7	9	9
37	36-2320-3640-1	0–61	1	1	1^n	9	1	1	2^n	6
68	36-2320-3640-2	61–213	10	10	1	9	8	6	0	8
2	36-1880-4080-1	0–61	0	0	0	5	0	0	0	5
1	36-1880-4080-2	61–213	10	8	6	7	5	5	8	9

[a] Emergence rating key: Emergence of seedlings rated on a scale of 0 – 10, where 0 = no emergence and 10 = emergence comparable to controls.

[b] Growth rating key: Growth rated on a scale of 0 – 10, where 0 = no growth and 10 = growth comparable to controls. Superscripts indicate as follows: n = necrosis, s = stunting, y = yellowing, and c = severe chlorosis.

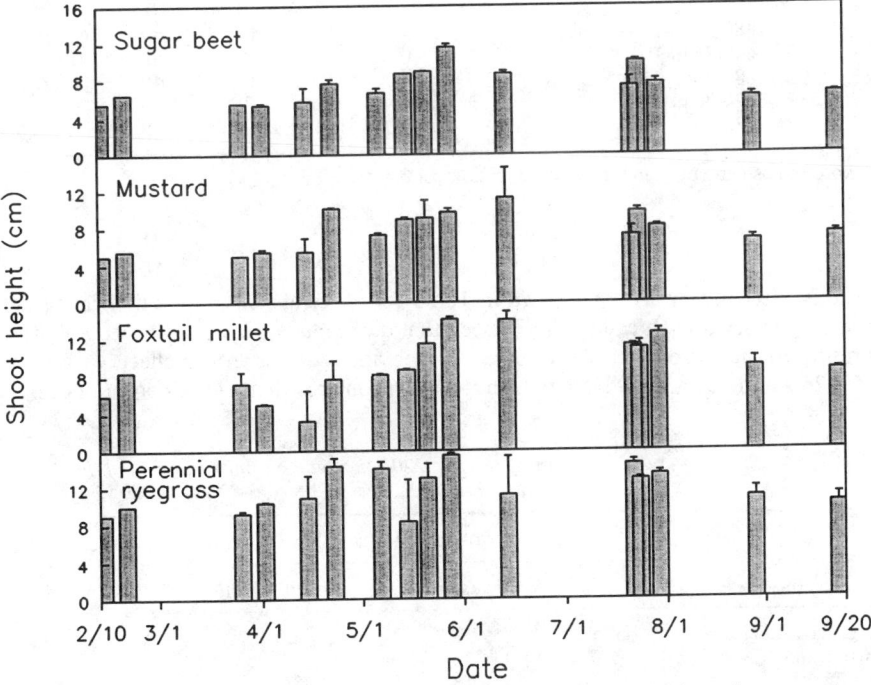

FIG. 2—*Seasonal variation in median top growth of control plants of 4 indicator species grown in fertile compost for different tests of soil samples from the Rocky Mountain Arsenal. Bars depict the mean and range of median shoot height of controls from each test measured 3 weeks after planting.*

TABLE 3—Growth of indicator plants in 26 of 120 soil samples from Section 36 in which shoot growth or root growth was reduced significantly on 2 or more plant species in primary tests.

		Length, cm[a]							
		Sugar Beet		Mustard		Foxtail Millet		Perennial Ryegrass	
Sample	Site Designation	Shoot	Root	Shoot	Root	Shoot	Root	Shoot	Root
513	36-1440-1030-1	62	57	104	75	32	24	38	18
273	36-2320-1030-1	59	86	80	107	44	19	23	16
111	36-2320-1030-2	49	46	73	113	42	34	56	9
707	36-3200-1030-4	46	100	48	55	0	0	61	141
378	36-1030-1440-2	34	152	53	117	42	67	32	46
278	36-1880-1440-1	29	49	36	45	3	4	5	3
277	36-1880-1440-2	23	26	42	79	3	4	2	2
459	36-1880-1880-1	100	62	100	81	54	61	31	6
59	36-2320-1440-1	5	0	0	0	0	0	0	0
46	36-2320-1440-2	9	6	16	9	7	3	11	4
317	36-2760-1440-1	63	67	84	81	0	0	38	6
303	36-2760-1440-2	64	65	61	76	27	9	44	13
331	36-2320-1880-1	0	0	0	0	0	0	0	0
260	36-2320-1880-2	0	0	0	0	0	0	0	0
311	36-2320-1880-3	43	28	50	42	4	3	31	8
113	36-2760-1880-1	43	37	64	37	10	0	40	18
74	36-2760-1880-2	69	147	92	70	45	15	34	16
627	36-1880-2760-1	26	33	1	5	13	15	19	2
613	36-1880-2760-2	67	123	57	79	34	47	31	33
299	36-1880-3200-1	65	60	22	12	35	19	9	12
31	36-2320-3200-1	31	18	27	5	0	0	30	5
63	36-1880-3640-2	64	96	24	30	14	26	45	32
9	36-1440-3640-1	27	27	0	0	0	0	11	11
67	36-1880-3640-2	34	49	34	19	17	10	28	38
37	36-2320-3640-1	15	...	18	...	12	...	75	...
2	36-1880-4080-1	0	0	0	0	0	0	50	56

[a] Expressed as percent of measurements on control plants.

Secondary Tests

Results of secondary tests are shown in Table 5 for several soil samples that gave positive results in primary tests. It can readily be seen that the secondary tests confirmed the primary test results for many soil samples. In particular, obvious adverse growth effects with Samples 46, 59, 260, and 331 are noted. Although the effects on growth are not as pronounced, Sam-

TABLE 4—Summary results of primary tests on 120 soil samples from Section 36 of the Rocky Mountain Arsenal.

	Number of Soil Samples Showing Positive Response			
Test Response	Sugar Beet	Mustard	Foxtail Millet	Perennial Ryegrass
Emergence rating	6	8	20	9
Growth rating	23	20	28	22
Top growth	15	16	21	18
Root growth	14	12	23	22

TABLE 5—Results of secondary tests on soil samples from Section 36 of Rocky Mountain Arsenal.

Sample	Site Designation	Emergence Rating[a]				Growth Rating[b]				Length, cm[c]							
										Shoot				Root			
		Sug	Msd	Fox	Rye	Sug	Msd	Fox	Rye	Sug	Msd	Fox	Rye	Sug	Msd	Fox	Rye
514	36-1440-1030-1	10	10	4	10	10	10	6	10	99	86	49	64	110	95	20	11
708	36-3200-1030-4	...	...	...	...	...	...	...	...	85	92	95	92	90	62	37	64
59	36-2320-1440-1	9	10	0	0	2^s	3^s	0	0	21	25	0	0	11	9	0	0
46	36-2320-1440-2	3	4	0	1	2^s	3^s	0	2^s	16	15	0	31	33	27	0	5
331	36-2320-1880-1	0	0	0	0	...	...	...	...	0	0	0	0	0	0	0	0
260	36-2320-1880-2	2	5	2	4	2^s	4^s	1^s	2^s	24	31	9	19	18	17	4	5
460	36-1880-1880-1	10	10	10	10	10	10	10	10	85	96	84	81	58	151	46	38
628	36-1880-2760-1	6	2	2	4	4^s	2^s	3^{s,n}	4^s	40	8	17	41	46	28	10	16
31	36-2320-3200-1	2	1	0	9	1	2^s	0	4^s	104	66	0	57	55	36	0	13
63	36-1880-3640-2	10	10	8	8	6	7^y	3^{y,s}	5^{y,n}	75	62	25	27	88	131	24	6
9	36-1440-3640-1	10	10	8	10	9	6^s	6^{s,n}	8^n	139	61	74	112	127	158	238	53
37	36-2320-3640-1	10	10	6	10	9	8	7^{s,n}	9	111	105	163	84	124	215	108	76
2	36-1880-4080-1	10	10	10	10	10	9	8	10	137	106	141	115	136	262	154	75

[a] Emergence rating key: Emergence of seedlings rated on a scale of 0 – 10, where 0 = no emergence and 10 = emergence comparable to controls.
[b] Growth rating key: Growth rated on a scale of 0 – 10, where 0 = no growth and 10 = growth of seedlings that is comparable to controls. Superscripts indicate as follows: n = necrosis, s = stunting, and y = yellowing.
[c] Expressed as percentage of measurements on control plants.

ples 31, 63, and 628 clearly gave positive results in the secondary tests. Contamination of these soil samples with phytotoxins was likely.

Secondary tests of other soil samples did not confirm the results of the primary tests. This was especially surprising with Samples 2 and 9. Pronounced growth inhibition was observed with these samples in primary tests, but only slight effects were recorded on indicator plants in the secondary tests. Presumably, poor physical condition or fertility may account for the positive results seen in primary tests of these soil samples.

Discussion

A survey of RMA for phytotoxic substances demonstrated the utility of modified screening procedures for evaluating herbicides to assess toxicity in soils suspected of contamination. Tests conducted on soil samples from several sections of RMA showed evidence of apparent contamination to be confined to Sections 26 and 36 [2]. Of course, correlations of results from the plant tests with contaminants in the soil samples should be examined. Unfortunately, this was not within the scope of the U.S. Army contract with the Boyce Thompson Institute, and coring data for compounds at RMA were not readily available to us. However, within Section 36 poor growth occurred in soil samples from areas of known gross contamination (i.e., the lime pits and Basin A regions) [2]. In addition, poor growth was encountered in samples from contiguous core samples. Thus it was concluded that "Section 36 is contaminated with potent phytotoxins in well defined areas" [2].

Although the test procedures used in the RMA study appeared adequate, several factors were seen to influence the sensitivity of the tests. Future use of these or similar procedures should address the factors that may influence the tests adversely. This consideration will be necessary to maximize the ability of the tests to assess the presence of phytotoxins in the environment.

Soil factors may have a pronounced impact on results. For example, if transport or storage of the soil samples is necessary before testing, changes may occur in the concentration of contaminants in the soil. Losses due to volatilization, microbial decomposition, photodecomposition, and chemical decomposition are possible [2]. Loss or decrease of a phytotoxic substance in the soil sample due to such causes could result in a "false-negative" response in the tests. Thus, the test might show a lack of phytotoxic substances in soils actually contaminated at the site sampled.

Infection by soil-borne pathogens may result in death of the indicator plants. Poor soil structure also may influence the growth of indicator plants adversely through mechanical impediment. Examples are: (1) interference with seedling emergence in crusted soils; (2) restriction of root development in hard surface soils; and (3) failure of roots to expand in compacted or fine-textured subsoils [4]. Low fertility of the soil samples might also be expected to result in suboptimum growth of the test plants. These factors could cause "false-positive" responses in the tests. The test results might indicate, therefore, contamination of a soil that, in fact, is free of phytotoxic substances.

Fluctuations in the environmental conditions under which the tests are conducted influence test results. This may be especially true in a glasshouse that is not air-conditioned or equipped with supplemental lighting. In this situation it is difficult to modulate the effects of seasonal changes in the ambient temperature and daily photoperiod. The use of controlled-environment facilities, therefore, is highly desirable when conducting tests for phytotoxicity.

Evaluation of plant responses should be based on several criteria. In primary tests visual ratings of seedling emergence and growth are expeditious. Such ratings are not only cost-effective but also appear reasonably reliable, especially when accompanied by observations on color, size, shape, or morphological effects. Quantitative measurements of weight or height provide information not discernible by visual observation. Although these criteria are more or less interrelated, all contribute to the overall evaluation of the test.

In the RMA survey cost, labor, space, and sample size constraints mandated the use of single replicates. It is encouraging that, despite a lack of replication, positive results were obtained. However, the level of contamination detectable is dependent on the precision of the test system. Control of random error by replication, in turn, enhances precision [5]. Also, replication facilitates statistical analysis. However, trade-offs between precision and practical constraints dictate the level of replication that can be employed.

Secondary tests are necessary on soil samples that show positive results in the primary screen. These secondary tests are essential to detect and discount reponses due to low fertility or poor physical characteristics of the soil samples. A degree of replication also is provided. Definitive chemical analysis of the soil samples suspected of contamination is imperative. However, chemical analysis does not provide direct evidence of phytotoxicity.

The survey of RMA for phytotoxins supports the concept of using herbicide screening techniques as the basis for delineating the extent of known or suspected soil contamination. Notwithstanding, refinement of the test system is necessary for full exploitation of the utility of these techniques. Efforts to control influences other than the effects of the contaminants themselves are mandatory. An appropriate experimental design is essential to reduce experimental error.

Acknowledgments

This work was supported by U.S. Army Medical Research and Development Command under Contract DAMD17-76-C-6039. The author wishes to thank P. A. Madigan, Chief, Research Data Management Branch, U.S. Army Medical Research and Development Command, for permission to publish this account.

References

[1] Torgeson, D. C. and Sirois, D. L., "Survey of the Rocky Mountain Arsenal for Phytotoxic Substances," Boyce Thompson Institute Report on U.S. Army Contract No. DAMD17-76-C-6039, 15 Oct. 1976.
[2] Cogley, D. R., Craker, L. E., Torgeson, D. C. and Sirois, D. L., "Biological Testing of Rocky Mountain Arsenal for Phytotoxic Substances," Report submitted to U.S. Army Medical Bioengineering Research and Development Laboratory, Environmental Research Requirements Branch, on Contract No. DAMD17-75-C-5059, Walden Division of Abcor, Inc., Wilmington, Mass., April 1979.
[3] Thomas, J. M., Skalski, J. R., Cline, J. F., McShane, M. C., Miller, W. E., Peterson, S. A., Callahan, C. A., and Greene, J. C., *Environmental Toxicology and Chemistry,* Vol. 5, 1986, pp. 487–501.
[4] Black, C. A., *Soil-Plant Relationships,* Wiley, New York, 1964, p. 11.
[5] Rubinstein, R., Cuirle, E., Cole, H., Ercegovich, C., Weinstein, L., and Smith, J., "Test Methods for Assessing the Effects of Chemicals on Plants," EPA Report 560/5-75-008, Office of Toxic Substances, U.S. EPA, Washington, D.C., 1975, p. 80.

Thomas H. Lillie[1] and Rebecca W. Bartine[1]

Protocol for Evaluating Soil Contaminated with Fuel or Herbicide

REFERENCE: Lillie, T. H. and Bartine, R. W., "**Protocol for Evaluating Soil Contaminated with Fuel or Herbicide,**" *Plants for Toxicity Assessment, ASTM STP 1091*, W. Wang, J. W. Gorsuch, and W. R. Lower, Eds., American Society for Testing and Materials, Philadelphia, 1990, pp. 198–203.

ABSTRACT: Sorghum and pinto bean plants were used as indicator organisms in a 14-day laboratory test to determine the impact of fuel and herbicide spills in agroecosystems. Plant survival and shoot length (from soil level to tip of highest leaf) were recorded and compared for plants grown in laboratory control soil and soil collected from the accident site. We also used soil from an uncontaminated area near the accident site as a control whenever possible. Chemical analysis was used to verify contaminants.

KEY WORDS: aviation fuel, JP-4, herbicide, soil bioassay

Fuel runoff from storage tanks or aircraft accidents can have a detrimental affect on agroecosystems. Likewise, drift and spills during herbicide application can sometimes contaminate surrounding areas for several months. The extent of damage is difficult to determine from simple observation at the site, and chemical analysis of the soil provides little data to determine the actual impact on plant growth.

The impact of accidental herbicide spills can be devastating if agricultural land is contaminated. Extensive studies have been performed to assess the toxicity of herbicides on target and nontarget plants. The type of test is dependent upon the mode of action of the herbicide, the route of contamination, and the environment in which contamination is likely to occur. The methods are either carried out on actual contaminated substrate (direct) or extracts of contaminated substrate (indirect) [1]. Plant survival [2], seed germination [3], shoot weight, plant heights, and root length [1] have all been used as variables for assessing the impact on plants.

A leaking fuel storage tank or aircraft accident may involve several thousand litres of flammable hazardous materials. Immediate fire danger exists as well as short- and long–term toxicity to plants and animals. Previous studies reported the toxicity and mutagenicity of jet fuels to dogs, monkeys, mice, rats, and humans [4,5]. Additional studies have addressed the impact of fuels on aquatic ecosystems [6]. Limited unpublished data are available on the effects of aviation fuel on agroecosystems, but no standardized laboratory test exists for assessing the impact of fuel in such areas.

The objectives of our study were: (1) to obtain baseline toxicity data for plants grown in

---

NOTE—Opinions and assertions contained herein are those of the authors. Mention of a proprietary product in this paper does not constitute an endorsement by the United States Air Force.

[1] United States Air Force Occupational and Environmental Health Laboratory, Brooks Air Force Base, San Antonio, TX 78235-5501. The first author is presently at Headquarters Space Systems Division/DEV, PO Box 92960, Los Angeles AFB, CA 90009-2960.

laboratory soil contaminated with either aviation fuel or herbicide; and (2) to develop a standardized laboratory test for evaluation of fuel or herbicide contamination of soil.

Procedure

Baseline Toxicity Tests

Sorghum (*Sorghum bicolor* L.) and pinto bean (*Phaseolus vulgaris* L.) plants were used as indicator organisms in a whole plant bioassay. The two plant species were selected as representative monocotyledon and dicotyledon, respectively. Seeds, obtained from Carolina Biological, were allowed to germinate for three days on moistened paper towels prior to planting to ensure that all seeds used in testing were viable. After germination, the seeds were planted in soil composed of a 1:1 mixture of Promix (seasoned Canadian sphagnum, peat moss, perlite, and vermiculite) and HYPONeX (peat moss, sand, and peat humus).

JP-4 aviation fuel was mixed into the soil to obtain concentrations of 6.25, 12.5, 17.5, 25, 35, and 50 mg of fuel per gram of soil (mg/g). The concentrations were made by mixing 720 g of soil with the appropriate amount of JP-4. The mixed soil was then divided evenly into eight 235-mL paper containers (i.e., 90 g per container). Four pinto bean seeds were planted in each of four containers of soil, and five sorghum seeds were planted in each of the other four containers of soil. The seeds were planted to a depth of approximately 2 cm below the surface of the soil. Separate containers of soil were contaminated with the herbicide Krovar® in concentrations of 0.01, 0.1, 0.5, and 1.0 mg/g. Uncontaminated soil was used as the control in all tests.

Immediately after planting the seeds, 50 mL of dechlorinated water was added to each container. Additional water was added as needed upon daily inspection, but all containers received the same amount of water throughout the test. The containers were maintained in an environmental chamber with climatic controls set at 16 h of light per diem (2.15×10^4 lux), $26.5 \pm 2°C$ day temperature, $10 \pm 2°C$ night temperature, and $70 \pm 10\%$ relative humidity. Plants were examined once a day and records were maintained of watering and climatic conditions. The test was terminated after 14 days; plants were cut at soil level, and each plant was measured from base of stem to tip of highest leaf. The entire procedure was replicated three times for each concentration of contaminant including the control.

Data were analyzed by analysis of variance and Duncan's multiple range test [7]. The specific test was "PROC ANOVA" with the "MEANS/DUNCAN" option. Sample sizes were 60 sorghum plants and 48 pinto bean plants for each treatment.

Evaluation of Herbicide and Fuel Spills

Herbicide and fuel spills associated with field incidents were evaluated according to procedures used in baseline toxicity tests. Environmental chambers were used with climatic conditions as described above. The experimental design differed in that 3 replicates were performed for each of three soil samples: (1) laboratory control soil, (2) field control soil, and (3) soil suspected of contamination. Field control soil was collected from an uncontaminated area near the spill or accident site. Plants grown in laboratory control soil were used to assess the impact of procedures and conditions during the test. Plants grown in field control soil were used to assess the ability of an uncontaminated sample to support plant growth. Plants grown in soil suspected of contamination were used as indicator organisms to determine the potential impact of herbicide or fuel in an agroecosystem. Each soil sample was also subjected to chemical analysis to verify contaminants. Data were analyzed by analysis of variance and Duncan's multiple range test [7].

Results

JP-4 aviation fuel significantly ($p < 0.01$) affected the growth of sorghum and pinto bean plants at all concentrations. Sorghum plants decreased in height as the JP-4 concentration increased in the soil (Fig. 1). Pinto bean plant growth produced a bimodal distribution with peaks at the control and at 35 mg/g (Fig. 2); however, a statistical test to determine the significance of bimodality was not performed.

Krovar affected plant growth at concentrations much lower than JP-4. Sorghum (Fig. 3) and pinto bean (Fig. 4) plants grew significantly ($p < 0.05$) shorter in soil containing 0.1 to 1.0 mg/g than in the control soil. Pinto bean growth in soil contaminated with Krovar produced a bimodal distribution similar to that observed for JP-4; however, peaks occurred at 0.01 and 0.5 mg/g (Fig. 4). The no effect level for pinto bean plants was 0.01 mg/g. A no effect level was not determined for sorghum.

From 1980 to 1989, the United States Air Force Occupational and Environmental Health Laboratory (USAFOEHL) performed over 40 whole plant bioassays in response to field incidents. Of these, roughly half were fuel related; the others were a combination of herbicide runoff and other contaminants. Results from the 40 bioassays have shown that aviation fuel affects plants at concentrations as low as 1 mg/g. Sorghum height tended to show a more direct relationship with fuel contamination than pinto bean height. However, the soil in which the fuel was spilled and the possibility of a fire associated with a crash may have affected the results as well. Follow-up tests were performed to monitor recovery of the soil. Recovery to a state of productivity equivalent to that observed in uncontaminated soil can require 2 to 4 years, depending on the concentration of JP-4 or herbicide in the contaminated soil. The recovery process can be reduced to about six months if mitigation measures are implemented.

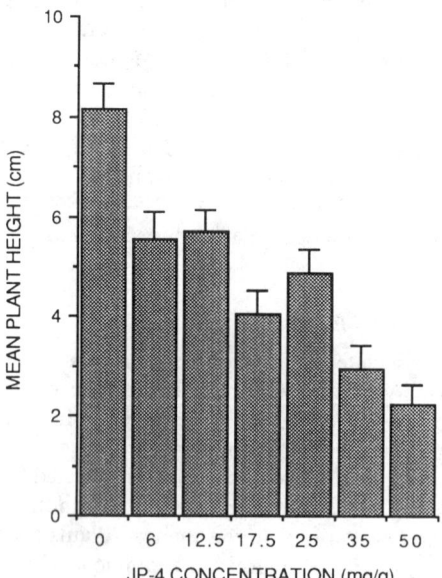

FIG. 1—*Mean height of sorghum plants grown in soil contaminated with known concentrations of JP-4 aviation fuel during a 14-day whole plant bioassay. Means are based on a sample size of 60 plants. Line above each column is the standard error of the mean.*

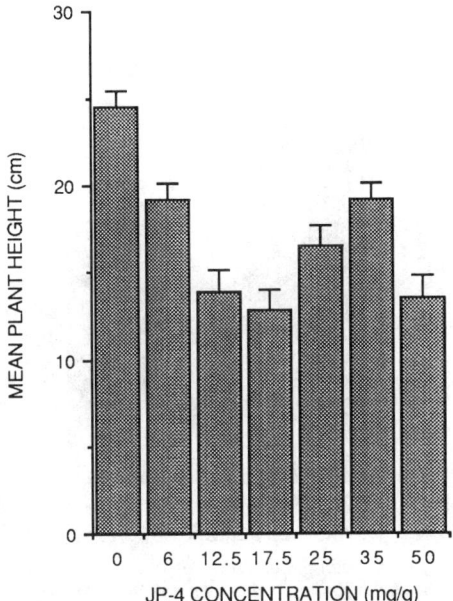

FIG. 2—*Mean height of pinto bean plants grown in soil contaminated with known concentrations of JP-4 aviation fuel during a 14-day whole plant bioassay. Means are based on a sample size of 48 plants. Line above each column is the standard error of the mean.*

Discussion

JP-4 aviation fuel was selected for evaluation because it is a primary fuel used in aircraft of the United States Air Force. It is composed of a complex mixture of about 300 aliphatic and aromatic hydrocarbon compounds varying in length up to C16. The herbicide Krovar was selected for testing because it is often the herbicide of choice in situations such as fence

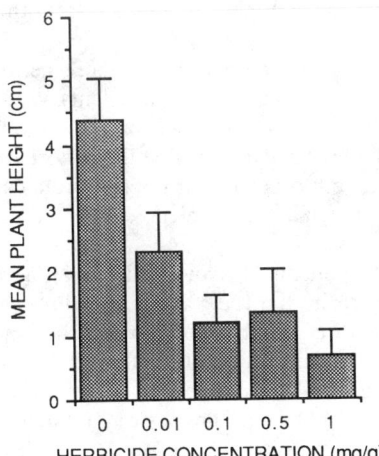

FIG. 3—*Mean height of sorghum plants grown in soil contaminated with known concentrations of the herbicide Krovar during a 14-day whole plant bioassay. Means are based on a sample size of 60 plants. Line above each column is the standard error of the mean.*

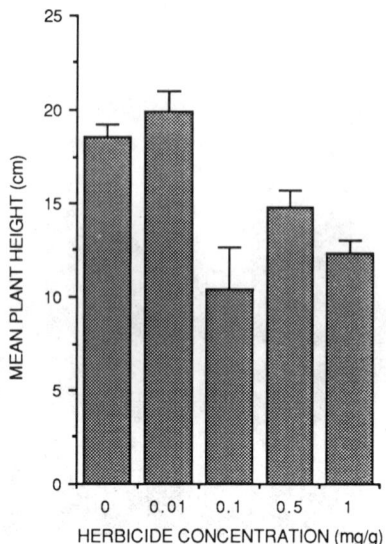

FIG. 4—*Mean height of pinto bean plants grown in soil contaminated with known concentrations of the herbicide Krovar during a 14-day whole plant bioassay. Means are based on a sample size of 48 plants. Line above each column is the standard error of the mean.*

rows and security areas where broad spectrum weed control is required. Krovar is composed of 40% bromacil (5-bromo-3-sec-butyl-6-methyluracil), 40% diuron (3-(3,4-dichlorophenyl)-1,1-dimethylurea), and 20% inert ingredients.

Previous tests done at USAFOEHL using the procedures described above have shown that plant growth is adversely affected by direct contact with soil containing JP-4. Soil will recover through natural processes 2 to 4 years after the crash date. The recovery processes can be assisted by aeration of the soil; turning it over and exposing lower layers to the air. This will encourage further evaporation of the volatile compounds from the jet fuel and enhance biodegradation. A no-effect level has not been established from these assays and further testing needs to be performed.

Results from the baseline assays are similar to those performed at USAFOEHL from 1980–1989 in response to field incidents. Sorghum plants were affected directly by toxins in the aviation fuel and herbicide. However, effects on the pinto bean plants were more complex. A component in JP-4 may be acting as a synthetic hormone. Because plant height was stimulated in the dicot (i.e., pinto bean) and not the monocot (i.e., sorghum) a synthetic auxin is most probable. Although auxin is one possibility, JP-4 is complex and other factors may be involved in the process.

Bromacil and diuron, components of Krovar, affect plant growth by inhibiting photosynthesis [8]. They are taken up by the roots and move through plants in the transpiration system. Bromacil is a potent direct inhibitor of photosynthesis acting at the chloroplast level. Diuron affects the light dependent portion of photosynthesis [9]. Both compounds provide general weed control at high concentrations, but diuron works as a selective weed killer at low concentrations. Residues of bromacil are persistent, the half life is about 5 to 6 months.

Conclusion

These test results are useful to organizations that deal with aircraft accidents, herbicide spills, and the legal implications associated with them. Although results from baseline tests could not replace actual studies done on soil contaminated during field incidents, they provide a reference point to which chemical analyses and bioassays could be compared. It is possible to perform the chemical analyses while the whole plant bioassay is under way, find the proportion of oils and greases or herbicide per gram of soil, and compare the chemical results to the baseline study. This would allow the researcher to validate the bioassay results. If fuel or herbicide is not the cause, then other contaminants and other factors can be examined, avoiding the possibility of unfounded legal claims and allowing the justified claims to be resolved as quickly as possible.

Acknowledgments

We wish to thank R. Brown and J. Jarrell for their assistance during Air Force-sponsored summer research programs. R. Brown also assisted with the literature search. We are also appreciative of the efforts of J. Crigler for chemical analyses and R. Binovi for consultation. We are extremely grateful to D. Ripley, J. Allen, S. Taffinder, R. Nakasone, J. Muilenberg, T. Koenig, C. Wilson, H. Casey, T. Doane, W. Rogers, P. Oxner, J. Schuermann, M. Spakowicz, and R. Rolon for their contributions to the USAFOEHL biomonitoring program.

References

[1] Clay, D. V., "Biological Assay Methods for Herbicide Residues," UK, Agricultural Development and Advisory Service, Ministry of Agriculture, Fisheries and Food No. 347, 1983, pp. 153–164.
[2] Wang, W., *Environmental Pollution* (Ser. B), Vol. 11, 1986, pp. 1–14.
[3] Wang, W., *Environmental Toxicology and Chemistry*, Vol. 6, 1987, pp. 409–414.
[4] Kapp, R. W. and Piper, C. E., "*In vitro* and *in vivo* Mutagenicity Studies of Jet Fuels, a Final Report," Hazelton Laboratories America, Vienna, VA, 1979.
[5] Knave, B., Mindus, P., and Struwe, G., *Acta Psychiatrica Scandinavica*, Vol. 60, 1979, pp. 39–49.
[6] Jenkins, D., Klein, S. A., and Cooper, R. C., *Water Research*, Vol. 11, 1977, pp. 1059–1067.
[7] *SAS User's Guide: Statistics*, Version 5, SAS Institute, Cary, N.C., 1985.
[8] Gardiner, J. A. in *Herbicides: Chemistry, Degradation and Mode of Action*, P. C. Kearney and D. D. Kaufman, Eds., Marcel Dekker, New York, 1975, pp. 294–321.
[9] Geissbuhler, H., Martin, H., and Voss, G. in *Herbicides: Chemistry, Degradation and Mode of Action*, P. C. Kearney and D. D. Kaufman, Eds., Marcel Dekker, New York, 1975, pp. 209–277.

M. Bassi,[1] *M. G. Corradi*,[1] *and M. A. Favali*[1]

Effects of Chromium in Freshwater Algae and Macrophytes

REFERENCE: Bassi, M., Corradi, M. G., and Favali, M. A., "**Effects of Chromium in Freshwater Algae and Macrophytes,**" *Plants for Toxicity Assessment, ASTM STP 1091,* W. Wang, J. W. Gorsuch, and W. R. Lower, Eds., American Society for Testing and Materials, Philadelphia, 1990, pp. 204–224.

ABSTRACT: We have tried to assess the effects of chromium(VI) in four freshwater green algae (*Coccomyxa minor, Scenedesmus armatus, S. dimorphus,* and *Haematococcus lacustris*) and two freshwater macrophytes (*Lemna minor* and *Pistia stratiotes*). The algae were kept in liquid or solid culture medium containing 1, 2, 5 and 10 mg/L of chromium (VI) supplied as $K_2Cr_2O_7$. *Lemna* and *Pistia* were grown in nutrient solution to which chromium was added as $K_2Cr_2O_7$ to give final concentrations of Cr(VI) from 0.1 to 20 mg/L.

Coccomyxa and *S. dimorphus* seem to be less tolerant than the other algae because 1 mg/L Cr(VI) is sufficient to induce alterations of chloroplasts and cell wall. *S. armatus* and the quiescent spores of *Haematococcus* show alterations if Cr(VI) concentration is at least 5 mg/L. With 10 mg/L spore germination is completely inhibited. All the algae show loss of chlorophyll and other pigments and cell vacuolation. The chloroplasts appear damaged, and electron-opaque deposits are visible in the cytoplasm and vacuoles.

In *Lemna* and *Pistia* the main macroscopical effects are retardation of growth, chlorosis, and wilting. Ultrastructural alterations consist in convolution and retraction of the plasma membrane, thylakoid disarrangement and swelling, transient increase of the number of mitochondria. Chlorophyll content diminishes, while protein content at first increases, then returns to average control values. These alterations are caused by even low doses of the metal (0.1 mg/L in *Lemna* and 1 mg/L in *Pistia*).

KEY WORDS: chromium toxicity, freshwater algae, freshwater macrophytes

Environmental contaminants act on cellular processes critical to algal growth such as photosynthesis and nucleic acid, protein, and lipid biosynthesis [1]. Chromium(VI), in particular, is a highly toxic water contaminant [2], which can alter the biological equilibrium of waters owing to its mutagenic or toxic effects in living organisms [3]. Therefore we tried to study the effects of Cr(VI) in the freshwater green algae *Coccomyxa minor, Scenedesmus armatus, S. dimorphus,* and *Haematococcus lacustris* to see if the effects of chromium vary according to the algal species.

We also have studied the effects of Cr(VI) in two common freshwater plants, *Lemna minor* and *Pistia stratiotes,* because they are both fast growing but of different size and because *Lemna* is a food source for fowl, fish, and mammals [4].

[1] Institute of Botany, University of Parma, 43100 Parma, Italy.

Materials and Methods

Algae

We used some of the tests suggested by Boyle [1] as useful in studies of algae as ecological indicators for water pollution:

- Changes in algae morphology and ultrastructure.
- Chlorophyll content as determined by fluorescence microscopy.
- Growth rate.
- Cell cycle: vegetative cells, colonies (cenobium formation), quiescent spores (akinetes).
- Mortality and/or tolerance.
- Recovery.

Culture conditions—*Coccomyxa minor, Scenedesmus armatus, S. dimorphus,* and *Haematococcus lacustris,* kindly supplied by IRSA, CNR, Milan, Italy, were kept in Hoagland culture medium [5]. *S. armatus* and *S. dimorphus* were grown together because they are generally found together in the same natural habitat. The liquid axenic cultures were routinely grown in climatic chambers in a static medium, at 22 to 24°C, with a 16-h photoperiod. The optimal pH for cell growth was 7; therefore this value was maintained by adding 0.05 M N-[tris(hydroxymethyl)-methyl] glycine (tricine) to the solutions.

The quiescent aplanospores of *Haematococcus* (akinetes) were maintained also in solid medium under continuous fluorescent light at room temperature. Chromium(VI) was supplied as potassium dichromate ($K_2Cr_2O_7$) to give final concentrations of 1, 2, 5, and 10 mg/L Cr(VI). All specimens were analyzed within one day of treatment and every other day for one month. The experiments were repeated four times.

Light (LM) and Fluorescence (FM) Microscopy—Living cells were observed and counted with a Zeiss Axioscope and their size distribution determined with a haemacytometer.

Chlorophyll content was determined subjectively by autofluorescence with a Zeiss Axioscope equipped with a fluorescent attachment.

Transmission Electron Microscopy (TEM)—The cells were centrifuged at 3000 g for 20 min, washed in 0.1 M sodium phosphate buffer, pH 7.0, then fixed with 3% glutaraldehyde in buffer for 3 h and post-fixed with 1% osmium tetroxide at 4°C for 2 h, dehydrated, and embedded in Epon-Araldite.

Ultrathin sections, unstained or stained with lead citrate, were examined with a Hitachi 300 electron microscope at 80 kV.

Recovery Tests—For recovery tests only *Haematococcus* akinetes were used. Akinetes grown on agar were transferred to Hoagland solution containing 1, 5, or 10 mg/L Cr(VI), and kept there for 4, 7, or 12 days. After Cr(VI) treatment all akinetes were transferred to fresh Hoagland medium and observed for 14 days more to see if they could germinate and produce vegetative cells as controls did. In addition, variations in cell structure and chlorophyll content were checked by LM and FM.

Freshwater Macrophytes

The plants were grown in half-strength Hoagland solution to which $K_2Cr_2O_7$ was added to give final concentrations of Cr(VI) ranging from 0.1 to 20 mg/L. Observations were made at regular intervals after 1 to 21 days of permanence in the solutions. Details of the experimental conditions have been published [6].

Fluorescence Microscopy—Whole leaves of *Lemna* and free-hand sections of *Pistia* leaves were observed directly with a Zeiss Axioscope equipped with a fluorescent attachment.

Transmission Electron Microscopy (TEM)—Specimens of leaves were fixed in phosphate-buffered 3% glutaraldeyde, post-fixed in osmium tetroxide, dehydrated in ethanol, and embedded in Araldite. Ultrathin sections were stained with uranyl acetate and lead citrate, and observed with a Siemens Elmiskop 1A electron microscope. Mitochondria were counted in sections of 20 control and 20 Cr(VI)-treated cells.

Protein Content—Protein content was determined by the Lowry's method [7] as modified by Hartree [8] and referred to dry weight. The data were subjected to statistical analysis.

Results and Discussion

LM and FM Microscopy

In *Coccomyxa*, hexavalent chromium induces morphological alterations very soon after the beginning of treatment (1 day), even at the lowest concentration (1 mg/L). The cells become significantly larger than controls (Table 1). Also, *S. dimorphus* and *S. armatus* show a slight increase in volume, which, however, is not significant. As seen in Figs. 2 and 3, volume increase is due to an increased cell vacuolation.

In control cultures of *S. dimorphus* and *S. armatus* many cenobia are always present (Fig. 1), while after Cr(VI) treatment the number of cenobia diminishes. In *S. dimorphus* 1 mg/L is enough to cause such diminution (Fig. 8); in *S. armatus* at least 5 mg/L are necessary.

In *Coccomyxa* chlorophyll content diminishes so drastically even with 1 mg/L that no fluorescence is detectable after 24 h. At this concentration in *S. dimorphus* and *S. armatus* chlorophyll content diminishes after three to four days of treatment, but some fluorescence is still detectable (cf. Fig. 5 and Fig. 4). With 10 mg/L no autofluorescence is observed.

Growth Rate

The growth rates of *Coccomyxa* (Fig. 6), *S. armatus* and *S. dimorphus* (Fig. 7) are reduced proportionally to the Cr(VI) concentrations used, but drastically reduced with 5 or 10 mg/L. After one week, growth is completely inhibited, and no recovery is visible even after one month.

TEM Observations

A typical control cell of *Coccomyxa* (Fig. 9) contains a mantle-shaped chloroplast with a well developed thylakoid system. No pyrenoid is present. In *Coccomyxa* Cr(VI) causes remarkable ultrastructural alterations (Fig. 10): with as little as 1 mg/L the cell organelles are completely damaged and large electron-opaque deposits are present in the cell. This picture suggests that *Coccomyxa* is highly sensitive to Cr(VI).

TABLE 1—*Dimensions (μm) of algal cells after 24 h of treatment with 1 mg/L Cr(VI) (means of 10 measurements $\pm$ s.e.n).*

Algae	Controls		Treated	
	Length	Width	Length	Width
Coccomyxa	3.5 ± 0.5	3.1 ± 0.3	7.4 ± 0.7	5.7 ± 0.1
S. dimorphus	10.2 ± 0.7	3.3 ± 0.3	11.5 ± 0.6	3.8 ± 0.4
S. armatus	7.6 ± 0.4	3.7 ± 0.2	8.1 ± 0.3	4.1 ± 0.2

[a] Significantly different from control ($p < 0.05$).

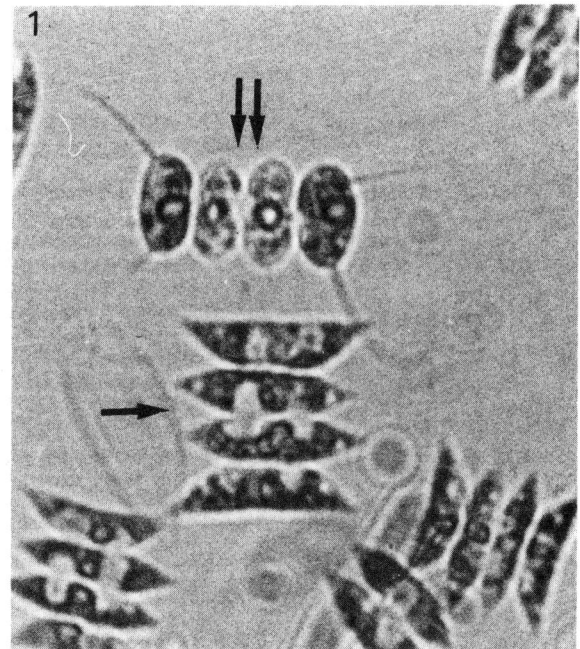

FIG. 1—*Typical cenobia of* S. dimorphus (arrow) *and* S. armatus (double arrow) *as seen in control coltures.* ×1100.

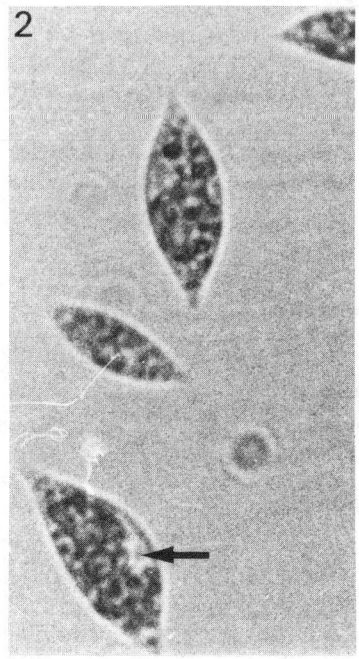

FIG. 2—S. dimorphus *kept in 1 mg/L Cr(VI); the cells become larger and vacuolated* (arrow). ×1100.

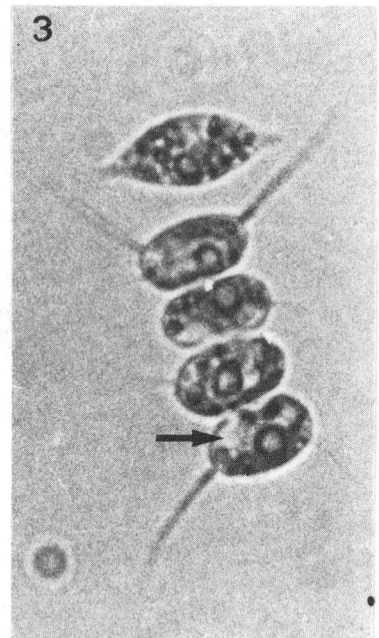

FIG. 3—*Cenobium of* S. armatus *grown in 5 mg/L Cr(VI); note cell vacuolation* (arrow). ×1100.

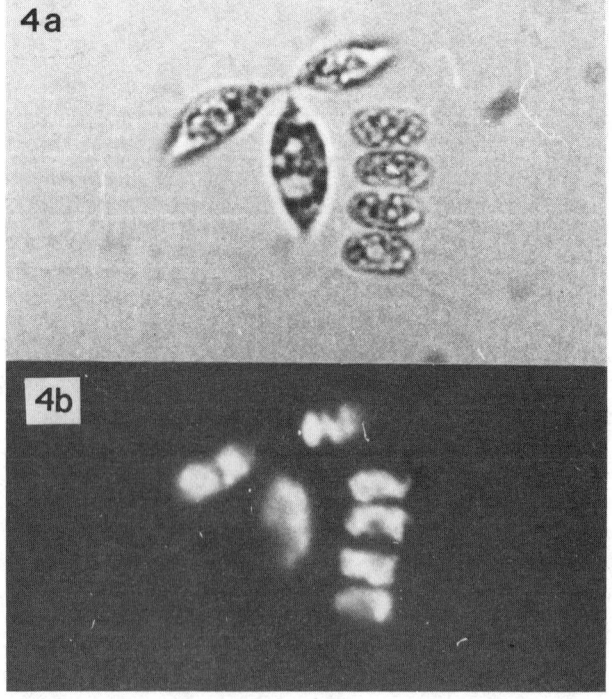

FIGS. 4a and 4b—*Light (LM) and fluorescence microscopy (FM) of control* Scenedesmus *cells; note the bright fluorescence of the cells.* ×900.

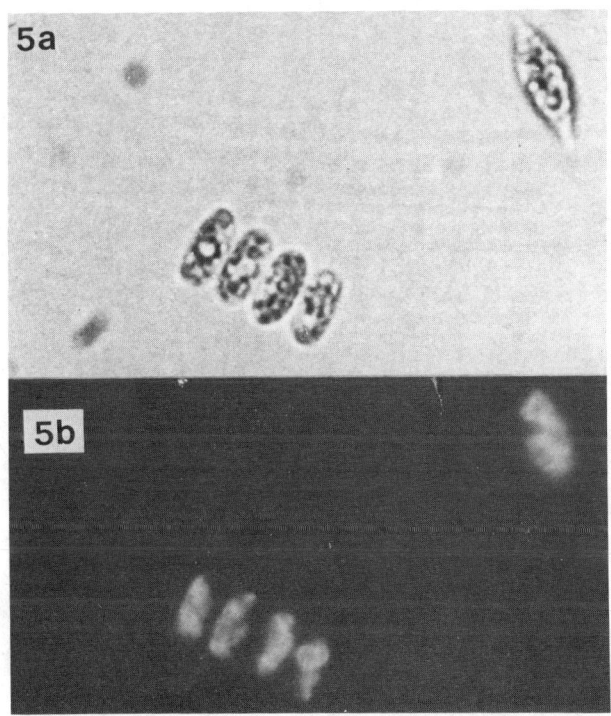

FIGS. 5a and 5b—*LM and FM of* Scenedesmus *cells treated with 1 mg/L Cr(VI); the algae are less fluorescent than controls (Fig. 4b).* ×900.

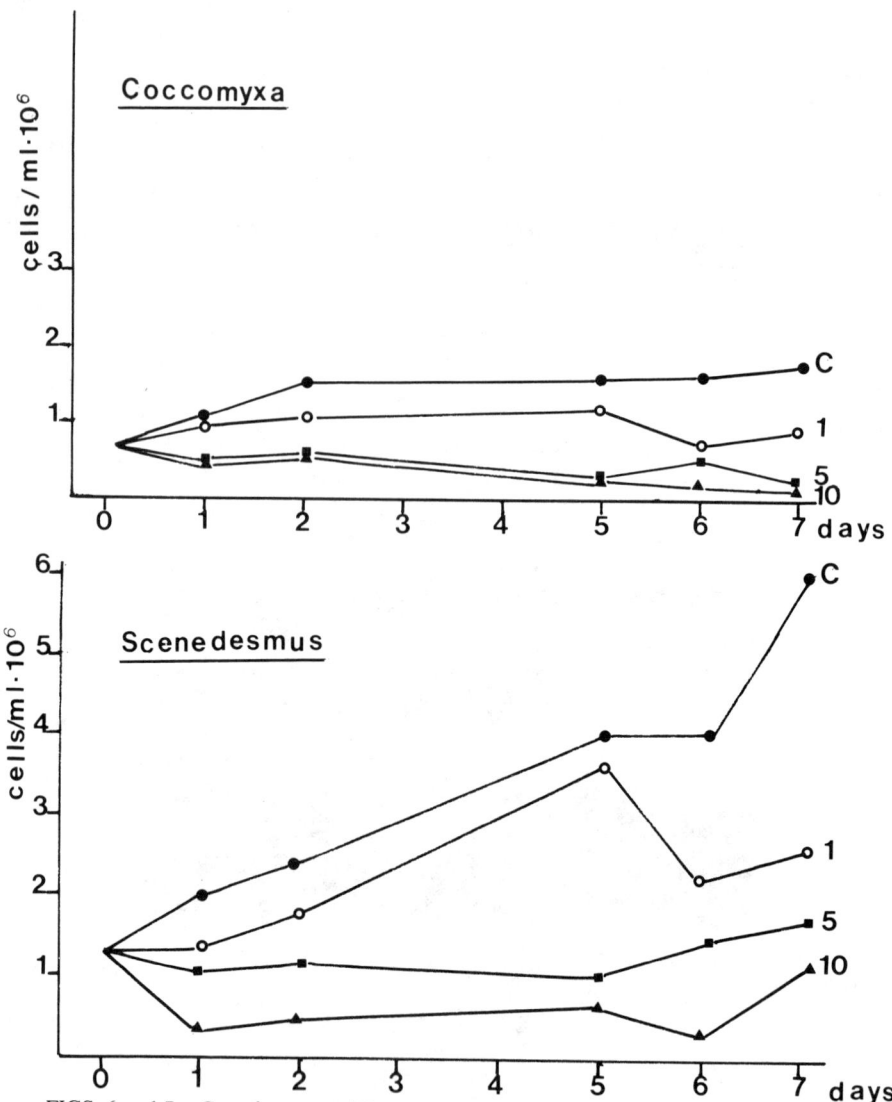

FIGS. 6 and 7—*Growth curves of* Coccomyxa *(Fig. 6) and* Scenedesmus *(Fig. 7) cultured cells treated with: 1 (○), 5 (■), or 10 mg/L (▲) Cr(VI). Controls (●). Means of three replicates. Differences between control and treated cells are highly significant after 2 days of treatment with 5 or 10 mg/L ($p < 0.05$).*

BASSI ET AL. ON EFFECTS OF CHROMIUM   211

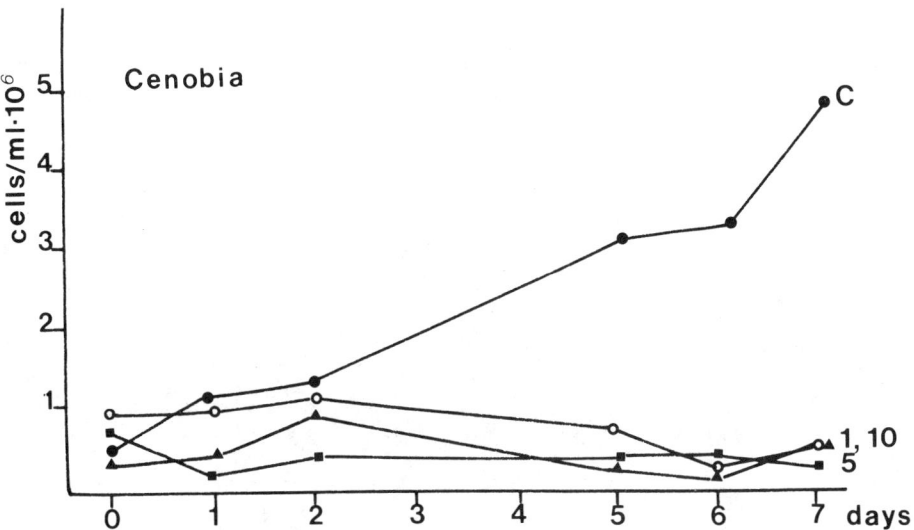

FIG. 8—*In* Scenedesmus *the number of cenobia decreases after Cr(VI) treatment. Means of 3 replicates. Differences between control and treated cells significant at* $p < 0.05$.

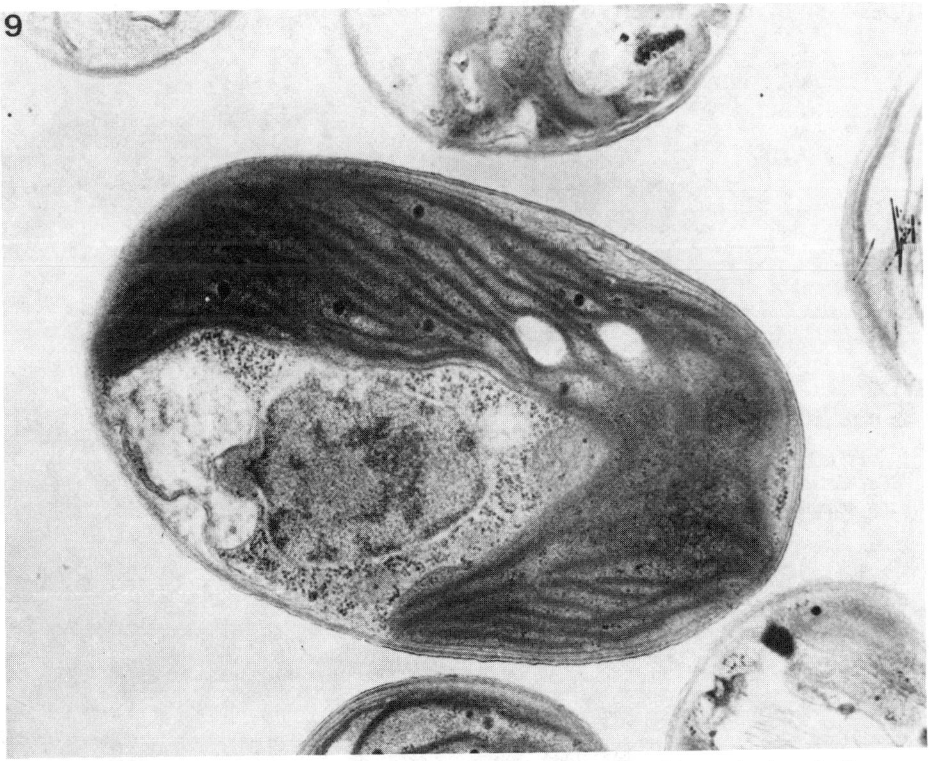

FIG. 9—*Electron micrograph of control cell of* Coccomyxa; *note the mantle-shaped chloroplast with a well-developed thylakoid system. No pyrenoid is present.* $\times 24\,000$.

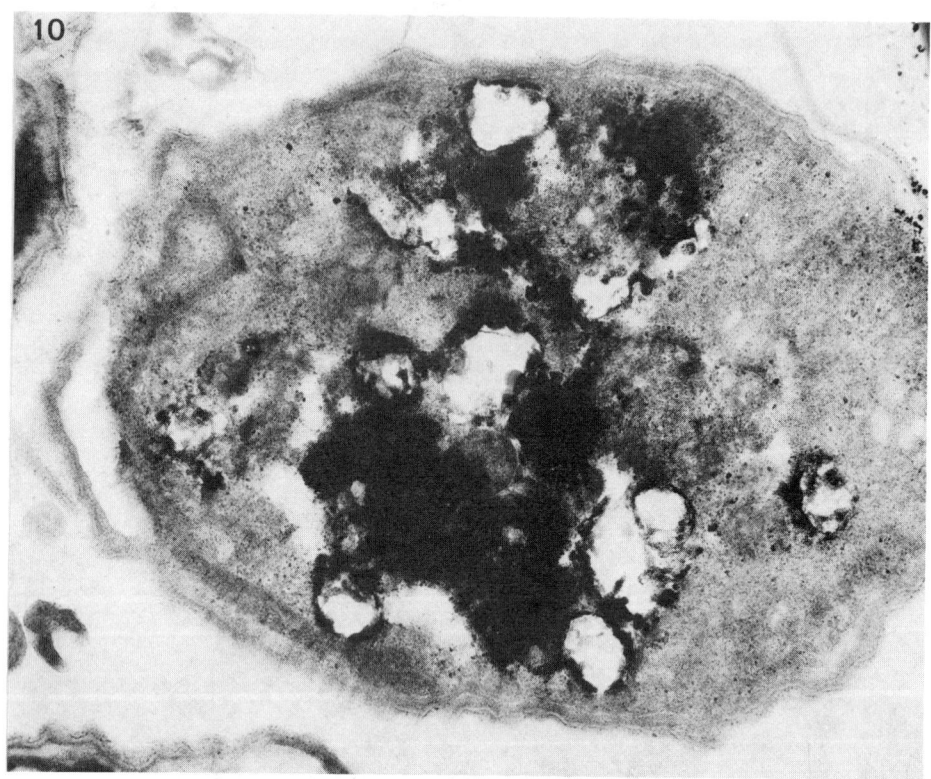

FIG. 10—*Electron micrograph of unstained section of* Coccomyxa *treated with 1 mg/L of Cr(VI); all cellular organelles are damaged and large electron-opaque deposits are visible in cytoplasmic vacuoles.* ×45,000.

Control cells of *S. dimorphus* (Fig. 11) and *S. armatus* (Fig. 12) contain chloroplasts with pyrenoids and starch plates. A typical cenobium of *S. dimorphus* is visible in Fig. 11. The cell wall of *S. armatus* shows spines and a verrucose layer (Fig. 12). The effects of Cr(VI) on the intracellular structure of *S. armatus* are less severe than in *Coccomyxa* (Figs. 13 and 14). With 1 mg/L some electron-opaque precipitates are visible in cytoplasmic vacuoles (Fig. 13), but the overall cellular structure is well preserved, while with 10 mg/L cell vacuolation is greatly increased and large electron-opaque precipitates are visible (Fig. 14). Similar precipitates have been observed in *Coccomyxa* and *Scenedesmus* grown in presence of copper [9] or other heavy metals [10].

Different mechanisms for Cr(VI) toxicity to algae have been proposed [11]. We suggest that the algae tested, mainly *S. armatus*, can tolerate the toxic effect of Cr(VI) by segregating the metal in cytoplasmic vacuoles. Detoxification is presumably achieved by complexing the metal with protein ligands, such as metallothioneins [12,13].

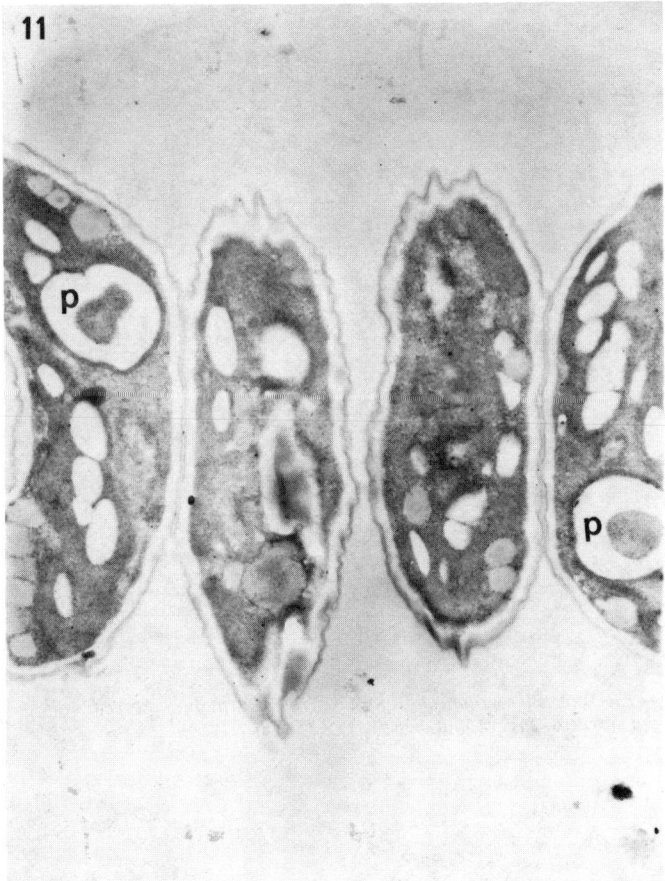

FIG. 11—*Electron micrograph of typical cenobium of untreated.* S. dimorphus. *The chloroplasts have pyrenoids with starch plates* (p). ×10,500.

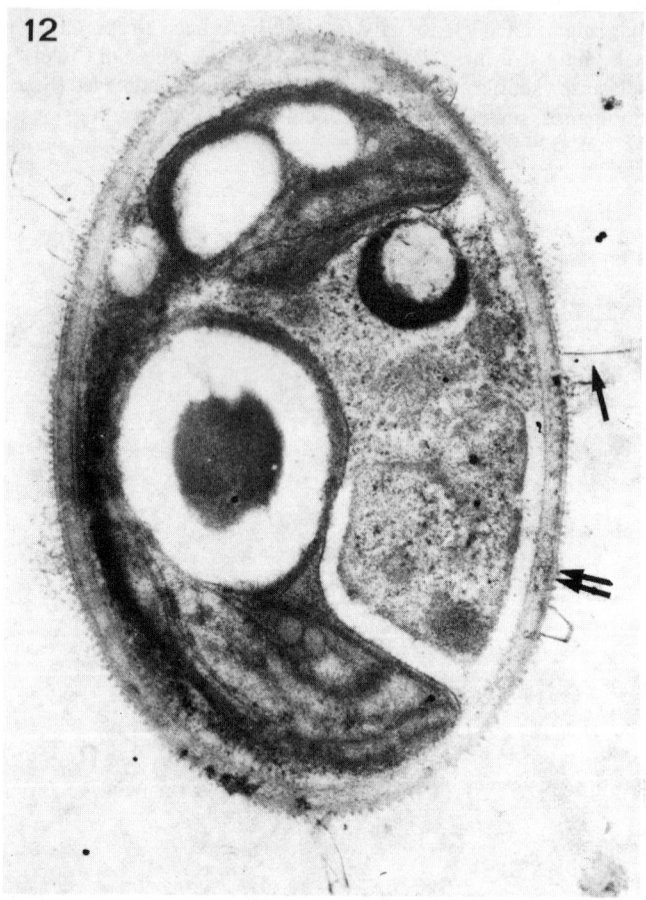

FIG. 12—*Electron micrograph of control cell of* S. armatus; *the cell wall shows spines* (arrow) *and a verrucose layer* (double arrow). ×15,000.

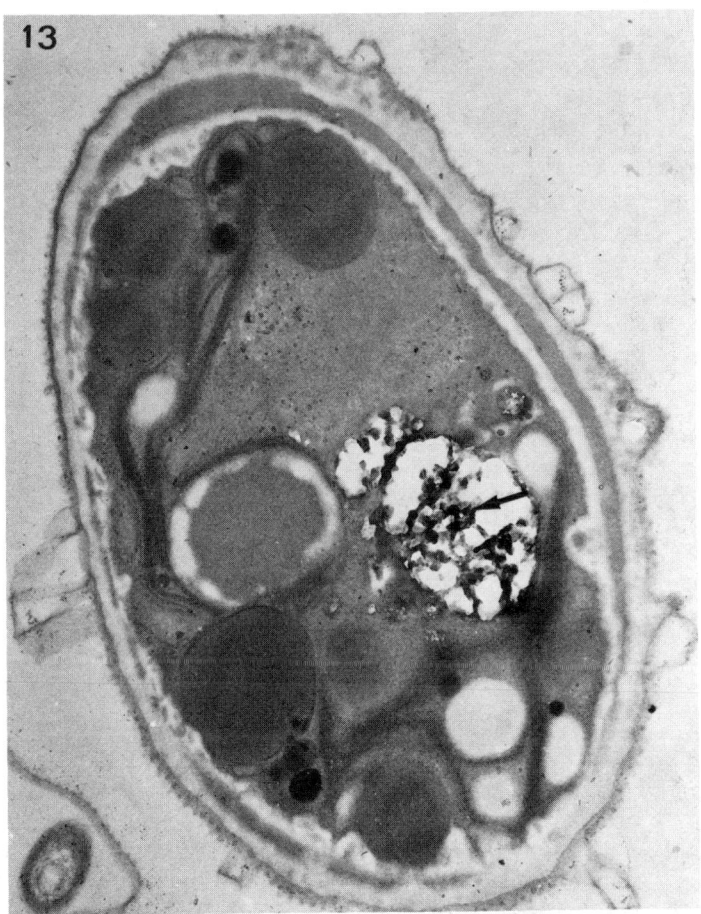

FIG. 13—*Electron micrograph of* S. armatus *kept in 1 mg/L Cr(VI); the overall cellular structure is preserved, even though the plasmalemma is convoluted and some electron-opaque deposits are visible in cytoplasmic vacuoles* (arrow). *Unstained.* ×27,000.

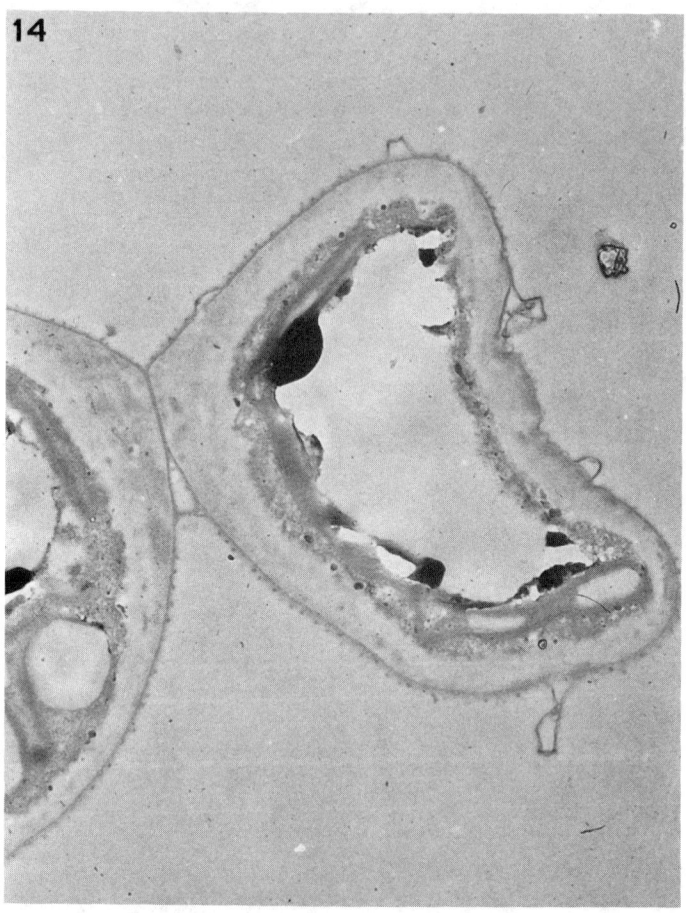

FIG. 14—*Electron micrograph of* S. armatus *treated with 10 mg/L Cr(VI); cell vacuolation is greatly increased and large electron-opaque deposits are visible. Unstained.* ×18,000.

Recovery Tests

Control akinetes do not show vacuolation, and astaxanthin, a red pigment characteristic of aging cells and akinetes, is uniformly distributed (Fig. 15). At the end of the experiment they are generally in the germinative stage (Fig. 20). Akinetes treated with 1 or 5 mg/L Cr(VI) show vacuolation and degradation of both astaxanthin and chlorophyll (Figs. 16 and 17). With 10 mg/L they become necrotic (Fig. 18). Akinetes kept in 1 ppm Cr(VI) (Fig. 16) and then retransferred to Cr-free Hoagland solution show a slight recovery after a few days because a few germinations occur. In those kept in 5 mg/L, recovery is poor, while those treated with 10 mg/L do not recover at all (Fig. 19).

With akinetes we could not use FM to test Cr effect on chlorophyll content because they are less autofluorescent than vegetative cells due to a masking of chlorophyll by the massive astaxanthin deposits.

Even very low concentrations of Cr(VI), such as 0.1 or 1 mg/L, induce severe damage to the macrophytes. The first target seems to be the plasma membrane, which becomes convoluted and focally detached from the cell wall (Fig. 22; compare with Fig. 21). Also, the chloroplasts become affected, their thylakoids appearing disarranged and swollen (Fig. 23). There is also a transient increase in the number of mitochondria (Fig. 24), which in normal cell sections is generally 2 to 4 per cell, in cells treated with Cr(VI) for 24 h is between 8 and 12. To these morphological alterations correspond a diminution of chlorophyll content and photosynthesis, and an increase in respiration [*14*].

Protein content increases significantly after 24 h, then returns to average control values (Table 2). This transient increase can probably be attributed to the increased mitochondrial population observed in the first days of poisoning, but could also be a response to the cellular stress induced by Cr(VI).

Macroscopically, the above changes lead to the appearance of chlorosis and finally to plant stunting and even death, if the metal concentration is strong enough. Many *Lemna* plants

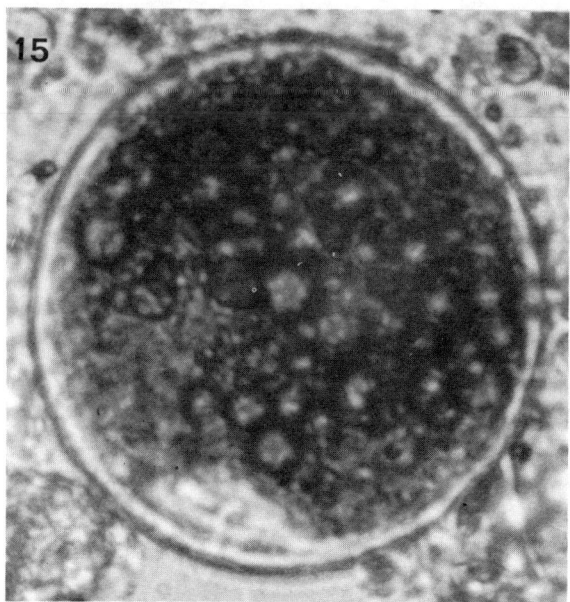

FIG. 15—*Control akinetes of* Haematococcus; *no vacuolation is visible.* ×250.

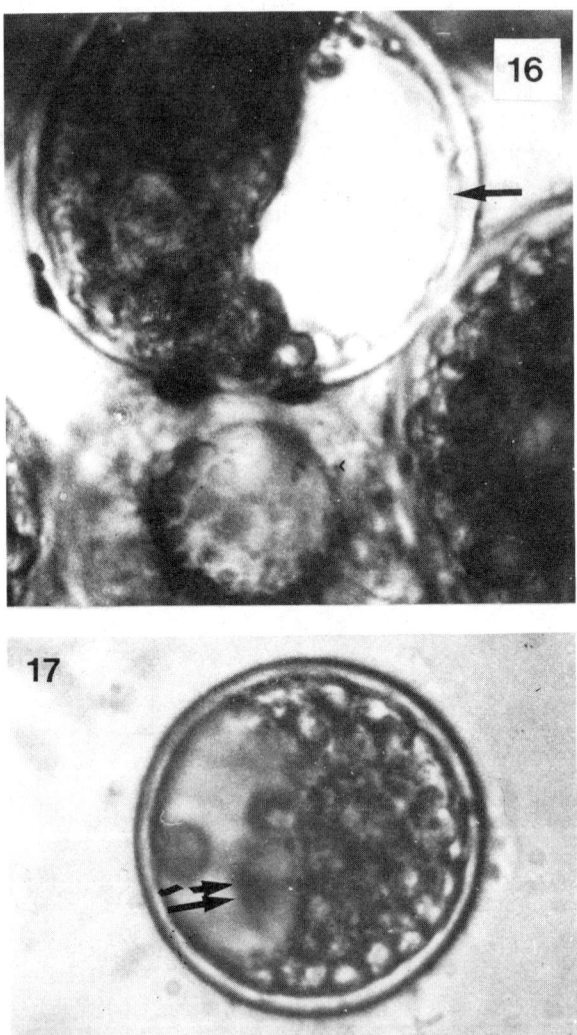

FIGS. 16 and 17—*Akinetes kept in 1 (Fig. 16) or 5 mg/L Cr(VI) (Fig. 17): note the progressive vacuolation* (arrow) *and the accumulation of astaxanthin in the vacuole* (double arrows). ×250.

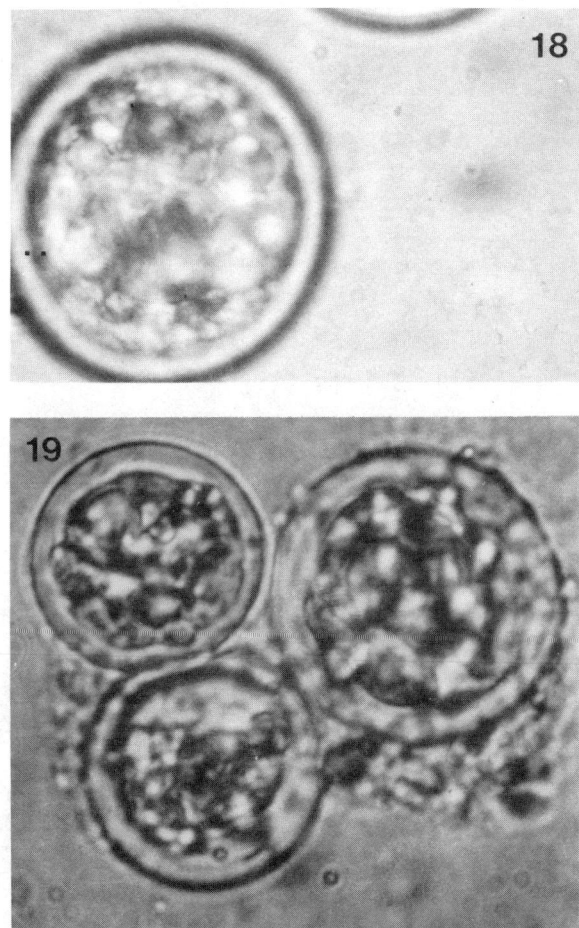

FIGS. 18 and 19—*Akinetes kept in 10 mg/L Cr(VI) before (Fig. 18) and after (Fig. 19) the recovery test; the cells are necrotic and can not germinate.* ×250.

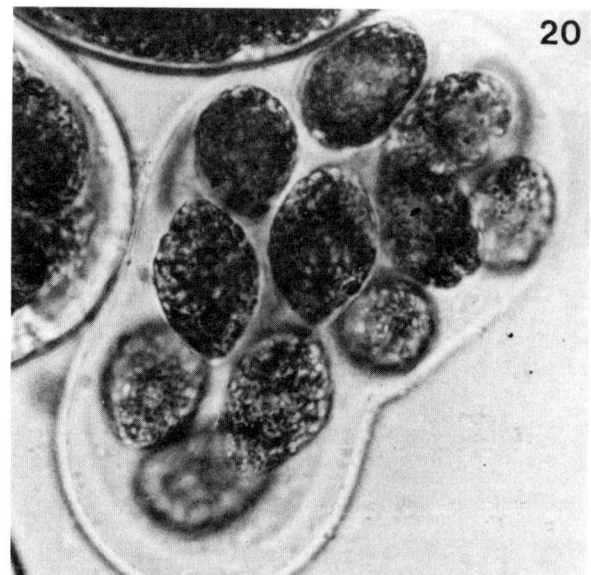

FIG. 20—*Control akinetes in the germinative stage.* ×420.

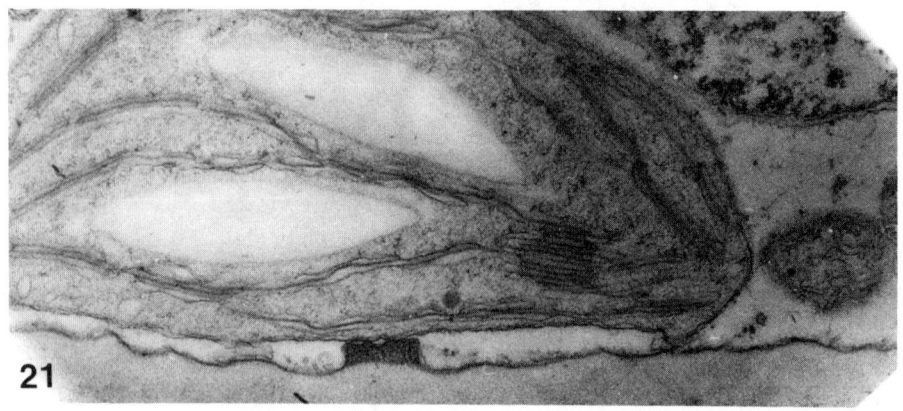

FIG. 21—*Electron micrograph of control leaf cell of* Pistia stratiotes. *The plasma membrane adheres perfectly to the cell wall and the chloroplast has a normal aspect, with starch grains.* ×32,000.

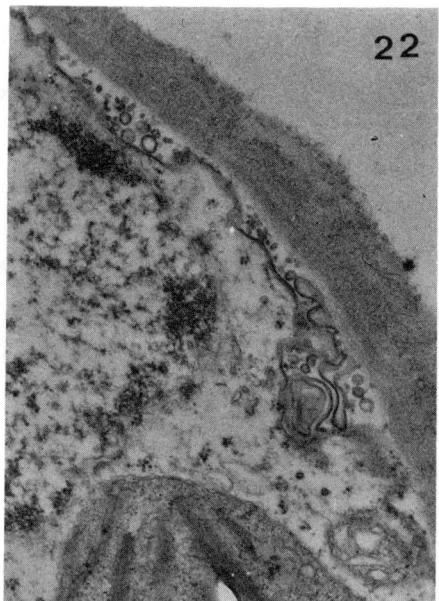

FIG. 22—*Electron micrograph of leaf cell of* P. stratiotes *kept for 48 h in the presence of 1 mg/L Cr(VI). The plasma membrane is highly convoluted and partially detached from the cell wall.* ×28,000.

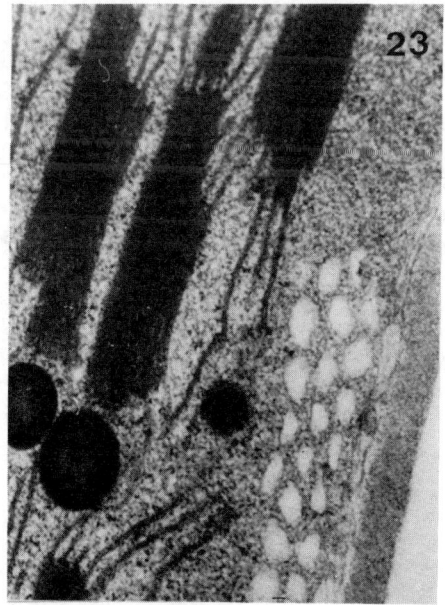

FIG. 23—*Electron micrograph of chloroplast of* Lemna minor *grown for 24 h in the presence of 1 mg/L Cr(VI). Tubule-like thylakoids give the stroma a fenestrated appearance, and the grana have a clumped appearance.* ×60,000.

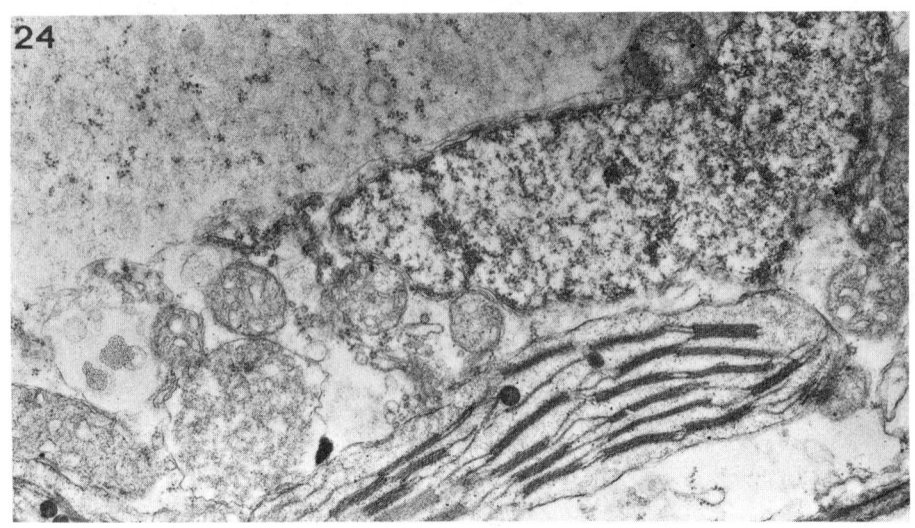

FIG. 24—*Electron micrograph of leaf cell of* P. stratiotes *kept for 24 h in the presence of 1 mg/L Cr(VI). Note the great number of mitochondria.* ×12,000.

died after 2 days of exposure to 5 mg/L, while all *Pistia* plants died after one week of exposure to 20 mg/L. Note that chlorosis, generally followed by necrosis, is not generalized to the whole leaf surface, but is restricted to circumscribed areas, which in *Lemna* take the form of either circular spots or straight bands across the leaf basis, and in *Pistia* are situated at the leaf upper margin, along the terminal veins. Apparently, chlorophyll is lost only from these areas and not from the entire leaf, as shown clearly by FM. Fluorescence diminishes or disappears only in such areas, while in the surrounding tissues normally fluorescent chloroplasts are present (Fig. 25). This observation confirms our previous hypothesis that the cellular alterations induced by Cr(VI) in the leaves are not consequent to a Cr(VI)-induced impaired function of the roots, but are probably due to a direct action of Cr(VI) on the leaf cells. In fact, we have demonstrated by X-ray microanalysis that the absorbed Cr(VI) is not confined to the roots but is translocated to the leaves [14].

The pH of the nutrient solution in which the Cr(VI)-treated plants are grown steadily increases with time if the solution is not renewed (Fig. 26). This correlates well with the diminution of K^+ uptake by these plants [14], which is probably consequent to an inhibition of proton extrusion [15,16]. These results confirm our previously expressed view that one of the major effects of Cr(VI) in freshwater macrophytes is the impairment of the plasma membrane function.

TABLE 2—*Protein content (mg/g dry weight) of* Pistia *leaves after treatment with 1 mg/L Cr(VI) (means of 6 replicates ± s. e. m.).*

	24 h	48 h	96 h
Control	*106.1 ± 3.00	112.6 ± 6.21	105.2 ± 5.87
Treated	*134.4 ± 5.75	128.8 ± 8.44	108.6 ± 6.42

* Difference significant at $p < 0.05$.

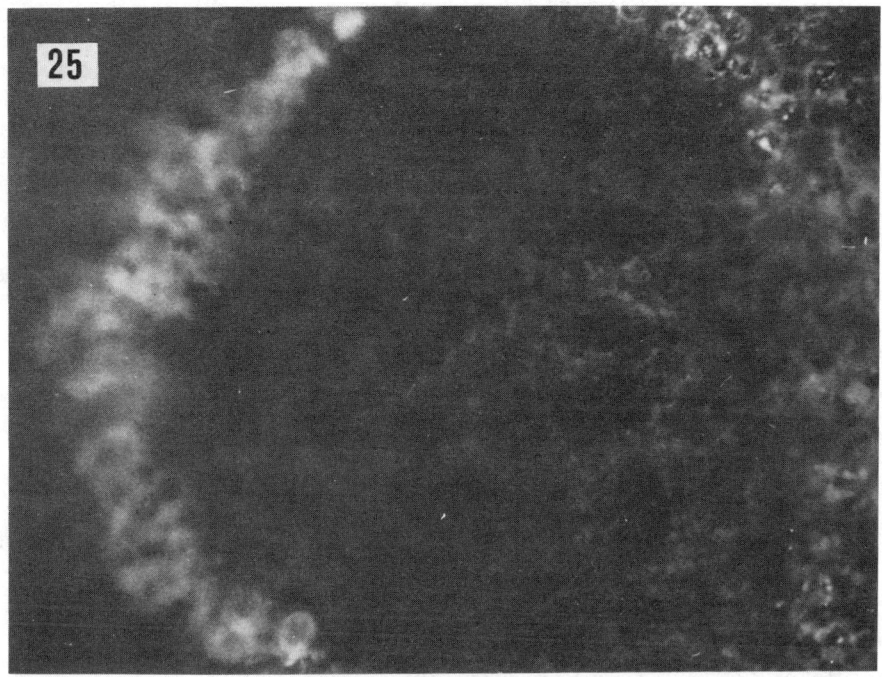

FIG. 25—*L. minor grown for 48 h in the presence of 1 mg/L Cr(VI). Fluorescence microscopy of a frond showing a chlorotic spot surrounded by normally fluorescent cells.* ×55.

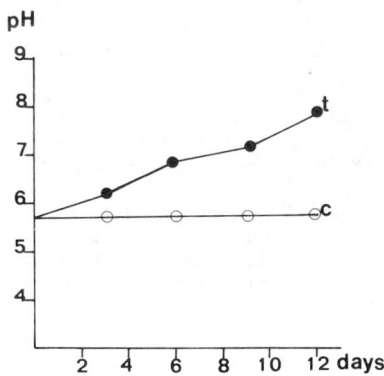

FIG. 26—*pH increase of the culture medium where* Pistia *plants are grown in the presence of Cr(VI); + = treated with 5 mg/L Cr(VI); c = control.*

Conclusions

Our data demonstrate that Cr(VI) toxicity varies in different algal species. Of the species tested *Coccomyxa* is less tolerant, together with *Haematococcus* in the vegetative stage, while *Scenedesmus* is more tolerant. Besides, of the two *Scenedesmus* species considered here, *S. armatus* is more resistant than *S. dimorphus*.

It seems that algal cells can neutralize the toxic action of Cr(VI) by segregating the metal in cytoplasmic vacuoles, in the same way as they segregate copper [9]. Since algae can be grown in strictly controlled conditions, they are particularly well-suited for studies of tolerance and/or regulatory mechanisms involving toxic metals.

Both freshwater macrophytes proved to be highly sensitive to Cr(VI), which seems to interfere with many metabolic processes. Even very low concentrations of the metal, such as can be easily found in polluted waters [2], cause retardation of growth and may lead in the end to plant death. These effects are more or less rapid according to plant size.

References

[1] Boyle, T. P., *Algae as Ecological Indicators,* Academic Press, London, 1984.
[2] Staves, R. P. and Knaus, R. M., "Chromium Removal from Waters by Three Species of Duckweeds," *Aquatic Botany,* Vol. 23, 1985, pp. 261–273.
[3] Bianchi, V. and Levis, A. G., "Recent Advances in Chromium Genotoxicity," *Environmental Toxicology and Chemistry,* Vol. 15, 1987, pp. 1–24.
[4] Culley, D. D., Jr., Rejmankova, E., Kuet, J., and Frye, J. B., "Production, Chemical Quality, and Use of Duckweeds *(Lemnaceae)* in Aquaculture, Waste Management, and Animal Feed," *Journal of the World Maricul. Soc.,* Vol. 12, 1981, pp. 27–49.
[5] Arnon, D. I. and Hoagland, D. R., "Crop Production in Artificial Culture Solution and Soils with Special Reference to Factors Influencing Yield and Absorption of Inorganic Nutrients," *Soil Science,* Vol. 50, 1940, pp. 463–483.
[6] Bassi, M., Corradi, M. G., and Realini, M., "Effects of Chromium(VI) in Two Freshwater Plants, *Lemna minor* and *Pistia stratiotes:* 1—Morphological Observations," *Cytobios,* Vol. 62, 1990, pp. 27–38.
[7] Lowry, O. M., Rosebrough, N. J., Fass, A. L., and Randall, A. J., "Protein Measurement with the Folin Phenol Reagent," *Journal of Biological Chemistry,* Vol. 193, 1951, pp. 265–275.
[8] Hartree, E. F., "Determination of Protein: A Modification of the Lowry Method that Gives a Linear Photometric Response," *Analytical Biochemistry,* Vol. 48, 1972, pp. 422–427.
[9] Favali, M. A., Mazzo, L., Ferrari, G., and Gerola, F. M., "Ultrastructural and Physiological Responses of the Green Algae *Coccomyxa* and *Scenedesmus* to Copper," *Proceedings,* Eleventh International Congress on Electron Microscopy, Kyoto, 1986, pp. 3259–3260.
[10] Silverberg, B. A., Stokes, P. M., and Ferstenberg, L. B., "Intranuclear Complexes in a Copper-Tolerant Green Alga," *Journal of Cell Biology,* Vol. 69, 1976, pp. 210–214.
[11] Riedel, G. F., "The Relationship between Chromium(VI) Uptake, Sulfate Uptake, and Chromium Toxicity in the Estuarine Diatom *Thalassiosira pseudonana,*" *Aquatic Toxicology,* Vol. 7, 1985, pp. 191–204.
[12] Albergoni, V., Piccinini, E., and Coppellotti, O., "Response to Heavy Metals in Organisms—I: Excretion and Accumulation of Physiological and Nonphysiological Metals in *Euglena gracilis,*" *Comparative Biochemistry and Physiology,* Vol. 62, 1980, pp. 121–127.
[13] Lerch, K., "Copper Metallothioneine in a Copper Binding Protein from *Neurospora crassa,*" *Nature,* Vol. 284, 1980, pp. 368–370.
[14] Bassi, M., Corradi, M. G., and Ricci, A., "Effects of Chromium(VI) in Two Freshwater Plants, *Lemna minor* and *Pistia stratiotes:* 2—Biochemical and Physiological Observations," *Cytobios,* 1990, in press.
[15] Zaccheo, P., Cocucci, M., and Cocucci, S., "Effects of Cr on Proton Extrusion, Potassium Uptake and Transmembrane Electron Potential in Maize Root Segments," *Plant, Cell, and Environment,* Vol. 8, 1985. pp. 721–726.
[16] White, C. and Gadd, G. N., "Inhibition of H^+ Efflux and K^+ Uptake, and Induction of K^+ Efflux in Yeast by Heavy Metals," *Toxicity Assessment,* Vol. 2, 1987, pp. 437–447.

D. L. Sirois[1]

Evaluation of Protocols for the Assessment of Phytotoxicity

REFERENCE: Sirois, D. L., "**Evaluation of Protocols for the Assessment of Phytotoxicity,**" *Plants for Toxicity Assessment, ASTM STP 1091*, W. Wang, J. W. Gorsuch, and W. R. Lower, Eds., American Society for Testing and Materials, Philadelphia, 1990, pp. 225–234.

ABSTRACT: Industrial chemicals and waste substances were tested in laboratory and growth chamber tests for toxicity to terrestrial and aquatic plants. Protocols for these tests were outlined in the 18 December 1978 and 16 March 1979 issues of the *Federal Register*. Tests on terrestrial plants included effects on seed germination and seedling growth as criteria of injury. Potential toxicity or stimulation of algae was assessed using *Anabaena flos-aquae*, *Chlorella vulgaris*, *Skeletonema costatum*, and *Selenastrum capricornutum*. Duckweed, *Lemna minor*, was used to assess potential phytotoxicity to higher aquatic plants. Evaluation of the results of tests showed that problems were encountered in implementing the test protocols. Unexplained variation in the data obstructed the calculation of confidence limits of the EC_{50} for the test substance with some species in the seedling growth test. The criteria for calculating the EC_{50} in *Lemna* and some algal tests were not met. Alternative use of analysis of variance (ANOVA) was hampered by heterogeneity of the variance in the test on *Anabaena*. In the algal tests chlorophyll appeared to respond differently than dry weight to the test substance. Although the protocols for these tests appear adequate, improvement is possible.

KEY WORDS: organic compounds, industrial wastes, indicator plants, algae, aquatic plants, bioassay, germination, plant, growth toxicity tests

Chemical contamination of the biosphere can be a consequence of commerce in an industrialized society. A chemical that becomes an environmental contaminant can seriously perturb both terrestrial and aquatic ecosystems if that chemical is toxic to the green plants that are essential to functional ecosystems. Protocols for assessing the potential phytotoxicity of new and existing industrial materials are a vital part of any effort to predict the impact of these materials on the environment. The objective of the present study was to evaluate certain study protocols using environmental samples.

We have conducted tests using study protocols outlined in the *Federal Register*. The test results have allowed us to critique these study protocols for planning future studies in our laboratory. Insights gained in the evaluation process may be of interest to other laboratories involved in assessing the effects of industrial chemicals or waste materials on plants.

Procedures

Protocols for the plant effects tests described in the 15 March 1979 issue of the *Federal Register* [1] were used to test industrial chemicals on terrestrial and aquatic plants. Seed germination tests were made on waste materials in accordance with procedures outlined in the Advance Notice of Proposed Rule Making (ANPR) that appeared in the 18 December

---

[1] Boyce Thompson Institute, Ithaca, NY 14853.

1978 issue of the *Federal Register* [2]. The proprietary nature of the chemicals tested preclude their description. Therefore letter designations are used instead of the actual test substances in presenting the data.

Seed Germination Bioassay

Radish (*Raphanus sativus* cv. Early Scarlet Globe) seeds (150) were placed in a Plexiglas germination chamber, and 100 mL of 1:10 dilution of the test substance (A, B, C, D, E, F, G) were added [2]. Radish seeds similarly prepared and treated with deionized water in a second germination chamber served as untreated controls. Both germination chambers were stored for 48 h in an unilluminated environmental chamber held at 25 ± 1°C. At the end of the incubation period, the length of the hypocotyl on each germinated seedling was measured and a standard t-test [3] used to compare treated and untreated seeds.

Seedling Growth Tests

Dicotyledonous seeds of cucumber, *Cucumis sativa* L. cv. Marketer, and pinto bean, *Phaseolus vulgaris* L. cv. Pinto, were placed in disposable growth pouches. Seeds of monocotyledonous species oat, *Avena sativa* L. cv. Clintford, and corn, *Zea mays* L. cv. Butter and Sugar, similarly were prepared. The seeds were then incubated in the dark for four days at 25 ± 1°C. The growth pouches with the germinated seeds were then removed from the incubator and exposed to ambient light in the laboratory for 24 h.

Washed horticultural vermiculite was treated with test substance X at concentrations of 62.5, 125, 250, 500, and 1000 µg/mL. Vermiculite treated with monuron [3-(p-chlorophenyl)-1,1-dimethylurea] at 0.015 and 1.5 µg/mL or deionized water was used for positive and untreated controls, respectively. The treated vermiculite was then placed in disposable Styrofoam cups (3 by 10 cm). Seedlings were transplanted into the treated vermiculite in four replicate cups for each treatment with one seedling per cup. The seedlings then were placed in a controlled environment chamber. A mixture of metal halide and high pressure sodium lamps maintained on a 12-h photoperiod provided a light intensity 400 µmol/s/m². Temperatures were held at 23 ± 1°C daytime temperature and 20 ± 1°C nighttime. The cups were arranged on a bench in a randomized complete block design and subirrigated with nutrient solution [4]. Drain holes drilled in the base of each cup provided access to nutrient solution held in Styrofoam saucers placed beneath each of the cups.

At the end of three weeks, the height and leaf area of the individual plants were measured. The roots and shoot portions of the plants were harvested, oven-dried at 80 ± 1°C for 24 h, and weighed separately. The mean values were determined and plotted against the log 10 of the test concentration to define dose-response curves [5]. The EC_{50}, defined in this test as the test substance concentration (µg/mL) required to reduce measurements on treated seedlings to 50% of untreated controls, was extrapolated from the dose-response curve. The 95% confidence limits for each EC_{50} were calculated where appropriate [6].

Lemna Test

The duckweed, *Lemna minor* L., was cultured under sterile conditions in 250-mL flasks containing 100 mL of nutrient solution. The basic nutrient solution used was a modified Hoagland's solution [7]. Solutions of the test substance were prepared by the addition of the material to the nutrient medium described above. Solutions containing concentrations of

100, 200, 400, 600, 800, and 1000 µg/mL of the test substance were formed. Technical grade monuron (3-p-[chlorophenyl]-1,1-dimethylurea) was included at a concentration of 1 µg/mL as the positive control for comparison. Any changes in the normal pH (5.6) of the nutrient medium induced by the addition of the chemical were corrected by addition of 0.1 N sodium hydroxide of 0.1 N hydrochloric acid. Each solution was added to each of three Erlenmeyer flasks. Each of the three replicate flasks was then inoculated with two microorganism-free rosettes of four fronds. The inoculated flasks were maintained in a growth chamber for the duration of the seven-day test. Temperature was held at 24 ± 1°C and light intensity at approximately 100 µmol/s/m^2 on a 16-h photoperiod.

The growth rate of the duckweed cultures was determined by measurement of the number of the oval, leaf-like fronds in each culture over time. A plot of the logarithm of the number of fronds against time yields a straight line with slope K. The growth rate (K) was thus calculated for each test culture and untreated control cultures. At the end of the test, the dry weight of the plants in each flask was determined and treatment means with their 95% confidence intervals computed. The coefficient of variation (CV) [3] also was calculated for the test.

Algae Tests

Cultures of *Selenastrum* and *Anabaena* were grown in a defined nutrient medium [4]. Filtered seawater fortified with vitamins and minerals [8] was the culture medium used for *Skeletonema*. The cultures were grown in shake culture on a 16-h photoperiod and a temperature of 24 ± 1°C.

Algal cells in the exponential growth phase were inoculated into 100 mL of sterile culture medium. The test substance was dissolved in water, diluted, and added to culture flasks to provide a dosage range. Additional cultures were treated with technical grade monuron [3-(p-chlorophenyl)-1,1-dimethylurea] at 1 µg/mL as positive controls. Each treatment was replicated five times. Five millilitre aliquots of cell suspension were removed periodically to provide extracts for pigment analysis. Spectrophotometric measurements of the extracts were used to calculate and monitor chlorophyll-*a* concentrations. At the end of the test the dry weight of algae in 25-mL aliquots from the cultures were determined [9]. The mean values of chlorophyll-*a* concentration, dry weight, and their 95% confidence intervals were calculated. The coefficient of variation (CV) [3] also was estimated for each response variable for each of the three algae.

Results

Seed Germination Bioassay

Data presented in Table 1 show that five of seven test substances reduced seed germination. Test substances A and C caused 32% and 44.6% reductions, respectively. These reductions were statistically significant ($p > 0.01$). Extracts E, F, and G elicited reductions of 14.4%, 15.0%, and 33.3%, respectively. These effects were statistically significant ($p > 0.05$). Growth reductions of the hypocotyls from the germinating seeds of less than 12% were not statistically significant.

These results indicate that this seed germination bioassay is an effective test system for assessing the potential phytotoxicity of test substances. The ability to detect a 15% inhibition suggests a sensitive test. Thus, release into the environment of materials that are active in this bioassay might be expected to have adverse effects on seeds of terrestrial plants. This bioassay, however, provides no information on possible interactions with the soil that could

TABLE 1—*Effects of diluted test substances on the germination of radish (*Raphanus sativus *L. cv. 'Early Scarlet Globe') seeds.*

Test Substance	Mean[a] Growth of Hypocotyl, mm		T_{Found}
	Test Substance	Untreated Control	
A	3.4	5.0	6.04[b]
B	4.5	4.8	1.49
C	2.0	4.4	10.97[b]
D	3.5	3.9	1.86
E	3.2	3.7	2.31[c]
F	3.4	4.0	2.74[c]
G	3.0	4.5	6.53[c]

[a] Mean of 150 seeds.
[b] Statistical significance ($p > 0.01$).
[c] Statistical significance ($p > 0.05$).

deactivate or alter the persistence of these materials in the environment. Also, no information is provided on growth parameters affected later in the life cycle or on the reproductive success of exposed plants.

Seedling Growth Tests

Only with corn was the data adequate enough to permit estimation of both the EC_{50} and 95% confidence limits in this test. With this species the estimated EC_{50} was comparable, at approximately 600 µg/mL, for root dry weight, shoot dry weight, and leaf area (Table 2). The confidence intervals varied somewhat in width but were calculable with the method used. A

TABLE 2—*Slope of the regression, coefficient of determination, and EC_{50} calculated from the dose-response relationship of corn (*Zea mays *L.), oat (*Avena sativa *L.), pinto bean (*Phaseolus vulgaris *L.), and cucumber (*Cucumis sativa *L.) seedlings treated with test substance X.*

Test Species	Growth Measurement	Slope (β) of Regression	Coefficient of Determination (R^2)	EC_{50} with 95% Confidence Limits, µg/mL[a]
Corn	Root Dry Weight	−88.246	98.2	669(286,2046)
	Shoot Dry Weight	−75.259	98.7	597(326,1267)
	Leaf Area	−70.090	96.2	646(217,3853)
	Height	−20.945	93.1	>1000
Cucumber	Root Dry Weight	−27.257	32.8	>1000
	Shoot Dry Weight	−23.099	25.8	882[b]
	Leaf Area	−22.268	19.7	>1000
	Height	−23.568	54.7	630[b]
Oat	Root Dry Weight	−45.874	30.7	733[b]
	Shoot Dry Weight	−54.093	62.6	>1000
	Leaf Area	−37.913	45.4	181[b]
	Height	−29.689	23.4	>1000
Pinto Bean	Root Dry Weight	−39.986	62.4	979[b]
	Shoot Dry Weight	−27.087	63.8	>1000
	Leaf Area	−27.592	76.0	>1000
	Height	−23.650	85.4	>1000

[a] Results from 4 replicate plants at each of 6 treatment levels.
[b] Calculation of 95% limits for these EC_{50} values not possible with the method used.

close fit of the dose-response curves to the data was indicated by the high values for the coefficient of determination, R^2.

With other test species the regressions describing the dose-response relationships of test substance X on dry weight and leaf area did not fit the data very well. Indeed, the low values of R^2 seen in the oat and cucumber tests (Table 2) indicate the presence of substantial residual variation in these data. This unexplained variation hampered calculation of confidence limits of the EC_{50} for leaf area and dry weight measurements on these plant species. Consequently, the validity of the EC_{50} values for the oat and cucumber data is questionable.

With pinto bean less than 50% reduction of all growth measurements except root dry weight was observed with the 1000 μg/mL treatment. Thus the EC_{50} for this species appeared to be greater than 1000 μg/mL; however, extrapolation of the dose-response curve beyond the limits of the data to estimate the EC_{50} was not warranted.

The EC_{50} for test substance X on plant height appeared greater than 1000 μg/mL on corn, oat, and pinto bean (Table 2). Only on cucumber did the EC_{50} for plant height appear less than 1000 μg/mL. This suggests that plant height may be less sensitive than other plant responses to treatment with test substance X.

It also simply may be that more error was present in the plant height data. Because error in the EC_{50} obtained from a dose-response curve is related to the slope of the regression line [5], comparison of the slopes is informative. A steep slope indicates relatively little error, while a shallow slope indicates large error. Examination of the slope of the regression lines (Table 2) used to calculate EC_{50} estimates for the corn data indicates a lesser slope for plant height than for other growth measurements.

Data shown in Table 2 imply that pinto bean may be more tolerant of test substance X than are the other test plants used. For instance, higher endpoints for dry weight and leaf area reduction are indicated for pinto bean than for corn.

Because only two concentrations of monuron were used, calculation of EC_{50}'s for the positive control treatments was not possible. At 1.5 μg/mL, the highest concentration used, severe toxicity was observed. At this level monuron caused more than a 50% reduction of all growth measurements on all test plant species.

The results of these seedling growth tests indicate that test substance X is only slightly phytotoxic to terrestrial plants; however, accurate specification of the threshold of toxicity to the test species was hampered by unexplained variation within the tests.

Lemna Test

Test substance X caused a 14 to 15% reduction in the growth of *Lemna* when applied at a concentration of 1000 μg/mL (Table 3). Because this was the highest concentration tested, calculation of EC_{50} values was not warranted. In contrast, the positive control substance, monuron, reduced growth by more than 70% when applied at a concentration of 1 μg/mL.

The test results indicate that concentrations greater than 1000 μg/mL of the test substance would be required to achieve a significant growth reduction in *Lemna* cultures; however, 1 kg/m^3 (2700 lb/acre-ft) of the test substance would be required to bring the concentration to 1000 μg/mL in an aquatic ecosystem [10]. This would occur only in cases of gross contamination of the environment. Therefore testing at concentrations greater than 1000 μg/mL appeared unnecessary.

With three replicates this test is adequate to detect a 25 to 30% inhibition of the growth of *Lemna*. Positive results with the herbicide, monuron, at 1 μg/mL imparts confidence that the test system would identify materials that are potential toxicants. Nevertheless, weak or marginally phytotoxic materials, such as test substance X, may prove difficult to evaluate.

TABLE 3—*Effects of test substance X on the growth of duckweed (*Lemna minor *L.) grown in nutrient solution.*[a]

Treatment	Concentration, µg/µL	Frond, No., (Day 7)	Frond Dry Weight, mg (Day 7)	Frond Growth Rate (K)
Test Substance X	1000	52.3(46.7,57.9)	4.3(3.6,5.0)	0.1202
	800	53.0(47.4,58.6)	4.9(4.2,5.6)	0.1186
	600	57.7(52.1,63.3)	5.2(4.4,5.9)	0.1253
	400	59.0(53.4,64.6)	4.8(4.1,5.5)	0.1293
	200	65.0(59.4,70.6)	5.1(4.4,5.8)	0.1332
	100	61.7(56.1,67.3)	5.5(4.8,6.2)	0.1317
Monuron	1	14.3(8.7,19.9)	1.5(0.8,2.2)	0.0348
Untreated Control	0	59.7(54.1,65.3)	5.0(4.3,5.7)	0.1271
CV (percent)		8.5	26.9	

[a] Data presented are averages of three replications. Values within parentheses represent lower and upper 95% confidence limits.

Algae Tests

Test guidelines call for testing at five or more concentrations, at least three of which inhibit growth by more than 20% and less than 90% of normal growth. In our tests on algae with test substance X, despite preliminary range-finding tests, dry weight was reduced by 50% or more with only one of the three algal species tested (Table 4). Consequently, calculation of the EC_{50} was not warranted and a two-way analysis of variance (ANOVA) was used as an alternative statistical procedure [6].

ANOVA requires that assumptions of normality and homogeneity of variance be met, however. A Kolomogorov-Smirnov test[2] for goodness of fit of dry weight data from tests of test substance X on algae did not indicate significant deviation from a normal distribution except with *Anabaena*. Bartlett's test for homogeneity of variance [6] showed homogeneity of variance in the dry weight data for *Selenastrum* and *Skeletonema* but not in the *Anabaena* data. Therefore transformations to make the data meet the assumptions of ANOVA or analogous non-parametric methods should be employed for the *Anabaena* data.

Dry weight measurements of *Skeletonema* cultures treated with test substance X at 1000 µg/mL showed a 64.3% growth reduction. A 33% reduction with *Selenastrum* cultures and a 23% reduction with *Anabaena* were seen at this concentration. Nevertheless, with these two species, the confidence intervals around the means overlap the confidence of the control mean; therefore a suitable multiple comparison test would not be likely to show statistical differences between means for test substance X and the mean for the untreated control.

A 1 µg/mL treatment with the control substance, monuron, caused a 93% reduction in dry weight of *Anabaena*. In *Selenastrum* and *Skeletonema* dry weight was reduced 64% and 79%, respectively, by treatment at this level of monuron.

Test substance X at 1000 µg/mL appeared to have little effect on the chlorophyll content of *Anabaena*. On the other hand, severe reductions of chlorophyll were measured in cultures of *Selenastrum* and *Skeletonema* at 1000 µg/mL. An 80.3% reduction was seen in the *Selenastrum* test and a 99[+]% reduction seen in the *Skeletonema* test. Chlorophyll also appeared less sensitive to inhibition in *Anabaena* than in the other two algae in response to the control substance, monuron. Monuron treatments at 1 µg/mL caused a 74% reduction of chloro-

---

[2] Statgraphics: Statistical Graphics System by Statistical Graphics Corporation, software available from STSC, Inc., 2115 East Jefferson St., Rockville, MD 20852.

TABLE 4—*Effects of test substance X on the growth of Anabaena flos-aquae (Lyngb.) Breb., Selenastrum capricornutum Printz., and Skeletonema costatum (C-rev.) Cl. in nutrient culture.*[a]

Treatment	Concentration, µg/mL	Anabaena		Selenastrum		Skeletonema	
		Dry Weight, µg/mL	Chl, µg/mL	Dry Weight, µg/mL	Chl, µg/mL	Dry Weight, µg/mL	Chl, µg/mL
Test Substance X	1000	64.2(52.1,76.2)	0.70(0.49,0.89)	11.3(7.8,14.7)	0.47(0.24,1.18)	15.9(8.8,23.1)	<0.01
	800	63.4(51.3,75.4)	0.74(0.54,0.94)	...	...	15.3(8.1,22.4)	<0.01
	600	75.3(63.2,87.3)	0.83(0.64,1.03)	...	...	8.3(1.2,15.4)	<0.01
	500	...	...	...	...	...	...
	400	80.6(68.5,92.6)	0.78(0.58,0.97)	14.4(10.9,17.8)	1.23(0.52,1.94)	25.4(18.2,32.4)	<0.01
	250	...	...	...	...	...	...
	200	84.8(72.7,96.8)	0.76(0.56,0.96)	10.0(6.4,13.3)	0.79(0.08,1.50)	38.9(31.7,46.0)	1.18(0.98,1.38)
	125	...	...	...	...	...	...
	100	76.3(64.3,88.3)	0.70(0.50,0.89)	16.8(13.3,20.2)	1.68(0.97,2.39)	48.4(41.2,55.5)	1.24(1.05,1.44)
	63	...	...	17.3(13.9,20.8)	2.14(1.43,2.85)	...	...
Monuron	1	5.7(−6.3,17.7)	0.20(0.04,0.40)	5.8(2.4,9.2)	0.03(−0.68,0.74)	9.4(2.3,16.6)	<0.01
Untreated Control	0	83.1(71.1,95.2)	0.76(0.57,0.96)	16.8(13.4,20.2)	2.38(1.67,3.09)	44.5(37.3,51.6)	1.06(0.86,1.26)
CV (percent)		19.7	43.4	28.4	61.7	30.2	49.4

[a] Data presented are the averages of five replications with values in parentheses representing the lower and upper 95% confidence limits.

phyll in *Anabaena*. In contrast, monuron at this level caused more than a 98% reduction of chlorophyll in both *Selenastrum* and *Skeletonema*.

Chlorophyll measurements thus show a different pattern than dry weight measurements. Chlorophyll appeared more reduced than dry weight in *Skeletonema* and *Selenastrum;* however, in *Anabaena* dry weight appeared more affected than chlorophyll. These test results suggest that dry weight and chlorophyll responses to test substances may vary among species. Therefore chlorophyll data may be a valuable complement to dry weight measurements in any algal test system.

Discussion

Determination of potential phytotoxicity by simple laboratory tests of substances that may be released into the environment requires valid test results. Achievement of valid test results, however, is dependent on tests with reasonable precision. Sensitivity, test replication, and temperature control all can affect test precision [5]. Sensitivity of the tests depends in part on the number of replicates, the probability level, and the type of statistical analysis [5].

Our experience suggests that some of the problems encountered in establishing the threshold of toxicity are related to the precision of the tests. This is illustrated by the results obtained with test substance X in the seedling growth test. For example, the poor fit of dose-response curves to the data with some species reflects large pot-to-pot variation within treatments. Consequently, it became difficult to determine reliable EC_{50} values for growth effects due to test substance X. In many cases, variability precluded calculation of the 95% confidence limits for the estimated EC_{50} values. The equation for estimating the confidence limits [6] failed when the standard error was high.

The small number of plants (4 per treatment dose) used in the seedling tests is a problem. This appears more a factor with some species (e.g., oat) than with other species (e.g., corn). Preselection of test plants for uniform size may be a possible solution. For example, one could grow a number of plants of each test species and select the most uniform plants for the test. Such efforts might be expected to enhance precision. Also, sensitivity can be expected to increase as the number of replicates is increased [5]. It also is essential to include enough dosage levels to establish at least four points on the linear portion of the dosage-response curve [5].

Heterogeneity of the variance in the test data as in the *Anabaena* test must be taken into account in making comparisons of treatment means. For a *Selenastrum* test, Horning and Weber [11] advocated the routine use of Dunnett's procedure to compare treatment means with the control mean. These authors suggested the use of Steel's Many-One Rank test when the assumptions of normality and homogeneity of variance are not met.

It may be desirable to compare treatment means for the test substance with the mean for the positive control, such as monuron used in our tests. Therefore it might be more appropriate to test potentially significant differences among all possible pairs of means. In this case, multiple comparisons might be made with a test such as Tukey's honestly significant difference procedure (*t*-method) or the Scheffe method [6]. When the assumptions of normality and homogeneity of variance cannot be satisfied, nonparametric tests likely would be more appropriate. Obviously, in selecting a test, judgments must be made that require input from a trained statistician.

Experimental design is also an important consideration. Although the radish germination test detects phytotoxins, for example, treatment replication is absent. The test would be greatly improved by placing 15 seeds in each of 10 germination chambers. While the number of seeds remains the same there now would be available some measure of the variability associated with the treatment procedure.

Choice of the plant species to be used also appears extremely important. Different mechanisms of toxicity may be expressed on various plants. The growth of one indicator plant may be inhibited, while another indicator plant may be tolerant to the test substance. For instance, in our seedling growth tests, pinto beans appeared more tolerant than other indicator plants to test substance X. Use of nonagricultural species as test subjects might show even greater species-to-species differences in response to test substances. Wild species retain genetic diversity in contrast to domesticated species where genetic diversity, generally, has been reduced through years of breeding. Consequently, varying response among individual plants to a test substance could be greater in the wild types and thus vary the assessed threshold of acute toxicity.

Likewise, in the algae tests, dry weight and chlorophyll content of the marine alga, *Skeletonema* was more affected than with the green alga, *Selenastrum*, or the blue-green alga, *Anabaena*, by treatment with test substance X. Although the test protocol [2] indicates that 24°C is a suitable temperature for *Skeletonema costatum*, this species perhaps is stressed at this temperature. Temperature stress thus might be the cause of the greater sensitivity apparent with this *S. costatum*. While inclusion of *Anabaena* affords data on the effects of the test substance on nitrogen fixation potential, it is unlikely that effects on nitrogen fixation account for the apparent difference seen between *Anabaena* and the other algae in our tests. In fact, it is not known why *Anabaena* was not as affected by test substance X. Nevertheless, the desirability of including several species in any test system is evident.

In the algal tests it appears desirable also to include more than one measure of growth or inhibition. The test protocol indicates that growth should be measured as dry weight increase or by a method calibrated with dry weight increase [1]. Our test results suggest that pigment content (i.e., chlorophyll-*a*), a principal measure of culture activity [4], responded differently than dry weight in algae treated with test substance X.

As a measure of biomass, chlorophyll measurements correlated poorly with dry weight measurements. Reduction of chlorophyll was more severe than reduction of dry weight in response both to the test substance and the control substance in tests with *Selenastrum* and *Skeletonema*. In the *Anabaena* test, however, the reverse situation was suggested. One has to question, therefore, whether the response of dry weight and chlorophyll were expressions of different physiological effects of the test substance.

The study protocols evaluated appear appropriate for assessing phytotoxicity to plants. Nevertheless, improvement of the protocols not only may be posssible but also, perhaps, should be a constant goal. The experience of the herbicide industry tells us that screening procedures for assessing phytotoxicity are not infallible. Indeed, some of the most successful commercial herbicides were once tested and discarded by one firm to be later discovered by another manufacturer using different screening procedures for physiological activity in plants [4]. Only through an ongoing process of evaluation and revision of study protocols is it likely that a battery of tests can be developed that will successfully identify all chemicals harmful to plant life.

Acknowledgments

The technical assistance of several individuals was employed for the tests reported. The contributions of P. Backus, J. M. Beglinger, C. Davis, J. E. Hvitfelt, C. M. Rusinko, and E. R. Sanstead are gratefully acknowledged.

References

[1] *Federal Register,* Vol. 44, No. 53, March 16, 1979, pp. 16285–16288.
[2] "Identification and Listing of Hazardous Waste," Advance Notice of Proposed Rule Making, *Federal Register,* Vol. 43, No. 243, Appendix VI, 18 Dec. 1978, p. 59026.

[3] Steel, R. G. D. and Torrie, J. H. in *Principles and Procedures of Statistics*, McGraw-Hill, New York, 1960, pp. 81–86, 109–110.
[4] Rubinstein, R., Cuirle, E., Cole, H., Ercegovich, C., Weinstein, L., and Smith, J., "Test Methods for Accessing the Effects of Chemicals on Plants," EPA Report 560/5-75-008, Office of Toxic Substances, U.S. Environmental Protection Agency, Washington, D.C., June 1975.
[5] Ratsch, H., "Interlaboratory Root Elongation Testing of Toxic Substances on Selected Plant Species" EPA Report 600/3-83-051, Office of Research and Development, U.S. Environmental Protection Agency, Corvallis, Ore., June 1983.
[6] Sokal, R. M. and Rohlf, F. J., *Biometry*, 2nd ed., W.H. Freeman, New York, 1981, p. 498, 241–262.
[7] Gerloff, G. C., in *Proceedings*, International Symposium on Eutrophication, National Academy of Sciences, Washington, D.C., 1969, p. 544.
[8] National Environmental Research Center, "Marine Algal Assay Procedure: Bottle Test," EPA Report 660/3-75-008, Pacific Northwest Environmental Laboratory, Eutrophication and Lake Restoration Branch, Office of Research and Development, U.S. Environmental Protection Agency, Corvallis, Ore., Dec. 1974.
[9] Strickland, J. D. H. and Parsons, T. R., "A Practical Handbook of Seawater Analysis," Bulletin 167, Fisheries Research Board of Canada, Ottawa, 1960, pp. 181–183.
[10] Klingman, G. C., Ashton, F. M., *Weed Sciences: Principles and Practices*, Wiley, New York, 1975, p. 376.
[11] Horning, W. B., II and Weber, C. I., Eds., "Short-Term Methods for Estimating the Chronic Toxicity of Effluents and Receiving Waters to Freshwater Organisms," EPA Report 600/4-85/014, Environmental Monitoring and Support Laboratory, Office of Research and Development, U.S. Environmental Protection Agency, Cincinnati, Ohio, Dec. 1985.

Mark W. Thomas,[1] Barbara M. Judy,[1] William R. Lower,[1] Gary F. Krause,[1] and William W. Sutton[2]

Time-Dependent Toxicity Assessment of Herbicide Contaminated Soil Using the Green Alga *Selenastrum capricornutum*

REFERENCE: Thomas, M. W., Judy, B. M., Lower, W. R., Krause, G. F., and Sutton, W. W., "**Time-Dependent Toxicity Assessment of Herbicide Contaminated Soil Using the Green Alga *Selenastrum capricornutum*,**" *Plants for Toxicity Assessment, ASTM STP 1091*, W. Wang, J. W. Gorsuch, and W. R. Lower, Eds., American Society for Testing and Materials, Philadelphia, 1990, pp. 235–254.

ABSTRACT: Bioassays utilizing the green alga *Selenastrum capricornutum* were performed on filtered eluates from soil treated with six commonly used forestry herbicides applied at label-recommended rates. The bioassays were conducted at three time periods after herbicide application—one hour, five days, and ten days. The 96-h EC_{50} values indicated growth inhibition (relative to control sample) for all treatments when assayed 1 h after herbicide application. Algal EC_{50} values of +100 (Control), +27.3 (Roundup®), −20.4 (Arsenal® [2 lb Acid Equivalent] [AE]/gal), −22.4 (Garlon® 4), −49.4 (Tordon 101M®), −100 (Velpar L®), and −100 (Velpar ULW®) were obtained. Assays conducted ten days after herbicide application to soil revealed substantially reduced toxicity of two herbicides. The 96-h EC_{50} values for Roundup and Arsenal were both +100. There was a significant enhancement effect observed with Roundup. A slight reduction in toxicity was noted for Garlon 4 (−15.9) and Tordon 101M (+9.9). No change in toxicity occurred for Velpar L or Velpar ULW. The herbicides were also applied to water and the following 96 h EC_{50} values in µg/ml were obtained: 5500 (Arsenal [2 lb AE]), 5300 (Arsenal [4 lb AE]), 5000 (Tordon 101M), 5000 (Garlon 4), 2600 (Roundup), 2.5 (Velpar L), and 1.2 (Velpar ULW).

KEY WORDS: bioassay, alga, *Selenastrum capricornutum*, toxicity, soil, herbicide, imazapyr, glyphosate, picloram, 2,4-D, triclopyr, hexazinone, EC_{50}

Bioassays have been useful tools for measuring the effects of potentially toxic substances for over 40 years [1,2]. The single cell green alga *Selenastrum capricornutum* (Printz) has long been recognized as a sensitive indicator of potential toxicants in aqueous media and standardized procedures have been developed to assess water quality [3,4]. Studies with *S. capricornutum* have investigated the toxic effects of nutrient loading (primarily phosphorus and nitrogen) on the aquatic environment [5–8], wastewater effluents [9–13], inorganic chemicals [14], oil refinery effluents [15], municipal sanitary landfills [16], effluents derived from the manufacturing of iron ore [17], and water contaminated with pesticides, specifically the chlorophenols [18] and herbicides (amitrol, atrazine, 2,4-D, glyphosate, picloram, simazine) [19]. Recently, a variation of the freshwater procedure has been used to investigate the effects of soil contamination by assaying the aqueous soil eluate [20] for phosphorus [21] and the herbicide bromacil [22,23], both in the field and the laboratory. Forestry applications

---
[1] Environmental Trace Substances Research Center, University of Missouri, Columbia, MO 65203.
[2] U.S. Environmental Protection Agency, Environmental Research Laboratory, Athens, GA 30613.

utilizing the *S. capricornutum* bioassay have documented the effects on water quality after the fertilization of forestry plantations [24]. Canada has recently developed a bioassay protocol for evaluating the toxic risk to regional fisheries resources posed by forest-use herbicides [25].

The present study documents the changes in toxicity over time after direct herbicide application to soil at commonly used rates. Aqueous eluates of the treated soils were assayed at intervals of one hour, five days, and ten days after treatment. Calculated effect values and EC_{50} values are given.

Materials and Methods

Soil Selection and Preparation

The soil used in the study was obtained from a farm owned by the University of Missouri, in Boone County, Missouri, and is designated as "East Farm" soil. The particular field had lain fallow approximately 15 years. The soil was removed at a depth of 15 cm and mixed for 16 h to ensure homogenization. An Agricultural Soil Test Analysis (Regional Soil Test Laboratory, University of Missouri, Columbia [UMC]) [26-31] was performed by the Missouri Cooperative Extension Service on ten 0.5 kg grab samples. Results of these tests revealed that the soil is classified as a clay loam (21.4% sand, 50.7% silt, and 27.9% clay) and contained 55 ppm nitrogen and 9.4 ppm sodium. The soil averaged 4.55% organic matter and the pH averaged 5.47. Other soil constituents averaged over the ten subsamples, standard deviations, and coefficients of variation are given in Table 1.

Herbicide Application

Herbicides tested in this study (Table 2) include: Arsenal® (Imazyapyr), American Cyanamid Chemical Company, at both the 2 lb Acid Equivalent [AE]/gal aqueous solution and

TABLE 1—*Average soil test results with standard deviations and coefficients of variation for ten subsamples of clay loam (East Farm) soil.*

	Mean	S.D.[d]	% C.V.[e]
pH	5.4–5.6	...	...
N.A.[a] (meq/100g)	3.50	0.0	0.0
O.M.[b] (%)	4.55	0.57	12.53
Bray I P (ppm)	1.075	0.079	7.35
Bray II P (ppm)	3.22	0.215	6.68
Ca (ppm)	185.25	11.29	6.09
Mg (ppm)	27.4	1.417	5.17
K (ppm)	20.86	1.18	5.66
Zn (ppm)	7.13	0.45	6.31
Fe (ppm)	70.49	4.38	6.21
Mn (ppm)	65.64	4.98	7.59
Cu (ppm)	0.87	0.43	49.42
SO₄S (ppm)	7.19	0.93	12.93
EC[c] (mmho/cm)	15.56	0.66	4.24

[a] Neutralizable Acidity (meq/100g).
[b] Organic Matter (%).
[c] Electrical Conductivity (mmho/cm).
[d] Standard Deviation.
[e] Percent Coefficient of Variation.

TABLE 2—*Trade name, active ingredient (AI), % AI, and toxicity data (LC_{50} in mg/L [ppm] 96-h static test) of four freshwater fish. Values in parentheses indicate number of tests conducted.*

Trade Name	AI	% AI	Toxicity Information (96-h Static), (LC_{50}) mg/L			
			Rainbow Trout	Fathead Minnow	Channel Catfish	Bluegill
Roundup®[a]	Glyphosate	41.0	5.6(11)	3.6(4)	15.9(5)	4.7(3)
Arsenal®[b]	Imazapyr	43.3	>100	>100	>100	>100
Garlon® 4[c]	Triclopyr	61.6	117	...	...	148
Tordon 101M[a]	Picloram	10.2	9.9(11)[d]	...	10.1(7)[d]	22.9(8)[e]
	2,4-D	39.6	>100(4)	682(20)	>180[f]	>150(8)
Velpar L						
ULW®	Hexazinone	25/75	>180[d]	...	...	>100[d]

[a] Source: *Manual of Acute Toxicity: Interpretation and Data Base for 410 Chemicals and 66 Species of Freshwater Animals*, F. L. Mayer and M. R. Ellersieck, U.S. Department of Interior, Fish and Wildlife Service, Res. Pub. 160.
[b] Source: American Cyanamid, Tech. Rpt. PE11967.
[c] Source: Dow Chemical Company, Tech. Rpt. 137-859-79.
[d] Based on 99% technical material.
[e] Based on 24.9% AI liquid.
[f] Based on 49.6% of liquid dimethylamine salt.

4 lb AE Applicators Concentrate; Roundup® (Glyphosate), Monsanto Chemical Company; Tordon 101M® (Picloram and 2,4-D), Dow Chemical Company; Garlon 4® (Ester formulation of Triclopyr), Dow Chemical Company; Velpar L® (liquid Hexazinone), DuPont Chemical Company; and Velpar ULW® (granular Hexazinone), DuPont Chemical Company. For the remainder of this paper, herbicide formulations will be referred to in the following manner: Arsenal (2 lb), Arsenal (4 lb), Roundup, Tordon 101M, Garlon 4, Velpar L, and Velpar ULW. Also, since all herbicides were applied at label-recommended rates, treatments should be considered as initially equal for comparative purposes. Algal EC_{50} values of each herbicide applied directly in water are also given.

The various formulations of herbicides used in the present study were composed of the following ingredients: Roundup—isopropylamine salt of N-(phosphonomethyl) glycine (glyphosate) 41.0% and 59.0% inert ingredients; Arsenal (2 and 4 lb AE)—2[4,5 dihydro-4-methyl-4-(1-methylethyl-5-oxo-1H imidazol-2-yl]-3 pyridinecarboxylic acid with 2-propanamine (1:1) salt; Tordon 101M—(picloram) [4-amino-3,5,6-trichloropicolinic acid] 10.2% + 2,4-dichlorophenoxyacetic acid [both triisopropanolamine salt] 39.6% and 50.2% inert ingredients; Garlon 4—triclopyr, butoxyethyl ester (3,5,6-trichloro-2-pyridinyloxyacetic acid) 61.6% and 38.4% inert ingredients; and Velpar L (25% AI) and Velpar ULW (75% AI)—hexazinone [3-cyclo-hexyl-6-(dimethylamino)-1-methyl-1,3,5-triazine-2,4 (1H, 3H)-dione].

Seven Pyrex dishes (265 by 375 by 55 mm) were partitioned into three equal areas with cardboard spacers. One hundred and twenty-five grams of East Farm soil were placed in each partition (approximately 20 mm depth) for a total of 375 g/dish.

The rate of herbicide application may be expressed in a number of ways. For the purpose of this publication, rates will be expressed as a percent for liquid formulated products and as lbs Active Ingredient/Acre (AI/AC) for the granular product. The corresponding conversion at the time of application in mg/L is as follows: 5013, Roundup; 5013, Arsenal (2 lb); 15 039, Tordon 101M; 7520, Garlon 4; 7520, Velpar L; and 1591, Velpar ULW.

Herbicides were evenly applied to the soil using a Dow Chemical Company herbicide applicator with the adjustable spray nozzle set for a fine mist. All herbicides, with the exception of Velpar ULW, were applied in enough reverse osmosis double-deionized water to achieve an initial soil moisture content of 25% (31.25-mL water/125-g soil). Herbicides were applied at the following rates: Roundup, 2%; Arsenal (2 lb), 2%; Tordon 101M, 6%; Garlon 4, 2.5%; and Velpar L, 2.5%. Velpar ULW, a granular formulation, was applied at a rate of 5.3 lb AI/AC. These rates can be considered as representative of average application rates for many silvicultural treatments. Reverse osmosis double-deionized water was applied alone to both the control and Velpar ULW treatments. All treatments were applied in a fume hood. The exhaust fan, however, was not turned on. Treatments were left in the fume hood for the duration of the study. The three treated 125-g subsamples were assayed at three time intervals: one hour after treatment and five and ten days after treatment.

Selenastrum capricornutum *Assay*

The three partitions of each treatment were designated "A", "B" and "C" and one was randomly chosen to be assayed for each time interval. The basic protocol is as follows: 125-g of treated soil were placed in a 1 L Teflon® container. Then, 500 mL of double-deionized reverse osmosis water were added, and the container was capped and inverted six times to thoroughly stir the contents. The container was then placed in a polystyrene holder and agitated for 48 h (120 excursions/min at 20°C) on an Eberbach Corporation shaker in the dark. This procedure was conducted in an environmentally controlled chamber (Unitherm). This is the extraction phase of the procedure. At the end of 48 h, the Teflon container was allowed to stand for 5 min, and the liquid was decanted and transferred to eight 40-mL Oak Ridge-type centrifuge tubes and centrifuged at 10 000 rpm for 10 min at 4°C. After centrifugation, the eluate was transferred into a 9 cm Buchner funnel with a 0.7 μm nonsterile glass fiber filter (Whatman GF/F) and filtered into a nonsterile side-arm flask under vacuum (25 in. Hg). A second nonsterile filtering was conducted using a nylon 0.45 μm MSI (9 cm) filter. The pH of the nonsterile-filtered soil eluates was then measured to determine if they were within the range of six and ten. No adjustment was necessary since they were above a pH of six. A final filtration was conducted using a sterile 0.45 μm, 47 mm MSI nylon filter and Millipore Sterilfit Holder into a sterile side-arm flask to obtain a sterile eluate.

A series of dilutions was then made at the following ratios: 1% to 99%, 10% to 90%, 50% to 50%, and 80% to 20% soil eluate to growth medium, respectively. The constituents of the growth medium are given in the Appendix. Two 125-mL Erlenmeyer flasks containing 50-mL aliquots of each dilution, along with two negative controls (100% growth medium) and two positive controls (100% growth medium plus 153.8 μg/L $ZnCl_2$ [74 μg/L Zn^{++}]), were inoculated with $1 \times 10^4 \pm 1 \times 10^3$ cells/mL of *Selenastrum capricornutum* (American Type Culture Collection–22662) and capped with nontoxic foam stoppers (Jaece Identi-Plugs) to allow for oxygen exchange during agitation. The inoculum was made from seven-day-old parent stocks in the logarithmic phase of growth. All work was performed in ultraviolet (UV) sterile chambers equipped with two UV lights (GE G15T8 Germicidal, 15 W) suspended 0.61 m (24 in.) from the surface. The twelve inoculated flasks of each treatment were then placed on an Eberbach shaker in a Conviron Controlled Environment Chamber kept at 24 $\pm$ 2°C. Relative humidity and temperature in both the Unitherm and Conviron were continuously monitored using hygrothermographs. The shaker was set at 100 excursions/min. Light intensity was kept at 4300 $\pm$ 430 lm (1.43 μmoL m^{-2} s^{-1}) and was monitored using a Li-Cor Quantum Radiometer Photometer (Model No. LI-185B). Test flasks were randomly placed on the shaker top.

The flasks were agitated for 96 h. At the end of 96 h, the test results were tabulated from the initial inoculum cell count and from the final 96-h cell count from each separate dilution

(2 replicates each) and for the negative control. Cell counts were obtained using a Coulter Counter Model ZBI equipped with a 100 μm aperture manometer. Results were calculated as follows:

$$\text{calculated effect} = \frac{(SM - IN) - P(NC - IN)}{P(NC - IN)} \times 100$$

where

SM_{80} = cells/mL of 80% soil eluate/20% growth medium,
SM_{50} = cells/mL of 50% soil eluate/50% growth medium,
SM_{10} = cells/mL of 10% soil eluate/90% growth medium,
SM_1 = cells/mL of 1% soil eluate/99% growth medium,
IN = cell/mL of inoculum ($1 \times 10^4 \pm 1 \times 10^3$ cells/mL),
P = proportion of growth media (0.20, 0.50, 0.90, 0.99 with 80%, 50%, 10%, 1% soil eluate, respectively), and
NC = cells/mL of negative control.

Algal EC_{50} values were calculated from the regression of the calculated effect at each concentration of filtered eluate as the dependent variable, Y, against the concentration of filtered eluate as the independent variable, X. This formula will periodically result in negative EC_{50} values. Generally, only positive values are reported. In this case, however, all treatments are considered equal at the time of application and EC_{50} values (positive or negative) distinguish between treatments and indicate a change in toxicity within a treatment over time.

Statistical Analysis

Raw data, as well as the log transformed means of the three cell counts per flask, were statistically analyzed using analysis of variance (ANOVA). A protected least significant difference (LSD) was computed at $p < 0.05$. Within-experiment variation was determined by calculating the difference between the cell growth in the two positive control (PC) flasks (with 74 μg/L ZN^{++}) using the formula

$$[1 - (PC \text{ high value}) \div (PC \text{ low value})] \times 100$$

Between-experiment variation was documented based on an ANOVA of log transformed cell count means for the interactions of treatment/time of assay/dilution and the experiment-wide mean square and coefficient of variation are reported.

Results and Discussion

Population growth (Table 3) at the end of 96 h expressed in cells $\times 10^6$/mL in all herbicide-treated soil eluates was less than in the control soil eluate at all four dilutions when assayed 1 h after herbicide application. This trend continued for the assay results conducted five days after treatment for five herbicides (Roundup, Garlon 4, Tordon 101M, Velpar L, Velpar ULW). There was reduced cell growth at all four dilutions. Arsenal (2 lb), however, induced a slightly stimulatory result at the 50% and 80% dilutions. This indicates a decrease in toxicity over the five-day interval the soil was in the fume hood. Temperature in the fume hood fluctuated between ten and 18°C which would limit microbial degradation. There was indirect light only during daylight hours which would limit photo-degradation. Ten days

TABLE 3—*Growth of* Selenastrum capricornutum *expressed in cells/mL × 10^6 in four dilutions of soil eluate (SE) at three time periods after application of six herbicides.*

	Percent Soil Eluate			
	1% SE	10% SE	50% SE	80% SE
Assay Initiated One Hour after Treatment				
Control[a]	3.1936	2.9363	1.5950	0.9191
Roundup	2.4695	2.4167	0.9129	0.4379
Arsenal (2 lb)	2.1200	1.4782	1.0679	0.5317
Garlon 4	1.5906	0.8057	0.0232	0.0120
Tordon 101M	1.4438	0.3015	0.0144	0.0100
Velpar L	0.0946	0.2742	0.0781	0.0781
Velpar ULW	0.0126	0.0103	0.0216	0.0166
Assay Initiated Five Days after Treatment				
Control[b]	3.2201	3.0381	1.5349	1.0249
Roundup	2.6027	3.0133	1.3587	0.3622
Arsenal (2 lb)	2.0171	2.2977	1.5980	1.1372
Garlon 4	1.8198	0.8456	0.0232	0.0128
Tordon 101M	2.5495	1.1201	0.0350	0.0209
Velpar L	0.0132	0.0112	0.0132	0.0137
Velpar ULW	0.0129	0.0130	0.0125	0.1534
Assay Initiated Ten Days after Treatment				
Control[c]	2.5015	2.4059	1.3645	0.7826
Roundup	2.4653	2.9431	2.1846	0.9576
Arsenal (2 lb)	1.4906	1.6899	1.2662	0.9647
Garlon 4	1.2803	0.5990	0.0893	0.1022
Tordon 101M	2.0967	0.6023	0.0129	0.0106
Velpar L	0.0118	0.0145	0.0199	0.0135
Velpar ULW	0.0156	0.0123	0.0133	0.0119

[a] Negative control = 3.3198 × 10^6 cells/mL in 100% growth medium.
[b] Negative control = 2.4457 × 10^6 cells/mL in 100% growth medium.
[c] Negative control = 2.5116 × 10^6 cells/mL in 100% growth medium.

after herbicide application, four herbicides continued to cause reduced cell growth at all dilutions (Garlon 4, Tordon 101M, Velpar L, Velpar ULW) when compared to the control soil eluate. Both Arsenal and Roundup soil eluates, however, induced a slightly stimulatory effect at the 80% dilution.

The fluctuation in *Selenastrum capricornutum* population growth from one time period to another is not of major concern, because the growth response in the four associated dilutions of soil eluate/growth media were similar. The negative controls (NC), for example, were 3.3198, 2.4456, and 2.5116 × 10^6 for the 1 Hour After Treatment (HAT), 5 Days After Treatment (DAT), and 10 DAT assays, respectively (Table 3). The CE formula presented earlier acts as a neutralizing mechanism, since the NC value is included and the amount of growth media is removed by appropriate multiplication (p = 0.99, 0.9, 0.5, and 0.2). To perform this calculation, one must assume a direct 1:1 correlation between population growth and the amount of growth media [63]. This assumption, though not precisely correct, is of sufficient precision that comparisons of CE and EC_{50} values between treatments can be made.

The raw cell numbers were normalized with log transformation. After transformation of the cell count means, values for the NC of 10.63, 10.32, and 10.35 were obtained for the 1 HAT, 5 DAT, and 10 DAT assays, respectively (Table 4). Experiment-wide between-exper-

TABLE 4—*Statistical significance of means of log transformed cell counts (LSD = 0.33).*[a]

Treatment	Percent Soil Eluate			
	1%	10%	50%	80%
Assay One Hour after Treatment (1 HAT)				
Control	10.59	10.51	9.90	9.35
Roundup	10.34	10.31	7.04^b	6.31^b
Arsenal	10.18^b	9.79^b	9.49^b	8.80^b
Garlon 4	9.89^b	9.20^b	5.67^b	5.03^b
Tordon 101M	9.79^b	8.24^b	5.20^b	4.84^b
Velpar L	4.78^b	5.84^b	4.61^b	4.61^b
Velpar ULW	5.07^b	4.87^b	5.02^b	5.33^b
Assay Five Days after Treatment (5 DAT)				
Control	10.60	10.54	9.86	9.46
Roundup	10.39	10.53	9.73	8.39
Arsenal	10.13^b	10.20^b	9.90	9.56
Garlon 4	10.03^b	9.16^b	5.56^b	5.09^b
Tordon 101M	10.37	9.55^b	6.04^b	5.57^b
Velpar L	5.12^b	4.96^b	5.11^b	5.15^b
Velpar ULW	5.09^b	5.10^b	5.06^b	5.26^b
Assay Ten Days after Treatment (10 DAT)				
Control	10.35	10.31	9.74	9.19
Roundup	10.34	10.51	10.21^c	9.39
Arsenal	9.80^b	9.96^b	9.67	9.40
Garlon 4	9.68^b	8.91^b	4.72^b	4.85^b
Tordon 101M	10.17	8.92^b	5.09^b	4.89^b
Velpar L	5.00^b	5.19^b	5.02^b	5.13^b
Velpar ULW	5.24^b	5.04^b	5.12^b	5.01^b

[a] 1 HAT Negative Control (NC) = 10.63, 5 DAT NC = 10.32, 10 DAT NC = 10.35.
[b] Significant inhibition ($p < 0.05$).
[c] Significant enhancement ($p < 0.05$).

iment variation based on an ANOVA within the interactions of treatment/time of assay/dilution resulted in an error mean square of 0.028 and a coefficient of variation of 2.03%. The average within-experiment variation of the two PC flasks for each time of assay was 10.67%. The stimulatory effect mentioned above with respect to Roundup was statistically significant ($p < 0.05$) at the 50% dilution.

Population growth can also be expressed as a calculated effect (CE) value (Table 5). The formula, designated as the Proportional Amendment Formula, has already been discussed. The calculated effect can have a negative or positive value. A negative value indicates a decrease in cell number (i.e., an inhibitory effect compared to the negative control). A positive value indicates an increase in cell number compared to the negative control and is regarded as an indication of a stimulatory effect. The change in calculated effect values over time indicates whether a substance becomes more toxic, remains the same, or becomes less toxic. EC_{50} values were calculated using the regression of the four calculated effect values against the four dilutions. Only negative calculated effect values are used to calculate EC_{50}.

Two rules were employed when encountering extremes with respect to EC_{50} values. If calculated effect values proved to be positive, EC_{50} values could not be calculated. Since positive calculated effect values indicate a stimulatory effect (i.e., an innocuous soil eluate), the EC_{50} was arbitrarily set at +100. If extremely toxic soil eluates were encountered that induced calculated effect values in all four dilutions > -98, the EC_{50} value was arbitrarily set at

TABLE 5—*Results of* Selenastrum capricornutum *bioassay utilizing soil eluate (SE) after herbicide application to soil at three time intervals (1 hour after treatment [HAT] and 5 and 10 days after treatment [DAT] expressed as calculated effect (CE) values at 1, 10, 50, and 80% SE dilutions and EC_{50} values.*

Time of Assay	Treatment	Percent Soil Eluate				EC_{50}[a]
		1%	10%	50%	80%	
		Calculated Effect				
1 HAT	Control	−2.8	−1.8	−4.2	37.4	+100[b]
5 DAT		33.1	38.1	25.2	108.4	+100[b]
10 DAT		0.6	6.4	8.3	54.5	+100[b]
1 HAT	Roundup	−24.9	−19.2	−95.0	−94.8	+27.3
5 DAT	(Glyphosate)	7.5	37.0	10.7	−27.7	—
10 DAT		−0.9	30.3	73.9	89.4	+100[b]
1 HAT	Arsenal	−35.6	−50.7	−36.0	−21.1	−20.4
5 DAT	(Imazyapyr)	−16.8	4.4	30.4	131.4	+100[b]
10 DAT		−40.3	−25.5	0.3	90.5	+100[b]
1 HAT	Garlon 4	−52.2	−76.0	−99.6	−99.9	−22.4
5 DAT	(Triclopyr)	−25.0	−61.9	−98.9	−99.4	+10.7
10 DAT		−48.7	−73.8	−100.1	−100.0	−15.9
1 HAT	Tordon 101M	−56.2	−90.2	−99.7	−99.9	−49.4
5 DAT	(Picloram 2,4-D)	5.3	−49.4	−98.0	−97.8	+26.8
10 DAT		−15.7	−73.7	−99.8	−99.9	+9.9
1 HAT	Velpar L	−100.0	−99.4	−100.1	−100.2	−100[c]
5 DAT	(Hexazinone-25%)	−99.9	−99.9	−99.7	−99.3	−100[c]
10 DAT		−99.9	−99.8	−99.9	−99.3	−100[c]
1 HAT	Velpar ULW	−99.9	−100.0	−99.8	−99.9	−100[c]
5 DAT	(Hexazinone-75% AI)	−99.9	−100.0	−99.8	−98.0	−100[c]
10 DAT		−100.0	−99.9	−99.8	−99.7	−100[c]

[a] Only negative CE values are used to calculate EC_{50} values.
[b] EC_{50} set to +100 when 80% CE values are positive.
[c] EC_{50} set to −100 when 1, 10, 50, and 80% CE values are > −98.

−100. This allows a range of EC_{50} values from +100 (innocuous soil) to −100 (toxic soil) with all other values falling between the two extremes. Direct comparisons between treatments can then be made.

The seven treatments can be divided into three groups based on their respective 10 DAT EC_{50} values. The Control, Roundup, and Arsenal treatments were the least toxic to *S. capricornutum* (EC_{50} = +100). Tordon 101M (EC_{50} = +9.9) and Garlon 4 (EC_{50} = −15.9) were of intermediate toxicity, while Velpar L and Velpar ULW were the most toxic (EC_{50} = −100).

Ranking treatments as to their changing toxicity to *S. capricornutum* over time becomes possible when graphing the percent soil eluate against their associated calculated effect (Fig. 1). Assay results initiated 1 h after herbicide application to soil indicate that the toxicity ranking (from least to most toxic) is as follows: Control < Roundup < Arsenal < Garlon 4 < Tordon 101M < Velpar L = Velpar ULW. The ranking changed slightly five days after application: Arsenal < Control < Roundup < Tordon 101M < Garlon 4 < Velpar L < Velpar ULW. The assay initiated ten days after herbicide application to soil revealed the following ranking: Roundup < Arsenal < Control < Tordon 101M < Garlon 4 < Velpar L = Velpar ULW.

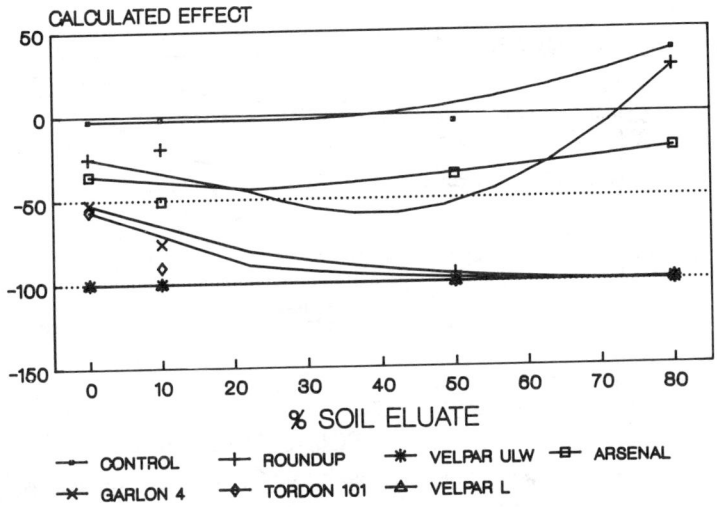

FIG. 1a—*Calculated effect values (assayed one hour after treatment) for 1, 10, 50, and 80% soil eluate dilutions.*

The modes of action of herbicides used in the present study to green alga is unknown but may be similar to the herbicides' effect on vascular plants. The modes of action of imazapyr and glyphosate are similar. Imazapyr induces mortality in plants by reducing the levels of three branched chain aliphatic amino acids (valine, leucine, isoleucine) through inhibiting the synthesis of an enzyme (acetohydroxyacid) common to the biosynthetic pathway of these

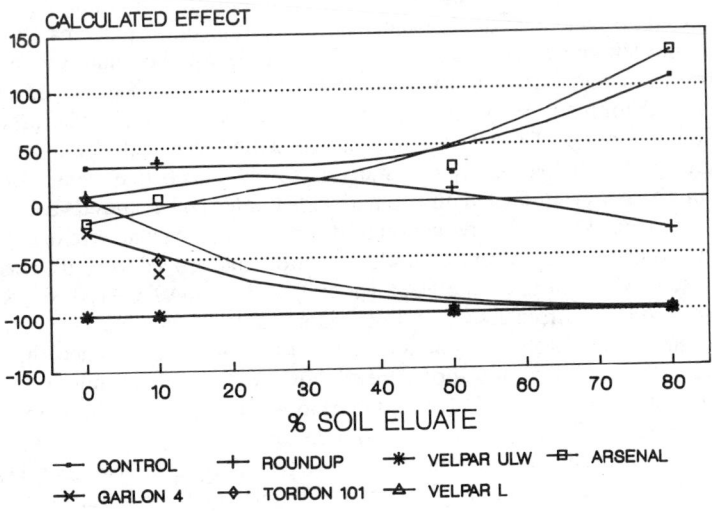

FIG. 1b—*Calculated effect values (assayed five days after treatment) for 1, 10, 50, and 80% soil eluate dilutions.*

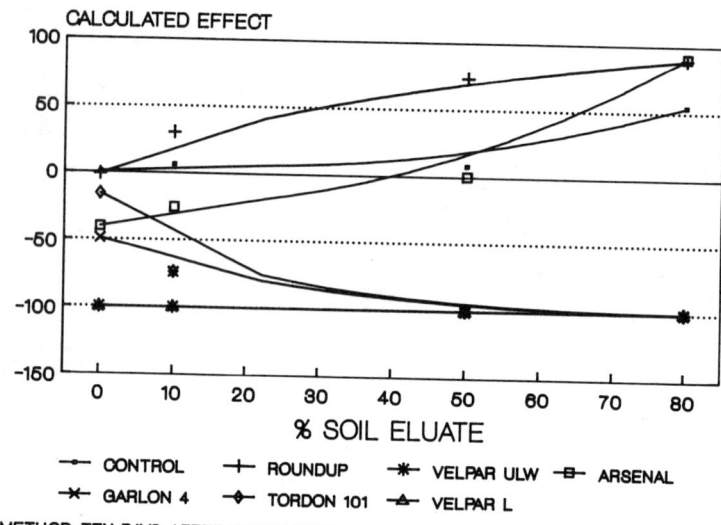

FIG. 1c—*Calculated effect values (assayed ten days after treatment) for 1, 10, 50, and 80% soil eluate dilutions.*

amino acids. Glyphosate inhibits the aromatic amino acid biosynthesis pathway and may also inhibit or suppress chlorismate mutase and/or prephenate dehydratase. Photosynthesis and respiration are impeded and the interruption of protein synthesis interferes with cell wall, cell membrane, chloroplast, and chlorophyll production. Picloram, 2,4-D, and triclopyr are "phenoxy" herbicides and induce characteristic auxin-type responses in plants: tissue proliferation, leaf curling or epinasty, tumor formation, cessation of root elongation, and the fruitless expenditure of carbohydrates in excessive respiration. Swelling parenchymal cells develop into callus tissue that induces mechanical crushing of the phloem as well. The mode of action of hexazinone is poorly understood, but it appears to be a photosynthetic inhibitor. The electron transport system of the Hill reaction is thought to be affected by specifically replacing a functioning protein with a nonfunctioning cytochrome carrier protein (possibly B-559). Note that herbicides with similar modes of action in vascular plants (i.e., Roundup/Arsenal and Tordon 101M/Garlon 4) resulted in similar effects to *S. capricornutum*.

The results of assays conducted on two herbicide treated soil eluates revealed reduced toxicity over time (Fig. 2). Toxicity reduction of Roundup and Arsenal were the most apparent. These two herbicides possess similar modes of action and proved to be the least toxic when applied to soil. EC_{50} values at one HAT (+27.3) and 10 DAT (+100) for Roundup and −20.4 and +100 for Arsenal support this finding. Both Arsenal and Roundup are highly biodegradable, with microbial degradation being the major cause of decomposition in the environment. Rates of decomposition depend, however, on soil and microflora populations [40]. Two treatments (Tordon 101M and Garlon 4) proved to be considerably more toxic to *S. capricornutum* (Fig. 3). A slight reduction in toxicity over time was observed with Tordon 101M and Garlon 4 but only at the 1 and 10% dilutions. These two herbicides also possess similar modes of action. EC_{50} values at 1 HAT and 10 DAT were −49.4 and 9.9 for Tordon 101M and −22.4 and −15.9 for Garlon 4, respectively. Both hexazinone formulations (Velpar L [liquid] and Velpar ULW [granular]) proved to be toxic to *S. capricornutum* in all dilutions (Fig. 3).

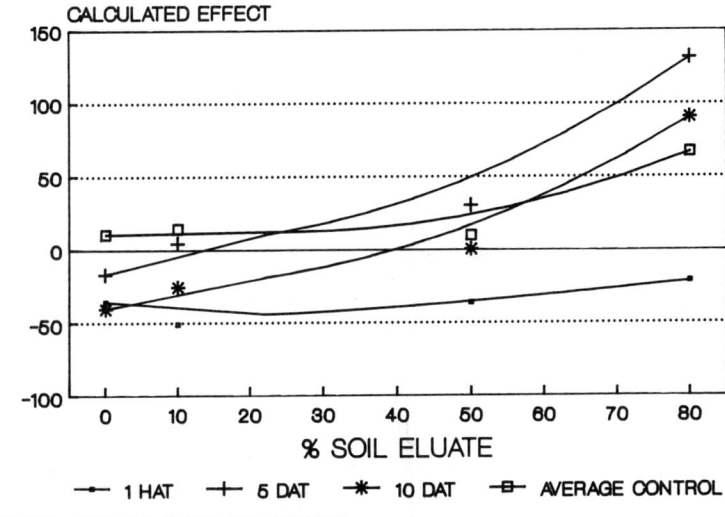

FIG. 2a—*Calculated effect values (assayed one hour, five days, and ten days after treatment) for Arsenal (2 lb) herbicide applied to soil.*

Each formulated herbicide was also mixed for 10 min with reverse osmosis double-deionized water at an initial concentration of 100 mg/L (Arsenal [2 lb and 4 lb], Roundup, Tordon 101M, Garlon 4) and 100 µg/L (Velpar L and Velpar ULW). A mortar and pestle was used to first grind the granular hexazinone into a fine powder. One, 10, 50, and 80% dilutions of

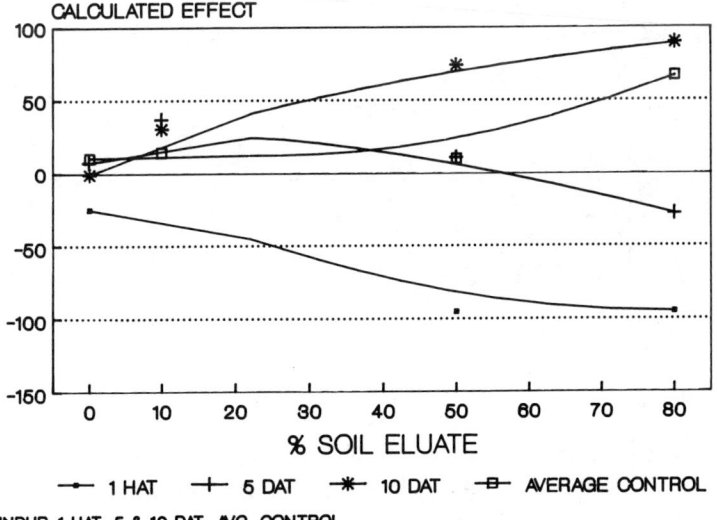

FIG. 2b—*Calculated effect values (assayed one hour, five days, and ten days after treatment) for Roundup herbicide applied to soil.*

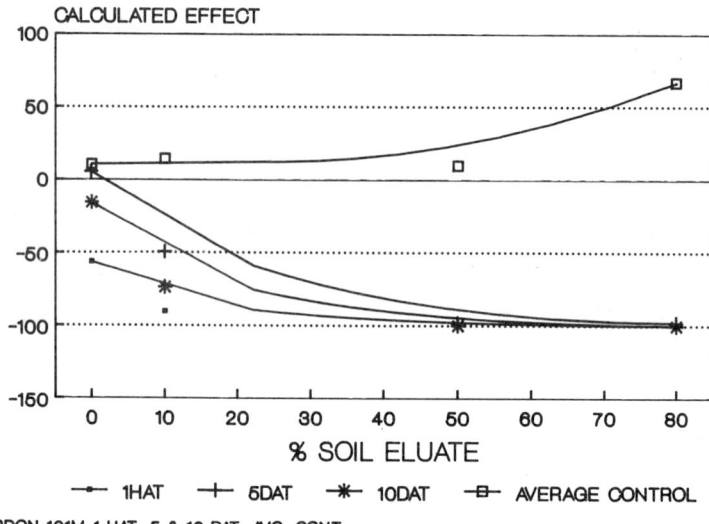

FIG. 3a—*Calculated effect values (assayed one hour, five days, and ten days after treatment) for Tordon 101M herbicide applied to soil.*

the sterile eluate were assayed. Both cell number (Table 6) and EC_{50} values (Fig. 4) are provided.

The order of toxicity in water was similar to that discussed previously in soil eluate except for glyphosate, which was more toxic. The toxicity ranking (from least to most toxic) expressed in µg/L was: Arsenal (2 lb) (EC_{50} = 5500) < Arsenal (4 lb) (EC_{50} = 5300) <

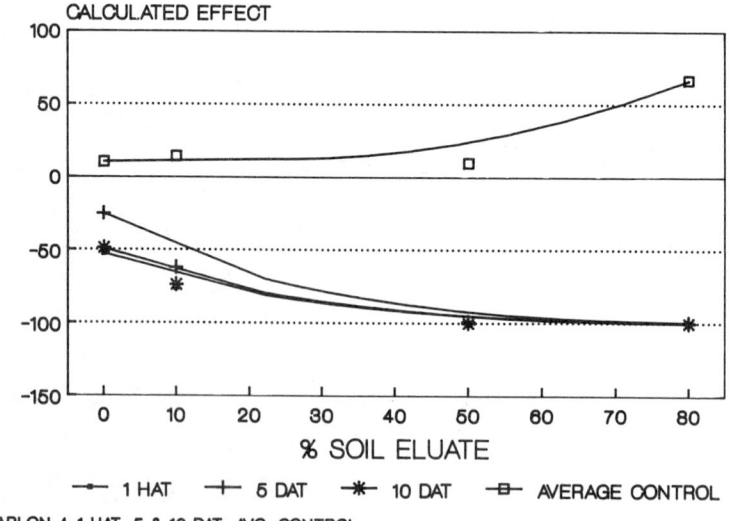

FIG. 3b—*Calculated effect values (assayed one hour, five days, and ten days after treatment) for Garlon 4 herbicide applied to soil.*

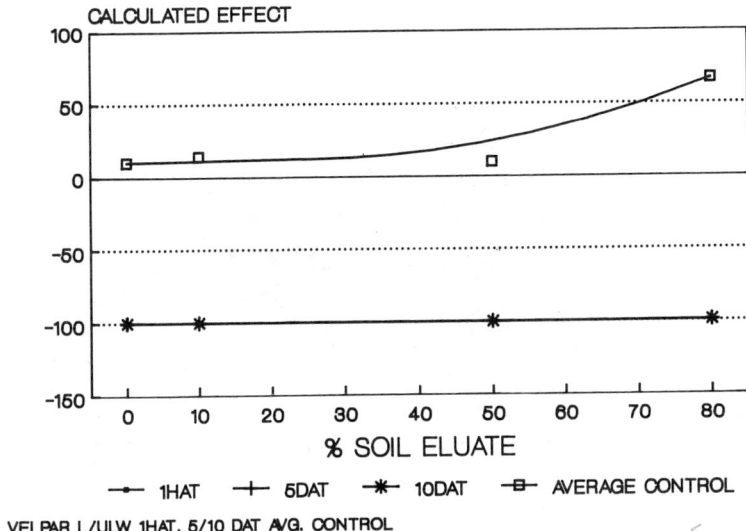

FIG. 3c—*Calculated effect values (assayed one hour, five days, and ten days after treatment) for Velpar L and Velpar ULW herbicide applied to soil.*

Tordon 101M = Garlon 4 (both EC_{50} values = 5000) < Roundup (EC_{50} = 2600) < Velpar L (EC_{50} = 2.5) < Velpar ULW (EC_{50} = 1.2). The increased toxicity of glyphosate noted in water (as compared to soil) may be the result of additional ingredients found in the Roundup formulation. Increased toxicity to aquatic organisms due to the surfactant (MON 0818) in the formulation has been reported [35]. The decreasing toxicity documented for Roundup after application to soil emphasizes the moderating or neutralizing properties of soil which result in rapid deactivation of some, but not all, herbicides.

The biological activity of a pesticide in soil decreases as the principal and inert ingredients are deactivated biologically and chemically over time [32]. Marking [32] developed an appropriate bioassay that estimates the total activity of a chemical and its degradation products but does not differentiate between chemical degradation, physical removal, and biolog-

TABLE 6—*Growth of* Selenastrum capricornutum *expressed in cells/mL* $\times$ *10^6 for seven herbicides applied to water (negative control = 3.8293 $\times$ 10^6 cells/mL).*

Treatment	1% (100 ppb)[a]	10% (1000 ppb)	50% (5000 ppb)	80% (8000 ppb)
Arsenal (2 lb)	3.6243	3.5668	2.0518	0.7727
Arsenal (4 lb)	3.3667	3.2546	2.0732	0.7395
Roundup	3.1523	2.7831	1.0703	0.0955
Tordon 101M	3.38265	3.47537	1.6260	0.7376
Garlon 4	3.0461	3.0773	1.8454	0.5946
	1% (100 ppt)[b]	10% (1000 ppt)	50% (5000 ppt)	80% (8000 ppt)
Velpar L	2.4039	2.1882	1.3425	0.4915
Velpar ULW	2.1992	2.1095	0.0925	0.0609

[a] Parts per billion (ppb) = μg/L.
[b] Parts per trillion (ppt) = ng/L.

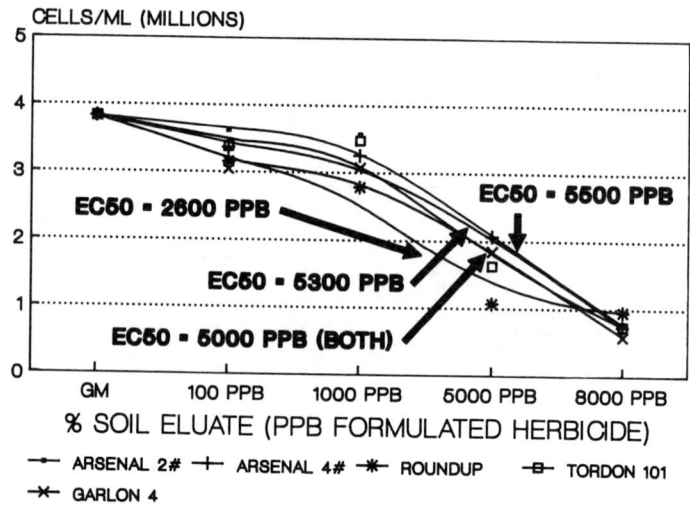

FIG. 4a—*Population growth of* S. capricornutum *in water containing 100 mg/L Arsenal (2 lb and 4 lb), Tordon 101M, Garlon 4, and Roundup at four dilutions.*

ical inactivation (microbial degradation). Deactivation indices have been developed [33] that determine changes in biological activity over time. Average values of between 0.5 and 1.5 are within the range of normal experimental variation [34]. A value of less than one indicates that a chemical becomes more toxic over time and a value greater than one indicates decreasing toxicity. "Aging" herbicide solutions for seven days resulted in decreasing toxicity 29% of the time, increasing toxicity 7% of the time and no change 64% of the time, based on 14 tests on 9 herbicides [34]. Average values for glyphosate of 1.12, picloram of 0.83, and 2,4-

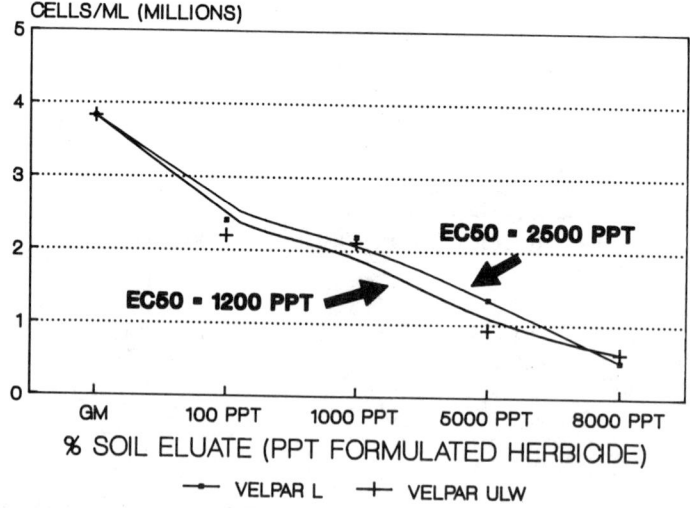

FIG. 4b—*Population growth of* S. capricornutum *in water containing 100 µg/L Velpar L and Velpar ULW at four dilutions.*

dimethylamine salt of 1.00 have been reported [35-38]. Mayer and Ellersieck [34] reported no change in toxicity after seven days for glyphosate while noting a slight increase with picloram. After four weeks of aging, a deactivation index of 1.02 for picloram was reported [34]. This suggests that picloram may be deactivated very slowly. Decreasing toxicity with 2,4-D herbicides was noted after 3 to 4 weeks with the exception of the dodecyl/tetradodecylamine salt of 2,4-D, which increased [34]. The data documented in the present study largely agree with previously published data; that is, the toxicity of glyphosate is rapidly reduced over time, while that of Tordon 101M (picloram and 2,4-D) degrades more slowly when applied to soil.

Herbicide formulations, not technical grade material, were used in the present experiment. On average, toxicity associated with formulated products remain unchanged 57% of the time, decreased 11% of the time, and increased 32% of the time [34]. The toxicity of granular formulated pesticides increased more often than liquid formulations: 40% as opposed to 31%, respectively [34]. This may partially explain why the granular formulation of hexazinone (Velpar ULW) proved to be so toxic in our study.

Even though the experimental conditions of the present study were not designed to and did not mirror field conditions (i.e., sunlight and precipitation), other factors may have been involved in deactivating the herbicides over time. Adsorption of glyphosate to soil colloids is well documented [39]. Microbial degradation appears to be the major cause of decomposition in soil [40]. The *S. capricornutum* bioassay of a glyphosate-treated soil eluate applied at a 2% rate supports these findings. Less is known concerning Arsenal, but the calculated half-life of imazapyr in soil ranges from 19 to 34 days and from 37 to 44 days in leaf litter [41]. The decreasing toxicity documented over the ten-day interval in the present study is consistent with the available literature. The degradation of picloram, however, is a complex and poorly understood process [42-46]. Lateral movement can occur to some extent [47,48] and only limited adsorption to soil colloids has been reported [49-52]. Degradation of picloram in the soil is slow [40]. Estimated half-lives for the Tordon formulation of picloram varies from 1 to 13 months [53] and from 4.5 months to 4.6 years in Kansas, South Dakota, Minnesota, and California [54]. Bovey et al. [55] reported half-lives of six to nine months in sandy soils in Texas and less than one year in Texas rangeland soils [56]. Triclopyr reportedly is less persistent than picloram in soil and degrades fairly rapidly with a half-life of 46 days in a Yolo loam soil from California, but is not strongly adsorbed onto soil colloids [40]. A half-life of ten days was reported in a Flanagan silty clay loam soil from Illinois [57]. Both of the above studies, however, were conducted at a temperature of 35°C with soil moisture at field capacity, two factors that would enhance deactivation. The present study was not particularly conducive to deactivation. There was a slight reduction in toxicity over time documented in the present study with Tordon 101M and Garlon 4, but this reduction was no so apparent as with Roundup and Arsenal.

A general rating scale based on 96-h EC_{50} values for aquatic species has been developed [58]. The relative toxicity scale includes various categories reported in mg/L (ppm): Relatively Harmless (>1000 ppm), Practically Nontoxic (100-1000 ppm), Slightly Toxic (10-100 ppm), Moderately Toxic (1-10 ppm), Highly Toxic (0.1-1 ppm), Extremely Toxic (0.01-0.1 ppm), and Super Toxic (<0.01 ppm). Using this scale to compare EC_{50} values of the seven herbicide formulations applied to water results in the following values (also reported in mg/L): Arsenal (2 lb) (5.5 ppm), Arsenal (4 lb) (5.3 ppm), Tordon 101M (5.0 ppm), Garlon 4 (5.0 ppm), and Roundup (2.6 ppm) are all considered to be Moderately Toxic, while Velpar L (0.0025 ppm) and Velpar ULW (0.0012 ppm) are Super Toxic. *S. capricornutum* is a more sensitive indicator of toxicity than rainbow trout, fathead minnow, channel catfish, and bluegill when comparing the values given in Table 1.

The method utilized in this study for extracting toxic constituents from soil is simple—the soil in question is mixed with water. As such, the toxic water soluble constituent(s) of the soil that are represented in the soil eluate would, theoretically, be present naturally in runoff after a storm event. These constituents can be thought of as being potentially environmentally available. This is in direct contrast to a chemical extraction of soil such as the TCLP-Toxicity Characteristic Leaching Procedure (elution with $0.1 N$ pH 5 sodium acetate buffer). Soil eluates derived from this procedure are not an accurate reflection of soil constituents that are normally biologically available. Soil eluates derived from water extraction, however, can be thought of as being more representative of potential availability subsequent to storm events.

Off-site movement of most herbicides (Arsenal, Tordon 101M, Garlon 4, Roundup) in stormflow and baseflow from treated watersheds would not be considered detrimental to aquatic fauna based on their respective EC_{50} values if label-recommended rates and directions are employed. Extreme care, however, should be used when applying hexazinone, especially the ULW formulation. Indigenous unicellular green algae may exhibit sensitivity to hexazinone similar to that documented with *S. capricornutum*.

Velpar ULW is a symmetrical triazine granular herbicide containing 75% AI hexazinone. Its high water solubility (33 000 ppmw at 25°C) makes it susceptible to off-site movement in storm runoff and in leaching and to lateral and vertical movement. Hexazinone resists photodegradation, but is subject to microbial degradation. The results of the present study indicate that extreme care should be taken and appropriate buffer strips used when applying Velpar ULW near waterways in order to protect the aquatic ecosystem. A direct application to a forest stream in Alabama resulted in hexazinone concentrations of 2400 μg/L (ppb) [59]. Detectable concentrations (<20 μg/L) were still present ten days later. A mean hexazinone concentration of 442 ± 53 μg/L was reported [60] in storm runoff from four upper piedmont watersheds three days after a granular application. The mean concentration had decreased to 14 ± 11 μg/L nine months after application. The 96-h EC_{50} value (in water) of Velpar ULW (75% AI hexazinone) documented in the present study is 1.2 μg/L.

Conclusion

Soil eluates of six herbicide formulations applied directly to soil were assayed using *S. capricornutum* one hour, five days, and ten days after herbicide application. A reduction in population growth occurred in all herbicide treatments compared to the control when assayed one hour after application. It should be noted that conditions were not conducive for herbicide deactivation (microbial degradation and photodegradation) in the present study. Nevertheless, a substantial decrease in toxicity was documented only five days after application with two herbicides, Roundup and Arsenal. A statistically significant enhancement was documented with Roundup ten days after treatment. An enhancement effect was also observed with Arsenal (2 lb). Two herbicides, Tordon 101M and Garlon 4, proved to be intermediate in toxicity. *S. capricornutum* exhibited extreme sensitivity to hexazinone (Velpar L and Velpar ULW).

Seven herbicide formulations were applied directly to water and assayed as well. Four herbicides were moderately toxic: Arsenal (2 lb), Arsenal (4 lb), Tordon 101M, and Garlon 4. Roundup was found to be more toxic in water than when applied to soil, but was still considered to be only moderately toxic. Velpar L and Velpar ULW, however, proved to be extremely toxic to *S. capricornutum*, with EC_{50} values of 2.5 μg/L and 1.2 μg/L, respectively.

S. capricornutum proved to be a sensitive bioassay organism for detecting herbicides both in soil eluates and water. Sensitivity was greater than in many taxa currently used for bioassay purposes (Table 1). The work of Kimerle [61], Blanck [62], and data generated in the present study suggest that *S. capricornutum*, *Daphnia*, and rainbow trout may in the future be the first three taxa of choice used for estimating the sensitivity of aquatic organisms to chemicals in a first-order tier of testing.

organism to herbicide concentrations for 96 h. Storm events over herbicide-treated land and the resultant runoff may not result in exposures of this length or concentration. Cautious interpretation of the results, therefore, is recommended when extrapolating the data given here to potential events.

Only one alga, *S. capricornutum,* was tested in the present study. The sensitivity of other algae to eluates obtained from herbicide-treated soil or herbicides applied directly to water is unknown. Additional research is required to determine the sensitivity of indigenous algae.

APPENDIX

The constituents of the growth medium are given in Appendix Table 1.

APPENDIX TABLE 1—*Composition of algal growth medium.*

	Macronutrients			
			Nutrient Composition of Prepared Medium	
Compound	Concentration (g/L)		Element	Concentration (mg/L)
	Stock Solution I[a]			
$NaNO_3$	25.500		N	4.200
$NaHCO_3$	15.000		Na[b]	11.001
			C	2.143
	Stock Solution II			
K_2HPOH_4	1.044		K	0.469
			P	0.186
	Stock Solution III			
$MgSO_4 \cdot 7H_2O$	14.700		S	1.911
	Stock Solution IV			
$MgCl_2 \cdot 6H_2O$	12.164		Mg[c]	2.904
$CaCl_2 \cdot 2H_2O$	4.410		Ca	1.202
	Micronutrients Stock Solution V			
			Nutrient Composition of Prepared Medium	
Compound	Concentration (mg/L)		Element	Concentration (µg/L)
H_3BO_3	185.520		B	32.460
$MnCl_{24} \cdot 4H_2O$	415.610		Mn	115.374
$ZnCl_2$	3.271		Zn	1.570
$CoCl_2 \cdot 6H_2O$[d]	1.428		Co	0.354
$CuCl_2 \cdot 2H_2O$[d]	0.012		Cu	0.004
$Na_2MoO_4 \cdot 2H_2O$[d]	7.260		Mo	2.878
$FeCl_3 \cdot 6H_2O$	160.000		Fe	33.051
$Na_2EDTA \cdot 2H_2O$	150.000		...	...

[a] Other forms of the stock solutions may be used as long as the resulting concentrations of elements are the same.
[b] Includes Na from $NaNO_3$.
[c] Includes Mg from $MgSO_4 \cdot 7H_2O$.
[d] Make up in 100-fold or more concentrations and add each in appropriate amount to Stock Solution V.

Acknowledgments

The authors are grateful to D. Schuster and K. Bick for their expertise in preparation of this manuscript. Dr. Armon Yanders and Dr. Ravi Puri provided assistance in reviewing the document and their help is greatly appreciated. The following companies supplied products: American Cyanamid Chemical Company, Dow Chemical Company, DuPont Chemical Company, and Monsanto Chemical Company.

References

[1] Hart, W. B., Doudoroff, P., and Greenbank, J., *The Evaluation of Toxicity of Industrial Wastes, Chemicals and Other Substances to Freshwater Fishes,* Atlantic Refining Company, Philadelphia, 1945.
[2] Doudoroff, P., Anderson, B. G., Burdick, G. E., Galtsoff, P. S., Hart, W. B., Patrick, R., Strong, E. R., Surber, E. W., and Van Horn, W. M., "Bio-assay Methods for the Evaluation of Acute Toxicity of Industrial Wastes to Fish," *Sewage Industrial Wastes,* Vol. 23, 1951, pp. 1380–1397.
[3] "Algal Assay Procedures: Bottle Test," National Eutrophication Research Program, Environmental Protection Agency, Pacific Northwest Environmental Research Laboratory, Corvallis, Ore., 1971.
[4] "The *Selenastrum capricornutum* Printz Algal Assay Bottle Test: Experimental Design, Application and Data Interpretation Protocol," EPA-6009-78-018, Environmental Protection Agency, Corvallis Environmental Research Laboratory, Office of Research and Development, Corvallis, Ore., 1978.
[5] Condit, R. J., "Phosphorus and Algal Growth in the Spokane River," *Northwest Science,* Vol. 46, No. 3, 1972, pp. 177–189.
[6] Shiroyama, T., Miller, W. E., and Greene, J. C., "Effect of Nitrogen and Phosphorus on the Growth of *Selenastrum capricornutum*," in *Proceedings,* Biostim. Nutr. Assess. Workshop, Oct. 1973, EPA-600-3-75-034, 1975, pp. 132–142.
[7] Parker, M., "The Use of Algal Bioassays to Predict the Short- and Long-Term Changes in Algal Standing Crop from Altered Phosphorus and Nitrogen Loadings," *Water Research,* Vol. 11, 1977, pp. 719–725.
[8] Brown, E. J. and Button, D. K., "Phosphate-Limited Growth Kinetics of *Selenastrum capricornutum* (Chlorophyceae)," *Journal of Phycology,* Vol. 15, 1979, pp. 305–311.
[9] Miller, W. E. and Maloney, T. E., "Effects of Secondary and Tertiary Wastewater Effluents on Algal Growth in a Lake-River System," *Journal of the Water Pollution Control Federation,* Vol. 43, 1971, pp. 2361–2365.
[10] Greene, J. C., Miller, W. E., Shiroyama, T., and Maloney, T. E., "Utilization of Algal Assays to Assess the Effect of Municipal, Industrial and Agricultural Wastewater Effluents upon Phytoplankton Production in the Snake River System," *Water, Air, and Soil Pollution,* Vol. 4, 1975, pp. 415–434.
[11] Maloney, T. E. and Miller, W. E., "Algal Assays: Development and Application," in *Water Quality Parameters, ASTM STP 573,* American Society for Testing and Materials, Philadelphia, 1975, pp. 344–355.
[12] Greene, J. C., Miller, W. E., and Shiroyama, T., "Use of Algal Assays to Assess the Effects of Municipal and Smelter Wastes upon Phytoplankton Production," in *Proceedings,* Symposium on Terrestrial and Aquatic Ecological Studies of the Northwest, 26–27 March, 1976, pp. 327–335.
[13] Joubert, G., "A Bioassay Application for Quantitative Toxicity Measurements Using the Green Algae *Selenastrum capricornutum*," *Water Research,* Vol. 14, 1980, pp. 1759–1763.
[14] Bartlett, L., Babe, F. W., and Funk, W. H., "Effects of Copper, Zinc and Cadmium on *Selenastrum capricornutum*," *Water Research,* Vol. 8, 1974, pp. 179–185.
[15] Gaur, J. P. and Kumar, H. D., "Effects of Oil Refinery Effluents on *Selenastrum capricornutum* Printz," *Internationale Revue Der Gesamten Hydrobiologie,* Vol. 71, No. 2, 1986, pp. 271–281.
[16] Plotkin, S. and Ram, N. M., "Multiple Bioassays to Assess the Toxicity of a Sanitary Landfill Leachate," *Archives of Environmental Contamination Toxicology,* Vol. 13, 1984, pp. 197–206.
[17] Blaise, C. and Couture, P., "Détection à l'aide d'un bio-essai *(Selenastrum capricornutum)* des répercussions d'un rèjet linier sur l'environnement aquatique: toxicité ou enrichissement en substance essentielle?" *Hydrodiologia,* Vol. 114, 1984, pp. 39–50.
[18] Shigeoka, T., Sato, Y., Takeda, Y., Yoshida, K., and Yamauchi, F., "Acute Toxicity of Chlorophenols to Green Algae, *Selenastrum capricornutum* and *Chlorella vulgaris,* and Quantitative

Structure-Activity Relationships," *Environmental Toxicology and Chemistry,* Vol. 7, 1988, pp. 847–854.
[19] Turbak, S. C., Olsen, S. B., and McFeters, G. A., "Comparison of Algal Assay Systems for Detecting Waterborne Herbicides and Metals," *Water Research,* Vol. 20, No. 1, 1985, pp. 91–96.
[20] "Chlorophyta Assay Using Soils as Sample Material-Method Protocol," USEPA, U.S. Environmental Monitoring Systems Laboratory, Las Vegas, Nev., and Environmental Trace Substances Research Center, University of Missouri, 1989.
[21] Hegemann, D. A. and Keenan, J. D., "Bioassay Techniques for Soil-Nutrient Availability," *Water, Air and Soil Pollution,* Vol. 19, 1983, pp. 259–276.
[22] Bennett, P.H. and Sandmann, E. R. I. C., "The Effect of Bromacil on the Net Photosynthetic Oxygen Production of *Selenastrum capricornutum*," *Chemosphere,* Vol. 12, Nos. 7–8, 1983, pp. 1047–1054.
[23] Bennett, P. H. and deBeer, P. R., "A Rapid Quantitative Bioassay for the Determination of Biologically Available Bromacil in Soils," *Pesticide Science,* Vol. 15, 1983, pp. 425–430.
[24] Gibson, C. E., "An Investigation into Effects of Forestry Plantations on the Water Quality of Upland Reservoirs in Northern Ireland," *Water Research,* Vol. 10, 1976, pp. 995–998.
[25] McLeay, D., "Development of a Bioassay Protocol for Evaluating the Toxic Risk to Regional Fisheries Resources Posed by Forest-Use Herbicides," FRDA Report ISSN 0835-0752;039, Canadian Forest Service/British Columbia Ministry of Forests, 1988, p. 48.
[26] Brown, J. R. and Rodriquez, R. R., "Soil Testing in Missouri: A Guide for Conducting Soil Tests in Missouri," Missouri Cooperative Extension Service, University of Missouri-Lincoln University, Report EC923, 1983.
[27] Debolt, D. C., "Readily Oxidizable Soil Organic Matter: A High Sample Volume Procedure for the Colorimetric Determination of Soil Organic Matter," communication in *Soil Science and Plant Analysis,* Vol. 5, 1974, pp. 131–137.
[28] Chapman, H. D., "Ammonium Acetate Extractable Calcium, Magnesium, Potassium and Sodium: Total Exchangeable Bases," Chapter 58 in *Methods of Soil Analysis, Part 2,* C. A. Black, Ed., Soil Science Society of America, Madison, Wisc., 1965.
[29] Bray, R. H. and Kurtz, L. T., "Extractable Soil Phosphorus—Bray I Method: Determination of Total, Organic, and Available Forms of Phosphorus in Soil," *Soil Science,* Vol. 59, 1945, pp. 39–45.
[30] McLean, E. O., "Soil pH in Salt Solution (pHs): Testing Soils for pH and Lime Requirements," Chapter 7 in *Soil Testing and Plant Analysis,* rev. ed., L. M. Walsh and J. D. Beaton, Eds., Soil Science Society of America, Madison, Wisc., 1973.
[31] Woodruff, C. M., "Determination of Neutralizable Acidity (NA): Determination of the Exchangeable Hydrogen and Lime Requirement of the Soil by Means of the Glass Electrode and a Buffered Solution," *Soil Science Society of America Proceedings,* Vol. 12, 1948, pp. 141–142.
[32] Marking, L. L., "Methods of Estimating the Half-Life of Biological Activity of Toxic Chemicals in Water," *Investigations in Fish Control,* 1972.
[33] Marking, L. L. and Dawson, V. K., "The Half-Life of Biological Activity of Antimycin Determined by Fish Bioassay," *Transactions of the American Fisheries Society,* Vol. 101, 1972, pp. 100–105.
[34] Mayer, F. L. and Ellersieck, M. R., *Manual of Acute Toxicity: Interpretation and Data Base for 410 Chemicals and 66 Species of Freshwater Animals,* Resource Publication No. 160, United States Department of the Interior, Fish and Wildlife Service, 1986.
[35] Folmer, L. C., Sanders, H. O., and Julin A. M., "Toxicity of the Herbicide Glyphosate and Several of Its Formulations on Fish and Aquatic Invertebrates," *Archives of Environmental Contamination Toxicology,* Vol. 8, 1979, pp. 269–278.
[36] Mauck, W. L., Olsen, L. E., and Marking, L. L., "Toxicity of Natural Pyrethrins and Five Pyrethroids to Fish," *Archives of Environmental Contamination Toxicology,* Vol. 4, 1976, pp. 18–29.
[37] Mauck, W. L., Olsen, L. E., and Megen, J. W., "Effects of Water Quality on Deactivation and Toxicity of Mexacarbate (Zectron) to Fish," *Archives of Environmental Contamination Toxicology,* Vol. 6, 1977, pp. 385–393.
[38] Woodard, D. F., "Toxicity of the Herbicides Dinoseb and Picloram to Cutthroat *(Salmo clarki)* and Lake Trout *(Salvelinus namaycush),*" *Journal of the Fisheries Research Board Canada,* Vol. 33, 1976, pp. 1671–1676.
[39] Hance, R. J., "Adsorption of Glyphosate by Soils," *Weed Science,* Vol. 7, 1976, pp. 363–366.
[40] *Herbicide Handbook of the Weed Science Society of America,* Weed Science Society of America, Champaign, Ill., 1983.
[41] American Cyanamid Chemical Company, Technical Information, "Fate of Arsenal in Forest Watersheds" (Abstract), Author: J. L. Michael, U.S. Forest Service, Auburn, Ala., 1988.
[42] Grover, R., "Studies on the Degradation of 4-amino-3,5,6-trichloropicolinic Acid in Soil," *Weed Research,* Vol. 7, 1967, pp. 61–67.

[43] Youngson, C. R., Goring, C. A. I., Meikle, R. W., Scott, H. H., and Griffith, J. D., "Factors Influencing the Decomposition of Tordon Herbicides in Soils," *Down to Earth*, Vol. 22, 1967, pp. 3–11.
[44] Hall, R. C., Giam, G. S., and Merkle, M. G., "The Photolytic Degradation of Picloram," *Weed Research*, Vol. 8, 1968, pp. 292–297.
[45] Haymaker, J. W., Youngson, C. R., and Goring, C. A. I., "Rate of Detoxification of 4-amino-3,5,6-trichloropicolinic Acid in Soils," *Weed Research*, Vol. 8, 1968, pp. 46–57.
[46] Guenzi, W. D. and Beard, W. E., "Picloram Degradation in Soils as Influenced by Soil Water Content and Temperature," *Journal of Environmental Quality*, Vol. 5, No. 2, 1970, pp. 189–192.
[47] Herr, D. E., Stroube, E. W., and Ray, D. A., "The Movement and Persistance of Picloram in Soil," *Weeds*, Vol. 14, 1967, pp. 248–250.
[48] Byrd, B. C., Williams, C. S., and Bjerke, E. L., "An Investigation of Lateral Surface Movement: Picloram," *Proceedings of the Southwest Weed Science Society of America*, Vol. 2, 1971, pp. 301–305.
[49] Keys, C. H. and Friesen, H. A., "Persistence of Picloram Activity in Soil," *Weed Science*, Vol. 16, 1968, pp. 34–42.
[50] Grover, R., "Adsorption of Picloram by Soils Colloids and Various Other Adsorbents," *Weeds*, Vol. 19, No. 4, 1971, pp. 417–418.
[51] Farmer, W. J. and Aochi, Y., "Picloram Sorption by Soils," *Soil Science Society of America Proceedings*, Vol. 38, 1974, pp. 418–423.
[52] Roa, P. S. C. and Davidson, J. M., "Adsorption and Movement of Selected Pesticides at High Concentrations in Soils," *Water Research*, Vol. 13, 1979, pp. 375–380.
[53] Goring, C. A. I., Youngson, C. R., and Haymaker, J. W., "Tordon Herbicide ... Disappearance from Soils," *Down to Earth*, Vol. 20, No. 4, 1965, pp. 3–5.
[54] Goring, C. A. I. and Haymaker, J. W., "The Degradation and Movement of Picloram in Soil and Water," *Down to Earth*, Vol. 27, No. 1, 1971, pp. 12–15.
[55] Bovey, R. W., Davis, F. S., and Morton, H. L., "Herbicide Combinations for Woody Plant Control," *Weed Science*, Vol. 16, 1969, pp. 332–335.
[56] Scifres, C. J., Hahn, R. R., Diaz-Colon, J., and Merkle, M. G., "Picloram Persistence in Semi-Arid Rangeland Soils and Water," *Weed Science*, Vol. 19, No. 4, 1971, pp. 381–384.
[57] "Technical Information on Triclopyr, the Active Ingredient of Garlon Herbicides," Technical Data Sheet, Form No. 137-859-79, Dow Chemical Company, 1979.
[58] U.S. Department of the Interior Fish and Wildlife Service, Research Information Bulletin No. 84-87, "Acute Toxicity Rating Scales," Contaminant Information Transfer Project, Columbia National Fisheries Research Laboratory, Columbia, Mo., 1984.
[59] Miller, J. H. and Bace, A. C., "Streamwater Contamination after Aerial Application of Pelletized Herbicide," U.S. Forest Service, Southern Forest Experiment Station, Research Note 50-255, 1980.
[60] Neary, D. G., Bush, P. B., and Douglas, J. E., "Off-Site Movement of Hexazinone in Stormflow and Baseflow from Forest Watersheds," *Weed Science*, Vol. 31, 1983, pp. 543–551.
[61] Kimerle, R. A., Werner, A. F., and Adams, W. J. "Aquatic Hazard Evaluation Principles Applied to the Development of Water Quality Criteria," *Aquatic Toxicology: Seventh Symposium*, ASTM STP 854, ASTM, Philadelphia, 1985, pp. 538–547.
[62] Blanck, H., Wallin, G., and Wangberg, S. A., "Species-Dependent Variation in Algal Sensitivity to Chemical Compounds," *Ectoxicology Environmental Safety*, Vol. 8, 1984, pp. 339–351.
[63] Greene, J. C., Bartles, C. L., Warren-Hicks, W. J., Parkhurst, B. R., Linder, G. L., Peterson, S. A., and Miller, W. E., "Protocols for Short-Term Toxicity Screening of Hazardous Waste Sites," U.S. Environmental Protection Agency, Corvallis, Ore., 1989.

John M. Macauley,[1] *James R. Clark,*[1] *and A. R. Pitts*[2]

Use of *Thalassia* and Its Epiphytes for Toxicity Assessment: Effects of a Drilling Fluid and Tributyltin

REFERENCE: Macauley, J. M., Clark, J. R., and Pitts, A. R., "**Use of *Thalassia* and Its Epiphytes for Toxicity Assessment: Effects of a Drilling Fluid and Tributyltin,**" *Plants for Toxicity Assessment, ASTM STP 1091,* W. Wang, J. W. Gorsuch, and W. R. Lower, Eds., American Society for Testing and Materials, Philadelphia, 1990, pp. 255–266.

ABSTRACT: Concurrent 12-week laboratory and field studies were conducted to determine toxicity of the suspended particulate phase (SPP) of drilling fluid to *Thalassia testudinum* and its epiphytes. Test systems were treated once per week to achieve nominal concentrations of 100 mg/L SPP. Chlorophyll content of *Thalassia* leaves and epiphyte biomass and chlorophyll content were monitored during each test. Laboratory exposures were conducted in 7-L, flow-through (7 L/h) microcosms consisting of Plexiglas cylinders containing intact cores of *Thalassia* from a local seagrass bed. Field exposures were conducted in water-tight Plexiglas chambers (2 by 2 by 1.5 m) placed over test plots in a seagrass bed for 24 h during SPP additions. The chamber base was buried several centimetres into the sediment to minimize water exchange. Drilling fluid exposure had no significant effect on chlorophyll a or b content of *Thalassia* leaves in laboratory or field tests. Epiphyte biomass was reduced after 6 weeks of intermittent exposure to SPP in laboratory and field tests. After 12 weeks, epiphyte biomass had increased to densities similar to control values.

Tributyltin chloride (TBT-Cl) was tested only in laboratory systems, using weekly treatments for 6 weeks. Nominal test concentrations ranged from 0.2 to 50 µg/L. Leaf protein and rhizome carbohydrate content of *Thalassia* were employed as effect measures in the TBT-Cl test. Leaf concentrations of chlorophyll a and b were not affected by exposure to TBT-Cl at nominal concentrations $\leq$ 50 µg/L. Leaf protein and rhizome carbohydrate concentrations were reduced by exposure to 50 µg/L TBT-Cl. Epiphyte biomass was reduced after exposure to 50 µg/L TBT-Cl for 6 weeks; concentrations $\leq$ 20 µg/L had no effect on epiphyte biomass.

KEY WORDS: *Thalassia,* epiphytes, drilling fluid, tributyltin

Seagrass beds have long been recognized as important contributors to the productivity of coastal waters. These areas are highly productive, faunally rich, and ecologically important habitats which are found worldwide [1]. Seagrasses offer shelter to juvenile forms of commercially important species of fish and shellfish, while leaf epiphytes are a source of food. Seagrass may also be grazed upon by herbivores such as parrotfish, sea turtles, sea urchins, and pinfish. In many areas seagrasses form the base of the food chain as energy from dead seagrasses is made available through decomposition.

Seagrasses are sensitive to different types of pollution. Thermal pollution from electrical generating plants adversely affects or kills *Thalassia* adjacent to the discharge [2]. Elevated turbidity and color from a polluted river drainage system severely reduced the biomass of

---

[1] U.S. EPA, Environmental Research Laboratory, Gulf Breeze, FL 32561.
[2] University of West Florida, Pensacola, FL 32561.

benthic macrophytes in an area of Apalachee Bay, Florida [3]. The herbicide atrazine and the preservative, pentachlorophenol, both depress photosynthesis/respiration ratios of *Thalassia* at concentrations of 1 ppm [4]. Two pollutants of recent concern, drilling fluids and tributyltin compounds, have been studied at our laboratory. The purpose of this effort was to determine possible effects of these pollutants on seagrasses and their epiphytes.

Most drilling fluids used in marine operations are water-based and usually consist of bentonite clays, barium sulfate as a weighting agent, lignosulfonates as thinners, and specialty agents for controlling pH and bacteria [5]. Diesel oil is commonly used as a lubricating agent [5]. All cause concern when disposed of in the marine environment. Drilling fluids discharged from a drill site may be carried by currents to shallow areas that support a seagrass community. Such discharges can reduce the light available for photosynthesis and possibly exert a toxic effect on epiphytes or associated organisms.

Tributyltin (TBT) is used as a wood preservative, disinfectant, and biocide [6]. It was used first as a pesticide in 1925 to control moths. It is a common ingredient in antifouling paints applied to the hulls of ships. As it slowly leaches out of its binding material, TBT eliminates fouling organisms. Thus, leaching of TBT from boat hulls is a source of exposure for aquatic and marine organisms, including seagrass communities in near coastal waters.

This report summarizes several studies conducted at the U.S. EPA Environmental Research Laboratory in Gulf Breeze, Florida (ERL/GB). It is meant to compare effects of drilling fluids in laboratory studies with experimental seagrass community dominated by *Thalassia testudinum* with those measured in the field, and to gain information on effects of TBT on an experimental seagrass community.

Materials and Methods

The effects of drilling fluids were tested in laboratory seagrass-community microcosms and in acrylic chambers placed in seagrass beds. Effects of tributyltin chloride were tested only in laboratory seagrass-community microcosm system. Treatments were based on nominal exposure concentrations of drilling fluid or TBT-Cl as determined from volume-to-volume dilutions of test material.

Laboratory Microcosms

A detailed description of the sample collection and laboratory testing system is given by Morton et al. [5]; only salient features are highlighted here. A *Thalassia testudinum* community located in Santa Rosa Sound approximately 3.0 km east of ERL/GB was the source of test material for microcosms. A 14.0 cm (ID) core was taken to a depth of approximately 10 cm. Plant material in the core was placed intact in a clear, acrylic cylinder (15.9 cm ID, 50 cm height) and transported to the laboratory for testing. The microcosms contained 10 cm of sediment with 7 L of water.

Microcosms were maintained in flowing seawater pumped from Santa Rosa Sound to each cylinder at the rate of 7 L/h. Theoretically, a 95% water exchange occurred every 3 h in the test cylinders. Four 400-W, halogen-halide lamps in the laboratory test system were supplemented with high-intensity fluorescent bulbs (GE Power Groove) rated at 250 W. Lights were suspended 46 cm above the test cylinders on a 12:12 photo/period and yielded an average rate of photon flux of 225 $\mu E/cm^2/s$ to the water surface of each cylinder, as measured by a LiCor LI 188B photometer. The light intensity was 35% of the 650 $\mu E/cm^2/s$ measured at noon during a sunny, summer day in the field plot at a depth of 1.25 m.

Field Chambers

A clear Plexiglas chamber (2 by 2 by 1.5 m) was used for *in situ* exposure of plants to drilling fluid.[3] This chamber was placed over the seagrasses and received test material from the surface through a 15-m garden hose attached to a fitting on the top of the chamber. A diffuser inside the top of the chamber distributed test material evenly. Material for the exposure was premixed in an 18 L bucket and pumped into the chamber with a small submersible pump. Water displaced from insider the chamber exited under the bottom edge of the chamber.

Experimental Design

Drilling Fluid—Tests were conducted with the suspended particulate phase (SPP) of a generic drilling fluid. The SPP is that portion of 1:10 dilution of a drilling fluid which remains in suspension 1 h after mixing [7]. The laboratory test system consisted of 48 microcosms (24 with drilling fluids, 24 control), to which concentrations were assigned randomly. Microcosms treated with drilling fluid were exposed to a nominal SPP concentration of 100 mg/L for 24 h weekly for 12 weeks to simulate a typical discharge pattern from a drilling rig. This was done by halting the water flow to the microcosms for 24 h after the addition of drilling fluid. Identical concentrations and durations were used for field exposures. The field site consisted of 9 plots: 3 reference, 3 control, and 3 drilling fluid. Reference plots were marked, unenclosed seagrass areas having the same dimensions as the test chambers. Control and drilling fluid exposures were performed in chambers; control chambers received natural seawater and the drilling fluid chambers received enough SPP to achieve a concentration of 100 mg/L. Chambers were only present during the 24-h exposure period. After exposure, chambers were removed and the sites opened to natural water movement.

One-half of the laboratory microcosms and one-half of each field plot were sampled at 6 weeks; the remaining microcosms and the other half of each field plot were sampled at 12 weeks. One *Thalassia* leaf with associated epiphytes was randomly taken for analysis of biomass, and chlorophyll from each laboratory microcosm and each of five field cores was randomly collected from each plot. Similarly one intact *Thalassia* plant was sampled for determination of protein and carbohydrate content. This produced 12 replicates per treatment for the laboratory test and 15 replicates per treatment from the field. These sample sizes were chosen based on variances measured from previous unpublished data.

Tributyltin Chloride—In Test I, 12 replicate microcosms per treatment for a total of 48 microcosms were exposed to concentrations of 0.0, 0.5, 5.0, and 50.0 µg/L of tributyltin chloride (TBT-Cl). Test II consisted of 16 replicates per treatment with test concentrations of 0.0, 0.2, 2.0, and 20.0 µg/L TBT-Cl, for a total of 64 microcosms. All treatments were randomly assigned to the test system.

Laboratory tests with TBT-Cl continued for 6 weeks. Half of the replicates in Test II were terminated at 3 weeks, the remainder at 6 weeks. At Test I termination, 12 *Thalassia* leaves with associated epiphytes were taken from each treatment group (1 per microcosm) for analysis of biomass and chlorophyll content, along with 12 *Thalassia* plants for protein and carbohydrate analysis. One-half of Test II (32 microcosms) was harvested at 3 weeks; two leaf blades with associated epiphytes and one *Thalassia* plant were collected from each microcosm for a total of 16 per treatment. The leaf samples were analyzed for epiphyte biomass and chlorophyll content, and *Thalassia* leaf chlorophyll content, whereas the whole plant

---

[3] Mention of trade names or commercial products does not constitute endorsement or recommendation for use.

was used for determination of leaf protein and rhizome carbohydrate. The remaining microcosms in Test II were harvested after 6 weeks and identical sampling and analysis performed.

Thalassia

Epiphytic material was removed from each side of *Thalassia* leaves with a razor blade and reserved for analysis. The length and width of *Thalassia* leaves were measured to the nearest millimetre to calculate leaf area, and the product was doubled to account for both sides of the leaf. The leaf was then placed into 20 mL of 90% aqueous spectrophotometric grade acetone and macerated with a Polytron tissue grinder. Chlorophyll was extracted from the macerated tissue for 24 h in the dark at 10°C; the acetone was then poured off and saved, another 20 mL of 90% acetone added, and extraction continued for another 24 h. Both 20 mL portions were combined and the sample filtered with a tared glass fiber filter (Gelman Type A/E). Filters were dried for 24 h at 100°C and reweighed to determine leaf tissue dry weight. Acetone extracts were adjusted to 40 mL with 90% acetone to account for volatilization, and optical densities were determined spectrophotometrically at 750, 663, 645, and 620 nm.

Chlorophyll a and b were quantified using the SCOR/UNESCO equations [8] and chlorophyll concentrations calculated per gram dry weight of leaf material. Whole *Thalassia* plants were separated into leaf and rhizome components, lyophilized, and ground in a Wiley tissue mill. Leaf protein and rhizome carbohydrate concentrations were determined by the Lowry procedure and carbohydrate analysis as described by Dawes [9]. Values are reported as milligram per gram of dry tissue.

Epiphytes

Epiphytic material which had been previously removed from *Thalassia* leaves were extracted in 5 mL of 90% spectrophotometric grade acetone in the dark for 48 h at 10°C. Sample material was then collected on tared glass fiber filters, dried at 100°C, and weighed. After drying, the filters were combusted at 500°C for 1 h, allowed to cool in a desiccator, and reweighed to determine epiphyte ash-free dry weight (AFDW). Epiphyte standing crop was estimated as AFDW per cm^2, calculated from the area of leaf tissue sampled.

After filtering, the volume of acetone in the chlorophyll extract was adjusted to 5 mL with 90% acetone to correct for volatilization of acetone during processing. After optical density of the chlorophyll extract was determined at 750 and 663 nm, a second extinction followed the addition of 0.1 mL $1N$ HCl to the spectrophotometric cuvette. Phaeophytin-corrected chlorophyll a was calculated [10] and expressed as μg per cm^2 of *Thalassia* leaf surface and per gram AFDW of epiphyte material.

Data Analysis

Means and standard deviations were computed for measurements from each treatment-time sampling. Data from each sample period of each experiment were treated independently. A one-way analysis of variance (ANOVA) was performed on the means to test for treatment effects. If the ANOVA was significant, the data was analyzed using Duncan's multiple range comparison [11]. Tests for significant differences were conducted at $\alpha = 0.05$.

Results

Thalassia

Drilling Fluid—No significant effect was detected for any parameter studied for *Thalassia* plants exposed intermittently to 100 mg drilling fluid SPP per litre for 6 or 12 weeks in laboratory microcosms (Fig. 1). There was a significant increase from 6 to 12 weeks in chlorophyll *a* per mg dry weight in the control plants. There also was a decrease in rhizome carbohydrate content for both the control plants and those plants exposed to drilling fluid (Fig. 1).

In situ Thalassia plants exposed intermittently to a nominal concentration of 100 mg/L drilling fluid SPP for 12 weeks also showed no significant treatment effects. However, all plants in the field showed a significant decrease in chlorophyll *a*/mg dry wt between weeks 6 to 12 (Fig. 1).

Tributyltin Chloride (Test I)—*Thalassia* chlorophyll content was not significantly affected by weekly exposures to a nominal concentration ≤ 50 µg/L of TBT-Cl for 6 weeks. Leaf

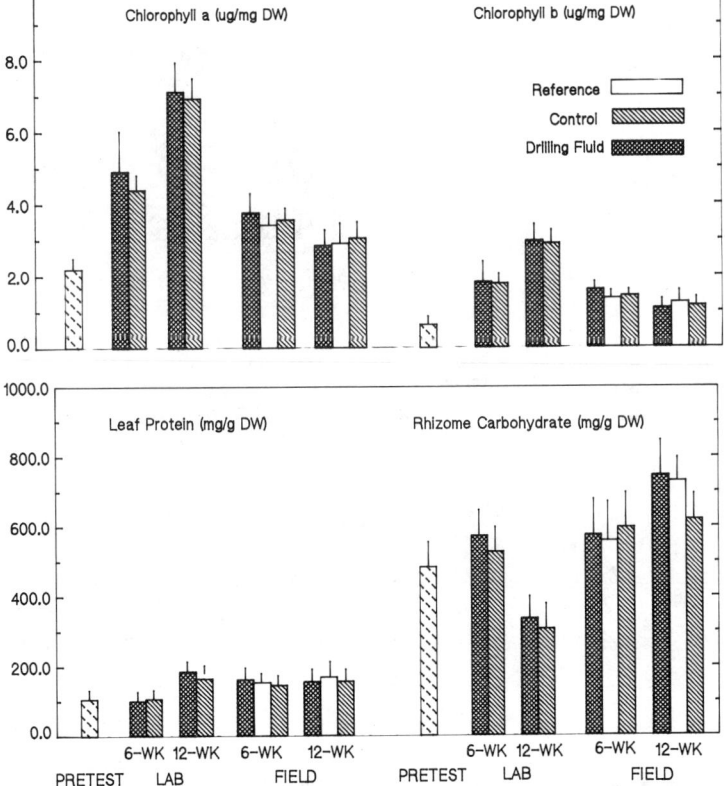

FIG. 1—*Mean values and 95% confidence intervals* (vertical bars) *for chlorophyll* a *and* b *content and the amount of leaf protein and rhizome carbohydrate in* T. testudinum *exposed weekly to 100 mg (SPP) drilling fluid. Exposures were conducted in laboratory microcosms or in field chambers for 12 weeks. The results of samples taken after 6 weeks of exposure also are shown.*

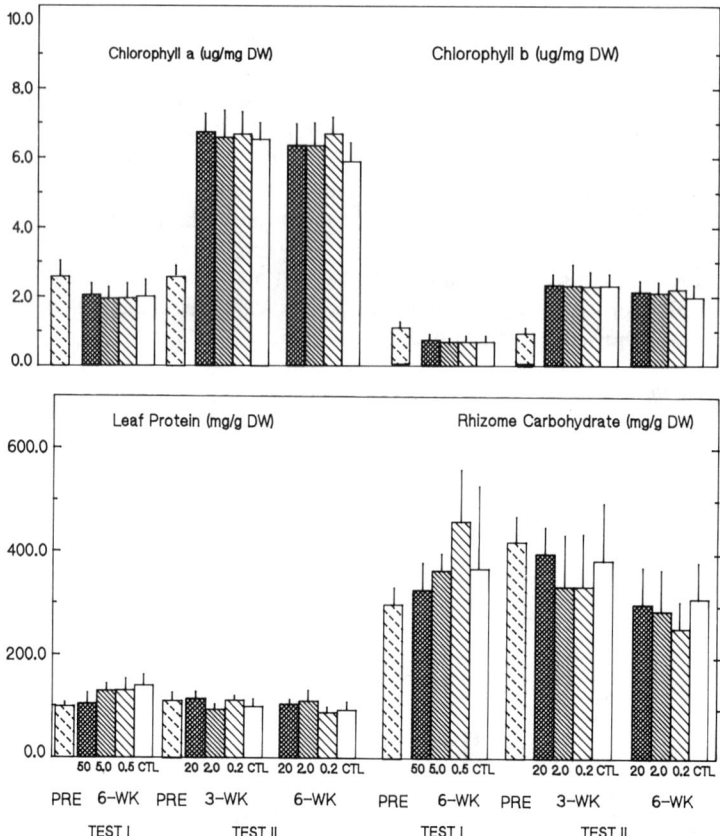

FIG. 2—*Mean values and 95% confidence intervals (vertical bars) for chlorophyll a and b content and the amount of leaf protein and rhizome carbohydrate in T. testudinum after exposure to tributyltin chloride (TBT-Cl). Two separate tests are illustrated, both of 6 weeks' duration. In Test I samples were only taken after 6 weeks of exposure to 0 (CTL), 0.5, 5.0, and 50.0 µg/L TBT-Cl. In Test II samples were taken after 3 weeks and again after 6 weeks of exposure to 0 (CTL), 0.2, 2.0, and 20.0 µg/L TBT-Cl.*

protein was significantly less in the 50 µg/L treatment when compared to the lower exposure concentrations and controls (Fig. 2).

Tributyltin Chloride (Test II)—*Thalassia* chlorophyll was not significantly affected by weekly exposures ≤ 20 µg/L for 3 and 6 week periods. No change was measured in the amount of carbohydrate stored in the rhizomes or protein in the leaves of the plants (Fig. 2).

Epiphytes

Drilling Fluid—When compared to the controls, exposure to 100 mg/L drilling fluid once a week for 6 weeks in laboratory microcosms significantly decreased the amount of chlorophyll *a* per mg of epiphyte tissue (µg chlorophyll *a*/mg AFDW) (Fig. 3). After 12 weeks of intermittent exposure to drilling fluid in the laboratory, chlorophyll *a*/mg AFDW of epiphytes remained less than controls.

In the field there was no significant difference measured between control plants and plants intermittently exposed to drilling fluids. When compared to reference plot values, the

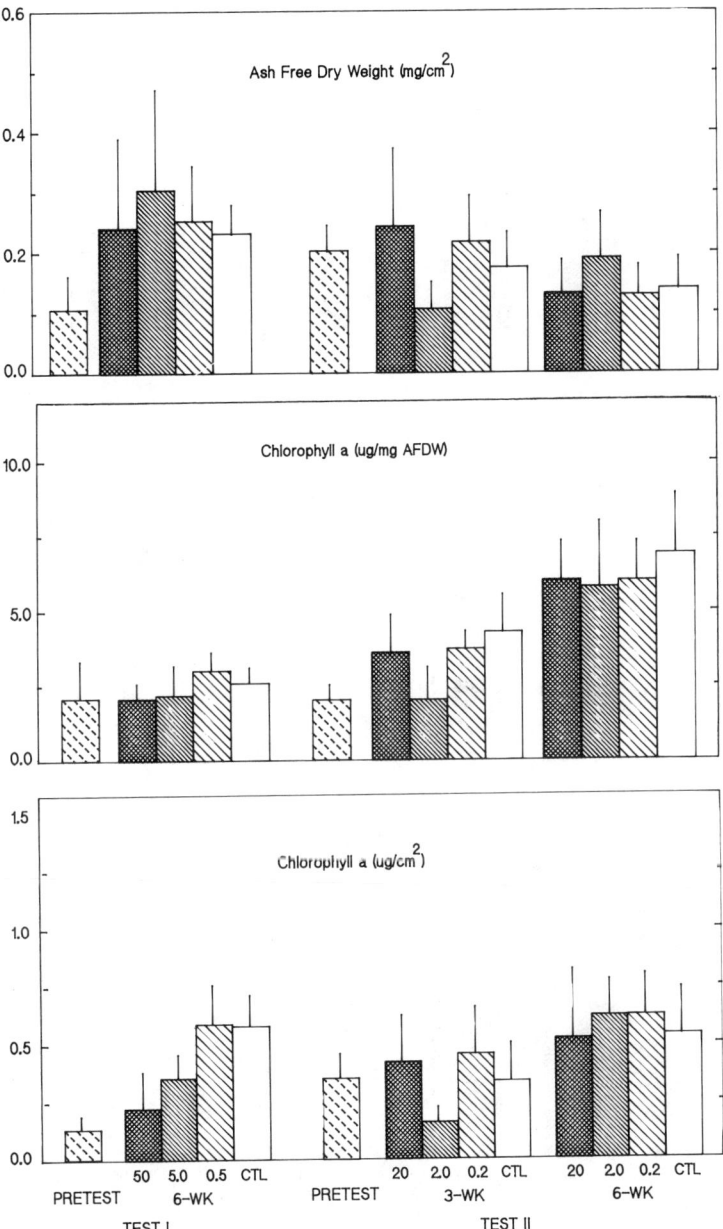

FIG. 3—*Means and 95% confidence intervals* (vertical bars) *for ash-free dry weight (AFDW), chlorophyll* a *content, and standing crop (μg chlorophyll* a*/mg AFDW) of epiphytes attached to the leaves of* T. testudinum. *Results illustrated are after 6 and 12 weeks of intermittent exposure to 100 mg/L (SPP) drilling fluid.*

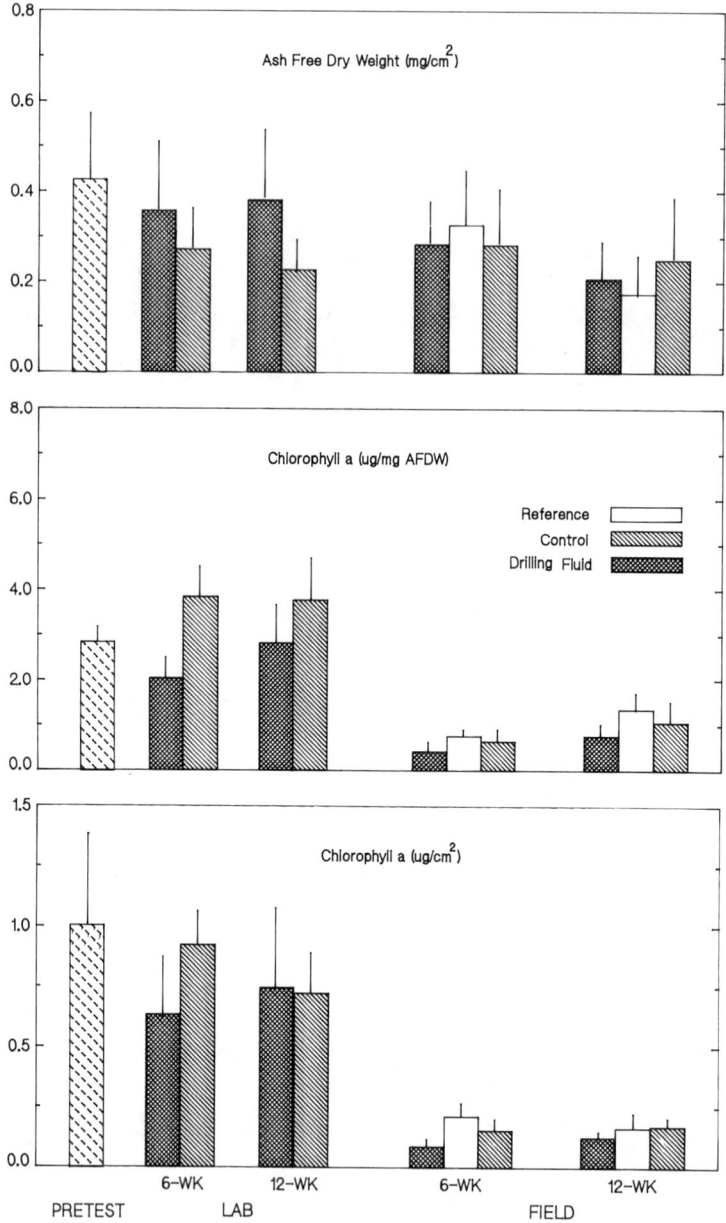

FIG. 4—*Mean and 95% confidence intervals* (vertical bars) *for ash-free dry weight (AFDW), chlorophyll* a *content, and standing crop (μg chlorophyll* a/mg AFDW) *of epiphytes on the leaves of* T. testudinum *exposed to tributyltin chloride (TBT-Cl). Two separate tests of 6 weeks' duration are illustrated. In Test I samples were taken after 6 weeks of exposure to 0 (CTL), 0.5, 5.0, and 50.0 μg/L TBT-Cl. In Test II samples were taken after 3 weeks and 6 weeks of exposure to 0 (CTL), 0.2, 2.0, and 20.0 μg/L TBT-Cl.*

amounts of epiphyte chlorophyll a/cm^2 and epiphyte chlorophyll a/mg AFDW were significantly reduced by intermittent exposure for 6 and 12 weeks to the drilling fluid. No significant difference was measured in the AFDW of epiphytes on *Thalassia* leaves exposed in the laboratory or the field.

Tributyltin Chloride (Test I)—Epiphytes exposed to 50 μg/L TBT-Cl for 6 weeks in laboratory microcosms had significantly less chlorophyll a/cm^2 *Thalassia* leaf area than controls (Fig. 4). Epiphyte chlorophyll a/mg AFDW followed a similar pattern, but the difference was not significant.

Tributyltin Chloride (Test II)—After exposure for 6 weeks to concentrations ≤ 20 μg/L of TBT-Cl, there were no significant effects on the AFDW or chlorophyll content of epiphytes on *Thalassia* leaves (Fig. 4). However, biomass and chlorophyll a concentrations, both per mg/AFDW and per cm^2 *Thalassia* leaf, had a significant increase overall from 3 to 6 weeks during the exposure period.

Discussion

Thalassia

Drilling Fluid—We detected no effect on *Thalassia testudinum* in laboratory microcosms or field enclosures from weekly exposure to 100 mg/L of drilling fluid SPP. Price et al. [*12*] reported that 6 weeks of continuous exposure to drilling fluids at concentrations ranging from 11 to 38 mg/L decreased *Thalassia* chlorophyll content in laboratory tests. However, because their tests were conducted at much lower light intensities than our tests (26 μE/cm^2/s versus 225 μE/cm^2/s), drilling fluid effects could not be differentiated from laboratory light effects. For treated and control groups, we observed an increase in chlorophyll content from 6 to 12 weeks. Barko [*13*] stated that chlorophyll in submersed macrophytes is inversely related to irradiance above that which is minimally required for tissue maintenance. The increase in chlorophyll a content measured in our laboratory from 6 to 12 weeks was a shading response of *Thalassia* plants to reduced light intensities compared to field irradiance. Similar higher chlorophyll content in response to low light intensity has been reported [*14,15*]. In the field, there was a concurrent, but lesser, decrease in chlorophyll content in all treatments from 6 to 12 weeks. This decrease was measured in July during the period of maximum temperature stress for *Thalassia*. This stress may account for reduction in chlorophyll content measured in plants exposed in the field, since it coincided with a seasonal reduction in *Thalassia* chlorophyll [*15*]. Seasonality of *Thalassia* is an inherent difficulty in conducting these studies. The baseline levels of chlorophyll and plant metabolites is different depending on the time of year the experiment is conducted. This is apparent in the difference in the amount of chlorophyll present in *Thalassia* leaves between TBT-Cl Test I and Test II (Fig. 2).

Increased chlorophyll content and utilization of carbohydrate reserves stored in rhizome tissues are signs of compensation for plants attempting to sustain growth under less than optimal light conditions [*9,14,15*]. These traits were present in control and treated *Thalassia* held in our laboratory microcosm system for 12 weeks during the drilling fluid tests. Thereafter we limited tests to 6 weeks' duration. Attempts to alter the design of our test system and increase artificial light intensity were thwarted by our inability to cope with a concomitant increase of heat from additional lights.

Tributyltin Chloride—The chlorophyll content of *Thalassia testudinum* was not significantly affected by exposures to concentrations ≤50 μg/L TBT-Cl. Leaf protein content was significantly reduced for 6-week exposure to 50 μg/L, whereas the 6-week exposure to 20 μg/L had no effect. Because leaves are generally the biosynthesis sites for seagrass, they contain large amounts of enzymes and membranes [*17*]. Therefore leaf structures of the plant contain the greatest amount of protein and were the focus of our analysis. In studies with terrestrial plants, such as pea *(Pisum sativum)* and spinach *(Spinacia oleracia)*, interference with chloroplast function has been observed after exposure to organotin compounds [18,19]. Similar effects could occur in *Thalassia*, thus accounting for the reduction in leaf protein.

Epiphytes

Drilling Fluid—Intermittent exposure to 100 mg/L drilling fluid SPP for 6 weeks significantly reduced the amount of chlorophyll in epiphytes in laboratory microcosms as measured by chlorophyll a/mg AFDW. Price et al. [12] reported reduced epiphyte biomass on *Thalassia* leaves exposed continuously for 6 weeks to drilling fluid SPP at an average concentration of 20.1 mg/L. Reduced light penetration and the toxicity of drilling fluids were cited as possible causes for this reduction. Results from our study for the laboratory system at Week 6 support the findings of Price et al.; however, we used considerably higher exposure concentrations and intermittent dosing. After 12 weeks of intermittent exposure the amount of chlorophyll was no longer significantly reduced. In the field there was no significant difference between control plant and plant exposed to the drilling fluid. When these values were compared to the reference plots the amount of chlorophyll in plants exposed to drilling fluid was significantly less. A possible explanation is that being placed in the enclosure affected the chlorophyll content and standing crop of epiphytes. This affect, enhanced by the addition of drilling fluid, produced a significant reduction in these values.

Tributyltin Chloride—Standing crop and chlorophyll content of epiphytes on the leaves of *Thalassia* plants were lowest in laboratory microcosms treated with 50 μg/L TBT-Cl. These results agree with reported data for single species of marine and freshwater algae and for field studies of populations of freshwater algae. Hopf et al. [20] reported that tributyl compounds are effective biocides against fouling organisms, including algae. Lethal concentrations of tributyltin chloride to *Skeletonema costatum* have been reported to be from 1 to 18 μg/L [21] to 11.5 μg/L [22]. Walsh et al. [22] also reported that cell death occurred in *Skeletonema* at 5.6 μg/L. Tributyltins were the most toxic group in their population growth studies with an average EC_{50} of 0.8 μg/L for *Skeletonema* and 1.19 μg/L for *Thalassiosira pseudonana*. Tributyl and triphenyl tin were the most effective algicides for *Scenedesmus quadricauda*, *Ankistrodesmus falcatus*, and *Anabaena flos-aquae* [23]. EC_{50}'s for reproduction inhibition occurred at concentrations of 2.0 to 5.0 μg/L along with 50% inhibition of primary productivity at 0.01 to 0.04 mg/L [23]. Also, Wong [23] tested natural populations of phytoplankton from Lake Ontario and found primary productivity inhibited by 50% at exposures to 2.0 to 3.0 μg/L of tributyltin. Our actual test concentrations probably were much lower than the nominal values because of the tendency for TBT to sorb to organic and inorganic matter rather than stay in solution. Therefore our effect concentration of 50 μg/L (nominal) may be closer to the range of toxic concentrations reported from single-species laboratory tests.

Summary

Intermittent exposure to a concentration of 100 mg/L of drilling fluid SPP had no effect on *Thalassia testudinum*. The only observed effect of increased leaf chlorophyll content was probably due to a shading response because of reduced light intensity in the laboratory test system compared to field light intensity [24]. After 6 weeks' exposure to 50 μg/L tributyltin, the protein content of the *Thalassia* leaves was reduced. A drilling fluid concentration of 100 mg/L significantly reduced the chlorophyll content of *Thalassia* epiphytes after 6 weeks of intermittent exposure, but the difference was no longer significant after 12 weeks. Six weekly exposures to 50 μg/L tributyltin reduced the standing crop and chlorophyll content of epiphytes.

The *Thalassia* plant appeared to be hardy and relatively insensitive to the toxicants we studied. Lack of sensitivity during our tests does not necessarily reflect on its sensitivity to other compounds or the potential to assess chronic effects. Because seagrasses form the base of many productive estuarine systems, it is important to develop and implement such test systems to assess potential impacts on these plants. Such testing represents a more realistic evaluation of potential impacts in the field than does simple laboratory testing. Epiphytes on the leaves of *Thalassia* plants were sensitive to both toxicants tested. Because epiphytes are important to a number of grazer organisms, their loss could have adverse effects on food resources in seagrass communities. More research on combined studies of seagrasses and epiphytes should be performed so that these studies may be incorporated into risk assessments for chemical releases into estuarine environments.

References

[1] Zieman, J. C., "The Ecology of the Seagrasses of South Florida: A Community Profile," FWS/OBS-82/25, U.S. Fish and Wildlife Service, Office of Biological Services, Washington, D.C., 1982.

[2] Zieman, J. C., "Tropical Seagrass Ecosystems and Pollution," Chapter 4 in *Tropical Marine Pollution*, E. J. F. Wood and R. E. Johannes, Eds., Elsevier Oceanography Series 12, Elsevier, New York, 1975.

[3] Zimmerman, M. S. and Livingston, R. J., "Effects of Kraft-Mill Effluents on Benthic Macrophyte Assemblages in a Shallow-Bay System (Apalachee Bay, North Florida, U.S.A.)," *Bulletin of Marine Science*, Vol. 29, No. 1, 1979, pp. 27–40.

[4] Walsh, G. E., Hansen, D. L., and Lawrence, D. A., "A Flow-Through System for Exposure of Seagrass to Pollution," *Marine Environmental Research*, Vol. 7, 1982, pp. 1–11.

[5] Morton, R. D., Duke, T. W., Macauley, J. M., Clark, J. R., Price, W. A., Hendricks, S. J., Owsley-Montgomery, S. L., and Plaia, G. R., "Impact of Drilling Fluids on Seagrasses: An Experimental Community Approach," in *Community Toxicity Testing, ASTM STP 920*, J. Cairns, Jr., Ed., ASTM, Philadelphia, 1986, pp. 199–212.

[6] Champ, M. A., Ed., *Organotin Symposium*, Vol. 4, Proceedings of Oceans '86, IEEE, New York, 1986.

[7] Petrazoullo, G., "Proposed Methodology: Drilling Fluids Toxicity Tests for Offshore Subcategory; Oil and Gas Industry," Technical Resources Inc., Rockville, Md., 1983, 45 pp.

[8] Strickland, J. D. H. and Parsons, T. R., *A Practical Handbook of Seawater Analysis*, 2nd ed., Bulletin 167, Fish. Res. Bd. Can. Ottawa, Canada, 1972, 310 pp.

[9] Dawes, C. J., *Marine Botany*, Wiley, New York, 1981.

[10] Weber, C. I., "Biological Field and Laboratory Methods for Measuring the Quality of Surface Waters and Effluents," U.S. EPA 670/4-73-001, Washington, D.C., 1973.

[11] *SAS User's Guide: Statistics*, SAS Institute, Cary, N.C., 1982.

[12] Price, W. A., Macauley, J. M., and Clark, J. R., "Effects of Drilling Fluids on *Thalassia testudinum* and Its Epiphytic Algae," *Experimental and Environmental Botany*, Vol. 26, No. 4, 1986, pp. 321–330.

[13] Barko, J. W., "Influences of Light and Temperature on Chlorophyll Composition in Submersed Freshwater Macrophytes," *Aquatic Botany*, Vol. 15, 1983, pp. 249–255.

[14] Furguson, R. L., Thayer, G. W., and Rie, R. R., "Marine Primary Producers," in *Functional Adaptations of Marine Organisms*, J. Smith, Ed., Academic Press, Ann Arbor, Mich., 1980, pp. 9–69.

[15] Wiginton, J. R. and McMillan, C., "Chlorophyll Composition under Controlled Light Conditions as Related to the Distribution of Seagrasses in Texas and the U.S. Virgin Islands," *Aquatic Botany*, Vol. 6, 1979, pp. 171–184.

[16] Macauley, J. M., Clark, J. R., and Price, W. A., "Seasonal Cycle of *Thalassia testudinum* and Its Epiphytes in Northwest Florida," *Aquatic Botany*, Vol. 31, 1988, pp. 277–287.

[17] Walsh, G. E., and Grow, T. W., "Composition of *Thalassia testudinum* and *Ruppia maritima*," *Journal of the Florida Academy of Science*, Vol. 35, No. 2, 1973, pp. 97–108.

[18] Kahn, J. S., "Chloro-*n*-butyltin: An Inhibitor of Photophosphorylation in Isolated Chloroplasts," *Biochimica et Biophysica Acta*, Vol. 153, No. 1, 1968, pp. 203–210.

[19] Gould, J. M., "Inhibition by Triphenyltin Chloride of a Tightly-Bound Membrane Component Involved in Photophosphorylation," *Env. J. Biochem.,* Vol. 62, No. 3, 1976, pp. 567–576.
[20] Hopf, H. S., Duncan, J., Beesley, J. S. S., Webley, D. J., and Sturrock, R. F., "Molluscidal Properties of Organotin and Organolead Compounds with Particular Reference to Triphenyllead Acetate," *Bulletin of W. H. O.,* Vol. 36, No. 6, pp. 955–961.
[21] Thain, J. E., "The Acute Toxicity of Bis(Tributyltin) Oxide to the Adults and Larvae of Some Marine Organisms," Int. Counc. Explor. Sea, Mar. Environ. Qual. Comm., CM 1983/E:13, 1983, 5 pp.
[22] Walsh, G. E., McLaughlan, L. L., Lores, E. M., Louie, M. K., and Deans, C.H., "Effects of Organotins on Growth and Survival of Two Marine Diatoms, *Skeletonema costatum* and *Thalassiosira pseudonana,*" *Chemosphere,* Vol. 14, No. 3/4, 1985, pp. 383–392.
[23] Wong, P. T. S., Chau, Y. K., Kramer, O., and Bengert, G. A., "Structure-Toxicity Relationship of Tin Compounds on Algae," *Can. J. Fish. Aquat. Sci.,* Vol. 39, 1982, pp. 483–488.
[24] Odum, E. P., *Fundamentals of Ecology,* W. B. Saunders, Philadelphia, 1971, 544 pp.

Eric L. Morgan,[1] *Yueh-Chin A. Wu,*[1] *and Richard C. Young*[2]

A Plant Toxicity Test with the Moss *Physcomitrella patens* (Hedw.) B.S.G.

REFERENCE: Morgan, E. L., Wu, Y.-C. A., and Young, R. C., "**A Plant Toxicity Test with the Moss *Physcomitrella patens* (Hedw.) B.S.G.,**" *Plants for Toxicity Assessment, ASTM STP 1091,* W. Wang, J. W. Gorsuch, and W. R. Lower, Eds., American Society for Testing and Materials, Philadelphia, 1990, pp. 267–279.

ABSTRACT: A method for detecting toxicity to a representative primary producer was developed and tested using the moss *Physcomitrella patens* (Hedw.) B.S.G. Cultures of the moss were maintained on solid minimal medium in a growth chamber with a day length of 17 h and a temperature of 23 ± 1°C. Two stages of the moss life cycle (spore and gametophore) were cultured for five weeks in the laboratory while being exposed to various toxic treatments.

Cultures were exposed separately to six different salt solutions (aluminum sulfate, barium chloride, boric acid, cadmium chloride, cobalt chloride, and lead nitrate), a mineralized-acidic leachate, and a coal combustion fly ash leachate. After five weeks' exposure, morphological changes, dry weight, chlorophylls *a* and *b*, and total chlorophyll measurements were used as criteria to reflect dose-effect relationships.

Regenerating clones were frequently morphologically aberrant, and no regenerants were observed at higher concentrations of most treatments. Increasing toxicant concentrations generally reduced the dry weight and chlorophyll contents of moss cultures ($P \leq 0.05$). No significant differences in response variables between spore and gametophore cultures were found, and dose-linked changes in fresh weight were similar to those for dry weight.

Cadium chloride and aluminum sulfate treatments had the most negative influence on moss dry weight and chlorophyll content; boric acid and barium chloride concentrations were the least toxic. The fly ash leachate did not appear to be toxic, possibly due to antagonistic chemical interactions in the fly ash/limestone scrubber leachate treatments.

KEY WORDS: plant toxicity test, moss *(Physcomitrella patens)* cultures, chlorophyll and standing crop (dry weight), phytotoxicity, metals, leachates

Various plant systems have been suggested for use in toxicity testing, including algal assays [1–5] for quantitative toxicity measurements and the radish and sorghum seedling germination test [6] for assessing the toxicity of solid waste leachate. The U.S. Environmental Protection Agency (EPA) Environmental Assessment Manual [7] lists two primary producer systems in routine use, the spiderwort *Tradescantia* and the maize waxy locus system. These autotrophic systems require large mature plants and cannot be treated by routine short-term microbiological methods. It has been proposed [7,8] that additional short-term test systems be developed, because existing tests have not been sensitive to the complex mixtures of chemicals that are often encountered in runoff from managed watersheds and agricultural areas.

Chemical analysis of tissues of submerged aquatic plants may provide valuable information about the extent of contamination of surrounding waters [9,10]. Other studies have clearly shown the potential value of several bryophytes for monitoring heavy metal contam-

---
[1] Center for the Management, Utilization and Protection of Water Resources and the Department of Biology, Tennessee Technological University, Cookeville, TN 38505.
[2] Young-Morgan & Associates, Inc., Franklin, TN 37064.

inants in a riverine environment [11,12]. The result of more recent studies suggests that the moss *Physcomitrella patens* could be suitable experimental material for biochemical, genetic, and developmental studies [13–16]. Mutants of *P. patens* have been obtained by exposing the moss to ionizing radiation or to chemical mutagens [16,17]. *P. patens* is known to complete its life cycle *in vitro* in two months [15], is haploid, and can be easily cultured in small test chambers. *P. patens* also produces many uninucleate, haploid spores and the formation of heterozygous diploid gametophytes by apospory, which are important characteristics to biochemical and genetic studies [16]. These attributes have favored the use of this moss in various laboratory studies.

The possibility of using the moss *P. patens* in toxicity assays has received little attention. Thus the objective of this research was to develop and evaluate a toxicity screening test that uses a primary producer, the moss *P. patens*. In meeting the objective of this study, emphasis was placed on the following: (1) development of a simplified moss culture procedure tailored to meet the requirements for toxic screening tests under routine laboratory conditions, (2) carrying out preliminary toxicity screening tests employing moss cultures subjected to a variety of waste substances, (3) identifying which culture stage of the moss (spore or gametophore) is more sensitive to toxicants, (4) selecting specific morphological and functional criteria that reflect moss dose-response relationships, (5) evaluating the practicality and applicability of moss culture toxicity screening tests, and (6) evaluating the reliability and cost effectiveness of the test system.

The moss toxicity test method was evaluated using a series of salts and two leachate solutions. Reagent grade salts were used to prepare the test solutions. One leachate solution was derived from a mineralized rock formation (Anakeesta) that is known to influence stream acidification processes in the Southern Appalachians. The second leachate solution was taken from fly ash sludges resulting from a coal combustion-limestone scrubber operation. The potential toxicants selected for testing were chosen not for identifying specific response limits or EC_{50} values, but were selected for the purpose of developing methodology and assessing the applicability of the moss toxicity assay.

Materials and Methods

Physcomitrella patens (Hedw.) B.S.G. was collected originally from a watershed in Gransden Wood, Huntingdonshire, England, by H. L. K. Whitehouse. Two subcultures of the strain used by Engel [16] were supplied to the authors by Dr. Paula Nakosteen, Department of Botany, University of Tennessee, Knoxville. Cultures were established from spores due to the difficulty in decontaminating moss gametophytic tissues. The two subcultures were incubated on solid minimal medium (Knop's salts, Nitches' minor elements, 10 mg/L ferric citrate, agar) under a 17 h day length at $23 \pm 1°C$ in a modified temperature-controlled algal growth chamber (Table 1). Light was supplied by cool white fluorescent bulbs positioned 35 cm from cultures with an intensity of approximately 36 000 ergs/cm^2 [14].

Establishment of Moss Cultures for Tests

Glassware and instruments used for transfer, blending, and preparing culture media were sterilized by using standard sterilization techniques or autoclaving. The following procedures were followed to obtain large numbers of inocula and to process two different life stages of moss culture for experimental treatments:

1. *Gametophore (Plant) Cultures*—From the two stock cultures, 4 g of plant tissue, were removed with forceps and transferred to a sterile blender containing 20 mL sterile liquid

TABLE 1—*Recommended test conditions for the moss* Physcomitrella patens *toxicity assay.*

Item	Recommendation
Temperature (°C)	23–24 ± 1
Light intensity (erg/cm^2)	36 000
Light source	Cool white fluorescent
Photoperiod (h)	17 light–7 dark
Medium:	
Liquid	Knop's salts, Nitches minor elements and 10 mg/L ferric citrate, pH 6.5
Solid	*Ibid.* plus agar
Hydrogen ion level (pH)	3.5 to 8.5
Septic state	Axenic
Test vessel size (mL)	100 glass petri plates
Volume of culture medium (mL):	
Stock	100
Test	25
Inoculum density:	
Gametophore	40 mg tissue per mL medium
Spore	1000 to 2000 spores per mL medium
Duration (weeks):	
Life-cycle (spore to spore)	8
Test (gametophore/spore)	Screening test 1 to 3
	Definitive test 4 to 6
Response criteria	Dry weight
	Total chlorophyll

medium (Knop's salts, Nitches' minor elements, and ferric citrate 10 mg/L). The blender was capped and the moss and medium blended for approximately 50 s with the speed varying between low and high. The resulting slurry contained moss fragments of various sizes (usually four to five cells) and cell debris. This slurry was filtered through 125 μm mesh nitex cloth, and the remaining debris was washed with 50 to 100 mL of sterile liquid medium.

The filtrate was resuspended in 100-mL sterile liquid medium. A 1-mL volume of the resuspended filtrate was transferred to a petri plate that contained 25-mL solid minimal medium. This process was repeated until 100 culture plates had been inoculated.

2. *Spore Cultures*—Before this procedure was initiated, capsule production was promoted by lowering the temperature in the growth chamber to 17°C [*18*]. Cultures were flooded with 2 mL of sterile distilled water to help induce the sexual phase of the moss life cycle (formation of antheridia and archegonia). Once induced, fertilization occurred and capsules were formed on the apex of the gametophore within one week. Under a dissecting microscope, a non-dehisced (unopened) capsule was selected. The isolated capsule was surface sterilized in 70% ethanol before being placed in a commercial bleach solution for 5 min. On the inside wall of a small test tube that contained 10 mL of sterile medium, the capsule was broken by pressure with a dissecting needle to liberate the spores. The tube was capped and tipped to allow the liquid medium to disperse the spores. The resulting spore suspension contained approximately 1 to 2 × 10^3 spores per mL of medium. A 1-mL volume of the spore suspension was immediately transferred to a sterile petri plate containing 25-mL solid medium. A discussion of the preparation and composition of the solid-phase culture medium used is given by Steeres et al. [*19*].

Experimental Treatments

Preliminary screening tests were used to establish suitable effective concentration ranges for the eight potential toxicants selected for this study. Once completed, a second aliquot of each toxicant was added to spore and gametophore cultures plates and mixed thoroughly. One group of untreated plates was maintained as controls for comparison. The following procedures were followed in selecting and preparing the eight test solutions.

Leachates

The Anakeesta formation comprises part of the thick sequence of metasedimentary rock that outcrops throughout the Southern Appalachian Mountain region. Herrmann et al. [20] and King et al. [21] described the Anakeesta formation as a sulfide-rich pyritic and pyrrhotitic strata, which is the source of acid drainage in regional streams. Recent studies in the Southern Appalachians reveal that widespread damage has occurred in many mountain streams, eliminating trout populations [22] and severely disrupting benthic macroinvertebrate communities [23,24]. Aquatic macrophyte communities are important components of the stream ecosystem in the Southern Blue Ridge Province, but information about Anakeesta leachate effects on these communities is lacking. Characteristically, the concentration of mineral elements in Anakeesta leachates varies considerably [25]. For example, the pH and specific conductance values of a laboratory derived leachate solution, taken as 100% leachate after seven days leaching, have been found to reach 3.9 and 140 (μmhos/cm) [26]. This leachate also contained high concentrations of Al, Cd, Co, Cu, Fe, Mn, Pb, and Zn. After several months of stabilization, the author's 100% leachate solution had a pH 1.7, and 1% leachate solution had a conductance of 100 (μmhos/cm). The pH values of test treatments (in liquid culture medium less agar) ranged from 3.4 to 7.5 with controls maintained at 6.5.

The Anakeesta leachate test solution was prepared by placing Anakeesta rock (115 kg) in a plastic-lined stock tank containing approximately 1100 L dechlorinated tap water for several months. Then 50 mL of the solution were taken as the 100% test solution.

Coal Combustion Fly Ash Leachate

The increasing utilization of coal for generation of electricity coupled with the reduction in stack emission requirements achieved by advanced precipitation designs have correspondingly increased the levels of fly ash and limestone scrubber residue. Fine fractions of fly ash are composed of spherical glass-like particles, formed by the melting of silicate minerals during coal combustion [27]. Characteristic coal fly ash contains several compounds, such as FeS, $3Al_2O_2$, Fe_2O_3, FeO_4, CaO, MgO, $CaSO_4$, and unburned pyrite [28], as well as toxic trace elements that sometimes exceed the U.S. EPA [29] recommended limits for surface waters.

The wet limestone scrubber process or "flue gas desulfurizer" is an additional process that extracts greater than 90% of the sulfur oxides after electrostatic precipitation. The oxides are scrubbed from the flue gas by a limestone suspension. The additional residues produced from the process are flue gas desulfurization sludges, fluidized bed boiler waste, and bottom ash. The chemical composition of the wet limestone scrubber sludge is typically 20 to 40% fly ash, 6 to 22% $CaSO_4$-$2H_2O$, 18 to 49% $CaSO_3$-½ H_2O, 3 to 39% $CaCO_3$, 2 to 5% $MgSO_4$, and 7 to 20% total S [30]. The slurry washed from the scrubber apparatus is discharged into a basin for the settling of solids. Barium, boron, nickel, silver, and thallium levels were higher in scrubber pond sludge extract when compared to scrubber pond return. Published values for fly ash chemical composition support evidence that trace elements are concentrated to

TABLE 2—*Test treatments for the moss* P. patens *toxicity assay.*[a]

Test Substance	Life Stage Tested*	Number of Treatments	Concentration Range	
			Treatment	Hydronium Ion (pH)**
Anakeesta Leachate	S,G	5	1 to 100%	2.7 to 6.9
Fly Ash Leachate	S,G	5	1 to 100%	7.0 to 8.6
Aluminum Sulfate ($Al_2(SO_4)_3 \cdot 16H_2O$)	S,G	8	10^{-7} to $0.5\ M$	3.3 to 7.5
Barium Chloride ($BaCl_2 \cdot 2H_2O$)	S,G	4	10^{-6} to $1.0\ M$	6.1 to 7.2
Boric Acid (H_3BO_3)	S,G	8	10^{-7} to $0.5\ M$	5.9 to 6.8
Cadmium Chloride ($CdCl_2 \cdot 2\frac{1}{2}H_2O$)	S,G	5	10^{-6} to $0.25\ M$	N.A.
Cobalt Chloride ($CoCl_2 \cdot 6H_2O$)	S,G	5	10^{-6} to $0.01\ M$	6.4 to 6.9
Lead Nitrate ($Pb(NO_3)_2$)	S,G	6	10^{-6} to $0.1\ M$	3.1 to 6.8

[a] With a control group for both life stages.
* Spore (S) and gametophore (G).
** Values taken in liquid culture medium without agar.

higher levels in ash pond sludge as compared to ash pond effluent [31]. High trace element levels associated with fly ash samples and scrubber pond core extracts have been shown to be associated with trace element sorptive processes and precipitation in the ash and scrubber ponds [32].

The fly ash leachate test solution consisted of a 3:1 volumes mixture of dechlorinated tap water with fly ash sludge. The mixture was aerated with a paddle-type aerator for 24 h and the ash sediment allowed to settle for two days. The supernatant obtained after settling was taken as the 100% test solution. Test treatment pH values measured in liquid culture medium without agar ranged from 7.0 to 8.6.

Stock solutions of the six reagent-grade chemicals were prepared in distilled water. The experimental design is summarized in Table 2. Treated petri plates were sealed with parafilm, and cultures were maintained in the environmental growth chamber for five weeks under the same regime as stock cultures.

Toxic Response Criteria

Three different observations were made to establish relationships between treatment level and moss response:

1. *Morphological Observations*—After five weeks of incubation, growth patterns, and morphological characteristics of treated cultures were observed and recorded with the aid of a dissecting microscope. These observations included assessments of germination success and measurements of gametophore length.

2. *Standing Crop Measurement*—Treated culture plates were heated on a hot plate to melt the solid agar. Plant tissue was then collected on filter paper and transferred to an aluminum cup. The collected material was then dried in an oven to a constant weight of 105°C for 24 h. Before and after each sample was dried, the total weights were taken. The weight of the aluminum cup was subtracted to obtain the fresh weight and dry weight [33] of the sample moss.

3. *Chlorophyll Content Measurement*—After dry weights had been taken, the sample was placed in a tissue grinder and covered with 2 to 3 mL of 80% aqueous acetone solution. The sample was homogenized and filtered, and the grinder was rinsed with several mL of 80% aqueous acetone. This rinse was added to the extraction slurry. Total volume was adjusted to 50 mL by adding 80% aqueous acetone. The optical density of the extracts was measured spectrophotometrically at 645 and 663 μm, and the chlorophyll contents of the extracts were calculated by using equations given by Arnon [*34*].

Statistical Treatments

The Model I (fixed-effects) experimental design [*35*] using a two-way layout in stage (spore or gametophore culture) and treatment (test concentrations) was applied for evaluating the null hypothesis that a group of means were all equal and that observed differences resulted from random variation. The alternative hypothesis was that some of the stage or treatment means differ significantly. In the simplest form, with no replication and with each stage sampled once in each concentration, the model was represented as

$$X_{ij} = \mu + S_i + C_j + [(SC)_{ij} + \epsilon_{ij}]$$

where

X_{ij} = individual observations (wet weight, dry weight, chlorophyll-*a*, chlorophyll-*b*, total chlorophyll),
μ = population mean,
S_i = stage effect,
C_j = concentration effect,
$(SC)_{ij}$ = stage $\times$ concentration interaction, and
ϵ_{ij} = error.

The assumptions for the analysis of variance (ANOVA) of normality, addivity, and homogeneity of variance were imposed [*35*]. When no replication was included, the design provided no independent estimate of the within-cell variance (ϵ_{ij}). Thus we assumed that there was no $S \times C$ interactions or that $(SC)_{ij} = 0$. Once this assumption had been made, the extent of the S or C main effect was examined with respect to the interaction mean square, as an estimate of the error variance. The $S \times C$ interaction mean square consisted only of the error (ϵ) and was used to assess the added variability resulting from both stage and concentration. It was realized that by assuming no interaction, an F-test that showed nonsignificant main effects may be difficult to interpret; namely, whether differences between stage or concentration means were being observed by error or $S \times C$ interactions [*36*]. However, by recalculating the ANOVA while assuming the two stages were replicates, no major differences were observed in the resulting F-values compared to those derived with no replication. By assuming replication, a significant concentration effect was substantiated for all toxicants except fly ash. Most computations were performed using the Statistical Package for the Social Sciences (SPSS) [*37*].

Results and Discussion

No significant differences among treatment concentrations and response criteria were observed for tests done with fly ash leachate. However, other toxicant treatments did result in significant reductions in all parameters measured (Table 3). Typically, as the concentra-

TABLE 3—*ANOVA statistics for moss gametophore and spore stage response to different concentrations of eight toxic treatments.*

Source of Variation	Degrees of Freedom	F-Value				
		Fresh Wt.	Dry Wt.	Chlor-a	Chlor-b	Chlor Total
Anakeesta Leachate:						
Stage[a]	1	3.543	0.064	0.009	0.236	0.103
Conc.[b]	6	19.168***	12.760**	12.599**	6.359*	8.395*
Fly Ash Leachate:						
Stage	1	0.005	0.971	0.036	0.035	0.036
Conc.	6	2.696	2.581	1.335	1.331	1.349
Aluminum Sulfate:						
Stage	1	0.353	4.716	17.225**	6.278*	14.515**
Conc.	7	25.340***	306.579***	96.119***	5.148*	24.050***
Barium Chloride:						
Stage	1	0.342	0.851	2.653	2.624	2.732
Conc.	4	16.566**	13.260*	80.704***	24.023**	37.011**
Boric Acid:						
Stage	1	0.749	1.031	9.379*	0.089	1.475
Conc.	6	7.758*	17.049**	28.017***	27.503***	36.254***
Cadmium Chloride:						
Stage	1	2.038	0.057	2.360	0.859	1.217
Conc.	5	8.408*	12.687**	63.677***	12.078**	21.488**
Cobalt Chloride:						
Stage	1	0.527	1.988	0.976	0.055	0.813
Conc.	5	14.245**	19.578**	53.399***	38.939***	52.693***
Lead Nitrate:						
Stage	1	54.193***	16.608**	0.008	0.327	0.150
Conc.	6	40.320***	12.400**	16.306**	9.326**	13.117**

* $P < 0.05$, ** $P < 0.01$, *** $P < 0.001$.
[a] Stage = moss gametophore and spore inoculated cultures.
[b] Conc. = treatment concentrations.

tions increased, the fresh weight, dry weight, and chlorophyll contents decreased concomitantly. Analysis of two-stage statistical comparisons reflected no significant differences in toxic response criteria between the majority of spore and gametophore produced cultures.

Responses of P. patens

Because responses of spore- and gametophyte-derived cultures to each toxicant were generally similar (Table 3), showing no significant differences between stages (except aluminum sulfate and lead nitrate), data for the gametophore stage were chosen for graphic display. Also chlorophyll a, b, and total were generally similar for a given treatment, so total chlorophyll values were selected for further evaluation. The dry weight and total chlorophyll contents were plotted against a logarithmic concentration range of the test treatment (Figs. 1 and 2). A summary of the ANOVA statistical treatments is presented in Table 3. In the following discussion, each experimental treatment has been addressed separately to provide clarity and consistency:

1. *Anakeesta Leachate*—The effects of Anakeesta leachate on moss growth and development were pronounced (Fig. 1). Morphological observations generally revealed single, dwarf-

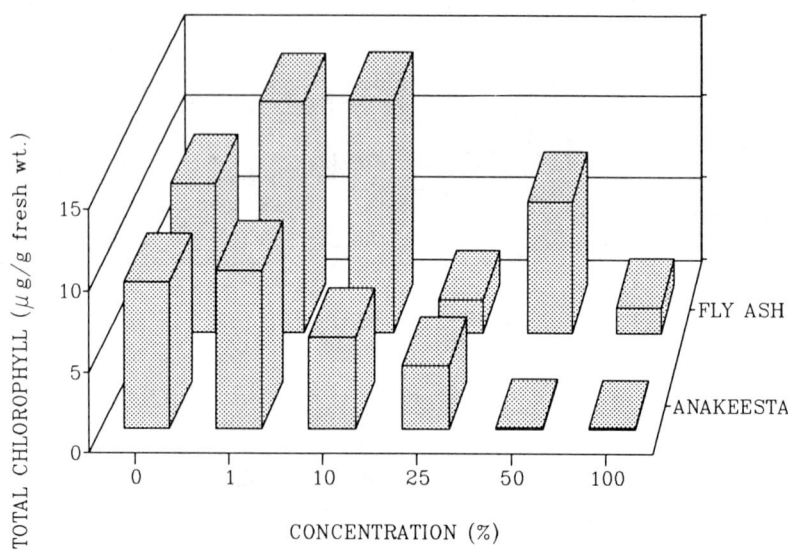
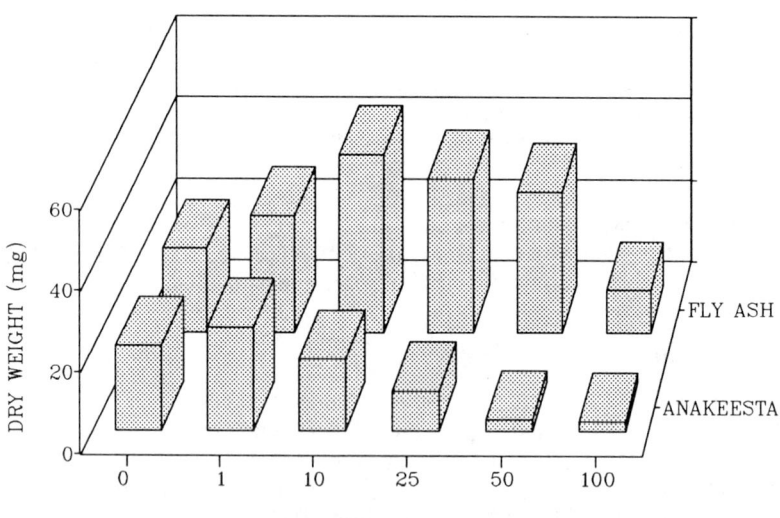

FIG. 1—*Influence of anakeesta and fly ash leachates on dry weight and total chlorophyll of moss gametophore cultures exposed 5 weeks.*

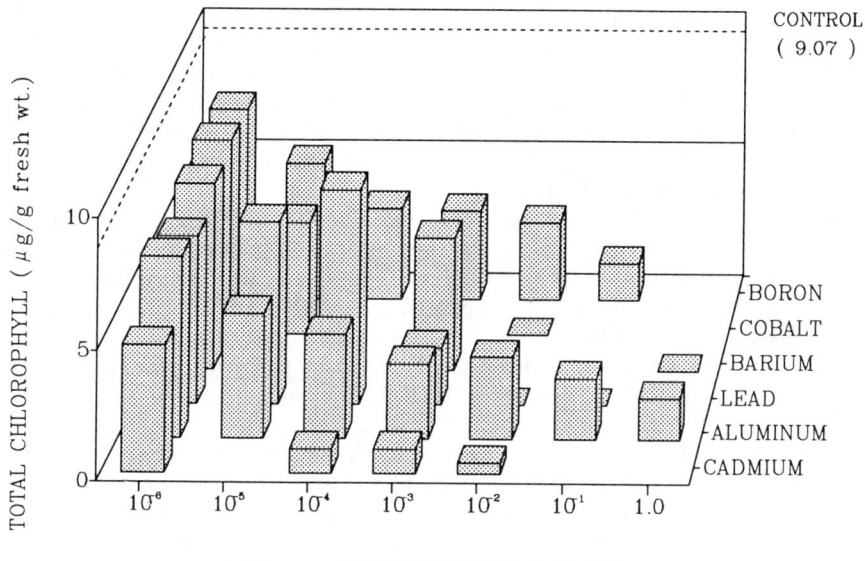

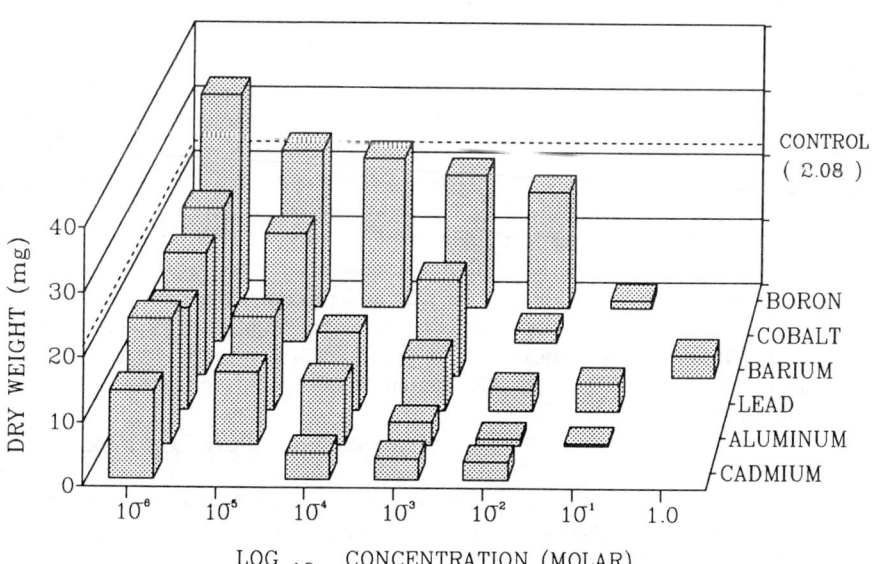

FIG. 2—*Influence of Al, Ba, B, Cd, Co, and Pb treatments on dry weight and total chlorophyll of moss gametophore cultures exposed 5 weeks; controls with average dry weight of 20.8 mg and total chlorophyll averaging 9.07 µg/g fresh weight.*

leafy shoots that grew on darker, more compact protonema. In the 50% leachate solution, moss protonemas were loosely arranged and no gametophore development was evident. On the media containing the test solution of 10% or less, protonemas had several young leafy shoots and a few rhizoids. In full-strength leachate treatments, the cultures were brownish-white and did not develop gametophores or rhizoids. Typically, as the treatment concentration increased, the cultures underwent a progressive significant reduction in dry weight and total chlorophyll contents (Table 3).

2. *Coal Combustion Fly Ash Leachate*—Moss cultures treated with the fly ash leachate showed both positive and negative responses relative to control cultures. The responses included low occurrence of the gametophore (in 100% fly ash leachate) and low levels of chlorophyll *a* and *b* (in 100% and 25% concentrations) (Fig. 1). No significant differences ($P \leq .001$) were found between standing crop and total chlorophyll levels (Table 3). The fly ash leachate at low concentrations did not inhibit gametophore regeneration and enhanced growth slightly, based on dry weight and total chlorophyll (Fig. 1).

The pH values for the two lowest fly ash treated (1 and 10% leachate concentration) moss cultures ranged between 7.0 and 7.1. This pH level, combined with excess sulfate concentrations in the presence of high calcium concentrations, may have induced chlorophyll production levels that exceeded those of controls. For example, gametophore cultures treated to 1 and 10% fly ash leachate concentrations produced 56 and 57% more total chlorophyll than controls while the spore cultures produced less total chlorophyll at these fly ash treatments. Conversely, the spore cultures had 44 and 74% more total chlorophyll production than spore-controls (average 7.81 μg/g fresh weight) at the higher fly ash treatments of 25 and 50% leachate.

3. *Aluminum Sulfate*—Aluminum has elicited wide environmental concern and has been shown to be environmentally problematic in some aquatic environments [40]. Aluminum sulfate treatments had a negative effect upon moss dry weight and chlorophyll contents (Table 3), showing progressive reductions as concentrations increase ($P < 0.001$). A decreased aluminum sulfate concentration favored the production of the gametophore. At an aluminum sulfate concentration of 10^{-2} M, both spore and gametophore cultures became predominately filamentous and produced no gametophores. The protonema in these solutions developed into compact, clustered masses. Abundant rhizoid development occurred in test concentrations exceeding 10^{-4} M aluminum sulfate.

4. *Barium Chloride*—Two distinct growth patterns were noted for moss cultures treated with barium chloride. When barium chloride concentrations were equal to or higher than 0.5 M, no gametophore development occurred and protomenatal branches of the young gametophores were aberrant. Rhizoid development was enhanced in some cultures treated with barium chloride concentraions of 0.1 M. When $BaCl_2$ concentrations were less than 0.5 M, the cultures were not morphologically different from the controls. Although barium chloride at concentrations of $\leq 10^{-3}$ M did not appear to be toxic based upon response criteria (Fig. 2), concentrations above this level significantly reduced dry weight and chlorophyll (Table 3).

5. *Boric Acid*—At low concentrations, boric acid induced the least pronounced effects of the substances tested on moss morphological characteristics, dry weight, and chlorophyll (Fig. 2). However, no gametophore development occurred in either spore or gametophore produced cultures at the highest boric acid concentration (0.1M). Moss dry weight and total chlorophyll levels were significantly depressed (Table 3) at the highest treatment concentration.

6. *Cadmium Chloride*—Morphological observations showed that cadmium chloride was the most toxic metal tested and concentrations as low as $10^{-4}M$ markedly reduced moss gametophore development and germination. Dry weight and chlorophyll content were sig-

nificantly depressed (Table 3) from that of controls at all concentrations tested. At a concentration of 10^{-6} M, the cultures consisted of brown compact protomenas with abundant rhizoids. Very few leafy shoots were observed attached to the protomenas, and these were small (single dwarf) and aberrent. At higher concentrations of cadmium chloride, more pronounced abnormalities were found. Cultures treated with $0.01 M$ cadmium chloride or higher were yellowish-white and appeared to be dead. At the highest concentration tested (0.025 M), the moss cultures contained no detectable amounts of chlorophyll.

7. *Cobalt Chloride*—Cultures exposed to $2.5 \times 10^{-3} M$ cobalt chloride or higher were morphologically abnormal. No gametophore production occurred at this concentration, and protonema clones were brownish to white in appearance. At cobalt chloride concentrations $\leq 10^{-5} M$ a few leafy shoots and rhizoids were noted. Results presented in Fig. 2 show that cobalt chloride is highly toxic to this primary producer and that its bryocidal effects are similar to those caused by Al and Cd salts.

8. *Lead Nitrate*—Moss generation was inhibited to within 0.1% of that for controls at the two highest lead nitrate treatments (10^{-1} and $10^{-2} M$), although abundant regeneration occurred (70 to 80% of controls) at the lower four treatments. The influence of lead nitrate on total chlorophyll content was especially pronounced (Fig. 2) and paralleled the dose-response relationship seen in moss regeneration. Although gametophore chlorophyll levels approached control concentrations at the $10^{-4} M$ treatment (8.13 µg/g fresh weight), nondetectable concentrations of chlorophyll a and b resulted at the two highest lead nitrate exposures (Table 3). An apparent combined-toxic effect occurred between pH (3.3 to 3.1) and lead nitrate concentrations above $10^{-4} M$. The concomitant reductions in pH to 5.0 (approaching modified Knop's inorganic culture medium, pH 5.5 to 5.8) and lead nitrate at $10^{-4} M$ may have contributed to the improved moss growth. However, the positive influence of the relatively high equivalent concentrations of nitrogen in the test treatments (as 1 part Pb to 3 parts NO_3) cannot be discounted, and this essential nutrient would be expected to promote growth at nontoxic concentrations of the lead cation.

Conclusions

The results of this study, which includes morphological observations and measurements of change in biomass and chlorophyll in response to various toxic treatments, support the assumption that the moss toxicity test can be used to screen environmentally hazardous substances. The assay is simple, inexpensive, and reliable. The quantifiable response parameters (biomass and chlorophyll content) were easily measured and sensitive to various plant toxicants. Because no significant difference between spore and gametophore culture types was found for most toxicants, a single culture life stage would be sufficient for routine testing. In order to evaluate the applicability of the moss toxicity test, this study was designed to test a wide range of toxicants, and assumptions were made that the growth potential of each culture was nearly the same and that no variance existed between the two stages that were used as replicates. However, in future studies, triplicate treatments at each test concentration would be useful.

The solid medium cultures were well suited for morphological observation and for obtaining measurements of biomass and chlorophyll. Moss growth had a strong dose-response relationship after 30 days treatment, and the cultures maintained similar growth patterns up for to 40 days. Recommended test conditions are summarized in Table 1.

Earlier studies have shown that *P. patens* is suitable for genetic, biochemical, and developmental studies. This study shows that the moss is also applicable to toxicity assays. Methodology developed in this study offers the following advantages:

1. *P. patens* provides several life stages for testing, being readily available in the laboratory from small quantities of stock culture, from which reliable spore and gametophore cultures can be grown.
2. By lowering the growth temperature to 17°C, the spore capsule is easily obtained. Spores remain viable even in dry specimens and can be stored indefinitely. In addition, spore and gametophore cultures are easily maintained.
3. Both morphological and physiological effects may be used in dose-response criteria.
4. *P. patens* appears to be sensitive to various toxicants and small volumes of toxic solution may be sufficient for testing.
5. No-observed-effects concentrations and the lowest observed effects concentrations can be calculated from the data obtained from the moss toxicity test.
6. The moss toxicity assay can provide useful information on an important component of the ecological community in a cost-effective manner.

Acknowledgments

Support for this study was provided by the Aquatic Biology Research Fund, Tennessee Technological University, and the U.S. Tennessee Valley Authority. The original research proposal designed for genotoxic study submitted by Dr. Paula Nakosteen, Department of Botany, University of Tennessee, to U.S. Tennessee Valley Authority for funding consideration was released to this study. Sincere appreciation is extended to her for providing consultation and *P. patens* cultures. Dr. Paul J. Tsai of Tennessee Tech served as statistical analyst. We gratefully acknowledge the computer-generated graphics done by Todd Hunt and Jane T. Red of Young-Morgan & Associates and the word processing and editing provided by The Center for the Management, Utilization and Protection of Water Resourses, Tennessee Technological University.

References

[1] Jourbert, G., *Water Research*, Vol. 14, 1980, pp. 1759–1763.
[2] Walsh, G. E., *Environmental Toxicology and Chemistry*, Vol. 7, No. 12, 1988, pp. 979–987.
[3] Cowgill, U. M., Milazzo, D. P., and Landenberger, B. D., *Environmental Toxicology and Chemistry*, Vol. 8, No. 5, 1989, pp. 451–455.
[4] Steele, R. L. and Thursby, G. B. in *Aquatic Toxicology and Hazard Assessment (Sixth Symposium), ASTM STP 802*, W. E. Bishop, R. Q. Cardwell and B. B. Heidolph, Eds., American Society for Testing and Materials, Philadelphia, 1983, pp. 71–88.
[5] "Guidance Document for Conducting Effluent Toxicity Test Using *Champia pavulia*," EPA Internal Report No. 600/X-85/242, U.S. Environmental Protection Agency, Cincinnati, OH, 1985.
[6] Edwards, N. T. and Ross-Todd, B. M., *Environmental and Experimental Botany*, Vol. 20, 1980, pp. 31–38.
[7] "Environmental Assessment: Short-Term Tests for Carcinogens, Mutagens, and Other Genotoxic Agents," EPA 625/9-79-003, Health Effect Research Laboratory, U.S. Environmental Protection Agency, Cincinnati, OH, 1979.
[8] "Standard Evaluation Procedure for Non-Target Plants: Growth and Reproduction of Aquatic Plants—Tiers 1 and 2," EPA 540/9-86-134, U.S. Environmental Protection Agency, Washington, DC, 1986.
[9] Adams, S., Cole, H., Jr., and Massic, L. B., *Environmental Pollution*, Vol. 5, 1973, pp. 119–147.
[10] Dietz, F., "The Enrichment Ratio of Heavy Metals in Submerged Plants," in *Advances in Water Pollution Research*, Proceedings of 6th International Conference, Pergamon Press, Oxford, 1973, pp. 55–62.
[11] Empain, A., *Bull. fr. Pissiculture*, Vol. 48, 1976, pp. 138–142.
[12] Empain, A., *Mem Soc. r. Bot. Belg.*, Vol. 7, 1976, pp. 141–156.
[13] Nakosteen, P. C., Dodson, D., and Hughes, K. W., *Environmental and Experimental Botany*, Vol. 19, 1978, pp. 93–97.

[14] Nakosteen, P. C. and Hofman, J., *The Bryologist,* Vol. 81, No. 1, 1978, pp. 162–166.
[15] Nakosteen, P. C. and Hughes, K. W., *The Bryologist,* Vol. 81, No. 2, 1978, pp. 307–314.
[16] Engel, P. P., *American Journal of Botany,* Vol. 55, 1968, pp. 438–446.
[17] Cove, D. J. and Ashton, N. W., *Hereditas,* Vol. 33, 1974, p. 135.
[18] Nakosteen, P. C., "Studies on the Induction, Selection, and Inheritance of Biochemical Mutants in Mosses," Ph.D. dissertation, University of Tennessee, 1978.
[19] Steeves, T. A., Susses, I. M., and Partanen, C. R., *American Journal of Botany,* Vol. 42, 1955, pp. 232–245.
[20] Herrmann, R., Morgan, E. L., and Green, R. L., "Aluminum Precipitation, Beech Flats and Walkers Prong Creeks, Great Smoky Mountains National Park," in *Proceedings, 1st Conference on Scientific Research in the National Parks,* USDI, National Park Service Transactions and Proceedings Services, 1979, p. 84.
[21] King, P. B., Neuman, R. B., and Hadley, J. B., "Geology of the Great Smoky Mountains National Park, TN and NC," U.S. Geological Survey, Prof. Pap. No. 587, 1968, p. 23.
[22] Morgan, E. L., Porak, W. F., and Arway, J. A., "Controlling Acid-Toxic Leachates from Southern Appalachian Construction Slopes: Mitigating Stream Damage," in *Wetlands, Floodplains, Erosion, and Storm Water Plumbing,* Transportation Research Board, Record 948, 1983, pp. 10–16.
[23] Green, R. L., "Benthic Macroinvertebrate Communities in Great Smoky Mountains National Park Streams Influenced by Acid Drainage," M.S. thesis, Tennessee Technological University, Cookeville, Tenn., 1975, pp. 122.
[24] Porak, W. F., "The Effects of Acid Drainage Mitigation upon Fish, Benthic Macroinvertebrates, and Water Quality in Streams of the Cherokee National Forest, Tennessee," M.S. thesis, Tennessee Technological University, Cookeville, Tenn., 1981.
[25] Huckabee, J. W., Goodyear, C., and Jones, R. D., *Transactions of the American Fisheries Society,* Vol. 104, 1975, pp. 677–684.
[26] Matthews, R. C., Jr. and Morgan, E. L., *Journal of Environmental Quality,* Vol. 11, No. 11, 1982, pp. 102–106.
[27] Adriano, C. C., Page, A. L. L., Chang, A. C., and Straugham, I., *Journal of Environmental Quality,* Vol. 9, 1980, pp. 333–334.
[28] Cavin, D. C., "A Study of Iron and Aluminum Recovery from Power Plant Fly Ash," M.S. thesis, Iowa State University, Ames, 1973, p. 115.
[29] "Quality Criteria for Water—1986," EPA 440/5-86-001, U.S. Environmental Protection Agency, Washington, D.C., 1986.
[30] Termin, G. L., "Solid Wastes from Coal-Fired Power Plants: Use or Disposal on Agricultural Lands," Bulletin Y 129, U.S. Tennessee Valley Authority, Muscle Shoals, Ala., 1978.
[31] Furr, A. K., Parkinson, T. F., Hinrichs, R. A., VanCampen, D. I., Bache, C. A., Gutenmann, W. H., St. John, L. E., Pakkala, I. S., and Lisk, D. J., *Environmental Soil Technology,* Vol. 11, 1977, pp. 1199–1201.
[32] Olem, H., Ship, J. W., Chu, T. J., and Ruane, R. V., "Toxic Pollutants in Conventional Wet Scrubber Wastes," in *Proceedings,* 8th Annual Water and Wastewater Equipment Manufacturers Association Industrial Pollution Conference, 1980.
[33] "Biological Field and Laboratory Methods for Measuring the Quality of Surface Waters and Effluents," EPA 670/4-73-001, Office of Research and Development, U.S. Environmental Protection Agency, Cincinnati, Ohio, 1973.
[34] Arnon, D. I., *Plant Physiology,* Vol. 24, 1949, pp. 1–15.
[35] Sokal, R. R. and Rohlf, F. J., *Biometry,* 2nd ed., W. H. Freeman, San Francisco, 1981.
[36] Schmitt, C. J. in *Aquatic Toxicology and Hazard Assessment (Fourth Conference), ASTM STP 737,* D. R. Branson and K. L. Dickson, Eds., American Society for Testing and Materials, Philadelphia, 1981, pp. 270–298.
[37] Nie, N. H., Hull, C. H., Jenkins, G. G., Steinbrenner, K., and Bent, L. M., *Statistical Package for the Social Sciences (SPSS),* 2nd ed., McGraw-Hill, New York, 1975.
[38] Morgan, E. L., Yehl, T. C., and Young, R. C., "Toxicity of Wet Limestone Scrubber Sludges and Effluent Waste from a Coal Combustion Process," in *Proceedings,* 37th Annual Purdue Industrial Waste Conference, Purdue University, West Lafayette, Ind., 1982.
[39] Morgan, E. L. and Young, R. C., "Toxicity of Wastes Generated by a Coal Combustion Power Plant Utilizing Limestone Scrubber Technology to Indigenous Aquatic Fauna," NTIS PB88-191416/WEP, July 1986, p. 133.
[40] Burrows, W. D., "Aquatic Aluminum: Chemistry, Toxicology, and Environmental Prevalence," in *Critical Reviews in Environmental Controls,* The Chemical Rubber Co., Cleveland, Ohio, 1977, pp. 167–216.

Wuncheng Wang[1] *and Gerald Elseth*[2]

Millet Root Elongation in Toxicological Studies of Heavy Metals: A Mathematical Model

REFERENCE: Wang, W. and Elseth, G., "**Millet Root Elongation in Toxicological Studies of Heavy Metals: A Mathematical Model,**" *Plants for Toxicity Assessment, ASTM STP 1091,* W. Wang, J. W. Gorsuch, and W. R. Lower, Eds., American Society for Testing and Materials, Philadelphia, 1990, pp. 280–294.

ABSTRACT: Plant seed germination and root elongation, the first phase of plant development, are sensitive to adverse environmental stress. The objectives of this study were (1) to use a millet root elongation method for measuring phytotoxicity of heavy metals and (2) to develop a mathematical model to relate the phytotoxicity of the metals to the root elongation effect and also to provide a basis for linear transformation of the data. The results show that the order of decreasing phytotoxicity of the metals is Cu, Ni, Cd, Cr(VI), Zn, Mn, and Ba. The metal toxicity is on the same order of magnitude in the millet, fish, and duckweed tests.

A mathematical model was developed:

$$\frac{P - P_M}{P_M} = A[M] \frac{BK + [M]}{K + [M]}$$

where A, B, and K are constants under conditions where metal concentration M is the only experimental variable. P_M is the root elongation in the presence of added metal ions and P is that in the control group.

A computer program has been developed to calculate the theoretical values of these metals. These values are in good agreement with experimental values. This model is likely to be applicable to other toxicity tests as well.

KEY WORDS: heavy metals, phytotoxicity, root elongation, millet, mathematical model

Seed germination and root elongation together represent the first phase of plant development. During this period, plant growth is highly sensitive to various environmental effects and can therefore be used as an indicator of environmental stress or adversity.

Recently there has been increasing attention given to aquatic and terrestrial toxicity tests using seed germination and the root elongation method. Luessem and Rahman [1] determined the phytotoxicity of water and wastewater by measuring their effects on root radical growth of cress seeds. Brusick and Young [2] developed the *Industrial Environmental Research Laboratory-Research Triangle Park Procedures Manual for Level 1 Environmental Assessment (Biological Tests).* The root elongation test was included as part of the terrestrial ecological effects bioassays. Wong and Bradshaw [3] used root elongation of perennial ryegrass to determine the phytotoxicity of Al, Cd, Cr(VI), Cu, Fe(II), Hg, Mn(II), Ni, Pb, and Zn. They reported that the order of toxicity was in agreement with the stability of metal-

---

[1] Water Quality Section, Illinois State Water Survey, Peoria, IL 61652.
[2] Department of Biology, Bradley University, Peoria, IL 61625.

organic complexes. They predicted that the method would be useful for detecting the toxicity of heavy metals in contaminated soils. Ratsch [4] conducted an interlaboratory study of five plant species (red clover, cucumber, lettuce, radish, and wheat) using ten chemical compounds. The results indicated that the effect on root elongation was a valid and sensitive test for environmental toxicity.

Wang [5] used plant biomass in the early life stage to detect the toxicity of phenol and seven chlorophenols. The biomass method, though simple, was found to be less sensitive than the root elongation method [6]. In a comparative study, results of the millet root elongation test showed that they were more sensitive to phenolic compounds than that of the cucumber and lettuce tests [7]. For inorganic ions such as Cd, Cr(VI), Cu, Mn, Ni, and Zn, lettuce tests appeared to be more sensitive to the metal toxicity than cucumber and millet tests [8].

The intent of this study is twofold. The first objective is to explore the potential of using the millet root elongation method for measuring the phytotoxicity of heavy metals. For this purpose, the toxicity of Ba, Cd, Cr(VI), Cu, Mn(II), Ni, and Zn on root radical growth of millet seeds was tested. The second objective is to develop a mathematical model that relates the phytotoxicity of the metals to effects at the molecular level and also provides a basis for a linear transformation of the data.

Methods

The method used in this study is similar to that reported earlier by Wang [6]. The seeds employed in this study were from the same stock of millet *(Panicum miliaceum)*. The seeds, which were stored in a freezer ($-10°C$), were treated with hypochlorite, imbibed, and washed according to the procedure reported in the earlier study. A series of 10 cm Petri dishes, each containing a Whatman No. 1 filter paper (9 cm diameter) and 5 mL deionized water or metal solution, was used. Six dishes were used per test solution and ten seeds were placed in each dish. After 120 h incubation in the dark at 24 to 26°C, the root length of each seed was measured to the nearest 1 mm. The statistical degrees of freedom were calculated based on 60 seeds per test solution. Experiments using Ba and Cu were each repeated three times.

The metal ions used in the study were prepared in solution from reagent-grade $BaCl_2$, $CdCl_2$, K_2CrO_4, $CuCl_2 \cdot 2H_2O$, $MnCl_2 \cdot 4H_2O$, and $ZnCl_2$. The pH of each solution was 6.5 using 0.1 mol/L NaOH or HCl. In every case, deionized water was used as the control and dilution water.

The different series of metal solutions were prepared in a 60% reduction scale. The highest concentration of each metal was selected after a preliminary experiment, giving at least an 80% toxic effect. Each solution containing the highest concentration of a metal was placed in a 100 mL volumetric flask. A 40 mL fraction was pipetted from the flask into a vessel. Six 5 mL subsamples were then pipetted into separate Petri dishes. The remaining 60 mL in the flask was diluted to the mark with deionized water to prepare the next lower concentration of a metal ion in solution. The diluted solution was well mixed manually and the procedure was repeated until there were seven different metal concentrations, giving the concentration ratios of 100, 60, 36, 22, 13, 7.8, and 4.7.

Results and Discussion

Concentration-Effect Relations

The inhibitory effects on root elongation produced by varying concentrations of the seven metal ions are shown in Fig. 1. The results are represented by semilogarithmic graphs in

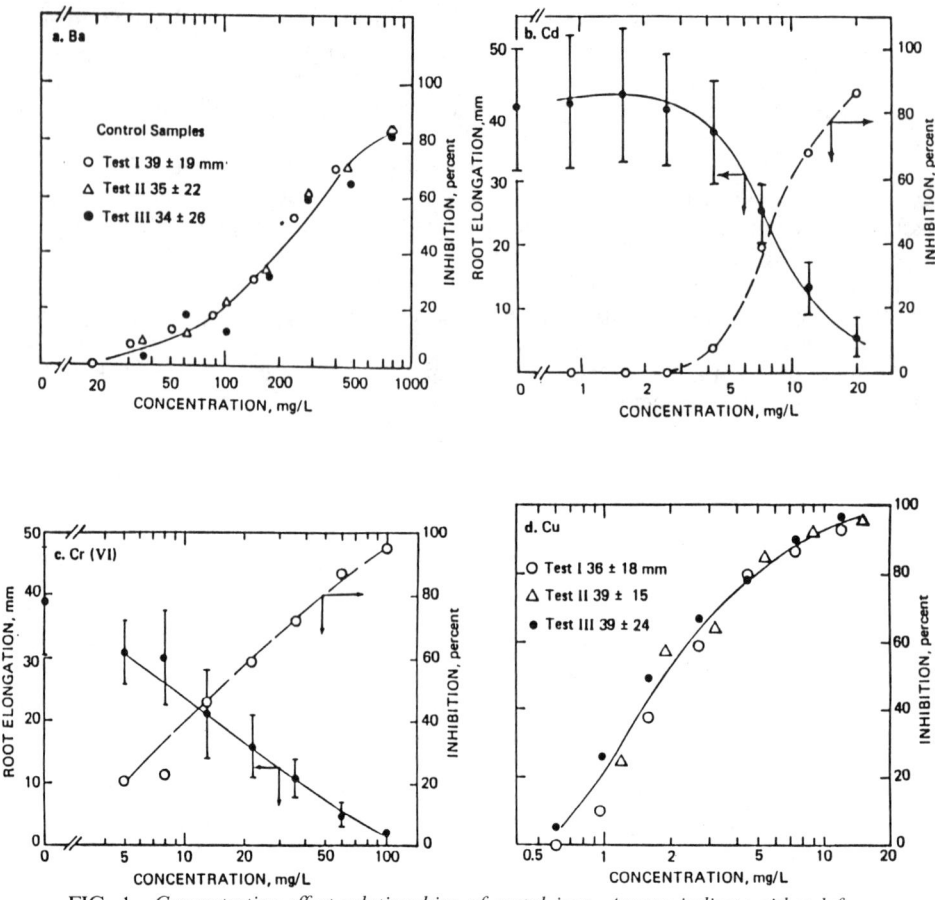

FIG. 1—*Concentration-effect relationships of metal ions. Arrows indicate either left- or right-hand scale to be used.*

which the logarithm of the metal ion concentration is plotted along the abscissa. The ordinates express root lengths in millimetres along the left-hand scale (elongation curves) and percent inhibition along the right-hand scale (inhibition curves). The left-hand scale provides the range of standard deviation, while the right-hand scale is used to locate each EC_{50} value (the 50% inhibitory effect concentration relative to the control sample). Ba and Cu were each tested three times in order to demonstrate repeatability of test results. Because of the triple tests of Ba and Cu, only the inhibition curves are plotted, as shown in Figs. 1a and 1d; in these two figures, the means and standard deviations of each test are given. The repetitions of the tests indicate that the results are generally reproducible.

In general, the inhibition curves in Fig. 1 appear as sigmoid relations. Ideally, the EC_{50} value can be obtained from the concentration along the abscissa that produces 50% inhibition; examples include the Ba and Ni results in Figs. 1a and 1f. The degree of uncertainty is known to be greater as the concentration is extended to both higher and lower ends of the curve [9].

The results in Fig. 1 show toxicity thresholds (i.e., no effect concentrations [NOECs]) [9] of the metals Ba, Cd, Cu, Ni, and Zn; they were 19, 2.6, 0.6, 0.93, and 27 mg/L, in that order. Below these thresholds, toxicities of these metals were undetectable.

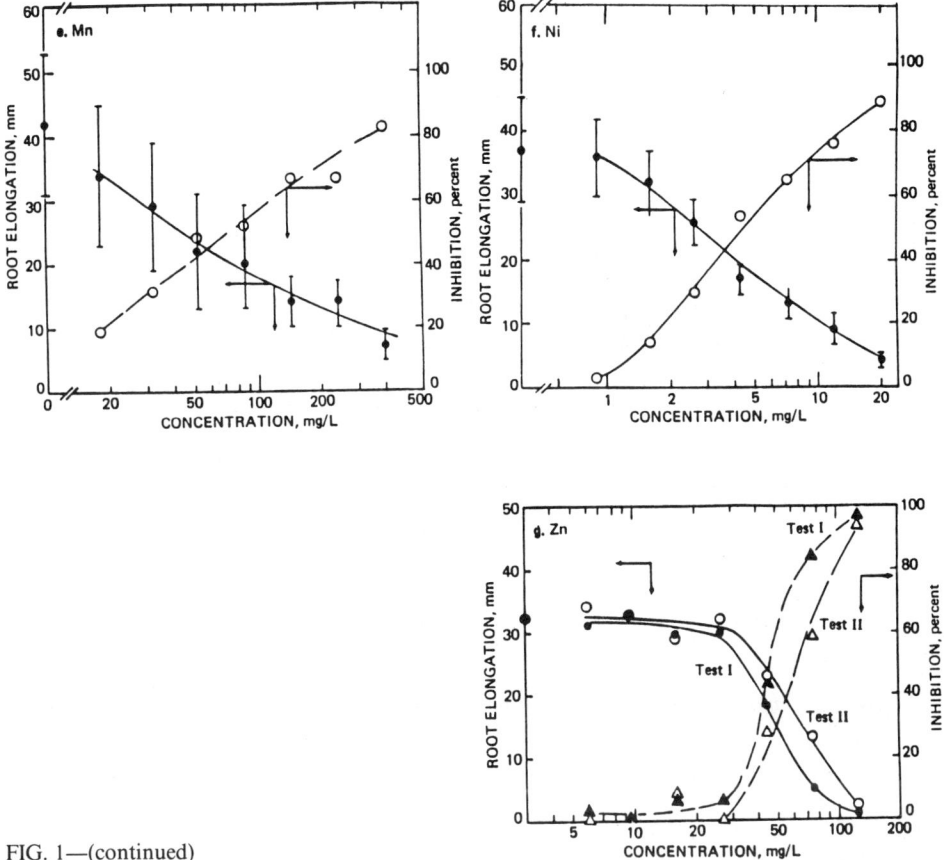

FIG. 1—(continued)

Although generally in a sigmoid form, the concentration-effect relations are not superimposed (Fig. 2). The concentration $1\times$ was 6 mg/L and 20 mg/L for Cd and Mn, respectively. The concentration ratios between EC_{80} and EC_{20} (80% inhibitory effect concentration and 20% inhibitory effect concentration) ranged from a low of approximately 3 for Cd to a high of slightly greater than 16 for Mn. In other words, Cd needed only a 3-fold increase in concentration to raise its toxic effect from 20 to 80%, while Mn required about a 16-fold increase in concentration. The significance of these results is that the metal with a greater toxicity potential (lower EC_{50}) tends to exert its toxicity in a narrower toxicity range (EC_{80}/EC_{20}).

Comparison with Other Tests

The toxicity of the seven metal ions in the millet root elongation tests is expressed in the form of EC_{50} values in the last column of Table 1 using the graphic method. The results indicate that among the seven heavy metals, Cu is the most toxic and Ba is the least toxic. The results are compared with the toxicity data obtained in other studies using the ryegrass root elongation, duckweed reproduction, and fish mortality tests. The ryegrass test took 14 d, the duckweed and fish tests, 96 h, and the millet test, 120 h. The millet test was held in the dark; ryegrass and duckweed, in constant light; and fish, in a light and dark regime as a common practice.

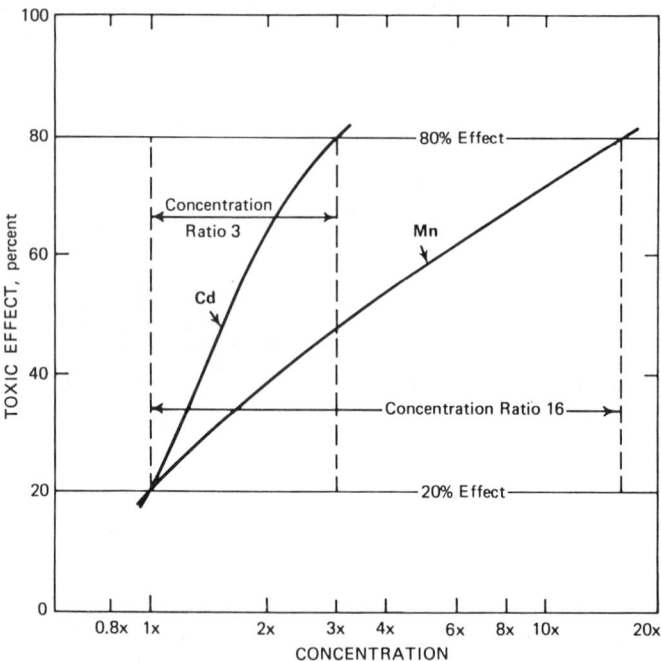

FIG. 2—*Concentration ratios (EC_{80}/EC_{20}) of cadmium and manganese. The concentration $1\times$ was 6 mg/L and 20 mg/L for cadmium and manganese, respectively.*

In summary, the results in Table 1 show that metal toxicity varied one to two orders of magnitude among the millet, duckweed, and fish tests. The ryegrass test appears to be the most sensitive; the test was one or two orders of magnitude more sensitive to metals except Cd than millet, duckweed, and fish tests.

Mechanism of Metal Action

The principle effects of metal ions on living systems result from the role of metals in enzyme expression and regulation. These effects, like the reactions between ions and proteins in general, can be quite complex. The metals may be essential for enzyme activity—an estimated one third of all enzymes require a metal ion in some phase of catalysis—or can replace other metal ions as nonspecific activators of enzymes. The metals also exert their effects in

TABLE 1—*Comparative metal toxicity to different test species (metal concentrations in mg/L).*

	Rye Grass [3] (14 d EC_{50})	Duckweed [10] (96 h EC_{50})	Fish [11] (96 h LC_{50})	Millet (This Study) (120 h EC_{50})
Ba	...	26	...	250
Cd	1.85	0.2	0.92	8.6
Cr(VI)	2.0	35	58.5	15
Cu	0.02	1.1	0.08–1.2 (3 species)	1.9
Mn(II)	0.45	31	...	72
Ni	0.18	0.45	13.6–48.8 (4 species)	4.4
Zn	1.60	10	0.4–55 (6 species)	46–64

two different ways: directly, by combining with the protein that they influence, or indirectly, by altering the activity of a normal activator or inhibitor of an enzyme or of an inducer or repressor of protein synthesis.

When the metal ion concentration is varied, many metals exhibit a concentration optimum for their enzyme effects, increasing enzyme activity at concentrations below the optimum and decreasing (inhibiting) enzyme activity at concentrations above it. This type of biphasic response can be used to account for the results in the present study. To describe this response mathematically, it is necessary to know only how the rate of an enzyme reaction in the presence of added metal ions (v_M) is related to the rate in their absence (v). Since metal toxicity is of primary concern, a general model can be expressed as a measure of enzyme inhibition (Z) defined as $Z = (v - v_M)/v_M$. Assuming that root elongation is limited by the rate of the affected reaction, Z can also be expressed as $(P - P_M)/P_M$, where P_M is the measured root elongation in the presence of the added metal ions and P is the elongation in the control group. In the case of simple competitive or noncompetitive inhibition, Z varies linearly with the concentration of inhibitor $[M]$ as $Z = A[M]$. In this case, $Z/[M]$ remains constant as long as only $[M]$ is varied. The relationship is more complex in the case where the metal is also an activator. For simple nonessential activation, a general model then takes the form (see Appendix I for a derivation)

$$Z = A[M] \frac{BK + [M]}{K + [M]} \quad (1)$$

where A, B, and K are parameters that depend only on the concentrations of substances other than M (substrates, other metal ions, etc.) and on the rate constants of the system. The values of the parameters are themselves constant when $[M]$ is the sole experimental variable, as it was in this study. The precise meaning of the parameters depends on the mechanism involved. In general, however, the parameter A measures the degree of inhibition per unit metal concentration, while the values of B and K provide information concerning the presence of a toxicity threshold and activation effects. When viewed strictly from the standpoint of curve fitting, B and K in the model can be regarded as "shape factors" that allow for a departure from linearity as shown by $Z = A[M]$.

Table 2 gives some conventional measures of metal toxicity by the parameters defined in the model. If $B = 1$, the metal will exhibit only simple inhibitory effects. In this case, $Z = A[M]$ and the metal is expected to give an EC_{50} of $1/A$ and an EC_{80}/EC_{20} of 16. Of the various metals tested in this study, only Mn gave these results (Table 3). In contrast, activation is evident when B has a negative value. Z will then equal 0 when $[M] = -BK$, giving a threshold value that $[M]$ must exceed before any inhibitory effect is realized. This toxicity threshold can account for the initial increase in root elongation that occurs with some metal ions at low values of $[M]$ (Fig. 3). Those metals showing a toxicity threshold are expected to give an EC_{80}/EC_{20} of less than 16.

Different methods can be used to evaluate the parameters in the model from experimental data. The approach taken in this study was to obtain initial estimates of A, B, and K from EC_{20}, EC_{50}, and EC_{80} values applying Eq 1 and the relationships derived in Table 2. These estimates were then used as initial parameter values for a regression analysis of each set of metal data expressed in the form of a log $[(P - P_M)/P_M]$ versus log $[M(BK + [M])/(K + [M])]$ plot. A computer program employed for the analysis used an iterative procedure to calculate an adjusted set of values for the parameters that gave the best fit to a line of unit slope. The parameter values obtained by this procedure are given in Table 3.

The goodness-of-fit to the model is shown by the log $[P - P_M/P_M]$ versus log $[M(BK +$

TABLE 2—*Metal effect concentrations derived from Eq 1.*

20% GROWTH INHIBITION (EC_{20})

$$M_{0.2} = \frac{1}{2A}(\tfrac{1}{4} - ABK)[1 + \sqrt{1 + AK/(\tfrac{1}{4} - ABK)^2}]$$

For $B = 1$: $M_{0.2} = \frac{1}{4A}$

50% GROWTH INHIBITION (EC_{50})

$$M_{0.5} = \frac{1}{2A}(1 - ABK)[1 + \sqrt{1 + 4AK/(1 - ABK)^2}]$$

For $B = 1$: $M_{0.5} = \frac{1}{A}$

80% GROWTH INHIBITION (EC_{80})

$$M_{0.8} = \frac{1}{2A}(4 - ABK)[1 + \sqrt{1 + 16AK/(4 - ABK)^2}]$$

For $B = 1$: $M_{0.8} = \frac{4}{A}$

MAXIMUM GROWTH

$M_{max} = K[\sqrt{1 - B} - 1]$
For $B = 1$: $M_{max} = -K$

$[M])/(K + [M])]$ plots for the different metals in Fig. 4. Clearly all sets of data show good agreement with theoretical expectations.

In an effort to simplify the analysis of metal effects that lacked a discernible toxicity threshold, Eq 1 was converted to a two-parameter model for the purpose of curve fitting. The two-parameter model was derived by setting $BK = 0$. The goodness-of-fit to this model is shown by the log $[(P - P_M)/P_M]$ versus log $[[M]^2/(K + [M])]$ plots in Fig. 5. A close fit is obtained in each case by selecting a value of $K = 12$ for the Cr, Cu, and Ni data and a value of $K =$

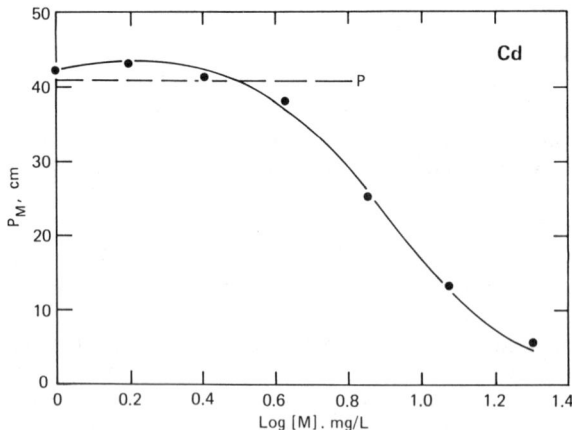

FIG. 3—*Theoretical concentration-effect relationship of cadmium with actual data points.*

TABLE 3—*Calculated constants* A *(L/mg),* B *(unitless), and* K *(mg/L) derived from Eq 1.*[a]

		A	B	K
Ba	I	0.0052	−0.426	122
	II	0.0055	−0.260	199
	III	0.0062	−0.258	179
Cd		0.9449	−0.100	33
Cr		0.0911	−0.238	13.3
Cu	I	0.608	−0.169	2.46
	II	0.934	−0.107	3.04
	III	1.04	−0.104	2.82
Mn		0.1819	1	...
Ni		0.386	0.143	5.43
Zn		0.0625	−0.036	140

[a] Parameter A measures the degree of inhibition per unit metal concentration; Parameters B and K provide information on toxicity threshold and activation effects.

120 for the Ba data. The effects of these metals in this system can therefore be described adequately with only two parameters, A and K. See Appendix II.

Application of Mathematical Model

Current environmental toxicology is heavily influenced by regulatory requirements. Most studies are concerned with median effect concentration and 95% confidence level. The results are then converted with a safety factor to derive a maximum permissible concentration. This rigid approach has been used to calculate water quality criteria [*12*]. The deficiency of this approach is that it ignores the diverse nature of various toxicants. For example, the ratios of EC_{80}/EC_{20} for Cd and Mn were 3 and 16, respectively. The safety factor of a pollutant should take this variability into consideration.

A less rigid and fundamentally more precise approach would be to arrive at water quality criteria with the aid of Eq 1. By knowing the experimental values of A, B, and K for each toxicant as given in Table 3, one can predict not only the conventional measures of toxicity (EC_{50}, EC_{80}/EC_{20}, etc.) but also the exact nature of the concentration-effect relationship. This approach could be particularly helpful in cases where a pollutant exhibits a toxicity threshold. The values of B and K would then provide the information needed to estimate threshold concentrations and also optimum concentrations for plant growth. The parameter A measures the degree of inhibition, while B and K in the model provide information about activation effects. Values of B and K are especially important in cases where the toxicant is essential for plant growth in small amounts but becomes inhibitory at higher concentrations.

The model as given in Eq 1 is not only applicable to higher plants but has also been applied to other test species. Studies in the laboratory of one of the authors (GE) have extended the model to *E. coli* for toxicity testing of Ba, Cd, Cu, Mn, and Ni. The metal toxic effect on bacterial growth was monitored up to 5 h. The results of these toxicity tests were all in close agreement with the model.

The mathematical model is likely to be applicable to other test species as well (e.g., algae, fathead minnow, *Daphnia magna*) because the biochemial processes of these species are similar and therefore, the same biochemical principles should apply. Furthermore, the model might be useful for a variety of toxic substances in addition to heavy metals. By further application of this model, a unified ecotoxicology might be developed.

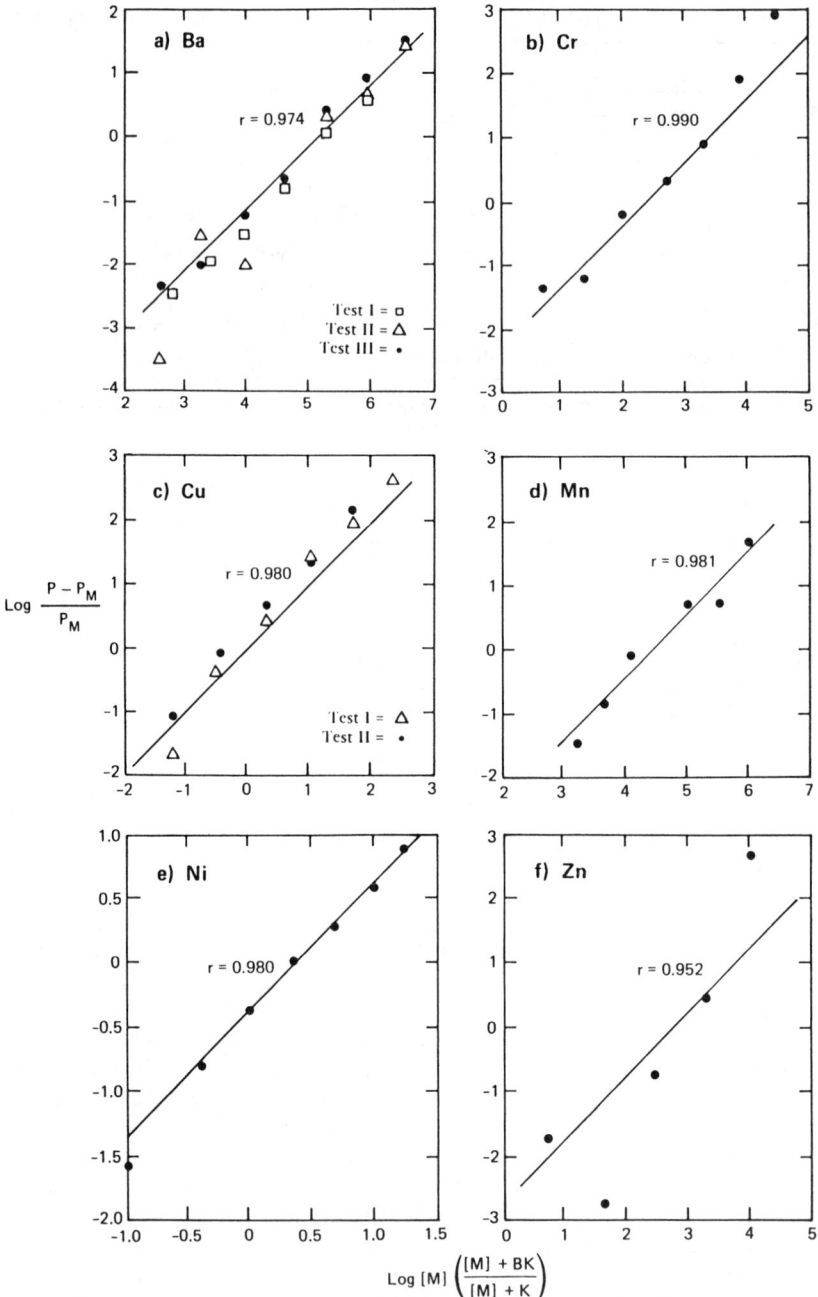

FIG. 4—*Fit to the model log $(P - P_M/P_M)$ versus log $[M]([M] + BK/[M] + K)$.*

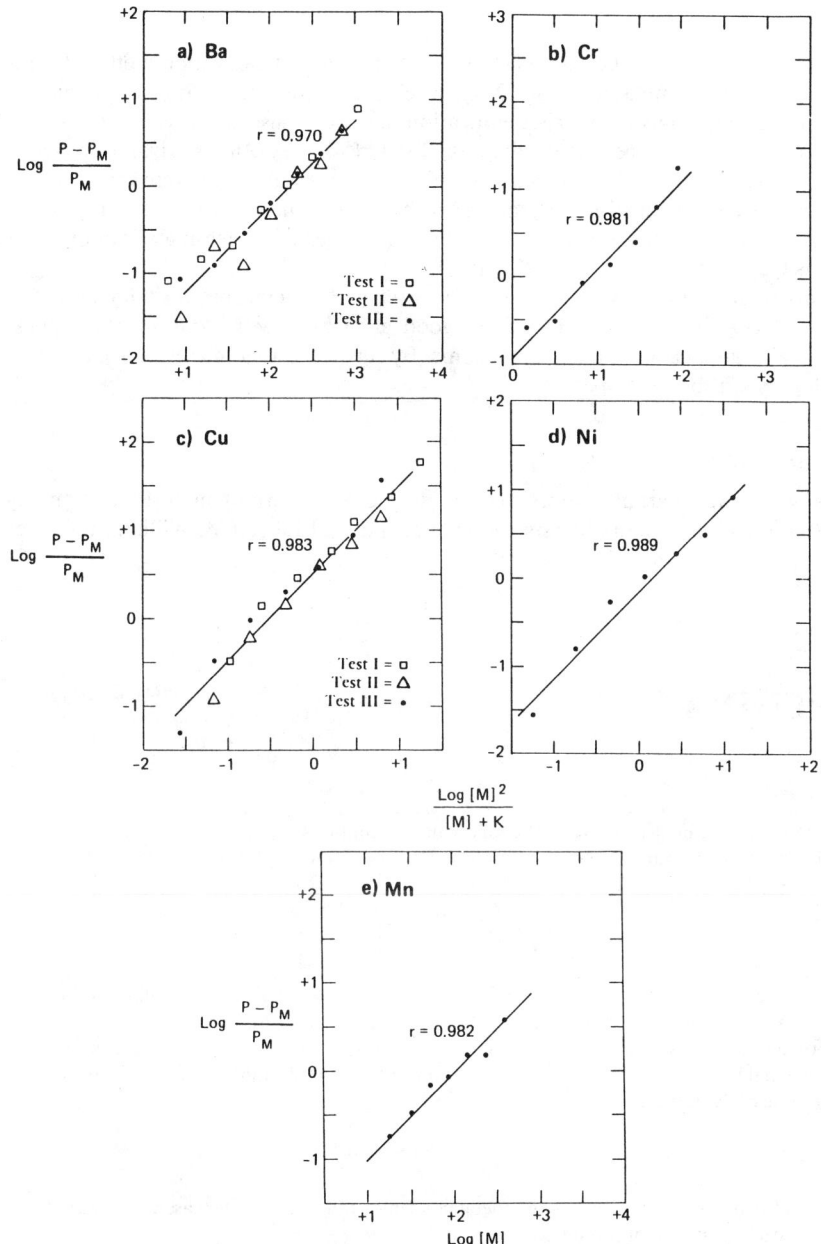

FIG. 5—*Fit of simplified two-parameter model log* $(P - P_M)/P_M$ *versus log* $[M]^2/(K + [M])$.

Summary and Conclusion

The objectives of this study were (1) to use the millet root elongation method for measuring phytotoxicity of metal ions and (2) to develop a mathematical model to relate the phytotoxicity of the metals to the root elongation effect and also to provide a basis for linear transformation of the experimental data. Ba, Cd, Cr(VI), Cu, Mn(II), Ni, and Zn were tested. Repeated tests of Ba and Cu indicated that the experimental results were reproducible.

The millet root elongation test measures the environmental stress affecting the earliest developmental stage of the plant. In this test, the order of decreasing phytotoxicity of the metals is Cu, Ni, Cd, Cr(VI), Zn, Mn, and Ba.

A mathematical model was developed to correlate the metal concentration and root elongation results. Theoretical values are in good agreement with experimental values. This model is also applicable to bacterial systems. By further testing of the model, a unified ecotoxicology might be developed.

Acknowledgments

This work was partially supported by the U.S. Environmental Protection Agency, R810834. We wish to thank Messrs. Don Schnepper and Judson M. Williams for developing the computer program and also Mr. Williams for technical assistance.

APPENDIX I

General Model

The derivation of Eq 1 is based on the following assumptions:
1. The inhibition parameter Z can be expressed as a function of $[M]$ by the relationship

$$Z = \frac{(a - b)[M]}{1 + b[M]}$$

This general equation gives the results of activation if $a < b$ or the results of inhibition if $a > b$. Inhibition is linear if $b = 0$ or is hyperbolic if $b > 0$. For further information regarding the basic types of activation and inhibition and their general effects on the velocity of an enzyme reaction, see Ref *13*.

2. For metal ions to exhibit activation at low concentrations and inhibition at high concentrations, a increases linearly with $[M]$ as

$$a = c[M] + d$$

where c and d are constants. Raising the metal concentration will thus increase the apparent chance of the metal reaching its site of inhibition.

Equation 1 can now be obtained by the following mathematical steps:

$$\begin{aligned} Z &= \frac{(c[M] + d - b)[M]}{1 + b[M]} \\ &= \frac{c[M](d/c - b/c + [M])}{b(1/b + [M])} \\ &= A[M] \frac{KB + [M]}{K + [M]} \end{aligned}$$

where $A = c/b$, $K = 1/b$, and $B = (d - b)b/c$.

A number of situations can lead to Eq 1. One example is provided by the following case.

Activation and Inhibition of a Single Enzyme

Since all metals used in this study are bivalent, they conceivably have two points of attachment to the proteins on which they act. Each divalent ion could facilitate enzyme activity when bound in chelant fashion to two groups at its attachment site. The metal could inhibit enzyme activity when two ions are bound to the attachment site, each combined by a single valence to the enzyme at a separate group.

The first step in metal binding would be the complexing of the metal ion M with the enzyme E to one of the groups at the attachment site

$$E + M \underset{-1}{\overset{1}{\rightleftharpoons}} C_1$$

The complex C_1 could then be converted to an active form (C_1') by the metal becoming anchored to both binding groups at the attachment site, or it could be converted to another inactive form (C_2) upon combining with a second metal ion at the other binding group:

$$C_1 \underset{-2}{\overset{2}{\rightleftharpoons}} C_1'$$

or

$$C_1 + M \underset{-3}{\overset{3}{\rightleftharpoons}} C_2$$

If this metal activator is not essential for enzyme activity, then substrate (S) can combine with E and C_1' and convert to product (P):

$$E + S \underset{-4}{\overset{4}{\rightleftharpoons}} C_3 \overset{5}{\rightarrow} E + P$$

and

$$C_1' + S \underset{-6}{\overset{6}{\rightleftharpoons}} C_4 \overset{7}{\rightarrow} E + M + P$$

The conservation equation for enzyme is then

$$E_t = E + C_1 + C_1' + C_2 + C_3 + C_4$$

and the initial rate of the reaction in the presence of the metal can be written as

$$v_M = k_5 C_3 + k_7 C_4$$

Assuming that steady-state conditions hold for the above reactions, the initial velocity becomes

$$v_M = v \frac{K + [M]}{K + [M] + A(KB + [M])[M]}$$

where

$$K = k_{-1}k_{-2}k_4k_5(k_{-6} + k_7)/k_1k_2k_6k_7(k_{-4} + k_5)$$
$$A = k_1k_3(k_{-4} + k_5)K/k_{-1}k_{-3}(k_{-4} + k_5 + k_4[S])$$

and

$$B = \frac{1}{AK}[Ak_{-3}/k_3 + Ak_2k_{-3}/k_{-2}k_3 + Ak_2k_{-3}k_6[S]/k_{-2}k_3(k_{-6}) + k_7) - 1]$$

Rearranging the above expression for initial velocity gives the model in the form of Eq 1.

APPENDIX II

Metals That Fit a Two-Parameter Model

Results of regression analysis:

Ba (Statistics based on 20 observations, assuming a value of $K = 120$)
- Estimate of $\log A = -2.10881$
- Standard error of estimate of $\log A = 0.158503$
- Value of $r = 0.97000$
- Estimate of slope = 0.954306
- Standard error of slope estimate = 0.056379

Cu (Statistics based on 20 observations, assuming a value of $K = 12$)
- Estimate of $\log A = -0.514578$
- Standard error of estimate of $\log A = 0.152308$
- Value of $r = 0.98348$
- Estimate of slope = 1.021508
- Standard error of slope estimate = 0.044319

Cr (Statistics based on 7 observations, assuming a value of $K = 12$)
- Estimate of $\log A = -0.93611$
- Standard error of estimate of $\log A = 0.144107$
- Value of $r = 0.981283$
- Estimate of slope = 1.040043
- Standard error of slope estimate = 0.091276

Ni (Statistics based on 7 observations, assuming a value of $K = 12$)
- Estimate of $\log A = -0.14951$
- Standard error of estimate of $\log A = 0.135874$
- Value of $r = 0.988764$
- Estimate of slope = 0.995428
- Standard error of slope estimate = 0.067302

The accompanying graphs of these metals shown in Fig. 5 are plots of log $(P - P_M)/P_M$ versus log $[M]^2/(K + [M])$.

Metal That Fits a One-Parameter Model ($BK = 1$)

Mn (Statistics based on 7 observations; no assumptions made of parameter values)
- Estimate of log $A = -1.70425$
- Standard error of estimate of log $A = 0.91776$
- Value of $r = 0.981754$
- Estimate of slope $= 0.96411$
- Standard error of slope estimate $= 0.078314$

The accompanying graph shown in Fig. 5 is a plot of log $(P - P_M)/P_M$ versus log $[M]$. (The slight departure of the slope estimate from the theoretical value of 1.0000 is due to one point that fell below the others.)

Metal That Fits a Three-Parameter Model

Cd (Statistics based on 4 observations, assuming that $K = 33$ and $BK = -3.3$)
- Estimate of log $A = -0.05664$
- Standard error of estimate of log $A = 0.030848$
- Value of $r = 0.99952$
- Standard error of slope estimate $= 0.023838$

No graph was made of this metal.

References

[1] Lussem, H. and Rahman, A., "Root Length Test with Garden Cress-Simple Ecotoxicological Test," *Vom Wasser,* Vol. 54, 1980, pp. 29–35.
[2] Brusick, D. J. and Young, R. R., *IERL-RTP Procedures Manual: Level 1, Environmental Assessment-Biological Tests,* U.S. Environmental Protection Agency EPA-600/8-81-024, Washington, D.C., 1981, 138 pp.
[3] Wong, M. H. and Bradshaw, A. D., "A Comparison of the Toxicity of Heavy Metals, Using Root Elongation of Ryegrass, *Lolium perenne,*" *New Phytologist,* Vol. 91, 1983, pp. 255–261.
[4] Ratsch, H. C., "Interlaboratory Root Elongation Testing of Toxic Substances on Selected Plant Species," U.S. Environmental Protection Agency EPA-600/S3-83-051, Washington, D.C., 1983.
[5] Wang, W., "The Use of Plant Seeds in Toxicity Tests of Phenolic Compounds," *Environment International,* Vol. 11, 1985, pp. 49–55.
[6] Wang, W., "Use of Millet Root Elongation for Toxicity Tests of Phenolic Compounds," *Environment International,* Vol. 11, 1985, pp. 95–98.
[7] Wang, W., "Comparative Toxicology of Phenolic Compounds Using Root Elongation Methods," *Environmental Toxicology and Chemistry,* Vol. 5, 1986, pp. 891–896.
[8] Wang, W., "Root Elongation Method for Toxicity Testing of Organic and Inorganic Pollutants," *Environmental Toxicology and Chemistry,* Vol. 6, 1987, pp. 409–414.
[9] *Environmental Risk Assessment,* A. V. Whyte and I. Burton, Eds., Wiley, Chichester, England, 157 pp.
[10] Wang, W., "Toxicity Tests of Aquatic Pollutants by Using Common Duckweed," *Environmental Pollution,* Vol. 11, 1986 (B), pp. 1–14.

[11] Spehar, R. L., Christensen, G. M., Curtis, C., Lemke, A. E., Norberg, T. J. and Pickering, Q. H. "Effects of Pollution on Freshwater Fish (Review)." *Journal of the Water Pollution Control Federation,* Vol. 54, 1982, pp. 877–922.
[12] Stephan, C. E., Mount, D. I., Hansen, D. J., Gentile, J. H., Chapman, G. A., and Brunge, W. A., "Guidelines for Deriving Numerical National Water Quality Criteria for Protection of Aquatic Life and its Uses," NTIS, Springfield, Va., 1985, 98 pp.
[13] Segal, I. H., *Enzyme Kinetics,* Wiley, New York, 1975.

New Approaches

Donald Miles[1]

The Role of Chlorophyll Fluorescence as a Bioassay for Assessment of Toxicity in Plants

REFERENCE: Miles, D., "**The Role of Chlorophyll Fluorescence as a Bioassay for Assessment of Toxicity in Plants,**" *Plants for Toxicity Assessment, ASTM STP 1091,* W. Wang, J. W. Gorsuch, and W. R. Lower, Eds., American Society for Testing and Materials, Philadelphia, 1990, pp. 297–307.

ABSTRACT: The typical green terrestrial plant is well adapted to sensing and reporting significant changes in its environment. This allows native plants growing in natural settings to be used to assess changes that might be toxic to plant or animal tissue. The basis of this bioassay is the chlorophyll molecule, which serves as an intrinsic fluorescent probe of the performance and capacity of photosynthesis. Under normal conditions, 97% of the light energy absorbed by chlorophyll is converted to biochemical forms of energy in photosynthesis. Stress conditions can reduce the rate of photosynthesis, disturb the pigment-protein apparatus, or block the light-driven photosynthetic electron transport in the chloroplast. This results in an increased loss of absorbed light energy of 6 to 10% via chlorophyll fluorescence with a peak in emission at 683 nm at physiological temperatures. The inverse relationship between *in vivo* chlorophyll fluorescence and photosynthesis has long been known as the *Kautsky Effect*.

Light-induced chlorophyll fluorescence from dark-adapted leaves can be recorded with portable, sensitive instruments using intact leaves. This nondestructive method essentially monitors the physiological well-being of the plant. Any stress including disease, nutritional stress, water, temperature, radiation, and chemical stress can be quickly and accurately recorded. The overall photosynthetic process can be thought of as a series of sensitive sites connected to the fluorescent photosynthetic reaction center, which respond to a large number of different insults and report these effects as a change in fluorescence. Chlorophyll fluorescence in intact native plants can be used to assess toxicity in the environment or in a laboratory bioassay.

KEY WORDS: photosynthesis, chlorophyll, fluorescence, chloroplasts

The use of *in vivo* chlorophyll-*a* fluorescence as a measurement of photosynthesis is being applied more frequently to a variety of fields of plant physiology [1,2]. The chlorophyll molecule can be considered an intrinsic fluorescent probe of the photosynthetic system in chloroplast. Fluorescent probes can report externally the physiological conditions occurring in the most basic biosynthetic process of plants. In the leaf of higher plants or in algal cells, the yield of fluorescent emissions is influenced in a number of complex ways by processes that are either directly related to photosynthesis or indirectly influence photosynthesis. This report will review the use of the fluorescence signal to monitor the physiological well-being of the individual plant—in particular, the fluorescence emission by isolated leaf sections or by intact chloroplasts, which have been intensely studied. These findings apply directly to

[1] Division of Biological Sciences, Tucker Hall, University of Missouri, Columbia, MO 65211.

the fluorescence emission observed by entire photosynthetic organs, such as stems or leaves. The fluorescent characteristics of the isolated chloroplast are much better controlled and more carefully studied than the entire leaf. The basic interpretations of the changes in the fluorescent signal of the chloroplast can be applied to the intact plant leaf or to a larger plant canopy, provided care is taken to include adequate controls. In order to compare fluorescent emission from one experimental situation to another the conditions must be very clearly defined. This article discusses the basic description of the light emission system of photosynthesis and what this fluorescence reveals to the investigator about the state of the photosynthetic process. Examples of effects of chemical stress and other well-known environmental stresses on changes in fluorescence are also given. There is an extensive literature available resulting from the basic study of the photosynthetic mechanism which can be applied to assessment of chemical toxicity. Researchers have utilized a large number of different types of inhibitors to dissect the photosynthetic system. In addition, there is a large literature developing on environmental stress effects on chlorophyll fluorescence and the use of fluorescence in characterizing these stresses. The majority of the effects described here will be characteristic fluorescent emission from plants in natural conditions of temperature on a slow time scale (15 to 30 s). The much faster microsecond or picosecond changes in chlorophyll fluorescence are more closely related to the primary photophysical and photochemical events of photosynthesis and will not be discussed here. The picosecond time scale of fluorescence is much more difficult to measure and would have less application to stress physiology or chemical toxicity studies.

General Description of Photosynthetic Apparatus

Light energy utilized in photosynthesis by higher plants and algal cells is absorbed by a number of photosynthetic pigments with absorption spectra covering a large range of the available light energy. The most prominent pigments that absorb this energy are chlorophyll-a and chlorophyll-b (Fig. 1). The light energy that is absorbed by the chloroplast first excites pigment molecules of the light harvesting chlorophyll proteins (LHC). These LHC proteins transfer their energy to either Photosystem I (PS I) or Photosystem II (PS II). These photosystems contain the reaction center pigments for the conversion of absorbed light energy to oxidation and reduction potential to drive dark electron transport. Light energy absorbed initially by the LHC and transferred to the reaction centers is lost by a number of different mechanisms. Approximately 3% of the light energy absorbed by chlorophyll pigments is re-emitted from the first excited state as fluorescence. Figure 2 shows the typical fluorescence emission spectrum of chloroplasts or whole photosynthetic cells. At low temperature this fluorescent emission has a major peak at 683 nm, a shoulder at 695 nm, and a broad second peak at 740 nm. At room temperature, light energy absorbed in photosynthesis is re-emitted and observed at the 683 nm emission peak. The light energy absorbed by the reaction center drives photosynthetic electron transport through PS II and PS I leading to the oxidation of water, oxygen evolution, the reduction of $NADP^+$ to NADPH, membrane proton transport, and eventually to ATP synthesis (Fig. 3).

The loss of light energy from the reaction center as fluorescence comes primarily from the PS II reaction. When the chloroplast or leaves have been dark-adapted, the pools of oxidation or reduction intermediates for the electron transport pathway return to a common level. Upon illumination of a dark adapted leaf, there is a rapid rise in light emission from PS II fluorescence followed by a series of slow oscillations. This is referred to as the *Kautsky Effect*. Figure 4 shows the usual onset kinetics of fluorescent emission from a typical dark adapted higher plant leaf. Changes in the fluorescent yield in the kinetics of fluorescent emission from

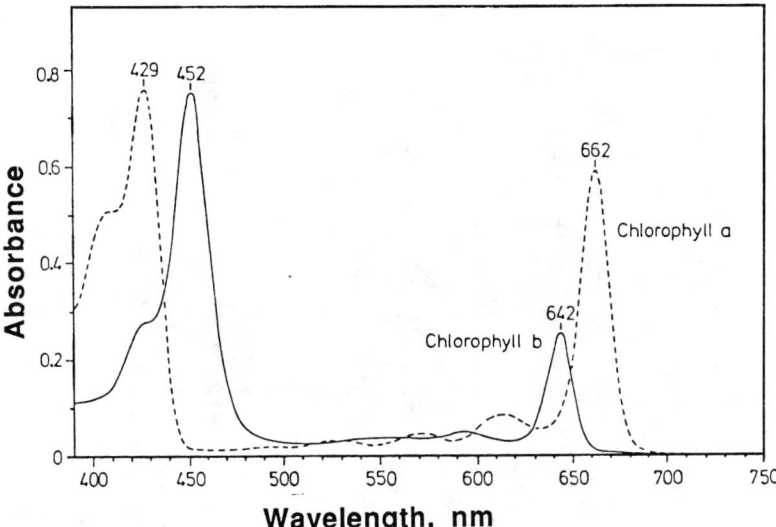

FIG. 1—*Absorption spectrum of solvent extracted and separated chlorophyll-a and chlorophyll-b.*

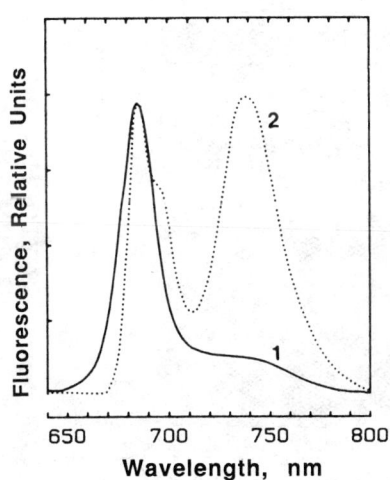

FIG. 2—*Fluorescence emission spectrum of isolated chloroplasts with excitation at 430 nm. (1) Solid line is typical spectrum at 25°C. (2) Broken line is emission spectrum at 77 K.*

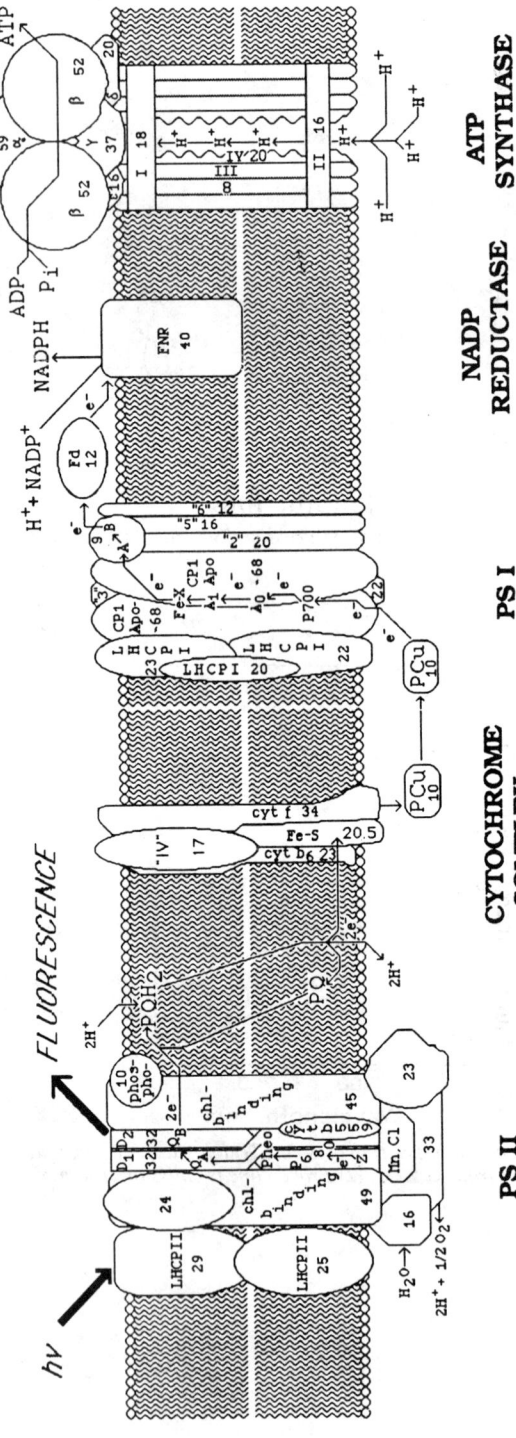

FIG. 3—*Model for organization of chloroplast inner membrane showing the relationship of PS I, PS II, cytochrome complex, and ATP synthetase. This model illustrates the path of electron flow from water to NADP. The apparent molecular mass for each polypeptide is indicated in kDa by the numbers.*

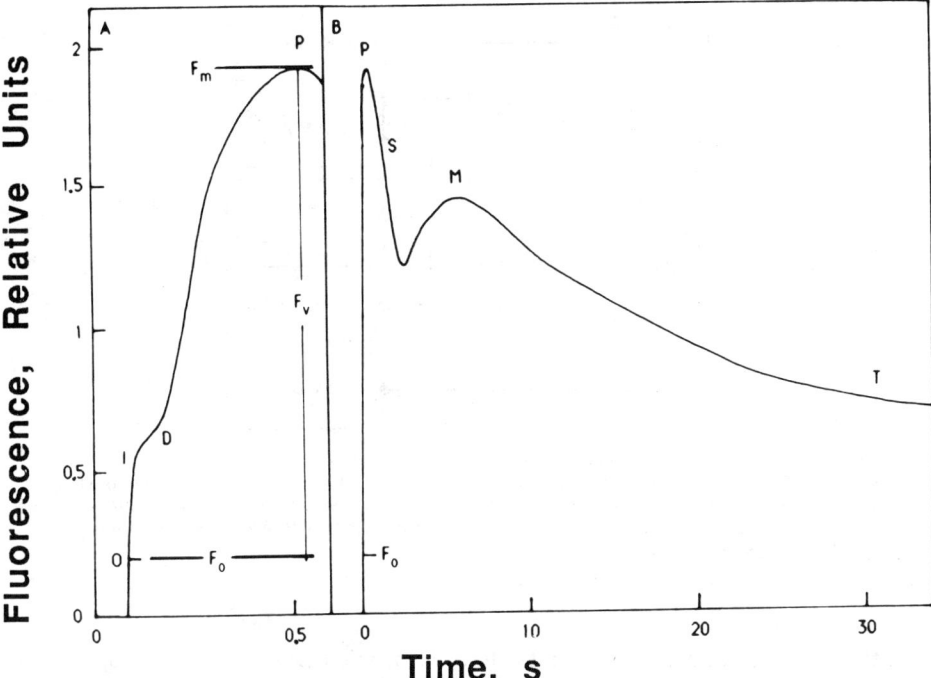

FIG. 4—*Typical response of chlorophyll fluorescence from a dark-adapted leaf from the time of illumination.*

dark-adapted leaves are sensitive to changes in the photosynthetic apparatus. Following many years of study of chlorophyll fluorescence to analyze its relationship to photosynthesis and to characterize photosynthesis, we know that any unusual change in overall bioenergetic status of the plant can be detected by a change in chlorophyll fluorescence [*3*]. This includes all the reactions from the oxidation of water through electron transport, development of the electrochemical gradient, ATP synthesis, and eventually the series of enzymatic reactions for CO_2 reduction to carbohydrate in the leaf. Even changes in the plant that affect stomata opening and gas exchange with the atmosphere are reflected by changes in the fluorescence characteristics as a leaf.

Measurement Methods

A variety of instruments has been used to record the fluorescent emission from chlorophyll in chloroplasts or plant leaves. The requirements for these instruments are (1) an actinic light source that will excite any photosynthetic pigment and (2) a method for measurement of the 683 or 740 nm emission peak of chlorophyll while excluding the actinic illumination. A typical instrument for measuring fluorescence kinetics of leaves or chloroplasts is shown in Fig. 5. In this instrument blue light is provided by a tungsten light source through a blue glass filter with a peak transmission of 430 nm. Fluorescence emission is measured with a photomultiplier tube or amplified photodiode blocked by a red glass cut-off filter (transmits 90% of the light over 670 nm). With this apparatus the dark-adapted leaf is oriented so that

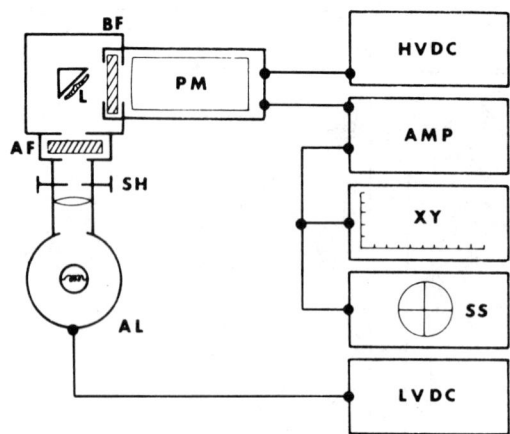

FIG. 5—*Diagram of kinetic fluorimeter.* (LVDC) *Low voltage power supply for the actinic lamp.* (AL) *Actinic lamp.* (SH) *Photographic shutter.* (AF) *Actinic filter (broad blue band).* (L) *Leaf.* (BF) *Blocking filter (red light transmitting).* (PM) *S-20 response photomultiplier (extended red sensitive).* (HVDC) *High voltage power supply.* (AMP) *Photocurrent amplifier.* (XY) *Plotter.* (SS) *Storage oscilloscope.*

when the photographic shutter is open to allow the actinic beam to excite chlorophyll, the yield of fluorescence emission from the leaf is recorded by the sensitive photomultiplier tube. The signal from the photomultiplier tube or photodiode is amplified and recorded on a chart recorder or, for faster recordings, a storage oscilloscope. In a modern instrument, the recordings can easily be made by interface boards and stored in a personal computer.

In addition to this laboratory instrument, which can be constructed simply, a small number of portable field instrument using photodiode light sources and solid-state photodetectors are now available commercially. These instruments are very useful for environmental field work, provided good controls are used to obtain accurate measurements [4,5].

The other general form of instrument used for recording fluorescence characteristics of photosynthetic organisms is the spectrofluorometer. The spectrofluorometer utilizes two monochromators in order to scan the exciting wavelengths of energy or to measure the emission wavelengths. A standard spectrofluorometer utilizes a high-intensity xenon light source through grating monochromators to provide precise wavelengths of actinic illumination to the sample. The emission is measured from the sample through a precision monochromator (usually double-grating) and detected on a wide-range, sensitive photomultiplier over the 400 to 750 nm range. With this instrument it is possible to measure excitation spectra for fluorescence at one wavelength or emission spectra of the photosynthetic tissue over a wide range. The most useful form of this spectrofluorometer contains a low temperature (liquid nitrogen) sample holder in order to measure high resolution fluorescence emission forms from the chloroplast (Fig. 2).

This report deals primarily with the simple fluorescence measurements of chlorophyll emission from dark-adapted leaves over 30 s. In the measurement shown in Fig. 4 the typical response has been identified by a series of phases. Immediately following excitation, the chlorophyll fluorescence rises to a point 0. From the initial point 0 there is a slower rise to a small peak (I) followed by a decline (D) and then the maximum level of fluorescence emission, referred to as P for the peak. This peak is reached in the average instrument at approximately

0.1 to 1.0 s after illumination. The timing for this series of oscillations to the peak depends upon a number of factors, including the intensity of the actinic light. After the fluorescence has risen to the peak in intact leaves, it now declines to a semi-steady state, S, and will rise in a second peak, commonly called M. Following the second smaller peak, there is a further decline to a level similar to S now referred to as T, the terminal level of fluorescence. In almost every photosynthetic system studied, this same series of oscillations occurs within the first 30 s of illumination of a dark adapted leaf. With chloroplasts the change in fluorescence ends with P.

After years of intensive study, we have information about each of these fluorescence changes. In order to compare the emission of one sample to another, a series of standard measurement is usually made. These measurements are referred to as F_0, for the initial level of fluorescence followed by F_M for the maximum level of fluorescence at (P). The difference between F_M and F_0 is the variable fluorescence (F_V), the most meaningful characteristic used to follow the physiological state and photosynthetic capacity of the photosynthetic apparatus. The variation of F_0, F_M, and F_V with light intensity is illustrated in Fig. 6. It is clear that F_V/F_M varies little with light intensity and this parameter can be used as a universal measurement of the physiological state of the chloroplast. Measurement of F_V or F_M alone is highly light intensity-dependent.

The electron transport reactions in the chloroplast, which are most important in determining the level of *in vivo* chloroplast fluorescence, have an effect on the oxidation-reduction state of the initial stable electron acceptor of PS II (Q_A). In the reaction center of PS II the primary chlorophyll, P-680, is excited by absorbed light energy. P-680 eventually reduces the Q_A electron acceptor in the PS II reaction center (Fig. 3). Q_A is a special plastoquinone bound to one of the reaction center polypeptides of PS II. If this acceptor is oxidized, it will receive the electron from the reaction center and the level of fluorescence will remain low (is quenched, therefore Q). If this electron acceptor is reduced (Q_A), there is no immediate place for the electron from the reaction center to go, and the excited states of the reaction center will collapse back, releasing their energy as fluorescent emission of the chlorophyll. The key to regulation of the level of fluorescence of PS II (and therefore the entire chloroplast or the photosynthetic apparatus) is the oxidation-reduction state of Q_A. Since Q_A can be oxidized by all of the electron carrier pool between PS II and PS I, then any change in the ability of the carriers between PS II and PS I to oxidize Q_A will affect the level of fluorescence of the leaf. This is why we can use *in vivo* fluorescence to monitor all the electron transport reactions from PS II through the cytochrome complex to PS I. Through those reactions that generate membrane potential, ATP synthesis, $NADP^+$ activation and reduction to NADPH (Fig. 3), and eventually the utilization of this reducing potential for CO_2 reduction, any change in the reactions will affect the redox level of Q_A. This can be monitored as changes in the characteristics of fluorescence from dark-adapted leaves. Limitations of electron transport on the oxidizing (water splitting) side of PS II between the PS II reaction center and water will have the opposite effect on fluorescence. The level of fluorescence will remain low rather than high. A limitation of electrons being donated to the reaction center of PS II causes the fluorescence to remain at a level near F_0.

Application of Fluorescence

The characteristics of inhibition of photosynthesis allow us to use fluorescence as a monitor of the overall rate of photosynthetic electron transport [6]. Any alterations of electron transport on either the the oxidizing or the reducing side of PS II will cause a detectable

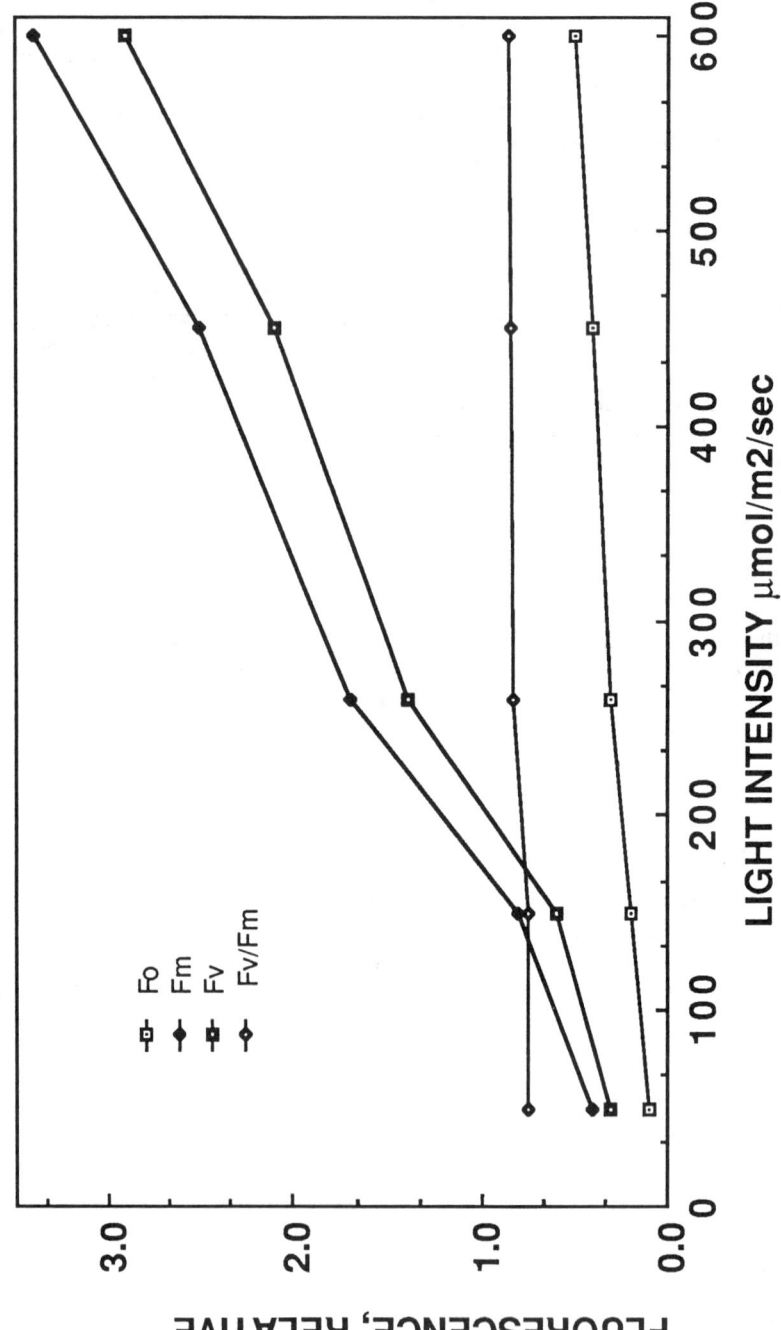

FIG. 6—*Comparison of the effect of light intensity on chlorophyll fluorescence parameters.*

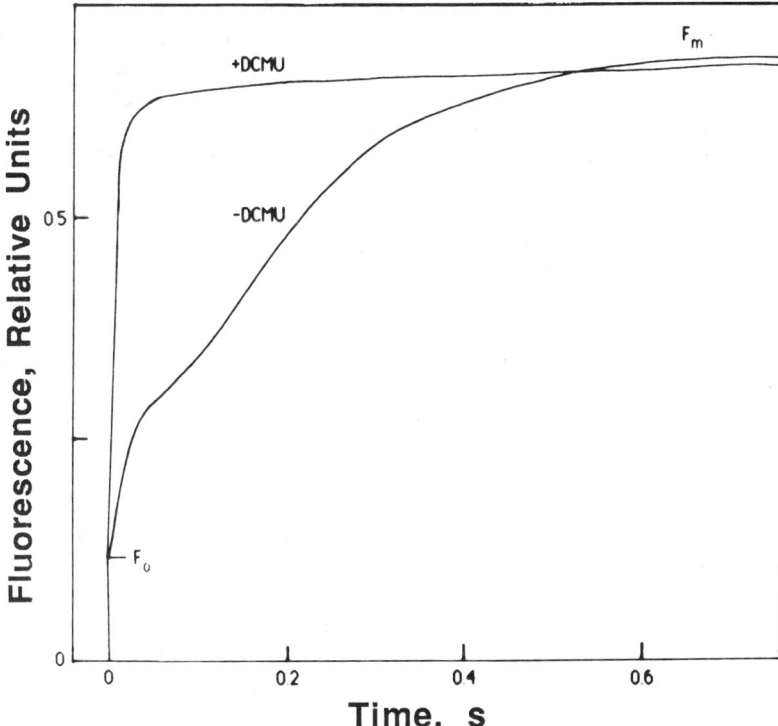

FIG. 7—*Chloroplast fluorescence changes in the presence of the inhibitor 3-(3,4-dichlorophenyl)-1,1-dimenthyl-urea (DCMU).*

change in the level and the emission spectrum of fluorescence. This system is an extremely useful intrinsic fluorescent probe of the bioenergetic status of the whole plant.

A typical effect of an inhibitor of photosynthesis is shown in Fig. 7. Whole plants or isolated chloroplasts exposed to a herbicide known to inhibit photosynthesis 3-(3,4-dichlorophenyl)-1,1-dimethyl-urea (DCMU) has a very dramatic effect on the fast level of fluorescent emission [7]. DCMU blocks electron transport just subsequent to the Q_A step. The only electron acceptors available are the limited pool of Q_A; therefore, when treated with DCMU, we find a very small change in variable fluorescence and a very high yield of fluorescence. This small change reflects the available Q_A's and the high yield reflects the blocked overall process. This increases the emission of fluorescence from the usual 3% level to 6 to 10% level. The specific site for DCMU inhibition of electron transport is well known. Any of the atrazine-type herbicides bind at or near this site, causing a block in electron transport and an immediate response in chlorophyll fluorescence. There have been several similar examples in the literature of the use of chlorophyll fluorescence to monitor the presence of herbicides in the environment [7,8].

Other inhibitors of electron transport that affect fluorescence are heavy metals [8]. We have studied the effects of lead, cadmium, and mercury on electron transport [9–11]. These elements either increase or decrease the level of F_M or F_V. With limited experimentations we can predict with some precision the site of interaction of these compounds with electron transport.

Specific genetic mutants of photosynthesis in the higher plants have also been very useful [*12*]. These genetic mutants have lesions in a variety of sites throughout the photosynthetic process, and each has a characteristic effect on the fluorescence emission. By knowing the locus of the mutation, we can now correlate change in specific photosystems with the emission characteristics of fluorescence. Working in reverse, it is possible to measure an effect of any type of stress on photosynthesis and, with our available knowledge, predict the reaction or sets of reactions in photosynthesis that may be responding to this stress.

This aspect of chlorophyll fluorescence has been used in a variety of environmental studies. In the study of stress, the effects have been quantitated by the use of chlorophyll fluorescence [*2,5,6,13*]. Water stress, nutrient stress, high or low temperature stress, the effect of high light intensity and/or ultraviolet light, all have been monitored through the changes in chlorophyll fluorescence emission. In addition, even gaseous pollutants affecting entire plants can have an effect on the emission of light energy in fluorescence. Studies of ozone damage in leaves have utilized fluorescence monitoring [*14*].

Previous work in which chlorophyll fluorescence has been used as a tool in general plant physiology [*15–17*] has measured emission kinetics and spectral changes of fluorescence at both room temperature and at low temperature (77 K). These changes can also be monitored to assess the effects of enviornmental pollutants or chemicals on the state of photosynthesis. In addition to fluorescence, there is a slow light admission (luminescence) from the reaction center with a half-time in the millisecond range [*4*]. This luminescence or delayed fluorescence indicates the recombination of electron acceptors and electron donors at the reaction center. Delayed fluorescence has been very useful in monitoring the funciton of reaction centers in photosynthesis and can also be useful in studying the effects of environmental changes on photosynthesis. However, delayed fluorescence is more difficult to measure, being only 1% of the fluorescence of PS II. Another method for monitoring changes in the bioenergetic of photosynthesis is measuring the carotenoid band-shift in the whole leaf. This bandshift occurs at 518 nm and is an important characteristic of PS I and PS II [*3*]. In addition to the measurement of fluorescence, both the delay fluorescence and the 518 nm absorbance change are further measurements that can be utilized to monitor photosynthesis in intact plants and provide further information.

Utilizing the Whole Plant Fluorescence

The development of our knowledge of chlorophyll fluorescence has been important to the biochemist and the physicist in understanding the basic reaction of photosynthesis. This information can be useful to environmentalist working in the opposite direction to determine how the photosynthetic systems have been altered. There are many sites in the electron transport chain related to photosynthesis that will sense a variety of different chemical compounds or stresses. Any changes in lipid soluble compounds or in highly reducing or oxidizing compounds will affect different sites in the electron transport system. The site can be almost immediately identified by monitoring the characteristics of chlorophyll fluorescence. In addition, any change in the series of carbon metabolism reactions of the chloroplast will eventually alter the level of the reduced NADP pool and this in turn will provide a characteristic change in the chlorophyll emission. Changes as remote as those affecting the gas exchange of the leaf will also be reflected in a change in fluorescence yield.

We can think of a higher land plant as a monitoring system of the environment. The plant is ideally suited, since it has a massive root system extending into the soil and ground water and taking up large amounts of water-soluble compounds. This extensive root system will allow the plant to collect and report on any chemical in its environment that is taken up. Any compound taken up by the root, transported through the stem xylem to the leaf, and finally to the leaf mesophyll cells can have an effect on photosynthesis. This would provide an immediate assessment, not only that a substance is limiting photosynthesis, but how it may be limiting photosynthesis and something about the chemical nature of the compound.

New instrumentation is being developed to image whole plants or groups of plants using solid-state video cameras. These instruments will record fluorescence emission characteristics in real-time using computer technology. This approach holds real promise for the use of chlorophyll fluorescence more widely to monitor any change in the characteristics of a plant. At present this monitoring is being accomplished at a 1 to 30 m range from the plant, but monitoring fluorescence from a much greater distance appears feasible [18].

We have the possibility of not only being able to use chlorophyll fluorescence in a well-controlled system in the laboratory to assess toxicity of chemical to biological systems but also moving that system into the environment. We should soon be able to use widely distributed key plants to assess changes in the environment by using either currently available portable instrumentation or remote sensing instrumentation.

Acknowledgments

The preparation of this manuscript was supported by research grant support of the U.S. Department of Agriculture, Grant CSRS 88-37234-3705.

References

[1] Krause, G. H. and Weis, E., *Photosynthesis Research*, Vol. 6, No. 1, 1985, pp. 139–157.
[2] Lichtenthaler, H. K. and Rinderle, U., *CRC Critical Reviews in Analytical Chemistry*, Vol. 19, 1988, pp. S29–S85.
[3] Hipkins, M. F. and Baker, N. R. in *Photosynthesis Energy Transduction; a Practical Approach*, M. F. Hipkins and N. R. Baker, Eds., IRL Press, Oxford, 4—Spectroscopy, 1986, pp. 51–102.
[4] Schreiber, U. and Schliwa, U., *Photosynthesis Research*, Vol. 11. No. 1. 1987, pp. 173–182.
[5] Oquist, G. and Wass, R., *Physiologia Plantarum*, Vol. 73, No. 2, 1988, pp. 211–217.
[6] Havaux, M., *Plant Physiology Biochemistry*, Vol. 26, No. 6, 1988, pp. 695–704.
[7] Harris, M. and Camlin, M. S., *Journal of Agricultural Science (Camb.)*, Vol. 110, No. 2, 1988, pp. 627–632.
[8] Samson, G. and Popovic, R., *Ecotoxicology and Environmental Safety*, Vol. 16, 1988, pp. 272–278.
[9] Miles, C. D., Brandle, J. R., Daniel, D. J., Chu-der, O., Schnare, P. D., and Uhlik, D. J., *Plant Physiology*, Vol. 49, No. 3, 1972, pp. 820–825.
[10] Lee, E. H. and Miles, C. D., *Plant Science Letters*, Vol. 5, 1975, pp. 33–40.
[11] Miles, D., Bolen, P., Farag, S., Goodin, R., Lutz, J., Moustafa, A., Rodriquez, B., and Weil, C., *Biochemical Biophysical Research Communications*, Vol. 50, No. 4, 1973, pp. 1113–1119.
[12] Miles, D., *Methods in Enzymology*, Vol. 69, 1980, pp. 3–23.
[13] Havaux, M., Ernez, M., and Lannoye, R., *Journal of Plant Physiology*, Vol. 133, No. 3, 1988, pp. 555–560.
[14] Rosema, A., Cecchi, G., Pantani, L., Radicatti, B., Romuli, M., Mazzinghi, P., van Kooten, O., and Kliffen, C., in *Applications of Chlorophyll Fluorescence*, H. K. Lichtenthaler, Ed., Kluwer Academic Publications, Dordrecht, The Netherlands, 1988, pp. 307–317.
[15] Peterson, R. B., Sivak, M. N., and Walker, D. A., *Plant Physiology*, Vol. 88, No. 1, 1988, pp. 158–163.
[16] Lichtenthaler, H. K., *Journal of Plant Physiology*, Vol. 131, No. 1, 1987, pp. 101–110.
[17] Stein, U., Buschmann, C., and Blaich, R., *Vitis*, Vol. 25, No. 2, 1986, pp. 129–141.
[18] McFarlane, J. C., Watson, R. D., Theisen, A. F., Jackson, R. D., Ehrler, W. L., Pinter, P. J., Idso, S. B., and Reginato, R. J., *Applied Optics*, Vol. 19, No. 10, 1980, pp. 3287–3289.

Barbara M. Judy,[1] William R. Lower,[1] C. Donald Miles,[2] Mark W. Thomas,[1] and Gary F. Krause[3]

Chlorophyll Fluorescence of a Higher Plant as an Assay for Toxicity Assessment of Soil and Water

REFERENCE: Judy, B. M., Lower, W. R., Miles, C. D., Thomas, M. W., and Krause, G. F., "**Chlorophyll Fluorescence of a Higher Plant as an Assay for Toxicity Assessment of Soil and Water,**" *Plants for Toxicity Assessment, ASTM STP 1091,* W. Wang, J. W. Gorsuch, and W. R. Lower, Eds., American Society for Testing and Materials, Philadelphia, 1990, pp. 308–318.

ABSTRACT: The chlorophyll fluorescence (CF) assay is recognized as an important tool for monitoring the electron transport system of photosynthesis. The increase or decrease of fluorescence, compared to control, can indicate a lesion in the chain of events in photosynthesis or damage to the chloroplasts. Inhibition of photosynthesis in leaf segments of *Tradescantia* by herbicides (glyphosate, picloram & 2,4-D, triclopyr and hexazinone), heavy metals (Zn, Cu, Cd), a surfactant (sodium lauryl sulfate), and sodium fluoride (NaF) has been demonstrated using this method. The strongest changes of fluorescence induction curves were caused by hexazinone. Photosynthetic inhibition was observed as an increase in initial fluorescence (F_o) and a decrease in variable fluorescence (F_v) and electron pool size (EP). A CF assay predicted herbicidal injury in *Tradescantia* leaves at least 24 h before leaf necrosis appeared.

Three heavy metals (Zn, Cu, Cd) are known as strong respiration and photosynthetic inhibitors in plant cells. In this study cadmium chloride, at the highest concentration (1000 ppm) tested, caused the strongest changes in F_v and EP size, which are characteristics of Photosystem II (PS II) photochemistry. Phytotoxicity of NaF caused an increase in F_v and EP compared to the control.

KEY WORDS: bioassay, chlorophyll, fluorescence, herbicides, heavy metals, *Tradescantia*, fluorometer

Chlorophyll fluorescence (CF) is a sensitive indicator of the energetic state of photosynthesis. This phenomenon can be used to monitor a plant's response to any factor that directly or indirectly affects photosynthetic metabolism [1,2].

Most of the fluorescence (light) is emitted from chlorophyll-*a* of Photosystem II (PS II) and from other protein-bound pigments near the reaction center (a complex of energy collecting pigments). Photosystems I and II are two separate groups of pigment that cooperate in photosynthesis. PS II absorbs wavelengths shorter than 690 nm and its reaction center is called P680. The photo-energy excites an electron in P680 that moves from P680 to the primary electron acceptor Q. Loss of the electron causes P680 to become positively charged and strongly attracts an electron from H_2O. The primary electron acceptor is historically called Q because of its ability to quench (i.e., reduce) CF emission.

[1] Environmental Trace Substances Research Center, University of Missouri, Columbia, MO 65203.
[2] Department of Biological Sciences, University of Missouri, Columbia, MO 65203.
[3] Department of Mathematics Science, University of Missouri, Columbia, MO 65203.

When electron transport is blocked in photosynthesis, an increased proportion of the absorbed excitation energy is reemitted as fluorescence. Much work has been done using CF as a method to evaluate the effects of the environment, including stress conditions, on plant growth [3-7]. The understanding of CF has been based to a large part on studies with isolated chloroplasts. In recent years instrumentation has been developed for the detection of CF in leaf segments of higher plants [8,9]. This method is relatively simple and can be applied conveniently to both laboratory and field conditions.

The present study assesses the toxicity of soil to plants after herbicide applications and evaluates the toxicity of several chemicals dissolved in water by measuring changes in CF of *Tradescantia* leaves (Clone 4430) as an indicator of photosynthetic inhibition.

Materials and Methods

Plant Material

Tradescantia plants (Clone 4430) were grown in Pro-Mix (a commercial potting mixture) under white artificial light (75 to 100μE) in a growth room maintained at 22 $\pm$ 2°C and 16/8 h light/dark cycle. Nutrients and water were applied to maintain optimal growing conditions. Fully developed leaves of *Tradescantia* were cut and immediately immersed, cut surface down, in 3 cm of treatment solution. In the herbicide study, leaves were exposed for 5 and 25 h in soil eluate (SE) under the same light and temperature conditions. In the study of inorganic chemicals and a surfactant, fluorescence was measured after 1, 3, and 5 h of exposure. Two 2-cm-long pieces from the middle part of five leaves were taken to measure fluorescence for each treatment and control for a total of ten measurements for each treatment-concentration.

Treatments

The soil used in the first part of this study was taken from a fallow field on a research farm owned by the University of Missouri. The soil was removed from a 15-cm plow layer and mixed for 16 h to ensure homogenization. The soil referred to as East Farm Soil is classified as a clay loam (21.4% sand, 50.7% silt, and 27.9% clay) and contains 4.5% organic matter.

The air dried soil samples (125 g each) were placed in Pyrex dishes at a depth of 20 mm and sprayed separately with four herbicides using a Dow Chemical Company herbicide applicator with the adjustable spray nozzle set for a fine mist (approximately equivalent to a TG 2.5 nozzle). All herbicides, with the exception of Velpar ULW, were applied in enough reverse osmosis double deionized (RODD) water to achieve an initial soil moisture content of 25% on a w/w basis. The application rates were: Roundup (Monsanto) 2%, Tordon 101M (Dow Chemical) 6%, and Garlon 4 (Dow Chemical) 2.5%. Velpar ULW (DuPont Chemical) was applied as a granular formulation at a rate of 5.3 lb of active ingredient per acre and RODD water was added to the soil to obtain a moisture content of 25% on a w/w basis.

One hour after herbicide application, the soil was mixed with 500 mL RODD water in a 1 L Teflon container and agitated at 120 rpm for 48 h at 20°C in the dark. This was the extraction phase of the procedure. At the end of 48 h, the liquid was transferred to centrifuge tubes and centrifuged at 10 000 rpm for 10 min at 4°C. After centrifugation, the liquid, now referred to as soil eluate (SE), was filtered through filters of decreasing pore size, starting with 0.7 μm nonsterile filter, with a final sterile filtration through a 0.45 μm filter. The SE was used at three dilutions made with RODD water of 1%, 10%, and 80% for the CF assay. The same three concentrations (1%, 10%, and 80%) plus 50% SE were used for the *Selenastrum* assay [9a].

The estimated concentrations of each active ingredient of herbicide in 100% of SE were: hexazinone (Velpar ULW), 6.7 ppm; glyphosate (Roundup), 4.4 ppm; triclopyr (Garlon 4), 3.0 ppm; and picloram +2,4-D (Tordon 101M), 1.2 ppm and 9.6 ppm, respectively.

In the second part of the study, inorganic chemicals (cadmium chloride, zinc chloride, sodium fluoride, cupric chloride) and the surfactant (sodium lauryl sulfate) were dissolved in RODD water at concentrations of 100 and 1000 ppm. All treatments were compared to control treatment of RODD water.

Chlorophyll Fluorescence

When a dark-adapted leaf at room temperature is suddenly exposed to light, the CF yield, as emitted light, shows characteristic changes (Fig. 1). The fluorescence first rises rapidly to a level O that is called the "constant" or "initial" fluorescence (F_o) and denotes fluorescence when the reaction center of PS II is fully oxidized. The quantum yield of F_o is independent of photochemical events and reflects the amount of chlorophyll. F_o is followed by variable fluorescence (F_v) of PS II. F_v is dependent on the redox state of the quencher Q, the first stable electron acceptor of PS II. Any action causing Q oxidation tends to result in fluorescence quenching (i.e., reduction in the emission of light). Oppositely, Q reduction increases the fluorescence yield. The rate of the fluorescence rise is proportional to the sum of Q reduction and oxidation rates [10]. The level P characterizes the "total" or "maximum" fluorescence yield (F_m) which is composed of F_o and F_v. The increase or decrease of fluorescence, compared to the control, can indicate a lesion in the chain of events in photosynthesis or damage to the chloroplasts. The "electron pool" (EP) is the area above the fluorescence curve between the O-P transients. This area characterizes the photochemical capacity of PS II, is proportional to the PS II electron acceptor pool, and determines the PS II photochemical activity [11].

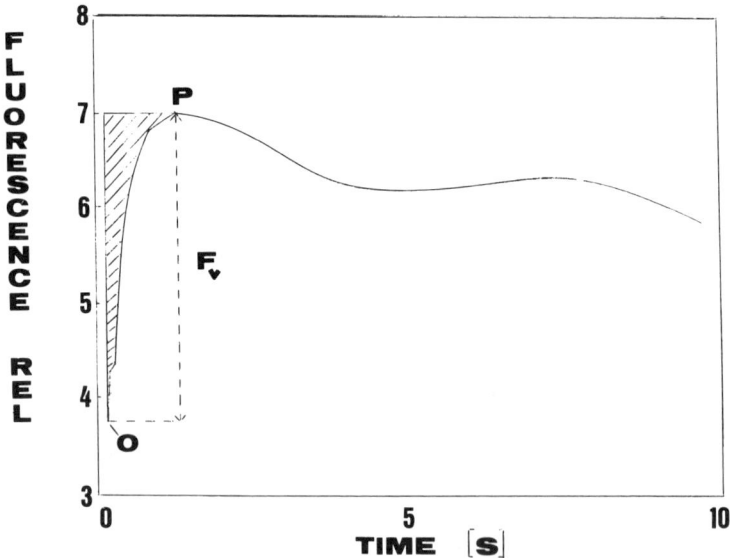

FIG. 1—*Schematic representation of chlorophyll fluorescence induction curve from* Trades-cantia *leaf.* (O) *Initial fluorescence* (F_o). (F_v) *Variable fluorescence.* ($F_o + F_v = P$) *Maximum fluorescence* (F_m). *Hatched area represents the electron pool (EP) size.*

Fluorescence induction curves were recorded by a fluorometer (Model SF-30, Richard Brancker Research Ltd., Ottawa, Canada) attached to a strip-chart recorder and to a personal computer. Fluorescence was measured at room temperature from the abaxial surface of *Tradescantia* leaves. The leaves were illuminated for 10 s with red light (670 nm, intensity 8.5 W/m^2) after a 20 min period of dark adaptation. The SF-30 chlorophyll fluorometer is able to store both the numeric data and the induction curves of up to 99 samples and provides interfaces for a digital plotter and a printer or personal computer. Parameters of particular interest in this study were the initial fluorescence level (F_o), variable fluorescence (F_v), and electron pool (EP). The shape of the curves was also considered.

Statistical Analysis

An analysis of variance with factorial sources for treatment, concentration and time was computed for each variable (F_o, F_v, and EP) for herbicides and chemicals separately. Means were compared using a protected least significant difference (LSD) rule (α = 0.05) [*12*]. Values in each table for each chemical followed by the same upper case letter do not significantly differ at the 5% level. The coefficient of variations for control are given in the Appendix for each of the fluorescence parameters.

Results and Discussion

Herbicides

Figures 2 and 3 show the fluorescence induction curves from *Tradescantia* leaves exposed to the aqueous eluate with varying concentrations of Velpar ULW. After 5 h of exposure

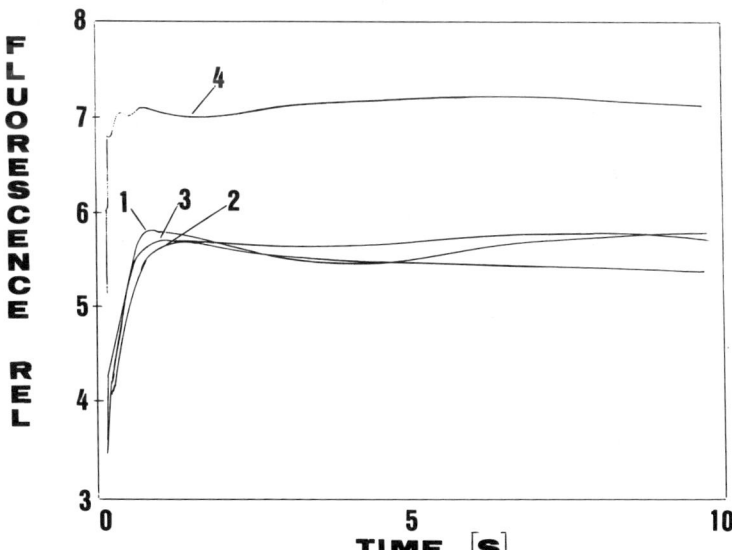

FIG. 2—*Example of chlorophyll fluorescence induction curves from* Tradescantia *leaves exposed to three concentrations of aqueous soil eluate (SE) with hexazinone for 5 h. (1) Control (H_2O). (2) 1% SE. (3) 10% SE. (4) 80% SE. The fluorescence is expressed in relative units.*

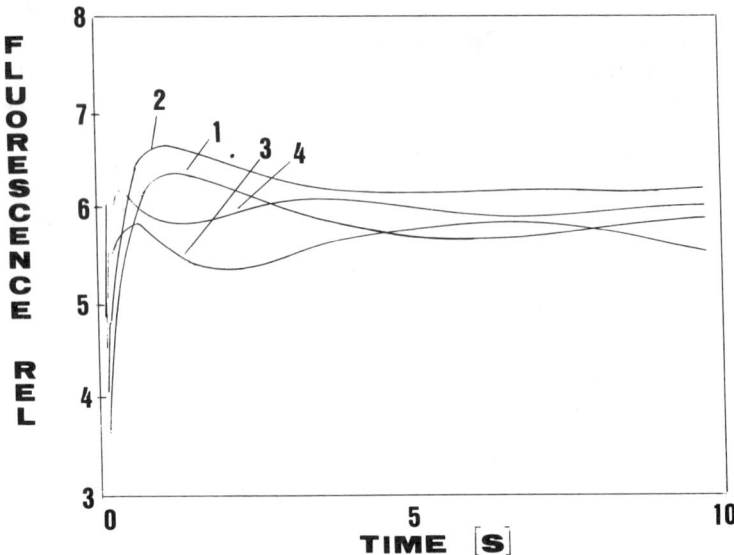

FIG. 3—*Example of chlorophyll fluorescence induction curves from* Tradescantia *leaves exposed to three concentrations of aqueous soil eluate (SE) with hexazinone for 25 h. (1) Control (H_2O). (2) 1% SE. (3) 10% SE. (4) 80% SE. The fluorescence is expressed in relative units.*

Tradescantia leaves show increased intensity of fluorescence of F_o level and changes in the shape of the curves, especially at highest concentration of SE (Fig. 2, Table 1). Longer exposure (25 h) to SE with Velpar ULW also caused significantly higher F_o at lower concentrations (1% and 10%) of SE. The F_o level is known to be affected by environmental stress which results in structural alterations at the PS II pigment level. Schreiber et al. [4,13] found that a dramatic increase in F_o was caused by thermal damage of PS II. Velpar ULW, especially

TABLE 1—*Percent changes in initial fluorescence* (F_o) *in* Tradescantia *leaves exposed to herbicides in soil eluate (SE).*

Time of Exposure, h	Concentration of SE, %	Herbicides in SE			
		Velpar ULW	Garlon 4	Tordon 101M	Roundup
		%			
5	0[a]	100 AC	100 A	100 A	100 A
	1	95 A	96 A	103 A	105 A
	10	108 C	97 A	98 A	110 A
	80	141 D	95 A	100 A	107 A
25	0[a]	100 AC	100 A	100 A	100 A
	1	120 B	102 A	97 A	109 A
	10	135 D	95 A	104 A	110 A
	80	138 D	98 A	106 A	109 A

[a] Control (H_2O).
[b] Values within a column for each herbicide followed by the same upper case letter do not differ at the 5% level according to the Protected LSD Rule, LSD = 11.0.

TABLE 2—*Percent changes in variable fluorescence (F_v) in* Tradescantia *leaves exposed to herbicides in soil eluate (SE).*

Time of Exposure, h	Concentration of SE, %	Herbicides in SE			
		Velpar ULW	Garlon 4	Tordon 101M	Roundup
		%			
5	0[a]	100 A	100 AB	100 A	100 A
	1	103 A	97 AB	105 AB	100 A
	10	96 A	108 A	106 AB	104 A
	80	104 A	99 AB	116 B	99 A
25	0[a]	100 A	100 AB	100 A	100 A
	1	93 A	104 A	102 A	99 A
	10	63 B	97 AB	107 AB	93 AB
	80	51 B	89 B	96 A	81 B

[a] Control (H_2O).
[b] Values within a column for each herbicide followed by the same upper case letter do not differ at the 5% level according to the Protected LSD Rule, LSD = 13.3.

the higher concentrations (10% and 80%) of SE, after 25 h of exposure, strongly reduced the F_v, about 43% (Table 2). After 25 h of exposure, a strong change in EP size were observed also. At 10% and 80% SE, the EP size was decreased about 81% (Fig. 3, Table 3). The mode of action of Velpar ULW (hexazinone) is not clearly understood, but it appears to be a photosynthetic inhibitor. The electron transport system of the Hill reaction is thought to be affected by specifically replacing a nonfunctioning cytochrome carrier protein (possibly B-559).

The F_m value is not always a useful indicator of herbicide toxicity due to the often opposite effects of F_o and F_v canceling each other. This was observed after 25 h of exposure of *Tradescantia* leaves to SE with Velpar ULW. The total fluorescence yield was not changed

TABLE 3—*Percent changes in electron pool (EP) size in* Tradescantia *leaves exposed to herbicides in soil eluate (SE).*

Time of Exposure, h	Concentration of SE, %	Herbicides in SE			
		Velpar ULW	Garlon 4	Tordon 101M	Roundup
		%			
5	0[a]	100 AD	100 AB	100 A	100 AB
	1	121 D	97 AB	108 A	112 A
	10	93 A	120 A	112 A	102 A
	80	48 B	104 AB	137 B	101 A
25	0[a]	100 AD	100 AB	100 A	100 AB
	1	65 B	103 AB	112 A	106 A
	10	18 C	92 B	116 AB	89 AB
	80	19 C	91 B	113 A	77 B

[a] Control (H_2O).
[b] Values within a column for each herbicide followed by the same upper case letter do not differ at the 5% level according to the Protected LSD Rule, LSD = 24.6.

because of F_o increased and F_v decreased. Samson and Popovic [6] also found that F_m yield is not a good indicator of pollutant phytotoxicity.

The other three herbicides do not influence photosynthesis directly. Roundup (glyphosate) inhibits the aromatic amino acid biosynthesis pathway and may also inhibit or repress chorismate mutase and/or prephenate dehydratase. Photosynthesis and respiration are impeded and protein synthesis is interrupted; this interferes with cell wall, cell membrane, chloroplast, and chlorophyll production. In this study Roundup did not change the F_o (Table 1). The F_v was significantly decreased but only at highest concentration after 25 h of exposure (Table 2). The causes of F_v lowering may be complex. Environmental stress that causes chloroplast damage (e.g., heat, cold, and photoinhibition) can lower the F_v yield [2]. Roundup had the tendency of also lowering the EP after 25 h of exposure at higher concentrations but differences were not significant compared to control (Table 3).

Tordon 101M (picloram & 2,4-D) and Garlon 4 (triclopyr) are auxin-type herbicides and induce characteristic responses in plants, such as tissue proliferation, leaf curling or epinasty, tumor formation, cessation of root elongation, and fruitless expenditure of carbohydrates in excessive respiration. Swelling parenchymal cells develop into callus tissue that induce mechanical crushing of the phloem as well. In this study both herbicides, especially Garlon 4, did not influence CF strongly. In the case of Tordon 101M, a tendency to higher F_v and EP was observed (Tables 2 and 3). After 5 h of exposure of *Tradescantia* leaves to 80% of SE F_v and EP were significantly increased about 16% and 37% respectively compared to control. After longer exposure changes were smaller and not significant compared to control.

Changes in photosynthetic activity of *Tradescantia* leaves in the CF Assay were noticed much earlier than the occurrence of visible injury. The first visible injury of *Tradescantia* leaves treated with Tordon 101M was seen after 48 h, with Garlon 4 after 5 days, and with Roundup and Velpar ULW after 8 days of exposure.

Heavy Metals

In the second part of the study, the three heavy metals (zinc, cadmium, and copper) were evaluated by the CF assay. These compounds are known as strong respiration and photosynthetic inhibitors in plant cells [14,15]. After 5 h of exposure of *Tradescantia* leaves to cadmium chloride and zinc chloride, in concentrations of 100 ppm and 1000 ppm, F_o was slightly decreased compared to control (Table 4). Only cadmium chloride caused stronger changes in F_v and EP (Tables 5 and 6). During the first 3 h of exposure F_v and EP were significantly higher. At the highest concentration (1000 ppm) of cadmium chloride, F_v was 22% higher and EP 31% higher than the control. The longer exposure to cadmium chloride (5 h) caused rapid decrease in F_v and EP, but these differences were not significant compared to control. De Filippis et al. [14] found that the three heavy metals (Zn, Cd, and Hg) strongly inhibit PS II activity, especially in young cultures of *Euglena*. Their observation demonstrated that the reduction in photosynthetic capacity in the presence of Zn, Cd, and Hg is due to interactions with the electron transport chain and the water splitting site, the Hill reaction. The reaction of *Tradescantia* leaves exposed to zinc chloride in this study was not as strong. After 5 h of exposure the F_o was slightly decreased and EP was slightly increased. These changes were not significantly different compared to control. Samson and Popovic [6] in an experiment with algae found that copper decreased the fluorescence intensity at the F_m transient without significant changes of the curve shape. These effects are attributed to an inhibition of photosynthetic electron transport on the oxidizing side of PS II. In this case the electron flow from the water-splitting system cannot reduce the PS II electron acceptor Q [16,17]. In this study *Tradescantia* leaves exposed to cupric chloride did not indicate significantly different CF compared to control (Tables 4, 5, and 6). The sensitivity of various

TABLE 4—*Percent changes in initial fluorescence (F_o) in Tradescantia leaves exposed to chemicals.*

Time of Exposure, h	Concentration, ppm	Chemicals				
		NaLS	$CuCl_2$	$CdCl_2$	$ZnCl_2$	NaF
				%		
1	0^a	100 A	100 AB	100 AB	100 AB	100 A
	100	120 B	91 A	105 A	106 A	103 A
	1000	128 B	101 AB	105 A	111 A	106 A
3	0^a	100 A	100 AB	100 AB	100 AB	100 A
	100	99 A	99 AB	107 A	100 AB	92 A
	1000	105 A	93 A	103 AB	102 AB	92 A
5	0^a	100 A	100 AB	100 AB	100 AB	100 A
	100	102 A	98 AB	90 B	91 B	99 A
	1000	100 A	112 B	88 B	87 B	97 A

^{a} Control (H_2O).
^{b} Values within a column for each chemical followed by the same upper case letter do not differ at the 5% level according to the Protected LSD Rule, LSD = 14.8.

TABLE 5—*Percent changes in variable fluoresence (F_v) in Tradescantia leaves exposed to chemicals.*

Time of Exposure, h	Concentration, ppm	Chemicals				
		NaLS	$CuCl_2$	$CdCl_2$	$ZnCl_2$	NaF
				%		
1	0^a	100 A	100 AB	100 A	100 A	100 A
	100	125 BC	90 A	122 B	92 A	117 AC
	1000	143 B	104 AB	124 B	99 A	122 BC
3	0^a	100 A	100 AB	100 A	100 A	100 A
	100	106 AC	87 A	122 B	99 A	134 BC
	1000	104 AC	86 A	122 B	106 A	137 BC
5	0^a	100 A	100 AB	100 A	100 A	100 A
	100	114 AC	85 A	86 A	98 A	120 AC
	1000	119 AC	119 B	90 A	100 A	121 C

^{a} Control (H_2O).
^{b} Values within a column for each chemical followed by the same upper case letter do not differ at the 5% level according to the Protected LSD Rule, LSD = 21.5.

TABLE 6—*Percent changes in electron pool (EP) in* Tradescantia *leaves exposed to chemicals.*

Time of Exposure, h	Concentration, ppm	Chemicals				
		NaLS	CuCl$_2$	CdCl$_2$	ZnCl$_2$	NaF
		←─────────────── % ───────────────→				
1	0[a]	100 A	100 A	100 AD	100 AB	100 A
	100	118 AC	95 A	127 AC	80 A	114 A
	1000	138 BC	100 A	131 BC	88 AB	121 A
3	0[a]	100 A	100 A	100 AD	100 AB	100 A
	100	177 AC	76 A	127 AC	94 AB	240 B
	1000	104 A	80 A	131 BC	104 AB	211 BD
5	0[a]	100 A	100 A	100 AD	100 AB	100 A
	100	115 AC	76 A	77 ED	114 B	170 C
	1000	126 AC	100 A	87 ED	110 B	194 CD

[a] Control (H$_2$O).
[b] Values within a column for each chemical followed by the same upper case letter do not differ at the 5% level according to the Protected LSD Rule, LSD = 30.8.

organisms can be different to the same toxicant. Also time of exposure and concentrations caused different results.

Other Chemicals

The third and last part of the study concerns the influence of two chemicals: sodium lauryl sulfate (a surfactant) and sodium fluoride on changes in CF. The surfactant caused strongest changes in fluorescence only after 1 h of exposure. At higher concentration of NaLS (1000 ppm), the F_o, F_v, and EP were higher compared to the control, about 28%, 43%, and 38% higher respectively (Tables 4, 5, and 6). Over time, fluorescence was similar to the control.

Tradescantia leaves treated with NaF showed higher F_v and EP. The strongest emission was observed after 3 h of exposure. A higher concentration (1000 ppm) of NaF caused 37% higher F_v and 111% higher EP (Tables 5 and 6). Significant changes in F_v and EP were observed also after 5 h of exposure.

Conclusions

The CF assay provides a simple and rapid method of studying the effects of toxicants in soil and water using higher plants. This method could also be useful in investigating the effects of various pollutants and the other stress factors that adversely affect photosynthesis. The advantage of this method is that results can be obtained very rapidly. In these experiments, toxic effects of herbicides were observed after only 25 h of exposure, much earlier than the appearance of visible injury.

Additional advantages of the fluorescence method is its solid foundation in the last 35 years of photosynthesis research. The kinetic and spectral characteristics of chloroplast and whole cell fluorescence is extremely well documented. Recording the changes in fluorescence spectral shifts, yields, and emission at different time points provides solid data on the rate and capacity for photosynthesis. This approach has been adapted by environmental physiologists to analyze the early stages of stress effects on plants. Natural changes in temperature, water availability, light quality, and light intensity are now being monitored through whole leaf fluorescence before any other changes in the plant occur. The present study extends the method to yet another important application of recording early changes of photosynthesis by minor quantities of soil toxicants.

Acknowledgments

The authors are grateful to Amha Asfair for writing an additional computer program and to D. Schuster and K. Bick for their expertise in preparation of this manuscript. Dr. Armon Yanders and Dr. Ravi Puri provided assistance in reviewing the document; their help is greatly appreciated. The following companies supplied products: Dow Chemical Company, DuPont Chemical Company, and the Monsanto Chemical Company.

APPENDIX

Appendix Tables 1 and 2 give the coefficients of variance for each of the fluorescence parameters.

APPENDIX TABLE 1—*Coefficients of variance (C.V.) for control (%).*

Fluorescence Parameters	Time of Exposure, h	Control for				
		NaLS	$CuCl_2$	$CdCl_2$	$ZnCl_2$	NaF
F_o	1	24.1	8.3	11.0	14.4	14.6
	3	12.8	13.0	10.7	8.7	11.8
	5	13.5	11.1	13.4	11.9	8.6
F_v	1	29.2	16.8	18.9	14.4	19.6
	3	11.7	17.8	12.0	5.4	13.7
	5	18.8	10.6	24.4	14.9	16.2
EP	1	30.0	26.3	24.3	13.0	37.2
	3	12.6	24.5	12.2	10.1	12.2
	5	26.5	17.7	29.9	19.1	24.3

APPENDIX TABLE 2—*Coefficients of variance (C.V.) for control (%).*

Fluorescence Parameters	Time of Exposure, h	Control for			
		Velpar ULW	Garlon 4	Tordon 101M	Roundup
F_o	5	8.8	8.5	8.0	11.8
	25	7.1	6.4	6.1	9.9
F_v	5	13.1	7.5	10.7	13.2
	25	6.3	8.4	11.4	10.4
EP	5	19.3	19.2	18.6	13.2
	25	19.5	11.6	18.8	20.5

References

[1] Smillie, R. M., *Plant Science Letters,* Vol. 28, 1982/83, pp. 283–289.
[2] Krause, H. G. and Weis, E., *Photosynthesis Research,* Vol. 5, 1984, pp. 139–157.
[3] Schreiber, U., Vidaver, W., Runeckles, V. C., and Rosen, P., *Plant Physiology,* Vol. 61, 1978, pp. 80–84.
[4] Schreiber, U. and Berry, J. A., *Planta,* Vol. 136, 1977, pp. 233–238.
[5] Hetherington, S. E., Smillie, R. M., Malagamba, P., and Huaman, Z., *Planta,* Vol. 159, 1983, pp. 119–124.
[6] Samson, G. and Popovic, R., *Ecotoxicology and Environmental Safety,* Vol. 16, 1988, pp. 272–278.
[7] Ali, A. and Machado, V. S., *Weed Research,* Vol. 21, 1981, pp. 191–197.
[8] Schreiber, U., *Photosynthesis Research,* Vol. 4, 1983, pp. 361–373.
[9] Lower, W. R., Yanders, A. F., Marrero, T. R., Underbrink, A. G., Drobney, V. K., and Collins, M. D., *Environmental Toxicology and Chemistry,* Vol. 4, 1985, pp. 13–19.
[9a] Thomas, M. W. et al., this publication, pp. 235–254.
[10] Vidaver, W., Popovic, R., Bruce, D., and Colbow, K., *Photochemistry and Photobiology,* Vol. 34, 1981, pp. 633–636.
[11] Lavorel, T., Brenton, J., and Lutz, M. in *Light Emission by Plants and Bacteria,* J. Govindjee and D. C. Fork, Eds., Academic Press, San Diego, Calif., 1986, pp. 57–98.
[12] Snedecor, G. W. and Cochran, W. G., *Statistical Methods,* ISU Press, Ames, Iowa, 1980.
[13] Schreiber, U. and Armond, P. A., *Biochimica et Biophysica Acta,* Vol. 502, 1978, pp. 138–151.
[14] De Filippis, L. F., Hampp, R., and Ziegler, U., *Archives of Microbiology,* Vol. 128, 1981, pp. 407–411.
[15] Li, E. H. and Miles, C. D., *Plant Science Letters,* Vol. 5, 1975, pp. 33–40.
[16] Samson, G., Morissette, J. C., and Popovic, R., *Photochemistry and Photobiology,* Vol. 48, 1988, pp. 329–332.
[17] Vierke, G. and Struckmeier, P. Z., *Naturforsch,* Vol. C33, 1978, pp. 266–270.

Yan Peng[1] and Te-Hsiu Ma[1]

Tradescantia Sister-Chromatid-Exchange (SCE) Bioassay for Environmental Mutagens

REFERENCE: Yan Peng and Te-Hsiu Ma, *"Tradescantia Sister-Chromatid-Exchange (SCE) Bioassay for Environmental Mutagens," Plants for Toxicity Assessment, ASTM STP 1091,* W. Wang, J. W. Gorsuch, and W. R. Lower, Eds., American Society for Testing and Materials, Philadelphia, 1990, pp. 319–323.

ABSTRACT: *Tradescantia paludosa* (spiderwort) has been one of the classical materials for cytogenetic studies since the late 1930s, and it possesses a number of endpoints to measure genotoxicity. Cytogenetic and genetic endpoints include chromosome/chromatid aberrations in its mitotic cells of microspores, root meristems, pollen tubes; meiotic pollen mother cells and gene mutation in the cells of staminal hairs. Although sister chromatid exchange (SCE) was first discovered in the plant root meristem of *Allium cepa,* the SCE technique has not yet been developed for *Tradescantia* root tip cells. Based upon the principles and technical procedures to measure SCEs in other plant systems, a standard SCE protocol was developed for *Tradescantia* in this study. Unlike the other plant system, *Tradescantia* roots developed from stem cuttings were used for SCE analysis. The major steps of this protocol include: root initiation, 5-bromodeoxyuridine (5-BrdU) treatment, thymidine chasing, colchicine treatment, fixation, pectinase digestion, squashing the meristematic cells under the coverglass, removal of coverglass, treatment with Hoechst and sodium salt solutions, UV-light treatment, staining with Giemsa. Well prepared slides were photographed under 400× magnification and analyzed for SCE frequencies. In addition, the spontaneous SCE frequency of *Tradescantia* was compared with those of *Vicia* and *Allium.*

KEY WORDS: *Tradescantia,* bioassay, sister-chromatid exchange, mutagen

Although the sister-chromatid-exchange (SCE) technique for mitotic chromosomes was first established in plant cells [1], more SCE studies have been carried out in animal cells than in plant systems at the present time. Animal cell [2,3] and human lymphocytes in culture [4] have been popular materials for SCE studies because mutagen-induced increases in the frequencies of SCEs in these materials could readily be related to mutagenic effects in humans. The actual mechanism of induced SCEs and the kinetics of this phenomenon are not well-understood, and SCE frequencies are usually used as a relative indicator of mutagenicity. Based upon the spontaneous frequencies of SCE, plant materials are, in general, more sensitive than animal chromosomes.

Tradescantia plants are currently used for a wide variety of bioassays that utilize various endpoints to detect the genotoxicity for environmental mutagen and clastogens [5]. These include cytogenetic and genetic responses in terms of chromosome/chromatid aberrations in microspores, root meristems, pollen tubes, and gene mutations in staminal hair and chromosome damage in the meiotic pollen mother cells. In this paper we report the development of a standard protocol for the *Tradescantia*-root-meristem SCE (Trad-SCE) assay as an additional bioassay that can be made in the same plant. This new assay may be specially useful for the studies of the kinetics of SCE induction.

[1]Department of Biological Sciences, Western Illinois University, Macomb, IL 61455.

Materials and Methods

Tradescantia paludosa clone #03 was used in this study. Old plant cuttings (about 5 to 10 cm long) were selected for initiating roots in tapwater in a 27°C incubator in the dark. A few drops of Malachite Green (5 ppm) were added to the fresh tapwater every day to inhibit fungus growth. In order to hasten growth and to obtain large quantities of roots, Rooton (naphthaleneacetamide indolbutyric acid and Tyiram, Union Carbide Agricultural Product Co.) was applied to the fresh cuts on each stem. The 5-bromodeoxyuridine (5-BrdU) treatment, slide preparation and staining procedures used in *Vicia* [6–8] and *Allium* [9,10] were modified for *Tradescantia* rooted cuttings. The following protocol was developed to standardize this Trad-SCE bioassay. 5-BrdU (50 to 100 mM) in 20 mL tapwater solution in a beaker was used for 5-BrdU incorporation during an 18-h exposure period. A 44-h recovery time in tapwater was then used to allow two rounds of mitotic cell divisions. Then 18 h of thymidine (100 mM) chasing was used to reduce the chance of further 5-BrdU incorporation and stimulate mitotic division. The intact roots were then treated with colchicine (0.2%) for 2 to 3 h to accumulate metaphase figures. The roots were then excised and fixed in aceto-ethanol (1:3 ratio) for 48 h. The meristematic tissue was digested with pectinase (9.7 units/mL) for removal of the cell wall, middle lamella, and dispersal of cells during slide preparation. For better adhesion of the metaphase cells to the slide surface, a gelatin (Sigma Chemical Co., St. Louis, Mo.) layer was added prior to the squashing process of the root tip cells. After removal of the root cap, the root meristematic cells were excised from the root tip and squashed in a drop of distilled water on the surface of the gelatin-coated slides under a coverglass. The coverglass was then removed after a quick-freezing treatment on a "cold plate" at -20°C. Slides with a high mitotic index of 10% or higher were selected for Hoechst 33258 (a fluorescent dye), RNase, and UV light treatments in the sodium citrate and sodium chloride (SSC) solution at 52°C for 30 min. This combination of treatments was designed to reduce the staining ability of 5-BrdU bearing chromatids for better differentiation between the sister-chromatids. Treated slides were then washed with distilled water and dried in an incubator (40°C) overnight. Thoroughly dried slides were then stained in Giemsa (a well-known DNA specific stain) for 15 min and washed in tapwater to destain the chromosomes for adequate color differentiation. The slides were dried in an incubator overnight and then mounted in Euparal as permanent preparations. Photomicrographs were made with Kodak Ektagraphic film and printed on polycontrast photographic paper for SCE frequency analysis. Frequencies of SCE were counted from 10 metaphase figures from each experimental series, and means and standard deviations were derived from 10 samples.

Results and Discussion

The standard protocol of the Trad-SCE bioassay developed in this study is a modified procedure of the Fluorescent Plus Giemsa technique developed for *Vicia faba* [11]. Adequate SCE preparations were obtained from the experiments conducted under various combinations of chemical concentration as shown in Table 1. The chemical treatments included 5-BrdU, gelatin, enzyme pectinase, colchicine, Hoechst, and RNase, and the adequacy of the SCE preparations were judged by mitotic index, percentage of metaphase figures, chromatid differentiation, and SCE frequency. The typical metaphase figures with clear SCE features were obtained. Photomicrographs of a cell with a normal karyotype ($2n = 12$), and a metaphase figure with SCEs is shown in Figs. 1 and 2 respectively.

Among commonly used plant materials for cytogenetic studies, *Vicia faba* (broad bean) and *Allium cepa* (onion) both have a relatively high spontaneous rate of SCE [7,9] and *Tradescantia paludosa* (spiderwort) has not yet had its spontaneous SCE frequency established. No reports or photomicrographs of SCEs in *Tradescantia* can be found in the current liter-

TABLE 1—*Results of SCE preparations under various combinations of chemical treatments.*

Chemical Agents (Dose Units)	Trial Numbers					
	1	2	3	4	5	6
5-BrdU (mM)	0.5	0.5	0.5	0.1	0.1[a]	0.1[a]
Gelatin	—	—	—	+	+	+
Pectinase (unit/mL)	2.4	4.9	9.7	19.4	9.7[a]	9.7[a]
Colchicine (h)	2	2	2.5	3	3[a]	3[a]
Hoechst (%)	0.001	0.001	0.001	0.004	0.004[a]	0.004[a]
RNase (%)	0.04	0.01	0.01	0.04	0.04[a]	0.04[a]
Mitotic index	1.3	0.8	14.6	5.8	15.1[a]	14.8[a]
Differentiation	poor	poor	poor	good	fair[a]	good[a]
SCE frequency	too high	—	—	2.0	29.4[a]	28.1[a]

[a] Favorable combination of chemical treatments.

ature. Relatively large quantities of 5-BrdU are required for both of these plant materials because the broad bean seeds or the onion bulbs absorb large amounts of this chemical and the seeds and bulbs provide large amounts of endogenous thymidine during the 5-BrdU incorporation. We expected that more efficient incorporation of 5-BrdU would be obtained with rooted cuttings of *T. paludosa*. The size of the stem and number of leaves on each cutting can be modified to regulate the nutrient and energy supply during the treatment and SCE slide preparation processes.

The average spontaneous SCE frequency of 29.4 per cell found in *Tradescantia* roots is higher than that of *Vicia* (21.0 per cell) [6] and lower than that of *Allium* (44.8 per cell) [9]. This difference is expected on the basis of the total length of chromosomes in each of these three plant species. The high frequency of SCE in *Allium* is the result of random exchange of 16 metacentric chromosomes that have a total length of 264 μm. Although the $2n$ number

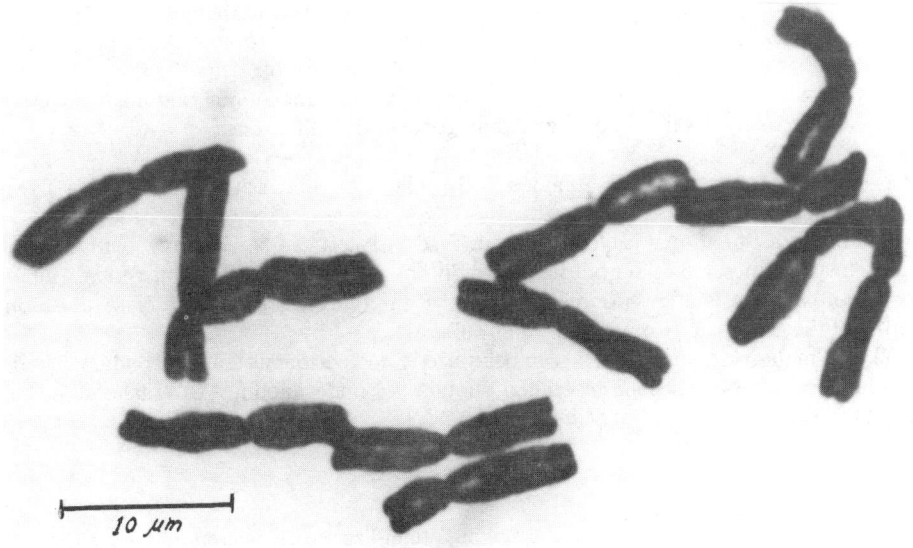

FIG. 1—*Karyotype of* Tradescantia paludosa *(2n = 12)*.

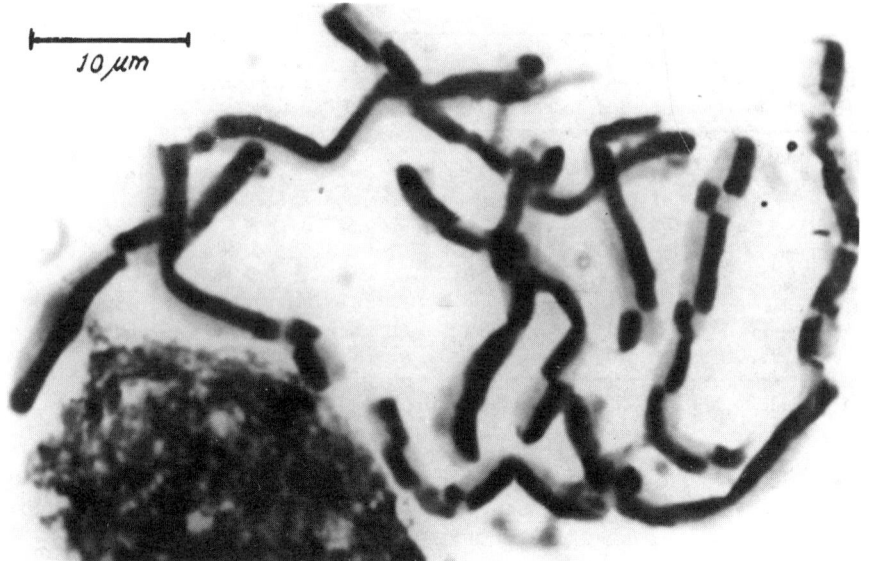

FIG. 2—*Metaphase figure with average frequency of SCE in* Tradescantia *root meristem.*

of the chromosomes of *Vicia* and *Tradescantia* are the same, the total length of two metacentrics plus 10 relatively short acrocentrics in *Vicia* is around 169.2 μm, while the 12 metacentrics in *Tradescantia* is around 234.3 μm. This is more obviously shown when the SCE frequencies are calculated on the basis of SCE per chromosome (i.e., *Vicia* has 1.8 SCE/chromosome, while *Tradescantia* has 2.5 SCE/chromosome). Average length of the acrocentrics in *Vicia* is around 11.4 μm. Average length of the metacentrics in *Tradescantia* is around 19.5 μm. Owing to the lower number and longer length of the metacentric chromosomes, *Tradescantia* root meristem is a more favorable material for SCE studies than *Allium*. The lower number makes it easier to prepare well-spread metaphase chromosome figures and to score SCEs. The average length of *Tradescantia* chromosomes is longer (19.52 μm) than that of *Vicia* chromosomes (16.5 μm), although the total length is shorter. Among the existing SCE bioassays for environmental mutagens using animal and plant materials, Trad-SCE bioassay has the following advantages over the other tests:

1. *Tradescantia* roots can be obtained easily in large quantities at very low cost for a year-round supply.
2. In contrast to animal cell cultures, the Trad-SCE test can be performed with minimal laboratory facilities and under nonsterile conditions. Thus the Trad-SCE bioassay is suitable for genotoxicity tests of environmental chemical mixtures and especially for *in situ* monitoring of the genotoxicity of air and water pollutants.
3. Test materials (plant cuttings) can be utilized for experiments that are conducted under various conditions of nutrient and energy supplies, or under modified external and internal factors to study the kinetics of SCE induction.

References

[1] Taylor, J. H., "Sister Chromatid Exchanges in Tritium-Labeled Chromosomes," *Genetics,* Vol. 43, 1958, pp. 515–529.
[2] Korenberg, J. R. and Freedlender, E. F., "Giemsa Technique for the Detection of Sister Chromatid Exchanges," *Chromosoma* (Berl.), Vol. 48, 1974, pp. 355–360.

[3] Davidson, K. L., Kaufman, E. R., Daugherty, C. P., Ouellette, A. M., DiFolco, C. M., and Latt, S. A., "Induction of Sister Chromatid Exchange by BrdU is Largely Independent of the BrdU Content of DNA," *Nature,* Vol. 284, 1980, pp. 74–76.

[4] DuFrain, R. and Garrand, T. J., "The Influence of Incorporated Halogenated Analogues of Thymidine on the SCE Frequencies in Human Lymphocytes," *Mutation Research,* Vol. 91, 1981, pp. 233–238.

[5] Ma, T. H., and Harris, M. M. "*In situ* Monitoring of Environmental Mutagens," in *Hazard Assessment of Chemicals: Current Development,* Vol. 4, Academic Press, New York, 1985, pp. 77–106.

[6] Kihlman, B. A. and Kronborg, D., "Sister Chromatid Exchanges in *Vicia faba:* I—Demonstration by a Modified FPG Technique," *Chromosoma,* Vol. 51, 1975, pp. 1–10.

[7] Scheid, W., "Mechanism for Differential Staining of BrdU Substituted *Vicia faba* Chromosomes," *Experimental Cell Research,* Vol. 101, 1976, pp. 55–58.

[8] Cortes, F. and Anderson, H. C., Analysis of SCEs on *Vicia faba* Chromosomes by a Simple Fluorescent Plus Giemsa Technique," *Hereditas,* Vol. 107, 1987, pp. 7–13.

[9] Schvartzman, J. B. and Cortes, F., "Sister Chromatid Exchange in *Allium cepa,*" *Chromosoma,* Vol. 62, 1977, pp. 119–131.

[10] Gutierrez, C., Gonzalez-Gil, G., and Hernandez, P., "Analysis of Baseline BrdU-Dependent SCEs at Different BrdU Concentrations," *Experimental Cell Research,* Vol. 149, 1983, pp. 461–469.

[11] Wolff, S. and Perry, P., "Differential Giemsa Staining of Sister Chromatids and the Study of Chromatid Exchanges without Autoradiography," *Chromosoma,* Vol. 48, 1974, pp. 341–353.

K. R. Sauser[1] and S. J. Klaine[1]

Activation of Promutagens by a Unicellular Green Alga

REFERENCE: Sauser, K. R. and Klaine, S. J., "**Activation of Promutagens by a Unicellular Green Alga,**" *Plants for Toxicity Assessment, ASTM STP 1091,* W. Wang, J. W. Gorsuch, and W. R. Lower, Eds., American Society for Testing Materials, Philadelphia, 1990, pp. 324–332.

ABSTRACT: Classic mutagenicity testing includes mammalian activation of promutagens. A relatively new approach to examining environmental mutagens involves the plant activation of promutagens. A plant system was developed using the aquatic unicellular green alga *Selenastrum capricornutum* as the activating mechanism. The genetic indicator organism used was the Ames strain TA98 of *Salmonella typhimurium.* Two treatment groups were examined: alga, TA98, and a test chemical incubated together; and TA98 and a test chemical incubated together without the alga. The test chemicals examined were *m*-phenylenediamine (MPD), aflatoxin B_1, and atrazine. Differences in the number of revertants were compared between the reaction mixtures of the two treatment groups. For MPD and aflatoxin, there was a significant difference between the two treatment groups with the algal treatment group producing a higher number of revertants. There was no effect seen with the atrazine at the concentrations tested or with the TA98 strain. It was shown that the chemically treated alga does not act as an exogenous source of histidine for the TA98. *Selenastrum* demonstrated a consistent trend of weak activation of the promutagens MPD and aflatoxin B_1.

KEY WORDS: plant activation, *Selenastrum capricornutum,* green alga, mutagenicity, environmental mutagenesis

Many chemicals, particularly those that act as mutagens, pose a possible threat to the environment as well as to public health. A mutagen is an agent that directly affects cellular DNA and macromolecules, while a promutagen is a chemical that must be metabolized to the active mutagenic form. While classic mutagenicity testing incorporates animal metabolism or "activation" of promutagens via a liver enzyme preparation, a relatively new approach to examining environmental mutagens includes the plant activation of promutagens. Development of this approach was based on several concerns. There is some evidence to support the assumption that xenobiotic metabolism by plants can be different from xenobiotic metabolism by animals [1]. Thus chemicals not metabolized in animal systems may act as promutagens when exposed to plant metabolism. Plants are the intentional as well as unintentional target of a wide range of extensively used chemical agents. Finally, plants form the base of the food chain in both terrestrial and aquatic systems. This is particularly important since plant generated mutagens might affect both animals and man. Although less well characterized than animal or microbial systems, several plant promutagen studies have provided genetic evidence that higher plants have the capacity to activate promutagens. Methods include extracting the activated metabolites from whole plants [2], from plant tissue homogenates [3,4] or from homogenates from plant cell cultures [5,6]. Another type of method, known as the plant cell/microbe coincubation assay, involves plant cells in suspension func-

[1] Memphis State University, Memphis, TN 38152.

tioning as the activating component coupled with microbial cells serving as the genetic indicator organism [7].

While higher plants are known to activate certain chemicals, few studies examine the potential of lower plants as activators of promutagenic chemicals [8,9]. This study was initiated to extrapolate existing plant activation assays to incorporate the freshwater unicellular green alga *Selenastrum capricornutum* Printz. Examining the role of algae in environmental mutagenesis is important for several reasons. Algae are ubiquitous in freshwater systems, are ultimately exposed to many chemicals via surface runoff, are the base of the aquatic food chain, and constitute a large metabolic pool in close contact with water. Many unicellular algae are also characterized by a large surface area, which would be a significant factor in the uptake of chemical residues present in the water column [10]. In addition, algae are known to metabolize certain chemicals [11]. These experiments represent the initial stages of investigating the role a unicellular green alga might have when in contact with chemical residues as measured by mutagenicity. Experiments were designed to examine the effects of chemical concentration as well as length of incubation time.

Materials and Methods

Bacterial Strain

Salmonella typhimurium strain TA98 *(hisD3052, uvrB, rfa, pKM101)* was used exclusively in all experiments. TA98 detects reagents that cause frameshift mutations. This strain was provided by Dr. Bruce N. Ames (University of California, Berkeley). Cells were harvested from overnight cultures by centrifugation (4000 g) for 10 min. The bacterial pellet was resuspended in 0.1 M potassium phosphate buffer and recentrifuged for 10 min. The pellet was resuspended in 10 mL 0.1 M potassium phosphate buffer, the titer was spectrophotometrically determined and adjusted according to the type of incubation protocol. For the coincubation assays the titer was 1.0 times 10^{10} CFU/mL (Colony-Forming Units); for the preincubation assays the titer was 1.0 times 10^9 CFU/mL. The strain charcteristics of TA98 were confirmed for each experiment by the method of Zeiger et al. [12]. The spontaneous reversion rate of TA98 ranged from 14 to 36 for these experiments.

Selenastrum capricornutum

Axenic cultures of *Selenastrum capricornutum* Printz were used in all experiments. The alga was cultured in an inorganic nutrient media described by Keating [13]. For each experiment, the alga was harvested from log growth phase, pooled, and concentrated to a cell density of at least 1.0×10^8 cells/mL.

Chemicals

The test chemicals examined were *m*-phenylenediamine (MPD), an aniline derivative; atrazine, an *s*-triazine herbicide; and aflatoxin B_1, a mycotoxin. The MPD and aflatoxin were obtained from Sigma Chemical (St. Louis, Mo.), and the atrazine was provided by Ciba-Geigy Corporation (Greensboro, N.C.). All other chemicals were obtained from Sigma. The test chemicals were dissolved in dimethylsulfoxide (DMSO) so that 50 μL or less was added to each reaction mixture. Each chemical was chosen because it is a promutagen and has been demonstrated to be activated to a mutagenic form in other plant systems [14–16].

TABLE 1—*Chemical concentrations added to reaction mixtures and final concentrations in plate.*

Chemical	Concentration Added to Reaction Mixture	Incubation	Final Concentration in Plate
MPD	0, 1.0, 10, 100 μg	Preincubation	0, 0.1, 1.0, 10 μg
MPD	0, 1.08, 2.7, 8.1, 10.8, 16.2 mg	Coincubation	0, 0.108, 0.27, 0.81, 1.08, 1.62 mg
Atrazine	0, 0.216, 2.16, 21.6 μg	Coincubation	0, 0.0216, 0.216, 2.16 μg
Aflatoxin B_1	0, 1.0, 2.0, 4.0, 20 mg	Preincubation	0, 10, 50, 200, 1000 ng
Aflatoxin B_1	0, 25, 100, 500, 1000 ng	Coincubation	0, 2.5, 10, 50, 100 ng

Treatment Protocols and Ames Assay

For each experiment two treatment groups were examined. One group consisted of reaction mixtures containing 4.5 mL of the algal concentrate, the appropriate concentration of test chemical, and 0.5 mL of the TA98 suspension. The other group consisted of reaction mixtures containing 4.5 mL of algal media containing no algae, the appropriate concentration of the test chemical, and 0.5 mL of the TA98 suspension. Controls for each experiment included using a positive direct mutagen (4-nitro-o-phenylenediamine) to confirm the phenotype of TA98, solvent controls, and negative controls where no solvent or test chemical was added to a reaction mixture. In some experiments the alga or algal media was preincubated with the test chemical for a length of time before the TA98 was added. After addition of the TA98, the preincubation reaction mixtures were coincubated for 30 min. In experiments that did not follow a preincubation scheme, all the components of the reaction mixtures were coincubated for a period of time. All reaction mixtures were incubated at 24°C under gold fluorescent light (F20T12/GO) to avoid photodegradation and shaken at 180 rpm. The light intensity was 7.9 μmol m^{-2} s^{-2}. Table 1 indicates the chemical concentrations added to each reaction mixture and the final concentration of test chemical/plate. Following incubation, the reaction mixtures were mixed, triplicate 0.5 mL aliquots were removed and added to 2 mL soft agar containing a trace amount of histidine and biotin. The soft agar was mixed and overlayed onto minimal glucose agar plates [17]. The plates were incubated at 37°C in the dark for 72 h and the number of revertants per plate were counted.

Data Analysis

The mean revertant rate of triplicate plates from each reaction mixture was analyzed for both treatment groups using the Student's *t*-test at the 0.05 level to determine if the treatment groups were statistically different.

Results and Discussion

Initial experiments with *m*-phenylenediamine (MPD) focused on coincubating the alga with the TA98 and the MPD. The length of coincubation was held constant at 4 h while the concentration of MPD was varied from 0.1 to 1.08 mg/plate. The number of revertants from the "with alga" group increased as a function of MPD concentration and was significantly different from the control at 1.08 mg/plate, indicating that an algal metabolic effect had been generated (Fig. 1). When 100 μg MPD/reaction mixture was preincubated with alga cells for various times the number of revertants from the "with alga" group becomes significantly different from the "without alga" group after 9 and 12 h of preincubation (Fig. 2). In order to test if it were necessary for the TA98 to be present while the alga was in contact with the chemical to have an effect, a 12 h coincubation with varying MPD concentrations (1.08 to

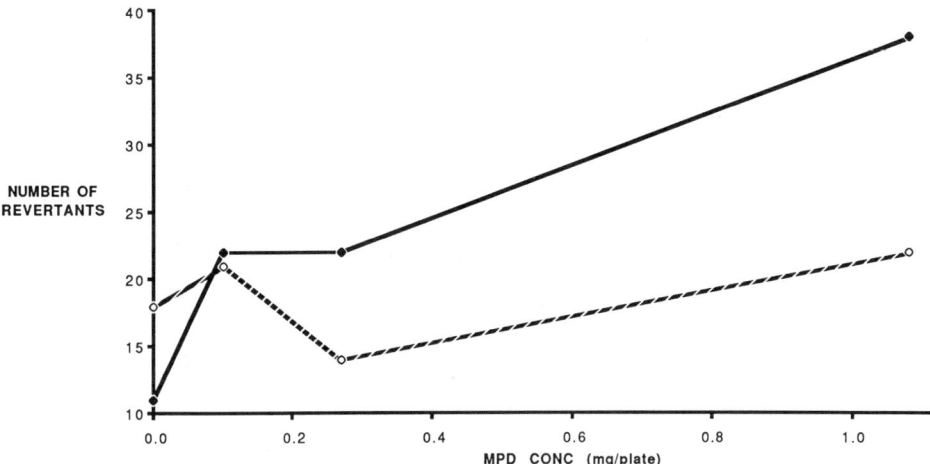

FIG. 1—*Coincubation test of* m-*phenylenediamine: Effect of varying the concentration. The solid line represents the algal treatment group. The broken line represents the without alga treatment group. The data represent the mean from three replicate plates.*

16.2 mg/reaction mixture) was conducted (Fig. 3). For the "with alga" group the number of revertants increased up to the 8.1 mg/reaction mixture (0.81 mg/plate) and subsequently declined. This decline in revertants might be attributed to chemical stress to the TA98 due to the length of exposure to the high MPD concentration. The number of revertants from the "without alga" treatment group remained essentially the same throughout the coincubation.

These results indicate that the concentration of MPD has a greater effect on the number of revertants generated by the alga than the length of either preincubation or incubation time. The number of revertants fell within a similar range regardless of the nature of treatment.

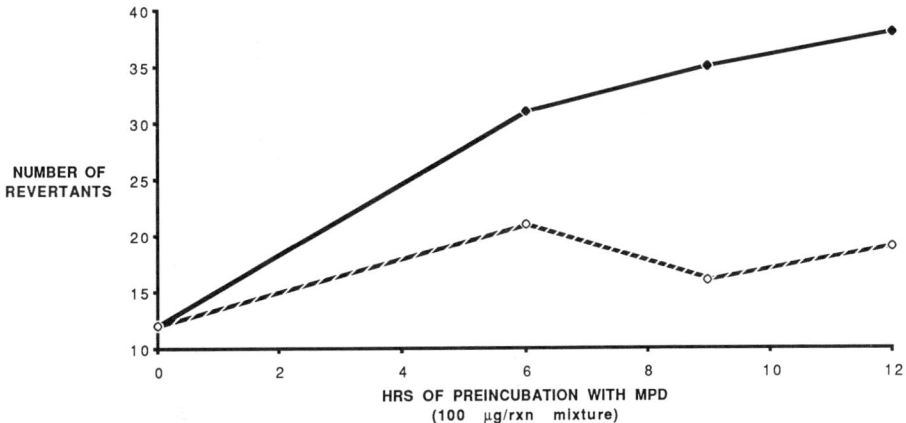

FIG. 2—*Preincubation test with* m-*phenylenediamine: Effect of varying the preincubation time. The solid line represents the algal treatment group. The broken line represents the without alga treatment group. The data represent the mean of three replicate plates.*

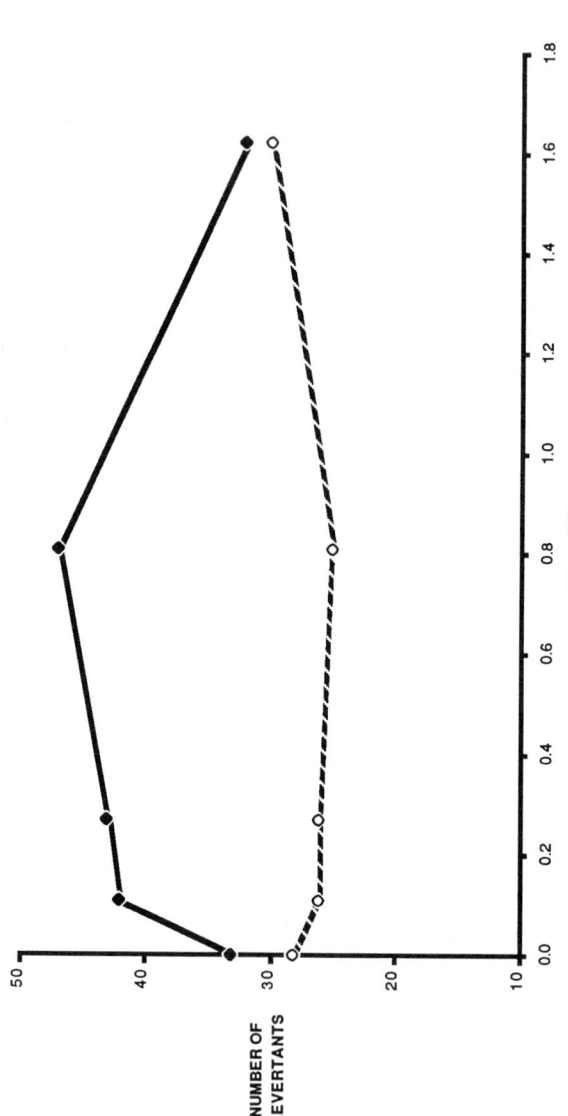

FIG. 3—*Coincubation test with m-phenylenediamine: Effect of varying the coincubation time. The solid line represents the algal treatment group. The broken line represents the without alga treatment group. The data represent the mean of three replicate plates.*

In comparison with other plant species that have activated MPD, using similar concentrations and the same Ames strain, *Selenastrum* is a weak activator. Other investigators have shown that MPD is differentially activated by plants ranging from *Tradescantia* to cultured tobacco cells at greater rates that that of *Selenastrum* [18,19].

Aflatoxin was examined by coincubating the two treatment groups with varying concentrations (25 to 1000 ng/reaction mixture) for 2 h. Throughout the range of concentrations examined there was little difference between the two treatment groups. Neither treatment group demonstrated an effect in response to the aflatoxin. Figure 4 represents the results of preincubating the treatment groups for 4 h with varying concentrations of aflatoxin (1 to 20 mg/reaction mixture). Higher concentrations were used for the actual preincubation, with the appropriate aliquot removed to achieve a range of 10 to 1000 ng/plate. The highest concentration of aflatoxin the TA98 could tolerate before significant cell death occurred was determined to be 1000 ng/plate. Both treatment groups exhibited an upward trend with the "with alga" group producing a statistically significant higher number of revertants. For aflatoxin it appears that both the concentration and the length of incubation time are effective for increasing the number of revertants produced by the algal treatment group.

Other studies show that aflatoxin B_1 is activated by plant systems. Using TA98, the S1 fraction from maize [20] and the microsomal fraction isolated from tulip bulbs [21] were effective in metabolizing aflatoxin B_1 to a mutagenic form.

Atrazine was examined by coincubating the treatment groups with varying concentrations (0.216 to 21.6 μg/reaction mixture) for 3 h. There was little difference between the treatment groups with neither group demonstrating an effect in response to the atrazine. Preincubation experiments were not attempted for atrazine.

Atrazine is unique among promutagens because it is considered by some to be a specific plant promutagen. That is, atrazine has been shown to be activated only in plants or with plant cell-free extracts [22]. No evidence of algal activation was seen in this study.

Because the algal treatment groups demonstrated a trend towards producing a higher number of revertants over the treatment group without algae, it became important to determine if the effect seen was a real metabolic effect rather than an experimental artifact. Because heritable mutations are dependent upon DNA replication, in the Ames assay the number of revertants are proportional to the number of cell divisions. Since the Ames strains are histidine dependent, the number of revertants can be directly related to the amount of exogenous histidine in the medium [23]. We wanted to see if the alga upon treatment with a test chemical were perhaps dying and releasing histidine to the reaction mixture where it would be available to the TA98 and thus account for the increase in the number of revertants for the "with alga" treatment group. Three different algal suspensions were prepared: live alga, heat killed alga, and alga treated with a lethal concentration of atrazine. These groups were compared with a control where no alga was present. Each group was tested with varying concentrations of a 0.5 mM histidine solution (0, 25, 50, 100 μL). At 0 μL there were no revertants for any group. This demonstrates that TA98 is histidine-dependent and that none of the groups acted as an exogenous source of histidine for the TA98 (Fig. 5). The number of revertants increased linearly with increasing histidine concentrations for all groups. The results also demonstrate that there is no difference between live alga and dead alga in regards to having an effect on the number of revertants. The number of revertants was slightly lower for the group without alga, but this variation falls within the normal fluctuations of the assay. To further demonstrate that there is not an inherent difference between a "with alga" group and a "without alga" group, a survey of the negative controls was conducted. The negative control of each experiment is where no solvent or test chemical is added. The number of revertants for the negative controls fluctuates from experiment to experiment, but there is

330   PLANTS FOR TOXICITY ASSESSMENT

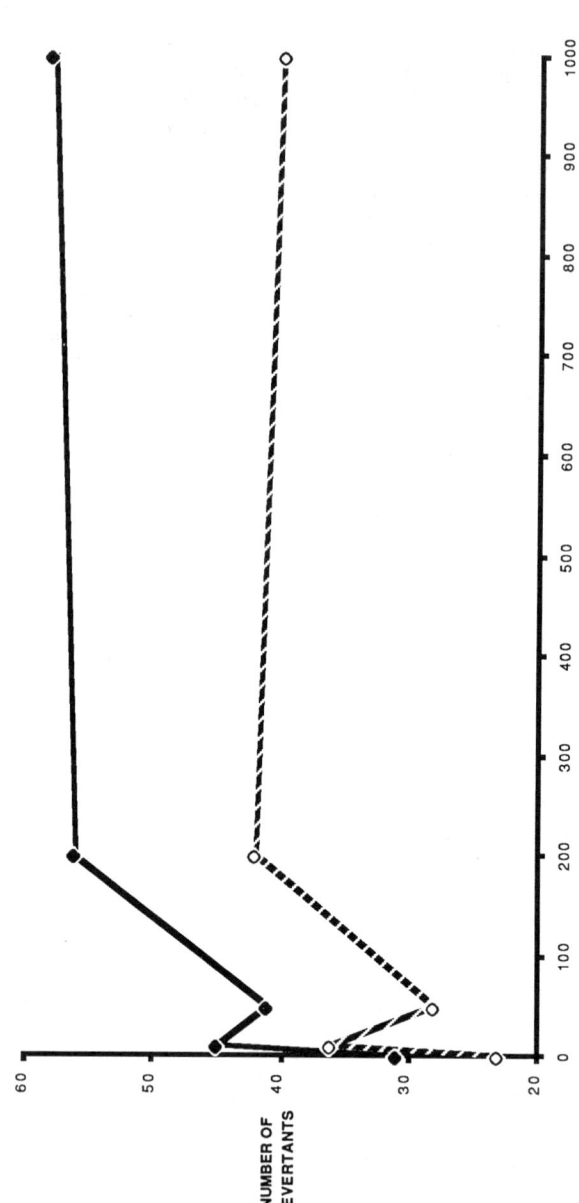

FIG. 4—*Preincubation test with aflatoxin B_1: Effect of varying the length of preincubation and concentration. The solid line represents the algal treatment group. The broken line represents the without alga treatment group. The data represent the mean of three replicate plates.*

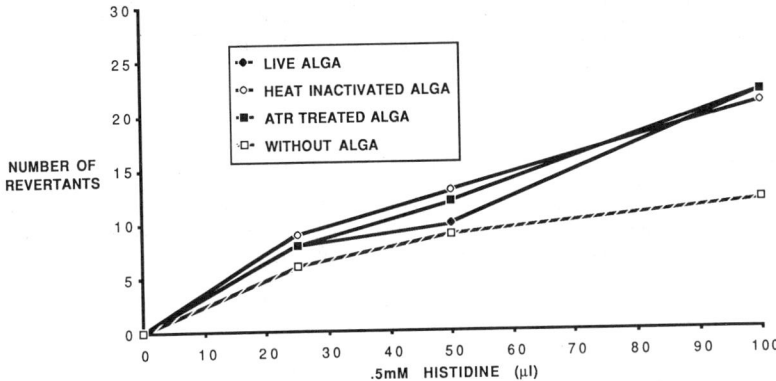

FIG. 5—*Histidine controls: Comparing the effect of treated algal groups and control in response to different concentrations of histidine.*

no significant difference between the groups within each experiment (Fig. 6). Therefore the trend of a higher number of revertants for the algal treatment groups over the without alga group can be viewed as an algal metabolic effect.

Conclusions

For m-phenylenediamine, there was a significant difference between the two treatment groups with the "with alga" group producing a higher number of revertants. The concentration of MPD seemed to be more effective than the length of incubation in producing a mutagenic effect. For aflatoxin B_1, there was a significant difference between the two treatment groups with the algal treatment group producing more revertants. Both the concentration of aflatoxin B_1 and the length of incubation time were effective in producing a mutagenic response. There was no effect seen with atrazine at the concentrations tested or with the TA98 strain. The chemically-treated alga does not appear to act as an exogenous source of

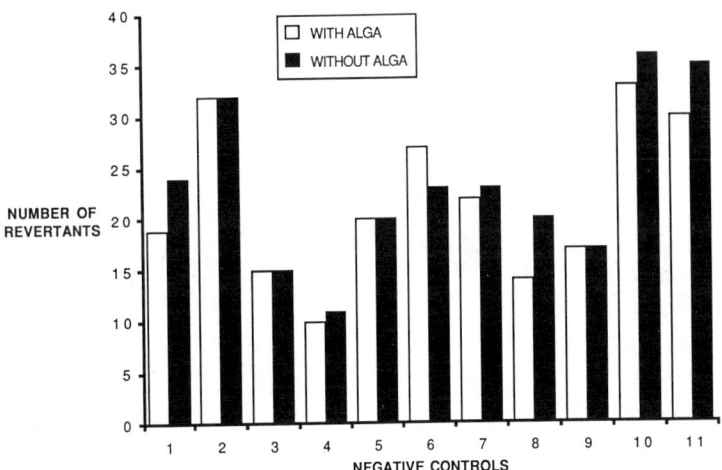

FIG. 6—*Survey of experimental negative controls.*

histidine for the TA98. Finally, *Selenastrum capricornutum* has demonstrated a consistent trend of weak activation of the promutagens MPD and aflatoxin B_1.

References

[1] Sandermann, H., in *Environmental Mutagenesis, Carcinogenesis and Plant Biology*, Praeger, New York, 1982, pp. 1-32.
[2] Plewa, M. J. and Gentile, J. M., *Mutation Research*, Vol. 38, 1976, pp. 287-292.
[3] Plewa, M. J. and Gentile, J. M., in *Chemical Mutagens*, Plenum, New York, 1982, pp. 401-420.
[4] Wildeman, A. G. and Nazar, R. N., *Canadian Journal of Genetics and Cytology*, Vol. 24, 1982, pp. 437-449.
[5] Benigni, R., Bignami, M., Camoni, Z. I., Carere, A., Conti, G., Iachetta, R., Morpurgo, G., and Ortali, V. A., *Journal of Toxicology and Environmental Health*, Vol. 5, 1979, pp. 808-819.
[6] Gentile, J. M., and Plewa, M. J., *Mutation Research*, in press.
[7] Plewa, M. J., Weaver, D. L., Blair, L. C., and Gentile, J. M., *Science*, Vol. 219, 1983, pp. 1427-1429.
[8] Schoeny, R., Cody, T., Radike, M., and Warshawsky, D., *Environmental Mutagenesis*, Vol. 7, 1985, pp. 839-855.
[9] Schafer, T., Gentile, J. M., and Plewa, M. J., *EMS Abstracts*, 1986, p. 5.
[10] Valentine, J. P., and Bingham, S. W., *Weed Science*, Vol. 22, No. 4, 1974, pp. 358-363.
[11] Wojtalik, T. A., Hall, T. F., and Hill, L. O., *Pesticide Monitoring Journal*, Vol. 4, 1971, pp. 184-203.
[12] Zeiger, E., Pagano, D. A., and Robertson, I. E. G., *Environmental Mutagenesis*, Vol. 3, 1981, pp. 205-209.
[13] Keating, K., *Water Research*, Vol. 19, No. 1, 1985, pp. 73-78.
[14] Means, J. C., Plewa, M. J., and Gentile, J. M., *Mutation Research*, Vol. 197, 1988, pp. 325-336.
[15] Lhotka, M. A., Plewa, M. J., and Gentile, J. M., *Environmental Molecular Mutagenesis*, Vol. 10, 1987, pp. 79-88.
[16] Higashi, K., Nakashima, K., Karasaki, Y., Fukanaga, M., and Mizuguda, Y., *Biochemistry International*, Vol. 2, 1981, pp. 373-380.
[17] Maron, D. M., and Ames, B. N., *Mutation Research*, Vol. 113, 1983, pp. 172-215.
[18] Anderson, V. A., Plewa, M. J., and Gentile, J. M., *Mutation Research*, Vol. 197, 1988, pp. 303-312.
[19] Plewa, M. J., Wagner, E. D., and Gentile, J. M., *Mutation Research*, Vol. 197, 1988, pp. 207-219.
[20] Means, J. C., Plewa, M. J., and Gentile, J. M., *Mutation Research*, Vol. 197, 1988, pp. 325-336.
[21] Higashi, K., Nakshima, K., Karasaki, Y., Fukanaga, M., and Mizugudu, Y., *Biochemistry International*, Vol. 2, 1981, pp. 373-380.
[22] Means, J. C., Plewa, M. J., and Gentile, J. M., *Mutation Research*, Vol. 197, 1988, pp. 325-326.
[23] Rasquinha, I. A., Wildeman, A. G., and Nazar, R. N., *Mutation Research*, Vol. 197, 1988, pp. 261-272.

R. R. Velagaleti,[1] D. Kramer,[1,2] S. S. Marsh,[1] N. G. Reichenbach,[1] and D. E. Fleischman[1,3]

Some Approaches to Rapid and Pre-Symptom Diagnosis of Chemical Stress in Plants

REFERENCE: Velagaleti, R. R., Kramer, D., Marsh, S. S., Reichenbach, N. G., and Fleischman, D. E., "**Some Approaches to Rapid and Pre-Symptom Diagnosis of Chemical Stress in Plants,**" *Plants for Toxicity Assessment, ASTM STP 1091,* W. Wang, J. W. Gorsuch, and W. R. Lower, Eds., American Society for Testing and Materials, Philadelphia, 1990, pp. 333–345.

ABSTRACT: The usefulness of ion accumulation patterns, leaf chlorophyll content, and delayed fluorescence of plant leaves as indicators of toxicity due to chemical stress is illustrated in this paper. Using plants as bioassay tools, these parameters can often be used to detect plant stresses rapidly and in advance of visible symptoms (pre-symptom diagnosis). They can also be used to distinguish and define the type and nature of chemical toxins present in a given substrate. Central to the use of plants as bioassay tools for toxicity assessment is the selection of appropriate species. The data presented here for the ion accumulation patterns and the chlorophyll content illustrate the genotypic variation in soybean indicated by the tolerance of the variety Manchu and the sensitivity of the Williams variety to sodium chloride (NaCl) and potassium chloride (KCl) stresses. Electrical conductivity measurements of leaf extracts have shown highly significant differences between Manchu and Williams; conductivity of leaf extracts in Manchu was significantly lower than Williams within the range of 10 to 40 mM of NaCl. Varietal and salt concentration differences were also highly significant for the concentration of chloride in the leaves beginning with stress levels of 20 mM NaCl or KCl. In Williams, chlorophyll concentration in leaves of two-week old plants declined significantly with increase in NaCl or KCl stress, the decline being concentration dependent. Highly significant differences in leaf chlorophyll were noticed between Williams and Manchu at various salt concentrations. The data presented for the delayed fluorescence technique show distinct patterns of leaf images under different stresses. The differences between stressed and nonstressed plants was evident for all the four parameters (conductivity, chlorophyll, chloride, delayed fluorescence), at certain stress levels, at a time when visible symptoms of chloride or sodium toxicity were not apparent, indicating the potential for pre-symptom diagnosis of chemical stresses in plants using these measures.

KEY WORDS: electrical conductivity, ion quantitation, chlorophyll content, delayed fluorescence, soybean, *Phaseolus* bean, NaCl, KCl, herbicides, toxicity assessment

Introduction

Easily adaptable and sensitive detection techniques and methods facilitate the use of plants for toxicity assessment. Implicit in the use of plants as bioassay tools is a selection process that should identify appropriate genera and/or varieties within a genus that respond and/or

---

[1] Health and Environment Group, Battelle, Columbus, OH 43201-2693.
[2] Present address: Department of Physiology and Biophysics, University of Illinois, Urbana, IL 61801.
[3] Present address: Biochemistry Department, Wright State University, Dayton, OH 45345.

are sensitive to a particular stress. As evident from numerous reports in the literature, genotypic variation in plants for a given stress is significant between genera and between the species within a genus, as well as between different varieties of a given species [1].

In an earlier study [2], several soybean varieties were screened for their response to 80 mM of NaCl stress. Severe symptoms of toxicity were observed in many of these varieties, which included leaf chlorosis and necrosis, loss of foliage, reduction in root growth, inhibition of root nodulation, and eventual death of the plant. Prominent among the sensitive varieties was Williams. A few varieties, such as Manchu, were tolerant, remained healthy, and continued to grow and fix nitrogen with no evidence of foliar injury. Similar patterns of sensitivity and tolerance were shown with KCl stress, although KCl was found to be more toxic than NaCl [3].

In addition to the visual morphological features described above, various approaches were investigated to understand the physiological differences between sensitive and tolerant varieties. Some of the approaches suggested here, such as stress detection through determination of ion accumulation patterns and chlorophyll content, are relatively simple. These techniques are well defined, and the instrumentation necessary is in the marketplace and can be easily adapted to field situations. Other approaches, such as identification and quantitation of specific ions and the delayed fluorescence technique, are more complex, involve the use of nonportable instruments, and cannot be easily adapted to on-site field evaluations. However, these other approaches provide a broader evaluation of the cause-and-effect relationship of stresses and provide a deeper insight into the scientific basis. A brief introduction to the four approaches for stress detection dealt with in this paper is provided below.

Ion Accumulation

Soil salinity is often measured as electrical conductivity (EC) of the saturation extract. Electrical conductivity measurements can be used for predicting yields over a range of salt concentrations. Crop yield response curves [1] and relative salt tolerance response curves [4] have been developed using EC measurements of the saturation extracts of soils. Increasing salt concentration has resulted in proportionate increase in the EC of the leaf extracts of plants grown under salt stress at concentrations where visible symptoms are difficult to discern (unpublished results). Since preparation of a leaf extract and measurement of conductivity using a portable conductivity meter is feasible for on-site field observations, we investigated the potential use of this technique for detection and measurement of toxicity due to low levels of salt stresses.

Identification and Quantitation of Ions

Differences in ion concentrations between genera, between species within a genus, and between plant parts within a species have been observed by several investigators, and such differences have been helpful in identifying adaptive characteristics to salt stress [5]. Earlier studies [2] have shown that the patterns of ion uptake and the expression of symptoms of toxicity differed significantly among different varieties of soybean when they were exposed to a constant stress of 80 mM NaCl or KCl. Rapid and excessive accumulation of ions in the leaf tissues resulted in chlorosis followed by drying and death of the sensitive varieties. In addition to shoot and root growth, the establishment and functioning of the symbiosis with soil rhizobia was inhibited. Tolerant varieties, on the other hand, grew well and fixed nitrogen under 80 mM NaCl and KCl stresses, although under the latter stress, plants showed signs of chlorosis after four weeks [6]. This prompted us to grow the plants of both sensitive and tolerant varieties at lower stress levels to identify tolerable salt levels for each variety. In experiments of this nature we noticed differences in ion accumulation patterns at a time when plants showed no visible symptoms of stress.

Chlorophyll Content

Chlorophyll concentration is a sensitive indicator of physiological response to salt stress in plants, and decline in chlorophyll content has been observed with increasing NaCl concentration [7]. A decline in the leaf chlorophyll concentration of soybean was observed with increasing concentration of either NaCl or KCl, more prominently in sensitive varieties [6]. Further examination revealed measurable decline in chlorophyll *a* and *b* concentrations at low levels of salt stress even though visual symptoms were not indicative of chlorophyll deficiency.

Delayed Fluorescence

The leaves of plants re-emit light for several minutes after they have been illuminated. This delayed fluorescence is different from ordinary ("prompt") fluorescence. Prompt fluorescence refers to the process in which a chlorophyll molecule, having absorbed a photon and been raised to an excited singlet state, re-emits light as it returns to its ground state. The chlorophyll excitation responsible for delayed fluorescence arises instead from a reversal of the primary photochemical event. The intensity of the emitted light at any instant is an indication of the amount of free energy stored in the chloroplast, for thermodynamic reasons that have been discussed by Marchiarullo and Ross [8]. Stresses that block the consumption of free energy for photosynthesis, such as unavailability of CO_2 due to stomatal closure or block of electron transport by an herbicide such as DCMU, or chloride or sulfate stress, increase delayed fluorescence intensity. The factors that determine delayed fluorescence induction and decay kinetics have been discussed in several reviews [9-11]. Sundbohm and Bjorn [12] and Bjorn and Forsberg [13] introduced the use of sensitive image intensifier tubes to record images of leaves using delayed fluorescence. They and subsequently Ellenson and his associates [14-16], and Blaich et al. [17] demonstrated that the delayed fluorescence images of leaves often change in dramatic and characteristic ways when the plants are subject to stresses such as herbicides, air pollutants, extreme temperature, and pathogens.

At Battelle, we constructed an instrument based roughly on the design of Sundbohm and Bjorn [12], which allowed us to record delayed fluorescence images of leaves of intact plants while simultaneously observing the total intensity of delayed and prompt fluorescence. Our intention was to use the instrument as a nonintrusive probe of photosynthesis. The results of some preliminary experiments with herbicide treatment along with a summary of a study of the photosynthetic behavior of salt-stressed soybeans are described here. Details of the photosynthetic behavior of the salt-stressed soybeans will be presented elsewhere.

Procedure

Plant Growth Conditions

Plants were grown in modified Leonard jar assemblies (each assembly comprising a pot and a jar) interconnected by a cotton wick. The soil used in the pots was an amended montmorillonite clay soil (manufactured by International Minerals Corporation under the trademark "Turface"). One-fifth strength Jensen's nutrient solution [18] was added to each jar.

The jars were refilled with nutrient solution whenever necessary. Salt stress was imposed from the time of planting by supplementing the nutrient solution with the desired stress level and type of salt.

Soybean varieties "Williams" and "Manchu" were selected as representative sensitive and tolerant varieties, respectively, based on earlier studies [2]. Seed of neither of these varieties germinated at a concentration of 120 mM NaCl; germination was poor and growth was severely restricted at 100 mM NaCl. Seeds of both varieties germinated well at 80 mM NaCl, and growth of the tolerant variety was satisfactory. The concentration of 80 mM NaCl was therefore selected as the upper stress limit for these experiments. All experiments were conducted below this stress limit. Electrical conductivity of the saturated extract of the soil stressed with 80 mM NaCl was 0.78 S m^{-1} or 7.8 mmhos cm^{-1}, while that of nonstressed soil (irrigated with only the nutrient solution) extract was 0.21 S m^{-1}.

Seed was inoculated with *Rhizobium japonicum* strain USDA 136 (from U.S. Department of Agriculture, Beltsville, Md.) at the time of planting. The plants were grown in walk-in growth rooms at approximately 27°C with relative humidity at approximately 80% and fluorescent and incandescent lighting providing 340 μmol m^{-2} s^{-1} at the level of plant canopy during a 14-h light and 10-h dark cycle.

Electrical Conductivity Measurement

Ion accumulation in the salt stressed (0, 10, 20, 30, and 40 mM) soybean plant leaves was determined by EC measurements using a digital conductivity meter (Yellowsprings Instrument Company Model 35). All leaves were collected from 2-week-old stressed soybean plants. Water extracts of the leaves were made by placing the leaves in deionized distilled water (10 mL/g of leaf tissue) heated to a temperature of 60°C for 5 min. The clear extract was decanted from the leaf debris, cooled, and the conductivity measured by immersing the electrode in the extract.

Identification and Quantitation of Chloride Ion

Plants were stressed with 0, 20, 40, and 60 mM NaCl and KCl concentration, and harvested after two weeks, and four weeks after seedling emergence. Individual ions were quantified by using a Waters 500 high-performance liquid chromatograph (HPLC). The leaf tissues were placed in wide-mouth test tubes, frozen in liquid nitrogen, and pulverized with a glass rod. After adding deionized distilled water (10 mL/g of leaf tissue), the tubes were covered with foil and placed in a 100°C water bath for 1 h to extract the inorganic ions. The insoluble residue was removed by centrifugation at 10 000 g for 15 min at 4°C and the supernatant solution analyzed for ions using HPLC. Chloride ion was separated on a Wescan Anion/R column using *p*-hydroxybenzoate (pH 8.2) as the eluent. A Wescan 213A conductivity detector was used, and measurement was based on peak areas using a Hewlett-Packard 3390A integrator.

Estimation of Chlorophyll Concentration

Leaves of soybean plants subjected to different salt stresses (0, 20, 40, and 60 mM NaCl and KCl) were harvested at 2 and 4 weeks after emergence. All the leaves of plants were separated from the branches and were extracted with ethanol. Chlorophyll ($a + b$) concentration of the extracts was measured using a Varian DMS 90, UV-visible spectrophotometer.

Delayed Fluorescence Measurements

Delayed fluorescence was measured in the most recently formed, fully expanded trifoliate leaves of 4-week-old salt-stressed plants. Two herbicides, diquat dibromide and DCMU (dichlorophenyl dimethyl urea), were applied at 10 μM concentration to the upper surfaces of the leaves as aqueous solutions containing 0.1% Triton X-100 as a spreading agent. The plants were left in the growth room overnight before their delayed fluorescence images were photographed. Salt stress was imposed following the procedure described under the Plant Growth Conditions section.

Delayed fluorescence was measured with a Becquerel-type phosphoroscope based on the design of Sundbohm and Bjorn [*12*], employing a Varo 510-5722-310 18 mm multichannel plate wafer image intensifier, and modified to detect the net instantaneous intensities of prompt and delayed fluorescence as the delayed fluorescence images were being recorded. Leaves, still attached to the plants, were placed horizontally in the instrument; the leaves were dark-adapted for 10 min, then their upper surfaces were illuminated and light emission from the upper surface was recorded. For most experiments the leaves were excited with approximately 10^4 ergs cm^2 s^1 of blue light. In each phosphoroscope cycle the leaves were illuminated for 10 ms, both the excitation and measuring shutters were closed for 6 ms, and the light emitted by the leaves was recorded for 30 ms. The images represent total light collected during the first 20 min of the light exposure of the leaves.

Statistical Analysis of Data

There were six replications for each experiment. Statistical analysis was performed using the PC version of the Statistical Analysis System [*19*]. A series of two-way analysis of variance (ANOVA) was used to test for differences in, for example, variety and salt concentration while holding salt type and week constant. Bonferroni multiple comparison procedures were used as follow-up procedures when the main effects and/or interactions were significant ($P < 0.05$). For the multiple comparison procedures an experiment-wide error rate of 0.05 was used. Variables were also transformed to natural logarithms when the variances were shown to be correlated with the means [*20*]. No statistical analysis was performed for the delayed fluorescence experiments.

Results and Discussion

Ion Accumulation

Salt accumulation in the leaf extracts of the sensitive variety Williams was more rapid than in the extracts of the tolerant variety Manchu. In the case of Williams the accumulation was approximately linear with increase in salt concentration (Fig. 1). Differences were statistically significant for the variety × concentration ($P = 0.0001$) interaction (Table 1), which indicated that as salt concentrations increased, so did the electrical conductivity and the variety differences for conductivity. Our earlier studies with numerous varieties [*2*] have shown that the ion accumulation in leaves distinguished the sensitive soybean varieties from the tolerant varieties.

The fact that the differences in electrical conductivity of the leaf extracts of salt-stressed and control plants of Williams were highly significant ($P = 0.0001$) supports the use of conductivity measurements for assessment of potential toxicity due to fertilizer or irrigation water related salt stresses in soybean. The availability of the portable battery operated con-

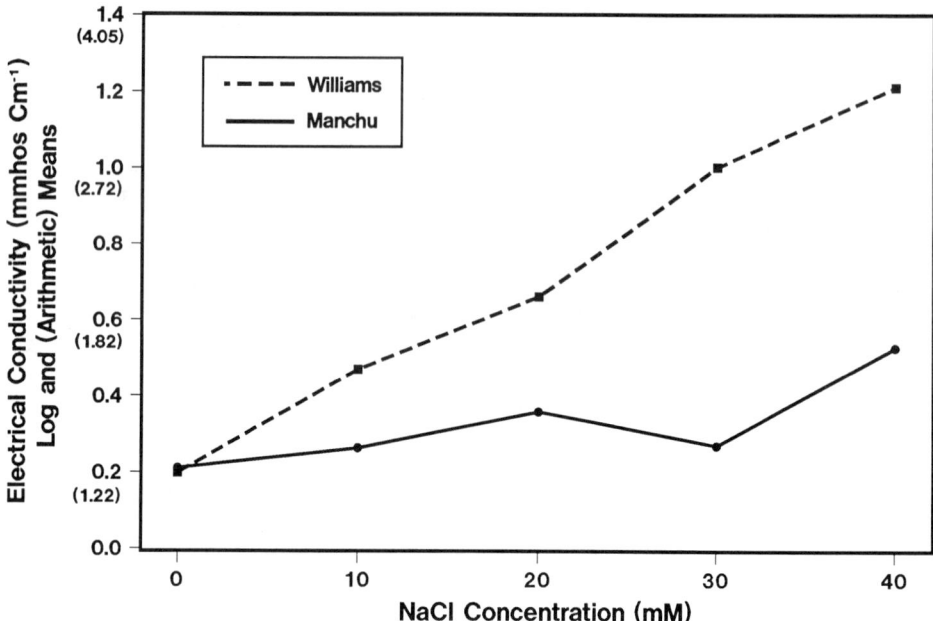

FIG. 1—*Electrical conductivity of leaf extracts of soybean varieties Williams and Manchu stressed with various levels of NaCl.*

ductivity meters further facilitates the use of this technique for field measurements. Experiments conducted with alfalfa and *Phaseolus vulgaris* (results not reported here) also showed trends similar to that of soybean. Since the conductivity measurements often reflect the total ionic strength of the leaf extracts, this technique could be used for a variety of chemicals where plant uptake is involved.

Identification and Quantitation of Chloride Ion

The accumulation of chloride ion in the leaves of Williams and Manchu plants stressed with NaCl and KCl showed trends consistent with those of the electrical conductivity measurements. The leaves of the variety Williams accumulated much larger quantities of chloride than those of Manchu at a given concentration, and the differences between the two varieties was highly significant ($P = 0.0001$) for both the salts (Fig. 2). Differences between concentrations were also highly significant ($P = 0.0001$) for each of the two salts. For NaCl stress, the variety × concentration interaction was highly significant ($P = 0.0001$), but this interaction was not significant for KCl stress (Table 1). This is consistent with our findings [3] that the variety Manchu was less tolerant to KCl stress than to NaCl stress, especially above 40 mM concentration. Such differences between NaCl and KCl stresses for Manchu were prominent only after four weeks of consistent exposure to these stresses. Thus chloride toxicity can be assessed in early phases of plant growth by the use of an appropriate variety. The varietal screening and the type of ionic stress are important considerations for utilizing the ion quantitation technique for assessing the plant toxicity.

TABLE 1—*Statistical summaries for electrical conductivity, chloride content, and total chlorophyll of leaf extracts under NaCl and KCl stresses.*

Salt and Week of Harvest	Source	Degrees of Freedom	F-Value	Significance Level
Electrical Conductivity of Leaf Extracts				
NaCl, 2 Weeks ($n = 60$)	Variety	1	208.38	0.0001
	Concentration	3	73.63	0.0001
	Variety × Concentration	3	28.98	0.0001
	CV = 19.81			
Chloride in Leaves				
NaCl, 2 Weeks ($n = 39$)	Variety	1	374.61	0.0001
	Concentration	3	177.21	0.0001
	Variety × Concentration	3	35.77	0.0001
	CV = 16.34			
KCl, 2 Weeks ($n = 38$)	Variety	1	64.80	0.0001
	Concentration	3	85.69	0.0001
	Variety × Concentration	3	2.11	0.1202
	CV = 21.27			
NaCl, 4 Weeks ($n = 47$)	Variety	1	439.13	0.0001
	Concentration	3	273.17	0.0001
	Variety × Concentration	3	11.96	0.0001
	CV = 10.95			
KCl, 4 Weeks ($n = 46$)	Variety	1	145.58	0.0001
	Concentration	3	372.82	0.0001
	Variety × Concentration	3	0.79	0.5084
	CV = 9.30			
Total Chlorophyll ($a + b$) in Leaves				
NaCl, 2 Weeks ($n = 39$)	Concentration	3	4.05	0.0153
	Variety	1	73.88	0.0001
	Concentration × Variety	3	18.00	0.0001
	CV = 1.98			
KCl, 2 Weeks ($n = 39$)	Concentration	3	13.49	0.0001
	Variety	1	106.59	0.0001
	Concentration × Variety	3	22.45	0.0001
	CV = 1.84			
NaCl, 4 Weeks ($n = 46$)	Concentration	3	53.97	0.0001
	Variety	1	242.81	0.0001
	Concentration × Variety	3	38.74	0.0001
	CV = 2.69			
KCl, 4 Weeks ($n = 46$)	Concentration	3	168.95	0.0001
	Variety	1	236.33	0.0001
	Concentration × Variety	3	24.50	0.0001
	CV = 1.89			

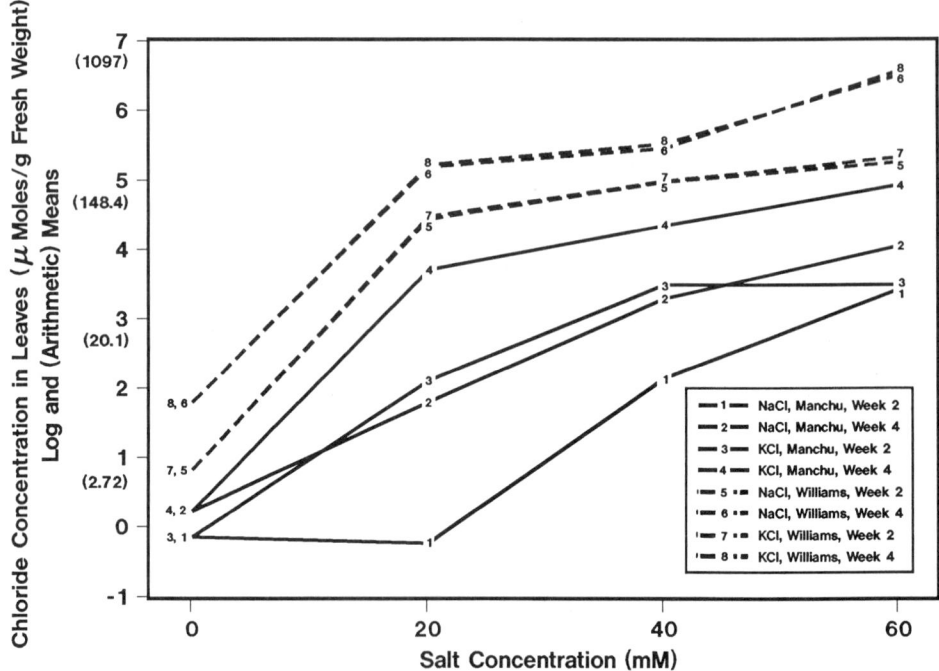

FIG. 2—*Concentration of chloride in the leaves of soybean varieties Williams and Manchu stressed with various levels of NaCl and KCl stresses.*

Chlorophyll Concentration in Leaves

Total chlorophyll of leaves harvested at 2 and 4 weeks declined with the increase in concentration of NaCl and KCl in both Williams and Manchu (Fig. 3). This decline was much more prominent for Williams at the 4-week harvest, although the leaf chlorophyll in NaCl and KCl stressed Williams plants harvested at 2 weeks also showed significant decline compared to that of the nonstressed controls. The variety Manchu, on the other hand, showed very little concentration-dependent decline in leaf chlorophyll concentration under NaCl and KCl stresses at the 2-week harvest. The decline in leaf chlorophyll was significant ($P = 0.0001$) in salt-stressed Manchu plants at the 4-week harvest (Fig. 3). Differences in the varietal response to the salt concentration was highly significant as indicated by the concentration $\times$ variety interactions which were highly significant ($P = 0.0001$) for all salts at both 2-week and 4-week harvests (Table 1).

The significant decline in the leaf chlorophyll of Williams under both NaCl and KCl stresses early in the plant growth reveals the potential use of leaf chlorophyll as an indicator for assessing the chemical toxicity in plants. Together with other approaches suggested above, these observations reaffirm the importance of varietal selection in the toxicity assessment process.

Delayed Fluorescence

Preliminary experiments were performed to determine whether delayed fluorescence measurements could distinguish between leaves that had been treated with herbicides whose mechanisms of action differed. Representative results are presented in Fig. 4. The upper sur-

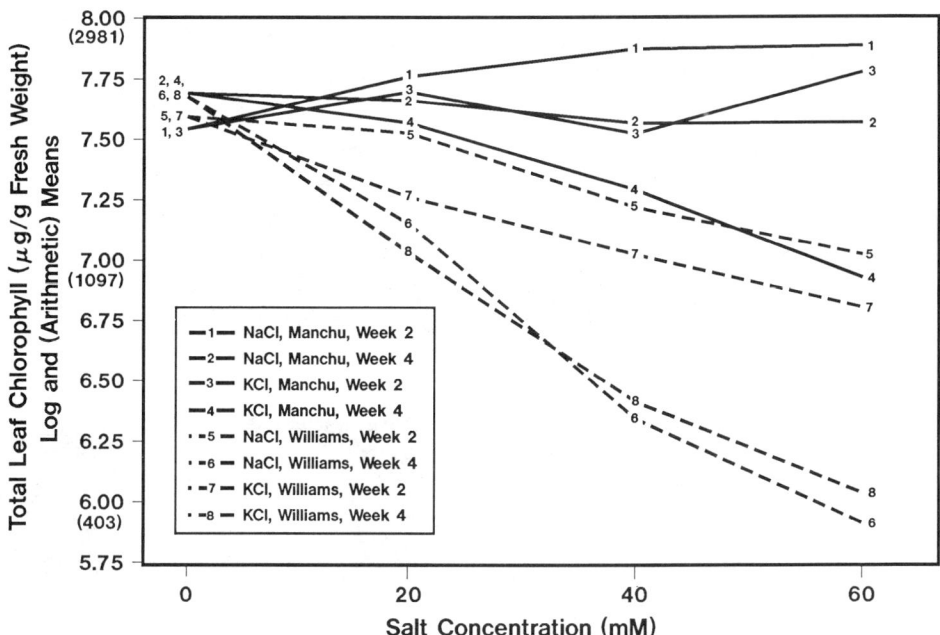

FIG. 3—*Concentration of chlorophyll* (a + b) *in the leaves of soybean varieties Williams and Manchu stressed with various levels of NaCl and KCl stresses.*

faces of bean leaves were treated with diquat (Fig. 4b) or DCMU (Fig. 4c) or with a control solution containing only Triton X-100, which served as a spreading agent in each treatment (Fig. 4a). The images of the diquat leaves are uniformly bright, while the veins of DCMU-treated leaves appear as dark areas surrounded by bright halos. For such experiments, instrumental parameters can be chosen which maximize the observed differences. Figure 4a to 4c show images of delayed fluorescence emitted between 6 and 36 ms after the leaves had been illuminated. Among parameters that can be varied in order to distinguish between different stressors are speed of the chopper motor (to select the best decay components), the period during the illumination of the leaves when the images are recorded (the first 20 min for the images in Figs. 4a to 4c), light or dark treatment of the leaves prior to the illumination (10 min of darkness for the images of Figs. 4a to 4c), intensity and the wavelength of the exciting light, temperature, and the composition of the gas surrounding the leaves.

A more detailed study of the effect of salt stress on the delayed fluorescence of soybean leaves was conducted. Salient features of findings of these studies are presented here. It was noticed that the images obtained from the first trifoliate leaves of plants grown in the presence of 80 mM NaCl were much brighter than those of control plants (Fig. 4d). The brightening first appeared 3 to 5 days after the opening of the trifoliate. Then the veins began to darken (Fig. 4e). Images of the leaves of Na_2SO_4-stressed plants also brightened a few days after the opening of the trifoliate, but the leaves then began to darken in the areas between the veins, resulting first in a scalloped appearance and finally in a pattern of bright veins against a dark background (Fig. 4f). We have not determined the physiological reason for the difference in the appearance of the leaves stressed by a chloride or sulfate salt. Nevertheless, the pattern differences shown in Figs. 4e and 4f between NaCl and Na_2SO_4 stressed plants are distinct and can be used for rapid and presymptom diagnosis of these stresses.

The source of brightening of the images of the stressed leaves is apparent when the time

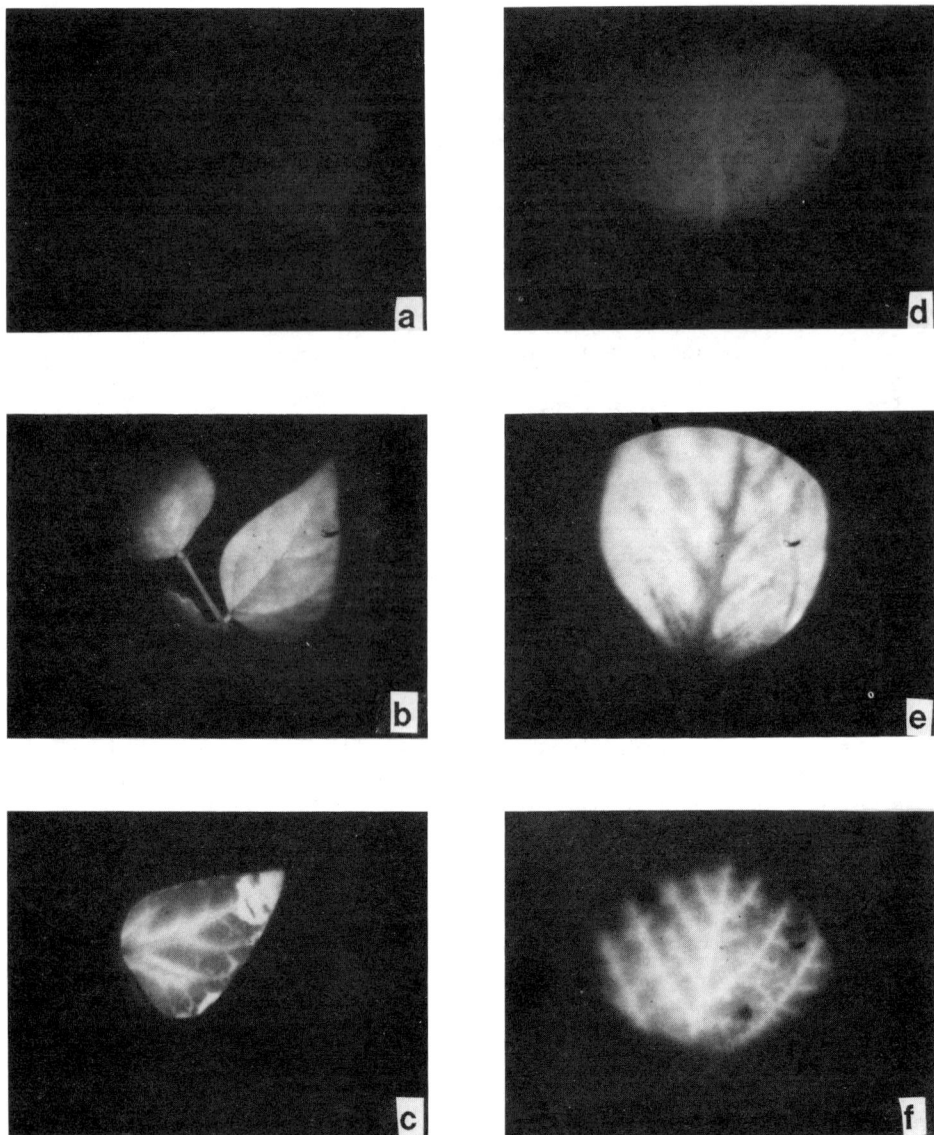

FIG. 4—*Delayed fluorescence images of leaves of Williams soybean plants stressed with herbicide and salt.* (a) *Control leaves of* Phaseolus *bean plants.* (b) *Leaves of* Phaseolus *bean treated with Diquat herbicide.* (c) *Leaves of* Phaseolus *bean treated with DCMU.* (d) *Control leaves of soybean plants.* (e) *Soybean leaves from plants treated with NaCl.* (f) *Soybean leaves from plants treated with Na_2SO_4.*

courses of the light emission by the stressed and control leaves are compared. The delayed fluorescence images were photographed during the first 20 min of illumination of the leaves. After illumination of leaves is begun, the intensity of the delayed fluorescence undergoes several waves of variation before settling to a steady-state level. The total amount of light emitted by the two leaves is about equal during the first 4 or 5 min. After that, however, light emitted by the control leaf continues to decline while that emitted by the stressed leaf does not. Thus the total amount of light emitted by the stressed leaf is much greater (Figs. 5a and 5b).

The higher steady-state delayed fluorescence intensity of the stressed leaf is due to an increase in magnitude of a delayed fluorescence decay component having a half-life of 0.3 to 0.5 s (data not shown). Delayed fluorescence in this time range is probably due to charge

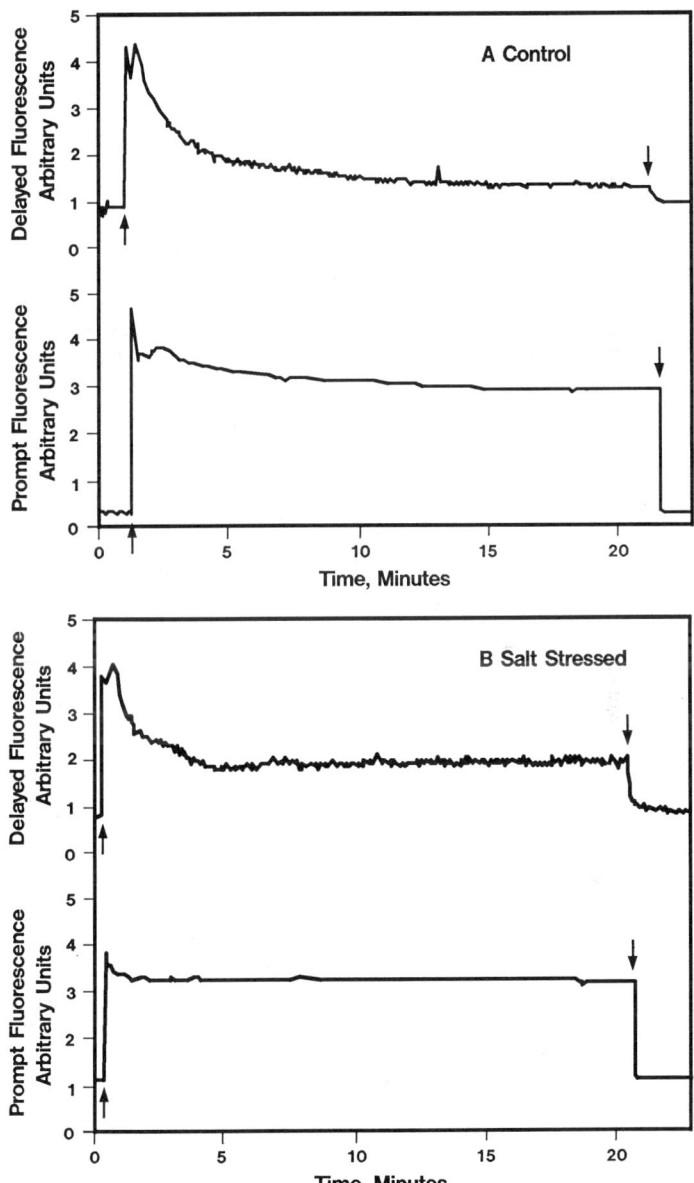

FIG. 5—*Kinetics of prompt and delayed fluorescence of leaves of soybean plants.* (a) *Leaf from control plant.* (b) *Leaf from NaCl stressed plant.*

recombination between the semiquinone form of the primary quinone electron acceptor, QA, and the S2 or S3 state of the oxygen-evolving system of the chloroplast [9-11]. Enhancement of this delayed fluorescence component could be due to increased protonmotive force across the thylakoid membrane or to an increased level of reduction of the quinone pool that lies between Photosystem I (PS I) and Photosystem II (PS II). The latter could be caused by stomatal closure or change in excitation energy distribution between PS I and PS II. A series of prompt fluorescence measurements (which will not be described here) lead us to the conclusion that the salt-stressed plant is unable to regulate excitation distribution between the photosystems. As a result PS II is overdriven, resulting in an overreduction of quinone pool and QA.

Conclusions

Results presented here demonstrate the usefulness of the electrical conductivity, and ion concentration measurements, chlorophyll content, and the delayed fluorescence of leaves for detecting plant stresses in advance of visible symptoms. These parameters can also be used to distinguish the tolerant and senstive genotypes under a given salt stress. The utility of these techniques for practical and field adaptations would depend on their simplicity and adaptability to a field situation. Electrical conductivity and chlorophyll content measurements are less equipment intensive and easily adaptable in comparison to the delayed fluorescence and ion quantitation measurements. However, the latter two techniques provide more elaborate data and contribute significantly to our understanding of toxicity assessment in plants. Clearly, the utility of these techniques for evaluating plants as bioassay tools would depend on the depth of information sought by an investigator, and the type of chemical stressor. For example, electrical conductivity measurements may be more valuable for salt stress measurement, while the chlorophyll content of the leaves may be the more useful measure of the pesticide stress. Thus a case-by-case approach is necessary for efficient use of these techniques for toxicity assessment.

Acknowledgments

We wish to thank Dr. M. F. Arthur and Mr. T. C. Zwick for their helpful comments and suggestions.

References

[1] Shannon, M. C. in *Salinity Tolerance in Plants, Strategies for Crop Improvement*, R. C. Staples and G. H. Toenniessen, Eds., Wiley, New York, 1984, pp. 231-254.
[2] Velagaleti, R. R., Marsh, S., Kramer, D., Fleischman, D., and Corbin, J., "Genotypic Variation in Growth and Nitrogen Fixation Among Soybean (*Glycine max* (L.) Merr) Cultivars Grown under Salt Stress," *Tropical Agriculture*, 1990.
[3] Velagaleti, R. R. and Marsh, S., "Influence of Host Cultivars and *Rhizobium* Strains on the Growth and Symbiotic Performance of Soybean under Salt Stress," *Plant and Soil*, Vol. 119, 1989, pp. 133-138.
[4] Van Genuchten, M. Th. and Hoffman, G. J. in *Soil Salinity under Irrigation, Processes and Management*, I. Shainberg and J. Shalhevet, Eds., Springer-Verlag, Berlin, 1984, pp. 258-271.
[5] Greenway, H. and Munns, R., *Annual Review of Plant Physiology*, Vol. 31, 1980, pp. 149-190.
[6] Velagaleti, R. R., Marsh, S., Corbin, J. L., and Fleischman, D. E., "Genotypic Variation for Growth and Nitrogen Fixation under KCl Stress and a Comparative Assessment of NaCl and KCl Stresses," *American Society of Agronomy Abstracts*, 1986, p. 124.
[7] Seeman, J. R. and Critchley, C., *Planta*, Vol. 164, 1985, pp. 151-162.
[8] Marchiarullo, M. A. and Ross, R. T., *Biochimica Biophysica Acta*, Vol. 636, 1981, pp. 254-257.

[9] Malkin, S., in *Encyclopedia of Plant Physiology,* A. Trebst and M. Avron, Eds., Springer-Verlag, Berlin, Vol. 5, 1977, pp. 473-491.
[10] Amesz, J. and Van Gorkom, H. J., *Annual Review of Plant Physiology,* Vol. 29, 1978, pp. 47-66.
[11] Jurisnic, P. in *Light Emission by Plants and Bacteria,* Govindjee, J. Amesz, and D. C. Fork, Eds., Academic Press, Orlando, Fla., 1986, pp. 291-327.
[12] Sundbohm, E. and Bjorn, L. D., *Physiologia Plantarum,* Vol. 49, 1979, pp. 39-41.
[13] Bjorn, L. O. and Forsberg, A. S., *Physiologia Plantarum,* Vol. 47, 1979, pp. 215-222.
[14] Ellenson, J. L. and Amundson, R. G., *Science,* Vol. 215, 1981, pp. 1104-1106.
[15] Ellenson, J. L. and Raba, R. M., *Plant Physiology,* Vol. 72, 1983, pp. 90-95.
[16] Ellenson, J. L., *Plant Physiology,* Vol. 78, 1985, pp. 94-98.
[17] Blaich, R., Bachman, O., and Baumberger, I., *Zeitschrift Naturforsch,* Vol. 37c, 1982, pp. 452-457.
[18] Vincent, J. M. in *Manual for the Practical Study of Root Nodule Bacteria,* Blackwell Scientific, Oxford, 1970.
[19] Johnson, N. and Leone, F., *Statistics and Experimental Design,* Vol. 2, Wiley, New York, 1977.
[20] *The Statistical Analysis System (SAS),* SAS Institute, Inc., Cary, N.C., 1986.

Jianhua Xu,[1] Wenjie Xia,[1] Xudong Jong,[1] Weichi Sun,[1] Guangheng Lin,[2] and Te-Hsiu Ma[3]

Image Analysis System for Rapid Data Processing in *Tradescantia*-Micronucleus Bioassay

REFERENCE: Jianhua Xu, Wenjie Xia, Xudong Jong, Weichi Sun, Guangheng Lin, and Te-Hsiu Ma, "**Image Analysis System for Rapid Data Processing in *Tradescantia*-Micronucleus Bioassay**," *Plants for Toxicity Assessment, ASTM STP 1091*, W. Wang, J. W. Gorsuch, and W. R. Lower, Eds., American Society for Testing and Materials, Philadelphia, 1990, pp. 346–356.

ABSTRACT: *Tradescantia*-micronucleus (Trad-MCN) bioassay is a well established short-term test for environmental mutagens in liquid or gaseous forms. For large-scale experiments that involve hundreds of slides, the micronuclei scoring process, data collection, and analysis are labor intensive and time consuming. Using computer image analysis process, a rapid data processing system was developed. This new facility is called the Trad-MCN Image Analysis System (Trad-MCN-IAS). The hardware of this system is composed of a photomicroscope attached to a video camera and a US-386 computer. The software for this system was written in FORTRAN language and run under an FG-105 real-time image processor. The operation begins with multi-threshold image segmentation, followed by mathematical morphology-based micronucleus extraction, and judging of normal tetrads with pseudo-color display. Using this system, a large number of tetrads (the four-cell stage of the meiotic pollen mother cells) image on the screen can be scored simultaneously for normal and micronucleus-bearing tetrads. The number of micronuclei in a large population of tetrads can be recorded and put in memory for data analysis. For high efficiency, four microscopes may be installed on a single Trad-MCN-IAS to gather tetrad images and count micronuclei; therefore a large number of samples can be scored and the experimental data processed in a short time. This system has several advantages: non-bias in the scoring process, automation, and minimizing the variance of scores introduced by different human observers.

KEY WORDS: image analysis, micronucleus, *Tradescantia*, data processing, computer

Tradescantia-micronucleus (Trad-MCN) bioassay is a widely adopted clastogenicity test for environmental pollutants and common genotoxic chemicals in both liquid and gaseous forms [1–3]. The exposure targets in this test are the early prophase I meiotic pollen mother cell chromosomes of *Tradescantia*, and the end points are the chromosome damage in the form of micronuclei (MCN) revealed in the tetrads (the four-cell stage at the end of meiosis) (Fig. 1). Scoring of the MCN frequencies from the microslides in this test is a labor intensive and time consuming process. Current investigation is designed to develop an automated computer image analysis system (IAS) to score the MCN frequencies and analyze the experimental data. Although the automated Mouse-erythrocyte-MCN [4] scoring system is cur-

---
[1] Department of Computer Science, Fudan University, Shanghai, People's Republic of China.
[2] Institute of Oceanology, Academia Sinica, Qingdao, People's Republic of China.
[3] Institute for Environmental Management and Department of Biological Sciences, Western Illinois University, Macomb, IL 61455.

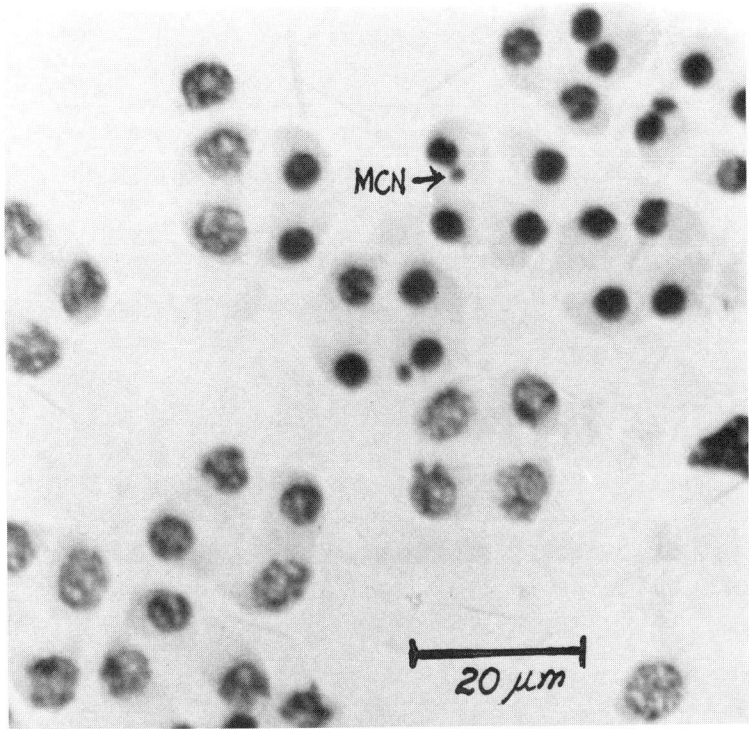

FIG. 1—*Photomicrographic images of tetrads. Some bear one or more micronuclei; some are irregular in shape or orientation and not qualified for image analysis.*

rently available [5] and attempts have been made to develop an automated scoring system for micronuclei in human lymphocytes [6,7], a special IAS is necessary for *Tradescantia* MCN bioassay because the pattern of the tetrads and MCN in this plant bioassay is quite different from that of the mouse erythrocyte system. A new approach in image analysis will be explored in the development of this system. The current IAS was designed under a different principle of computer technology. Application of this Trad-MCN-IAS will increase the efficiency in the scoring process and eliminate the inherent variance among human observers and among laboratories in a large-scale testing network.

Materials and Methods

The external features of the system are shown in Fig. 2. The major components of this *Tradescantia*-MCN-IAS (Trad-MCN-IAS) are a 32-bit US-386 microcomputer with a real-time vision board, colored monitor, and mouse-controller, and a video camera attached to a microscope. The detailed composition and operational procedure are presented below.

Hardware System

To allow the Trad-MCN-IAS to have world-wide adoption, most of the components of the hardware system comply with international standards. The major processor is composed of a US-386 super microcomputer (CPU is a 32-bit INTEL 80386), FG-105 vision board

FIG. 2—*Major components of Trad-MCN-IAS: photomicroscope, video camera, and computer systems for image analysis and computation.*

with 1024 by 1024 by 12 bits of frame memory. A block diagram of the hardware system is shown in Fig. 3.

System Optimization

Background Compensation—The optical system of the microscope and the video camera lens often create spherical aberrations. Therefore the background grey level of the blank

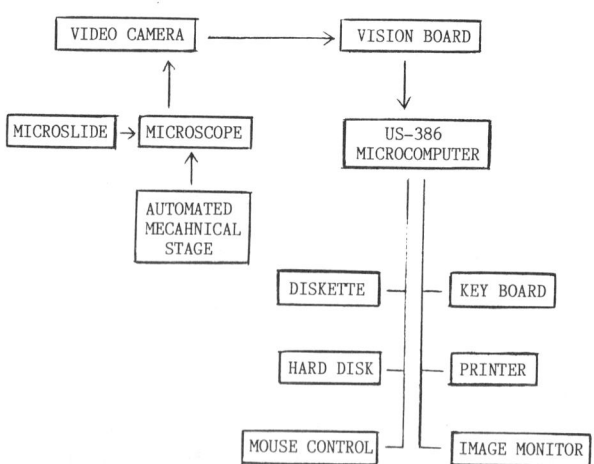

FIG. 3—*Block diagram of the hardware of the system.*

screen is usually uneven and the image of the objects distorted. A compensation procedure was applied using the formula

$$F(x, y) = f(x, y) - g(x, y) + C \qquad (1)$$

where C is a constant, $g(x, y)$ is the reference image of the background compensation (i.e., an empty field), $f(x, y)$ is the original image of the object, and $F(x, y)$ is the resulting image after compensation.

Application of Frame Memory as External Memory—To improve analytical ability by enhancement of the operational speed and transportable range of the computer system, the frame memory procedure was used to process the information point-by-point, read-and-write form, and to keep the information in external memory. This procedure was followed because the FORTRAN 77 4.0 version program has limited internal memory. When Trad-MCN and the accompanying tetrad images were processed with 512 by 512 by 8 bit array, only 60 s were needed to put the images in the frame memory system. This speed was about eight times faster than using the internal memory system.

Automation of Scanning Movement of Mechanical Stage of the Microscope—A series of commands was programmed in the software to activate the stepping motors that were built on the mechanical stage of the microscope to move the slides on the x and y axes automatically. For the protection of this delicate mechanism, programs were also built in the system to stop the movement when conflicting driving forces were initiated by mistake.

Software System

The software was programmed in the modular fashion and in the form of running subroutine in order to utilize the full potential of the US-386 microcomputer and FG-105 vision board. This system not only provides excellent automation and image detection but is user-friendly and transportable. A block diagram of the software system is shown in Fig. 4.

The Trad-MCN-IAS was specially designed for the MCN formed in the tetrads of the meiotic pollen mother cells of *Tradescantia*. The system can automatically process the following information:

1. Total number of tetrads and micronuclei.
2. Total number of tetrads, M, which contains a given number of micronuclei, k.
3. Total number of tetrads containing dead nuclei.
4. Ratio of total number of micronuclei over tetrads and the MCN frequency.
5. Data output of the bioassay on pollutants.

The software system designed specifically for Trad-MCN includes a program of compensation of the background grey level, image subtraction for local image regions, and mathematical morphology-based shape analysis. These programs were designed for the extraction of the features of the tetrads and micronuclei.

Extraction of Nucleus and Micronucleus from Tetrads

Image Segmentation—Since the video camera image of the tetrad may occasionally show the connection between the nucleus and MCN, the tetrad may appear as a single image. Thus the image subtraction-based entire image segmentation approach was used. This process can

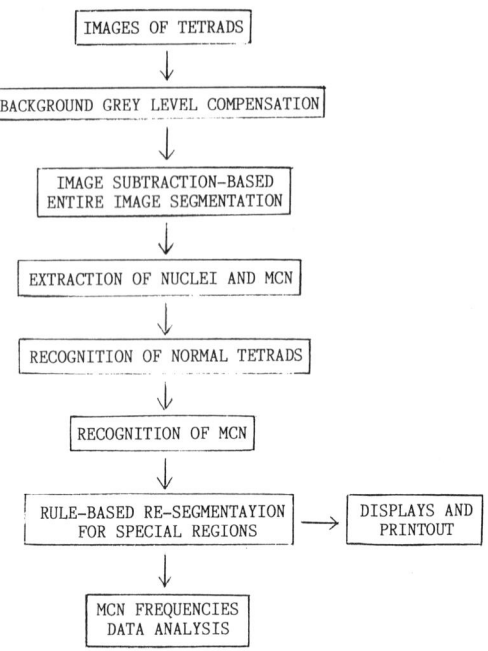

FIG. 4—*Block diagram of the software of the system.*

separate clearly the nucleus from MCN. The principle of this segmentation technique is that of comparing the tetrad image $f(x, y)$ with the background $g(x, y)$.

Let

$$F(x, y) = \begin{cases} 1, & \text{when } f(x, y) - g(x, y) < \text{Ths} \\ 0, & \text{Otherwise} \end{cases} \quad (2)$$

where *Ths* is the threshold for grey level, and $F(x, y)$ is the binary image for segmentation. The binary image is shown in Fig. 5; the video image of a population of tetrads is shown in Fig. 1 for comparison.

Extraction of Nuclei and Micronuclei—For the binary image $F(x, y)$, a labeled image can be obtained by image isolation processing (i.e., in the labeled image, every labeled region has only one label, and for different labeled regions, their labels are also different). Making the histogram for the labeled image, one obtains the area S_k of kth ($k = 1, 2, 3, \ldots, n$) labeled region. Then the nuclei and micronuclei may be separated according to the area threshold as follows:

$$\text{Labeled region} = \begin{cases} \text{nuclei}, & \text{if } Ta1 < S_k < Ta2 \\ \text{MCN}, & \text{if } Ta3 < S_k < Ta4 \end{cases} \quad (3)$$

where Tai ($i = 1, 2, 3, 4$) are area thresholds, and S_k is the area of the kth labeled region. Moreover, for the labeled image one obtains the central position (x_c, y_c) and distributed range of the kth region as follows:

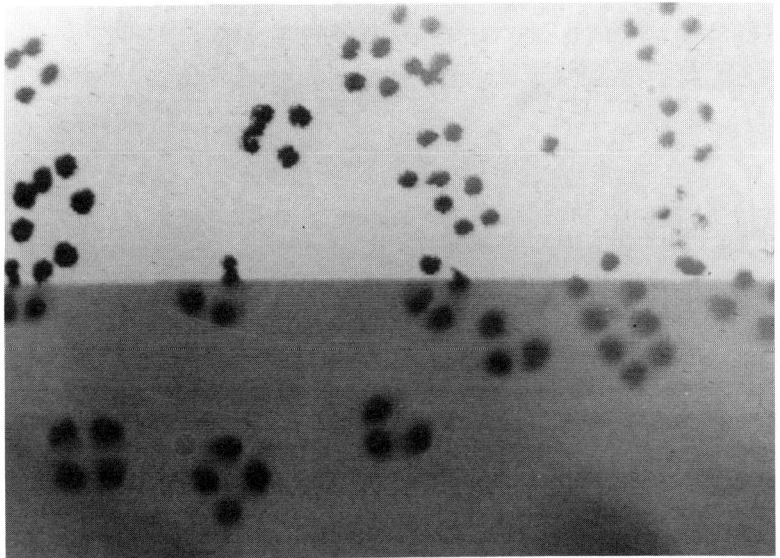

FIG. 5—*Scanning process during extraction of micronuclei from the binary images of the tetrads.*

Central Position:

$$x_c = \text{Sum } [x: f(x, y) = S_k]/S_k \quad (4)$$
$$y_c = \text{Sum } [y: f(x, y) = S_k]/S_k \quad (5)$$

Distributed Range:

$$[x_{k\min}, x_{k\max}] \text{ and } [y_{k\min}, y_{k\max}]$$

where $x_{k\min}$ denotes the minimum position of x-direction of the region labeled by k, $x_{k\max}$ denotes the maximum position of x-direction of the region labeled by k, and $y_{k\min}$ and $y_{k\max}$ denote the minimum and maximum positions of y-direction of the regions labeled by k respectively.

Recognition of Normal Nuclei and Micronuclei—All normal nuclei are circular and most abnormal nuclei have irregular shapes; normal nuclei have similar surface areas; therefore it is possible to extract the normal nuclei using the concept of circular degree. In a plane, a simple closed curve has a constant length and a circle surface has the largest area. That is:

$$L - 4\pi S > 0 \quad (6)$$

The condition that satisfies Eq 6 the most is that the closed curve be a circle. From this, one can use the concept of circular degree to recognize the normal nuclei and micronuclei. The circular degree is defined by

$$K = L^2/S \quad (7)$$

where L is the perimeter of the region, and S is the area of the region. The closer the shape of the region is to a circle, the less is its circular degree (near to 4π).

In the labeled image, if the label of a certain region is k, the area of the region is S_k; then the perimeter L_k of this region is

$$L_k = \tfrac{1}{4}[n(0, k) + n(k, 0)] + \tfrac{1}{2}[n(^k0) + (0^k)] \tag{8}$$

where $n(0, k)$ represents the number of the labels changed from 0 to k in the horizontal direction (from left to right) in this region, so that

$$K_k = L^2/S_k \tag{9}$$

According to Eq 9, the normal nuclei can be identified by the circular degree threshold, Thc. If $K_k < Thc$, then this nucleus is normal; otherwise it is abnormal.

In some cases, the distance between the nucleus and MCN is very short, and among the labeled images there may be some small labeled regions resulting from background noise. For accurate extraction of MCN, a distance threshold, Thd, was established. When the distance from the central position of a certain small labeled region to the normal nuclei is greater than the Thd, the small labeled region was not considered to be a true MCN.

Recognition of Normal Tetrads and Corresponding MCN

The central position of four nuclei of every normal tetrads should be located at the corners of a square. Therefore the rule of normal tetrads may be determined according to the geometrical properties of the square. For the binary segmented image of the tetrads, the labels of the four nuclei of the same normal tetrads are not numbered orderly because the labeled image was obtained by the isolation processing in a line scanning form. The central coordinates of four nuclei of a given tetrad are not always arranged in order. The normal tetrads were recognized from a set of central coordinates of four nuclei according to the following identification criteria rules:

Rule 1—The quadrilateral constructed by the central coordinates of four nuclei of a given tetrad should be approximately a square.

(*a*) On six connected lines constructed by four central coordinates, select a connected line having the longest distance as one diagonal. The connected line constructed by the other two central coordinates is taken as another diagonal. Four adjacent sides are produced at the same time.

(*b*) Distances of four adjacent sides are approximately equal.

(*c*) Distances of two diagonals are approximately equal.

(*d*) Two diagonals in the quadrilateral have an intercept. Distances from the intercept point to the four corners of the quadrilateral are approximately equal.

Rule 2—In the approximate square constructed by four nuclei of a normal tetrad, there should not be any other nuclei.

Rule 3—The normal tetrads recognized from a set of central coordinates constructed by four nuclei should be the best match among the other possible matches and checked repeatedly through Rules 1 and 2.

The MCN frequency is defined as the number of MCN found in the tetrads as recognized according to the rules stated above. At the same time, the following parameters are obtained:

- $M(0)$ = number of tetrads without MCN.
- $M(k)$ = number of tetrads with k number of MCNs ($k = 1, 2, \ldots, n$).

Some nuclei in the tetrads may be dead nuclei, and the ratio of the number of tetrads with dead nuclei to the total number of tetrads is also a very important parameter. Compared with live nuclei (at the time of treatment), dead nuclei have the following features:

- Dead nuclei are rounder and smoother (i.e., its circular degree is smaller).
- Dead nuclei show darker grey levels.

For these reasons, one can set a new circular degree threshold and a threshold of the region grey level mean value. When the circular degree and the grey level mean value of certain nuclei are smaller than the threshold, the nucleus is claimed to be the dead nucleus and the number of tetrads with dead nuclei can be obtained.

Rule-Based Image Resegmentation

With the limitation of the resolution of the present image input system, some MCN and nuclei appear to be in close contact. These MCN are not detected in the first round of subtraction-based segmentation. A second recursive threshold resegmentation processing for specific localized region was applied. The MCN were extracted on the basis of the rules of circular degree, area, and distance. The rules and the procedure of the recursive resegmentation process are as follows:

1. The attached MCN in a set of nuclei of a tetrad was detected by using the new circular degree threshold. Since the boundary of the nucleus shows a protruding part similar to the MCN, the circular degree is therefore relatively larger.

2. In the local rectangular window with the nucleus, the tetrad images have their original grey level after background compensation. Starting from the given initial grey level segmentation threshold, the resegmentation process was conducted for the specific region based upon the recursive reductive grey level thresholds. The attached MCN can be isolated.

3. A region having the larger area in newly isolated regions was selected first for resegmentation. If this area is larger than the area threshold of the nucleus extracted, and at the same time the boundary of the this region is sufficiently round, then this newly isolated region is identified as the newly isolated nucleus worth a MCN. If this new region does not satisfy the foregoing two conditions, the original nucleus does not have the attached MCN.

4. After the resegmentation of the specific region and forming the new nucleus, the other isolated new regions with small area should be inspected. If these small regions fit the conditions of area, boundary, and distance between nuclei and these small regions, they are identified as the new MCN. Otherwise, the original nucleus was declared to be a nucleus without attached MCN.

The image of tetrads and MCN can also be displayed by a pseudo-color processing technique in which the background color is green, the normal tetrads are red, and the MCN-bearing tetrads and the MCN are blue.

Determining MCN Frequencies

With the computer IAS continuously picking up tetrads and MCN images from the microscope through the video camera, about 300 tetrads are analyzed automatically. The following formulas are used to determine the MCN frequencies:

1. $M(k)$ = number of tetrads with k number of MCN:

$$k = 0, 1, 2, 3, \ldots, n \qquad (10)$$

2. Total number of tetrads, M:

$$M = \sum_{k=1}^{n} M(k) \qquad (11)$$

3. Total number of MCN, N:

$$N = \sum_{k=1}^{n} kM(k) \qquad (12)$$

4. Total number of tetrads with dead nucleus.
5. Ratio of total number of MCN to total number of tetrads:

$$\text{MCN frequency} = N/M \qquad (13)$$

Validation of Trad-MCN-IAS

To verify the reliability of MCN frequencies scored by the IAS, three series of experiments were performed. *Tradescantia* MCN-in-tetrads slides were prepared in accordance with the standard protocol described in an earlier publication [4]. For the purpose of comparing the accuracy of the scores between the IAS and human observation, two series of Trad-MCN slides were prepared for scoring. One group of slides was prepared from the flower buds treated with 40 R X-rays for the high MCN frequency (around 50 MCN/100 tetrads); the other group of slides was prepared from flower buds treated with 136 ppm of sodium azide for low MCN frequency (around 25 MCN/100 tetrads). These slides were repeatedly scored by IAS and human observation. The results are shown in Tables 1 and 2.

TABLE 1—*Proficiency of Trad-MCN-IAS for scoring MCN frequencies induced by X-rays (40 R) as compared with human observation.*

	Trad-MCN-IAS Scoring				Human Scoring			
	X-Ray (40 R)		Control		X-ray (40 R)		Control	
Exp. Groups	MCN/100 Tetrads	Time (min)	MCN/100 Tetrads	Time (min)	MCN/100 Tetrads	Time (min)	MCN/100 Tetrads	Time (min)
1	68.70	10.0	3.35	12.3	63.90	43	5.70	12.0
2	39.90	8.0	2.80	7.0	79.50	40	5.30	12.0
3	38.59	11.2	5.80	10.5	67.10	30	6.40	14.0
4	40.30	12.0	0.30	14.0	65.30	35	2.60	11.0
Mean	46.69	10.3	3.06	11.2	68.95	37[a]	4.50	12.3[a]
S. E.	6.30		0.98		3.09		0.79	

[a] Average time consumed did not include the time needed for data analysis.

TABLE 2—*Proficiency of Trad-MCN-IAS for scoring MCN frequencies induced by sodium azide (136 ppm) as compared with human observation.*

Exp. Groups	Trad-MCN-IAS Scoring				Human Scoring			
	NaN$_3$ (136 ppm)		Control		NaN$_3$ (136 ppm)		Control	
	MCN/100 Tetrads	Time (min)	MCN/100 Tetrads	Time (min)	MCN/100 Tetrads	Time (min)	MCN/100 Tetrads	Time (min)
1	14.90	5.30	3.35	12.3	25.30	15.0	5.70	12.0
2	15.25	5.70	2.80	7.9	25.30	17.3	5.30	12.0
3	12.01	5.23	5.80	10.5	29.00	14.9	6.40	14.0
4	20.38	6.09	0.30	14.0	23.00	18.7	2.60	11.0
5	13.83	5.85			26.00	16.4		
Mean	15.27	5.65	3.06	11.2	25.72	15.7	4.50	12.3
S.E.	1.56		0.98		0.96		0.79	

According to the results of this series of experiments, IAS scores were about 60 to 68% of the scores of human observation. This value could be used to express the IAS accuracy:

$$\text{Accuracy} = \frac{\text{MCN scored by IAS}}{\text{MCN scored by human observation}} \quad (14)$$

The speed of scoring under IAS and human observation was also recorded. The mean values are given in Table 1 for comparison.

Conclusions and Discussion

Trad-MCN-IAS developed in this investigation was capable of recognizing the standard tetrads of the meiotic pollen mother cells of *Tradescantia* and detecting the MCNs with 60 to 68% accuracy as compared with human observation. The speed of Trad-MCN-IAS scoring is about 3.5 times faster than that of human observation in the high frequency level (induced by 40 R of X-rays) slides, and about twice as fast when scoring the medium level (induced by chemical mutagens) slides. Since the automated IAS can analyze the experimental data simultaneously during the scoring process through the computer system, the speed of data collection is always faster than manual scoring. The great advantage in using this automated IAS is the reduction of inherent variance among the human observers and among the laboratories in a large-scale environmental monitoring project involving different geographic areas.

Acknowledgments

The authors would like to thank Long-xiang Bian, Lei Xuan, Wei-ming Yan, and Rui-li Lin for their comments and suggestions during the course of this investigation.

References

[1] Ma, T. H., Sparrow, A. H., Schairer, L. A., and Nauman, A. F. "Effect of 1,2-Dibromoethane (DBE) on Meiotic Chromosomes of *Tradescantia*," *Mutation Research,* Vol. 58, 1978, pp. 251–258.

[2] Ma, T. H., Harris, M. M., Anderson, V. A., Ahmed, I., Mohammad, K., Bare, J. L., and Lin, G., "*Tradescantia*-Micronucleus (Trad-MCN) Tests on 140 Health Related Agents," *Mutation Research,* Vol. 138, 1984, pp. 147–167.

[3] Ma, T. H., Anderson, V. A., Harris, M. M., Neas, R. E., and Lee, T. S., "Mutagenicity of Drinking Water Detected by the *Tradescantia* Micronucleus Test," *Canadian Journal of Genetics and Cytology,* Vol. 27, 1985, pp. 143–150.

[4] MacGregaor, J. T., Schlegel, R., Choy, W. N., and Wehr, C. M., "Micronuclei in Circulating Erythrocytes: A Rapid Screen for Chromosomal Damage during Routine Toxicity Testing in Mice," in *Development in the Science and Practice of Toxicology,* Haves, Ed., Elsevier Science Publications B. V., 1983, pp. 555–558.

[5] Ma, T. H. "*Tradescantia* Micronuclei (Trad-MCN) Test for Environmental Clastogens," in *In vitro Toxicity Testing of Environmental Agents,* Part A., Kolber, Wong, Grant, DeWoskin, and Hughes, Eds., Plenum, New York, 1983, pp. 191–214.

[6] Callisen, H. H., Pincu, M., and Norman, A., "Feasibility of Automating the Micronucleus Assay," *Analytical Quantitative Cytology and Histology,* Vol. 8, 1986, pp. 219–223.

[7] Pincu, M., Callisen, H. H., and Norman, A., "Micronuclear DNA Distributions," *International Journal of Radiation Biology,* Vol. 47, 1985, pp. 423–432.

Author Index

A

Amundson, G. H., 156
Arthur, M. F., 127

B

Bartine, R. W., 198
Bassi, M., 204
Benenati, F., 5
Bo, J., 170
Boudou, A., 97

C

Clark, J. R., 59, 255
Corradi, M. G., 204

D

DiGiulio, R. T., 143
Duke, K. M., 127

E

Elseth, G., 280

F

Favali, M. A., 204
Fleischman, D. E., 333
Fletcher, J. S., 33
Freemark, K., 14
French, D. R., 156

G

Garten, Jr., C. T., 69
Gorsuch, J. W., 1, 49
Greene, J. C., 177
Grigal, D. F., 156
Guangrong, C., 170

J

Jong, X., 346
Judy, B. M., 235, 308

K

Klaine, S. J., 324
Kramer, D., 333
Krause, G. F., 235, 308
Kringle, R. O., 49
Kromroy, K. W., 156
Kristich, M. A., 87

L

Laurence, J. A., 117
Lillie, T. H., 198
Lin, G., 346
Linder, G., 117
Lower, W. R., 1, 235, 308

M

Ma, T.-H., 319, 346
Macauley, J. M., 59, 255
MacQuarrie, P., 14
Mandl, R. H., 117
Marsh, S. S., 333
Miles, C. D., 308
Miles, D., 308
Ming, L., 170
Morgan, E. L., 267

N

Nwosu, J., 177

O

Olson, M. F., 156

P

Peng, Y., 319
Peterson, H., 14
Pitts, A. R., 255

R

Ratsch, H., 177
Reichenbach, N. G., 333
Ribeyre, F., 97
Richardson, C. J., 143
Robillard, K. A., 49

S

Sasek, T. W., 143
Sauser, K. R., 324
Schwarz, O. J., 87
Sirois, D. L., 188, 225
Smith, S., 177
Sun, W., 346
Sutton, W. W., 235
Swanson, S., 14

T

Teng, P. S., 156
Thomas, M. W., 235, 308
Tolle, D. A., 127

V

Velagaleti, R. R., 333
Versteeg, D. J., 40

W

Walti, K., 117
Wang, W., 1, 280
Weinstein, L. H., 117
Wilborn, D., 177
Wu, Y. -C. A., 267

X

Xia, W., 346
Xingguo, W., 170
Xu, J., 346

Y

Young, R. C., 267

Subject Index

A

Acetylene reduction, in nitrogen fixation, 69
Acute algal methods, 7, 40, 69
Advance Notice of Proposed Rule Making (ANPR), Federal Register, 225
Aerosal deposition, 127
Agronomic species (*See* Crops and Forage)
Air pollution, 1, 117, 156, 170
 due to aerosol deposition, 127
 and forest declines, 1, 143
Alfalfa, 156
Alflatoxin B1, 324
Algae
 Anabaena Flos-aquae, 225
 Champia parvula, 10
 Chlorella vulgaris, 10, 34, 69, 225
 Coccomyxa minor, 42, 204, 324
 Haematococcus lacutris, 204
 Scenedesmus armatus, 204
 Scenedesmus dimorphus, 204
 Selenastrum capricornutum, 6, 40, 69, 225, 235, 324
 Skeletonema costatum, 225
Algal assays, 2, 5, 10, 23, 205
 evaluation of protocols, 225
 for herbicides, 33, 69, 235
 as indicators for vascular plants, 10, 33, 69
 for pesticides, 14, 40, 69
 for promutagens, 324
 for waste water, 40, 225
 for metal pollution, 204
 in tier testing schemes, 33
Aluminum sulfate, 267
Anabaena flos-aquae, 225
Analysis of Variance (ANOVA), 225
Antioxidants, 143
Antracene, 87
Apparent free space (AFS), 87
Aquatic exotoxicology (*See also* Algae, *Daphnia*, Epiphytes, *Lemna minor* (Duckweed), and *Thalassia testidinum* (Seagrass)), 8, 97, 204, 225

Arabidopsis thaliana, 10
Archiving samples, 117, 156
Arsenal, 235
Ash-free dry-weight, 61
ASTM Committee E-47, 1
ASTM Subcommittee E-47.11, 1
Atrazine, 40, 324
Avena sativa (Oats), 226
Aviation fuel, 198

B

Bacteria, for mutagenicity testing, 324
Barium, 280
Barium chloride, 267
Barley (*Hordeum vulgare*), 69, 73
Baseline toxicity tests, 199
Beets (*Beta vulgaris*), 190
Bioaccumulation, 87, 97
Bioassays, 1, 49, 188, 225
 chlorophyll fluorescence, 297, 308
 micronucleus assays, 170
 of promutagens, 324
 sister-chromatid-exchange, 319
 time-dependent toxicity tests, 235
Biochemical markers, for air pollution tests, 143
Bioindicators, 117, 156, 225
 diagnostic gas exchange technique, 143
 of soil toxicity, 190
Biomass, 6, 69, 127, 255
 in Minnesota Bioindicator Study, 157
Biomonitors, 117
Boric acid, 267
Brassica nigra (Mustard), 190
Bush beans (*Phaseolus vulgaris*), 69, 73

C

Cadmium, 40, 118, 281, 308
Cadmium chloride, 267
Canada, testing in, 14
Carbaryl, 40
Carbohydrate content of rhizomes, 59
Carbon dioxide fixation test, 41

Carcinogenicity testing, 11
Champia parvula, 10
Chemical stress on plants, 1, 7, 11, 33, 49
  presymptom diagnosis, 333
Chemical Information System (CIS), 14
Chinese National Environmental Protection Agency, 170
Chlorella vulgaris, 10, 34, 69, 225
Chlorides, 118
Chlorophyll, 59, 297, 308
  content, 59, 255, 267, 333
Chlorophyll fluorescence, 6, 297, 308
  Kautsky effect, 297
Chloroplasts, 297
Chlorosis, 188
Chromium, 204, 281
Chronic algal methods, 40
Clover (*Melilotus alba*), 69, 87
Coal combustion fly ash leachate, 267
Cobalt chloride, 267
Coccomyxa minor, 204
Coleoptile growth, 49
Computer database (PHYTOTOX), 33
Computer image analysis system (IAS), 333, 346
Conductance, stomatal, 145
Contaminants, 97, 188
Copper, 40, 76, 118, 281, 308
Copper sulfate, 69
Corn (*Zea mays*), 226
Crops and forage, 5, 11, 157, 226
Cucumber (*Cucumis sativa*), 226
Cultivated plants, as bioindicators, 117
Cytogenetic studies, 319

D

Danish National Agency for Environmental Protection, 15
Daphnia, 6
Delayed fluorescence, 333
Development of testing, 198
Di-*tert*-butyl phenol, 8
Diagnostic gas exchange techniques, 143
Diethyl phthalate, 40
Dose-response, 156
Drilling fluids, 255
Dry weight tests, 267

E

Early seedling growth tests, 6, 8, 10–11, 49
Ecological assessment, 177
  of obscurant smokes, 127
Ecosystem-level effects, 127

Ecotoxicological studies, 1, 97
Electrical conductivity (EC), 333
Electron pool size, 308
Element uptake, 127
Elodea densa, 97
Enzymatic antioxidants, 143
Environment Canada, 14
Environmental hazards, 69, 117, 204
  data in PMNs, 6
Environmental mutagens, 319, 324, 346
EPA (*See* U.S. Environmental Protection Agency)
Epiphytes, 59, 61, 255
European Economic Community (EEC), 15
Exposure benches, 117

F

Faunal studies, 1
Federal Register, 225
Festuca arundinaceae (Fescue), 87
Field studies, 7, 59, 117, 297
  in China, 170
  Minnesota Bioindicator Study, 156
  at Rocky Mountain Arsenal, 188
  in Switzerland, 117
Field validation, 59
Fish acute toxicity, 6
Fluorescence (*See* Chlorophyll fluorescence)
Fluorides, 11, 117
Fluorometer, 308
Fly ash leachate, 267
Foliar symptoms, 157, 188
Food chain contamination, 1, 5, 87, 324
Forest declines, 1, 117, 143
Formative effects, 188
Foxtail millet (*Setaria italica*), 190, 280
Free radicals, 143
Fresh and saltwater algae, 9, 204
Freshwater macrophytes, 204
Fuel contamination, of soil, 198

G

Garlon 4, 235, 308
Genetic indicator, 1, 324
Genotoxicity testing, 170, 319, 346
Germination, 49, 188, 225
Glycine max. (Soybeans), 69, 73, 156, 333
Glyphosate, 235, 308
Grass cultures, 117
Growth chamber tests, 225, 267

H

Haematococcus lacustris, 204
Hazardous waste site evaluation, 177, 188
Health risk management, 11
Heavy metals, 280, 308
Herbicides, 69, 198, 235, 308, 333
  in algal assays, 10–11
  in Canada, 14
  and industrial wastes, 189
  in multispecies testing, 33, 69, 73
Hexazinone, 235, 308
Hordeum vulgare (Barley), 73
Hydrogen peroxide, 143
Hydrogen sulfide, 170
Hypocotyl growth, 49

I

Image analysis system (IAS), 333, 346
Imazapyr, 235
In situ testing (See Field studies)
In vivo testing, 297
Indicator gardens, 117
Indicator plants (See Bioindicators)
Industrial chemicals and wastes, 188, 225, 255
Inorganic mercuric chloride ($HgCl_2$), 97
Interaction of factors, 97
Intercellular CO_2 concentration, 143
Ion accumulation patterns, 333
Isolated root uptake test (IRUT), 87

J

JP-4 aviation fuel, 198

K

Kautsky effect, 297

L

Laboratory studies, 49, 97, 198, 225
  comparison to field tests, 59
  evaluation of protocols, 225
  growth chamber tests, 7, 25, 267
  mathematical models, 280
Leachates, 267
Lead, 118
Lead nitrate, 267
Leaf chlorophyll content, 333
Leaf protein content, 255
Legumes, 73, 156, 198, 226, 333
  nitrogen fixation in, 7, 70, 255

Lemna minor (Duckweed), 6, 9, 204, 225
Lettuce (*Lactuca sativa*), 49
Lichen, 117
Light emission, and photosynthesis, 297
Light intensity, 97
Lipid peroxidation, 143
Loblolly pine (*Pinus taeda L.*), 143
Long-term tests, 40, 69
Ludwigia natans, 97

M

m-phenylenediamine (MPD), 324
Macrophytes, 97, 204
Malondialdehyde, 143
Manganese, 281
Mathematical model, for toxicological studies, 280
Melilotus alba (White sweet clover), 128
Mercury compounds, 97
Metabolic processes, 33
Metabolism, of xenobiotic chemicals, 88
Methylmercury chloride (CH_3HgCl), 97
Microcosm technique, 128
Micronucleus bioassays (MCN), 170, 346
Milkweed, 156
Millet (*Panicum miliaceum*), 280
Millet, foxtail (*Setaria italica*), 190, 280
Mineralized-acidic leachate, 267
Minnesota Bioindicator Study, 156
Moss (*Physcomitrella patens*), 267, 280
Multispecies toxicity testing, 69, 73, 97
Mustard (*Brassica nigra*), 190
Mutagens, 11, 319, 324

N

Naphthalene, 87
Naphthol, 87
Native plants, as bioindicators, 117
Nickel, 118, 281
Nitrogen fixation, 7, 69
Nitrogen oxides (NO_x), 143
Nontarget plant testing, 14
No Observed Effect Concentration (NOEC), 50, 53, 82
Nutrient loss, 127

O

Oat (*Avena sativa*), 226
Obscurant smokes, 127
Octamethyl cyclotetrasiloxane, 8
Organization for Economic Cooperation and Development (OECD), 15, 33

Organic methylmercury chloride (CH$_3$HgCl), 97
Organic compounds, 34, 188, 225
Oxidant biomarkers, 143
Oxygen generation test, 41
Ozone, 143, 156

P

Pentachlorophenol, 40
Perennial ryegrass (*Lolium perenne*), 49, 127, 190
Peroxidase, 143
Pest Control Products (PCP) Act, R.S.C. 1970, C.P-10, 14
Pesticides, 14, 40, 69–70, 73
 in Canada, 14
 in food chain, 5
pH, 97
Phaeseolus beans, 73, 198, 226, 333
Phosphorus smoke-producing compounds, 127
Photoperiod (PH), 97
Photosynthesis, 143, 297
 Kautsky effect, 297, 308
Photosystems I and II, 308
PHYTOTOX database, 33
Phytotoxicity studies, 1, 10, 49, 177, 267
 in Canada, 14
 evaluation of, 225
 of heavy metals, 280
 of herbicides, 10
 of industrial wastes, 188
 in Minnesota Bioindicator Study, 157
 of organic chemicals, 34, 188 225
 of pesticides, 40, 69, 73
Picloram, 235, 308
Pinto beans (*Phaeseolus vulgaris*), 198, 226
Pistia stratiotes (Freshwater macrophytes), 204
Plant(s) (*See* Algae, Aquatic plants, Crops and Forage, Trees)
Plant data, and risk assessment, 5, 11
Plant testing, (*See* Bioassays, Early seedling growth tests, Field studies, Laboratory studies, Multispecies toxicity testing, Nitrogen fixation, Toxicity)
Pollution damage, 117
Polynuclear aromatics, 87
Population growth tests, 40, 42
Portable exposure bench, 118
Potassium chloride, 333
Potato, 156

Premanufacture Notifications (PMN), 2, 5, 11
Presymptom diagnosis, 333
Promutagens, 324
Protein content, of leaves, 59

R

Radish (*Raphanus sativus*), 49, 69, 226
Red phosphorus/butyl rubber (RP/BR), 127
Regenerating clones, 267
Rhizobium, genetically altered, 7
Rhizobium-legume symbiosis, 69
Rhizome carbohydrate content, 255
Rocky Mountain Arsenal (RMA), 188
Root concentration factor (RCF), 88
Root elongation, 10, 49, 177, 280
Root nodules, 69
Rooted macrophytes, 97
Roundup, 235, 308
Ryegrass (*Lolium perenne*), 49, 118, 190

S

Salmonella typhimurium, as a genetic indicator, 324
Salt concentrations, 267, 334
Saskatchewan Research Council, 14
Scenedesmus armatus, 204
Scenedesmus dimorphus, 204
Screening tests, 5, 33
Seasonal growth cycles, 59, 157
Sediment, 97
Seed germination and development, 10, 188, 225
 and root elongation, 8, 117
Selenastrum capricornutum, 6, 40, 225, 235, 324
 in herbicide testing, 69, 235
 in pesticide testing, 14, 40, 69
 promutagens in, 324
 in waste water testing, 40
Short term tests, 40, 177, 33
 multispecies testing, 69
Simazine, 40
Single-species toxicity tests, 177
Sister-chromatid-exchange (SCE), 319
Smoke exposure chamber, 127
Sodium chloride, 333
Sodium fluoride, 308
Sodium lauryl sulfate, 308
Soil pollution, 1, 189, 198, 235
 contamination, 1, 177, 190, 198, 235
 salinity of, 335

Soil-microcosm system, 127
Sorghum (*Sorghum bicolor*), 198
Soybean (*Glycine max.*), 69, 73, 156, 333
Spiderwort (*See Tradescantia*)
Standard grass cultures, 117
Standing crop measurements, 59
Static exposure system, 127
Stomatal conductance, 143
Streptomycin sulfate, 69
Stress-ethylene test, 127
Sugar beet (*Beta vulgaris*), 190
Sulfur, 118
Sulfur dioxide, 143, 156
Superoxidase dismutase (SOD), 143
Surrogate species, 33
Symbiosis, 69
Synergy (*See* Interaction of factors)

T

Temperature, 97
Terrestrial microcosm, 127
Terrestrial plant(s), 69
  and algal assays, 10, 33, 69
  testing, 6, 9, 49, 127, 177, 225
Test methods (*See also* Algal assays, Bioassays, Bioindicators, Field Studies, Herbicides, Laboratory studies, Long term tests, Pesticides, Phytotoxicity studies, Short term tests), 1, 5
  validity of, 10, 40, 225
Thalassia testidinum (Seagrass), 59, 255
Tier testing schemes, 33
Time dependent toxicity tests, 235
Tordon 101M, 235, 308
Toxic Substances Control Act (TSCA), 2–10
Toxicity (*See also* Air pollution, Herbicides, Industrial Chemicals and Wastes, Organic compounds, Pesticides, Soil pollution, Waste water, and Water pollution), 33, 40, 177, 188, 235
  algal assays for, 40, 69
  bioassays for, 297
  and multispecies approaches, 69, 308
  presymptom diagnosis of, 333

Tradescantia paludosa (Spiderwort), 308, 319
  micronucleus test (Trad-MCN), 170, 346
  sister-chromatid-exchange (SCE), 319
Transpiration stream concentration factor, 87
Trees, 1, 117, 143
Tributyltin, 255
Triclopyr, 235, 308
Trifolium pratense (Clover), 87
Trifolium aestivum (Wheat), 128
2,4-dichlorophenooxyacetic acid (2, 4-D), 71

U

U.S. Army Medical Research and Development Command, 127
U.S. Environmental Protection Agency, 15

V

Vanadium, 118
Variable fluorescence, 308
Vascular plants, 33
Velpar L, 235
Velpar ULW, 235, 308
Vicia faba leaf tip micronucleus test (Vicia-LTMT), 170
Visual assessment, 5

W

Waste water, 40, 225
Water pollution, 41, 97, 170
Wheat (*Triticium aestivum*), 127
White phosphorus/felt (WP/F), 127
White sweet clover (*Melilotus alba*), 127
Whole plant uptake test (WPUT), 87

X

Xenobiotics, 1, 59, 87
Xylene, 170

Z

Zinc, 118, 281, 308